Fluid Mechanics and Its Applications

Volume 105

For further volumes:
http://www.springer.com/series/5980

Aims and Scope of the Series

The purpose of this series is to focus on subjects in which fluid mechanics plays a fundamental role.

As well as the more traditional applications of aeronautics, hydraulics, heat and mass transfer etc., books will be published dealing with topics which are currently in a state of rapid development, such as turbulence, suspensions and multiphase fluids, super and hypersonic flows and numerical modeling techniques.

It is a widely held view that it is the interdisciplinary subjects that will receive intense scientific attention, bringing them to the forefront of technological advancement. Fluids have the ability to transport matter and its properties as well as to transmit force, therefore fluid mechanics is a subject that is particularly open to cross fertilization with other sciences and disciplines of engineering. The subject of fluid mechanics will be highly relevant in domains such as chemical, metallurgical, biological and ecological engineering. This series is particularly open to such new multidisciplinary domains.

The median level of presentation is the first year graduate student. Some texts are monographs defining the current state of a field; others are accessible to final year undergraduates; but essentially the emphasis is on readability and clarity.

Fernando Concha A.

Solid–Liquid Separation in the Mining Industry

Springer

Fernando Concha A.
Department of Metallurgical Engineering
University of Concepción
Concepción
Chile

ISSN 0926-5112 ISSN 2215-0056 (electronic)
ISBN 978-3-319-02483-7 ISBN 978-3-319-02484-4 (eBook)
DOI 10.1007/978-3-319-02484-4
Springer Cham Heidelberg New York Dordrecht London

Library of Congress Control Number: 2013953608

Printed on acid-free paper

Springer is part of Springer Science+Business Media (www.springer.com)

Preface

This book seeks to fill a gap in the bibliography of the mineral processing industry. Subjects such as comminution and flotation have received much attention by researchers in the field of Mineral Processing, while thickening and filtration have not.

It seems that the economic importance of size reduction, obviously the most costly stage in Mineral Processing, and the strategic position of flotation as the main concentration process, have relegated the last stages of mineral beneficiation, solid-liquid separation, to a lower level of importance. It is true that when a mineral processing plant is operating normally operators tend to regard thickening and filtration as auxiliary rather than fundamental processes in the plant. But the situation changes when problems appear in sedimentation of tailings or filtration of concentrates and it is not possible to recover all the water necessary for the process, or when it is not possible to obtain the required level of moisture in the final product. In those cases, solid–fluid separation acquires a fundamental importance.

Process engineers find themselves ill-prepared to face problems such as those described above. Possibly they ask themselves why their university gave less attention to these areas. The truth is that this academic "carelessness" has deeper roots related to the low level that mechanics and fluid mechanics has been given in mining, mineral, and metallurgical engineering programs. As a result very little research has been done in the areas of thickening, filtration, and pulp transport and therefore instructors lack knowledge and experience to share with their students.

Currently, there are research groups worldwide working on solid–liquid separation, especially in the field of Chemical Engineering, where the works of authors such as Tarleton and Wakeman (1999, 2005) are available. There has been less work in the field of Mineral Processing. It is necessary to change this situation by providing an adequate framework to solid–liquid separation in the mineral industry. I hope that this book will help in this respect.

This book is divided into 11 chapters. Chapter 1 introduces the field of Mineral Processing and the importance of water in processing minerals. The consumption of water in the various stages of concentration and the need to recover most of that water by recycling are discussed.

Chapter 2 lays the conceptual basis for the study of processes of solid–liquid separation. A rigorous but limited account of the Theory of Mixtures of continuum mechanics is given. It was considered unnecessary to deal with thermodynamic aspects. An introduction discusses the conditions that a multi-component body must fulfill to be considered a continuum. The concepts of body, component and mixture are then introduced and the concepts of deformation and rate of deformation are discussed. Mass and momentum balance equations are formulated for each component of the mixture and the need to establish constitutive equations to complete a dynamic process is discussed.

Mixtures of finely divided solid particles in water are the subject of Chap. 3. Here the equations derived in Chap. 2 are applied to particulate systems.

Chapter 4 deals with sedimentation of particulate systems considered as discrete media. Starting from the sedimentation of a sphere in an unbound fluid, a complete analysis is made of the settling of individual particles and suspensions, establishing their settling and fluidization velocities.

Sedimentation of suspensions, treated as continuous media, is studied in Chap. 5. The concept of an ideal suspension and ideal thickeners is established. Kynch's theory and its extension to a continuous process are presented. The solution to Kynch's problem is deduced through the theory of characteristics. Also the concept of Modes of Sedimentation is introduced.

Flow through porous media is dealt with in Chap. 6. Equations for particulate systems are reduced for the case of the flow of a fluid through a rigid porous material. Darcy's and Forchheimer's equations are used as constitutive equations for the relative solid–fluid force. Permeability and its geometric concept are studied. For the case of two-phase flow through a rigid porous medium, the concepts of relative permeability, saturation, and capillary pressure are introduced.

Chapter 7 considers particle aggregation. When agglomerated particles in a suspension increase in size they acquire greater sedimentation velocity essential to obtain a good separation by sedimentation. The agglomerates also form more permeable cakes, which accelerates the filtration process. Different methods to increase the size of solid particles are studied in this chapter, these being *coagulation*, by reduction of the interparticle electrostatic repulsion and *flocculation* by bridging particles with polymeric agents.

Chapter 8 history of thickening is laid out from the Stone Age to the present, emphasizing people and institutions that have been important actors. The chapter then reviews the thickeners used in the mining-mineral industry. The theory of sedimentation–consolidation is deduced from the equations for a particulate system and constitutive equations for the solid–fluid interaction force and sediment compressibility are postulated. Batch and continuous sedimentation are analyzed and simulations are compared to data from the literature. Experimental determination of thickening parameters and instruments for their determination are presented. Old and new methods for thickening design are reviewed and software for the design and simulation of batch and continuous thickening are presented. Finally, strategies for the operation and control of industrial thickening are discussed.

Chapter 9 deals with filtration. Following the same scheme as in the previous chapter, equipment, the theory of filtration, constitutive equations, and parameter determination are discussed and the operations of vacuum and pressure filters are simulated.

Chapter 10 discusses Rheology. The fluid mechanics of Newtonian and non-Newtonian materials are briefly presented and the different constitutive equations for shear stress are discussed. The measurement of viscosity and yield stress in the laboratory is described and models for the relationship of these parameters with concentration are deduced.

The last chapter of the book is related to the transport of pulps in mineral processing plants. Starting from the continuity equation and the equation of motion for a continuous medium, the expression for the pressure drop during fluid flow in a tube is obtained. Newtonian fluids are then treated for cases of laminar and turbulent flows. The concepts of friction factor and Reynolds number are introduced and the distribution of velocity, flow rate, and pressure drop in a tube is obtained. The transport of suspensions in pipelines is then treated, defining the different regimes separated by the limiting deposit velocity. First, the flow of heterogeneous suspensions is introduced and the form to calculate head loss is presented. Next, homogeneous suspensions modeled by different rheological approaches are discussed. Finally, equations for the transport of suspensions in open channel are dealt with.

All the chapters present problems with solutions to aid the reader in understanding the subjects.

A substantial part of this book, especially the chapters on sedimentation and thickening, are the results of research by the author and his research group at the Department of Metallurgical Engineering and the work of Prof. M. C. Bustos and R. Bürger at the Department of Mathematical Engineering at the University of Concepción, Chile. We had the important collaboration of Profs. Wolfgang Wendland at the University of Stuttgart, Germany, Kenneth Karlsen at the University of Bergen, Norway, and Elmer Tory at the University of Mount Allison in Canada. Special thanks to all of them.

The collaboration of Graduate and Engineering students at the universities of Concepción in Chile, Stuttgart in Germany, and COPPE/UFRJ in Brazil was especially important. Many thanks to P. Garrido, F. Betancourt, M. Kunik, O. Bascur, A. Barrientos, R. Becker, A. Quiero, R. Valenzuela, A. Christiansen, A. Rojas, F. Melo, H. Droguett, V. Soto, R. Ruiz, R. Burgos, R. Pradenas, P. Leonelli, and E. Almendra.

Many thanks to George Montgomery for correcting my English and to engineering students P. Silva and D. Vidal for checking the problems.

I would like to give special thanks to the late Prof. Giulio Massarani from the Graduate School (COPPE) at the Federal University of Río de Janeiro, Brazil (UFRJ) for his influence in the conceptual development of the theory of thickening. The several years that I spent in COPPE as visiting professor in the early 1970s, discussing the fundamental aspects of sedimentation, gave me the conviction that the Theory of Mixtures was the appropriate framework to study the thickening process.

I would also thank the late Prof. L.S. Sriven from the University of Minnesota who introduced me to the area of Transport Phenomena during my graduate studies at that university. He showed me the way to apply fluid mechanics to industrial processes.

Finally, I would like to thank several institutions which gave support to my academic and research work, especially the **University of Concepción; Conicyt:** *Fondecyt projects* 89/352; 1.097/83; 841/82; 1040597; *Fondef projects* MI-08; 2–43; F-I064;D97I2042, D00-T1027, D00-T2027, *DAAD project* 2003154; *Fondap projects* 15130015; **Andes Foundation**:; C-13634; C-13670; Humboldt N1 13131; **NSF INT**-8610400; **Stuttgart:** We-659/7–1; **OEA** 1971–1975; **Innova-Corfo**: projects 08 CM91–17; 08 CM01–18.

Contents

Abstract

This book shows the need for research and more emphasis on dewatering processes in the preparation of metallurgical, mining, and chemical engineers. Most of the research in mineral processing in recent decades has been directed at comminution and flotation, with very little emphasis on thickening, filtration, and suspension transport. This, of course, results from comminution being vital to mineral liberation and because it consumes about 80 % of the energy in mineral processing. Flotation research has always been vital to mineral processing because flotation is the backbone of separating desired minerals from waste material, as well as because of its costliness in terms of reagents and energy. These two operations each cost many dollars per ton of ore treated, whereas the cost per ton associated with dewatering is low (except for polymer flocculants). Thus, dewatering has not received a great deal of attention by most mineral processing researchers. With water shortages becoming a major problem globally, dewatering in mineral processing is becoming more important.

Chapter 1
Introduction

Abstract This chapter introduces the field of Mineral Processing. A mineral processing plant can be divided into four distinct operational units: *comminution, concentration, dewatering* and *pulp transport*. Comminution is the process of reducing the particle size of an ore until the free particles of minerals can be separated by available methods. Froth flotation is the most important mineral processing technique to recover sulphide minerals, such as copper, zinc and lead. It uses the differences in physicochemical surface properties of particles of different minerals and gangue to recover a concentrate and leave the gangue as *tailing* that is discarded. Dewatering is a process of *solid–liquid separation* achieved by thickening and filtration. *Thickening* uses the force of gravity to separate the particles from the water by sedimentation in large cylindrical tanks called thickeners, while *filtration* uses pressure forces to pass the slurry through a cloth and separate it into a *filter cake* on the cloth and clean water called *filtrate*. Safe and environmentally friendly deposition of mining waste is a major concern in the mining industry worldwide. Almost all the treated minerals in sulphide concentration plants are deposited as tailings since the recovered product represents a very small percentage of the total tonnage. Water has become a major concern in processing raw materials, in terms of water conservation and reuse. The chapter discusses water consumption in the various stages of concentration and the need to recover most of that water by recycling.

In the introduction to his book "Mineral Processing Technology", Barry Wills (1997) gives an excellent and concise account of the process to recover minerals from ores and the technical steps necessary to produce metals or industrial minerals. I will borrow some of his ideas to introduce the subject of solid–liquid separation in the minerals industry.

The extraction of valuable minerals from unwanted rock (gangue) is possible only if the mineral grains, which are to be separated, exist as liberated discrete particles. As the grade of mineral deposits decreases, the valuable minerals often appear in the form of finely disseminated grains in a solid rock (ore). Liberation is achieved by crushing and grinding them down to small sizes, that is, by *size reduction.*

F. Concha A, *Solid–Liquid Separation in the Mining Industry*,
Fluid Mechanics and Its Applications 105, DOI: 10.1007/978-3-319-02484-4_1,

Separation of mineral and gangue is strongly dependent on the size of the treated ore particles and there is an optimum size range for mineral recovery in all separation processes. Recoveries of both too fine and too coarse particles, that are outside this optimum size range, are low. For a satisfactory overall recovery, the size of the processed particles must therefore be carefully controlled. Because of the reduction in particle size that is required to achieve proper liberation, some over grinding is unavoidable.

A mineral processing plant, usually called the *concentrator* or *mill*, may be broadly subdivided into four distinct unit operations: *comminution, concentration, dewatering* and *pulp transport*. Comminution is the process of reducing the particle size of an ore until the free particles of minerals can be separated by available methods. Separation of the constituents depends strongly on the degree of liberation of the ore. Since the run-of-mine ore may have sizes in the range of fractions of meters and liberation may be in the range of microns, no unique operation of comminution is possible and several stages of size reductions are necessary.

Explosives are used in mining to remove ores from ore bodies and *blasting* can be considered as the first stage in comminution. See Fig. 1.1. Figure 1.2 shows the *flowsheet* of a Mineral Processing Plant. In this figure, the blue ellipse represents comminution, the green ellipse shows concentrations and the red ellipse represents dewatering processes. The separation of unit operations is not completely possible in a plant since, during the concentration process, some comminution and some dewatering are necessary.

Fig. 1.1 Blasting in an open pit mine as the first stage in comminution

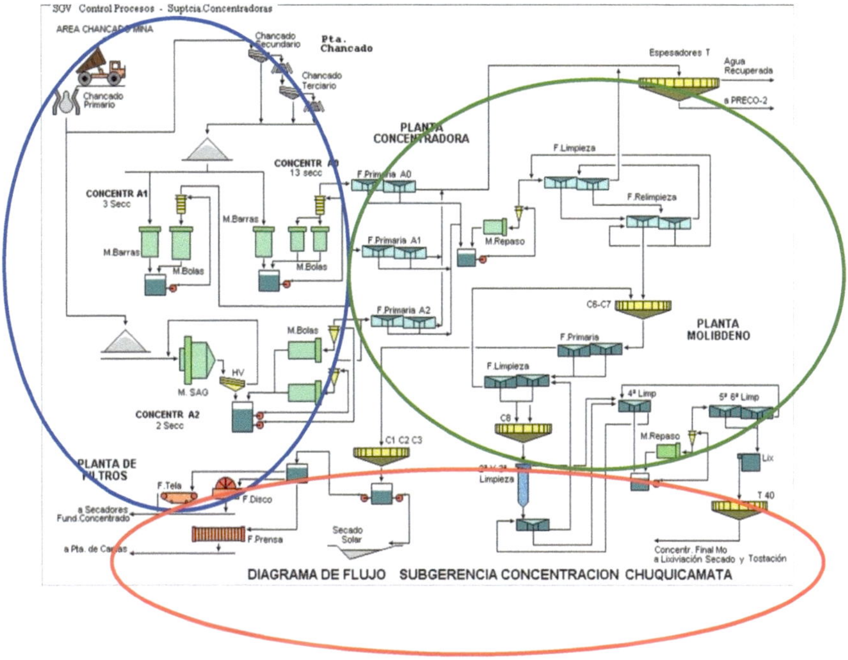

Fig. 1.2 Simplified flowsheet of the concentrator in Chuquicamata

1.1 Comminution

In a concentrator, comminution takes place as a sequence of crushing and grinding circuits. A *circuit* consists of a *size reduction stage* followed by a particle *classification stage*, a necessary scheme to avoid crushing or grinding to a smaller size than necessary. Within a circuit, particles smaller than a target size leave the circuit and particles larger than that size return to the size reduction stage. Crushing reduces the rock sizes from the run-of-mine ore to such a level, that grinding can be carried out until the mineral and gangue is substantially separated.

Reduction ratio of a comminution stage is the ratio of the maximum particle size in the feed to the maximum size in the product. Crushing is usually a dry process and is performed in several stages, with small reduction ratios, ranging from three to six in each stage. Figures 1.3 and 1.4 show equipment for crushing stages.

Product from crushing enters the tumbling mill section of the concentrator. A *tumbling mill* is a rotary drum with rods or balls, called *grinding media*, that reduce the size of the ore particle by impact and by attrition between the grinding media and the ore. Grinding is usually performed wet to provide a feed to the concentration process as slurry. Since the middle of last century, a new type of comminution equipment, the Semi Autogenous Mill or SAG mill, is being used in

Fig. 1.3 Primary crushing stage: gyratory crusher

Fig. 1.4 Secondary or tertiary crushing: cone crusher

Fig. 1.5 SAG mill in the foreground, with a secondary ball mill at the side

the mineral industry. This mill, see Fig. 1.5, which is much larger than a conventional ball mill, Fig. 1.6, can replace part of the crushing circuit, the secondary and tertiary crushing, and the primary grinding, with great success.

Fig. 1.6 Secondary grinding circuit with a battery of hydrocyclones

1.2 Concentration

Froth flotation is the most important mineral processing technique to recover sulphide minerals, such as copper, zinc and led. It utilizes the difference in physico-chemical surface properties of particles of different minerals and gangue. After treatment with reagents, these differences in surface properties makes air bubbles attach to valuable minerals and lift them to the pulp surface, where they are recovered and leave the gangue as slurry, called *tailing*, which is discarded. Figure 1.7 shows a flotation machine for copper flotation.

Fig. 1.7 Flotation cell in a copper concentrator

In most sulfide minerals concentration plants, the general feed to flotation comes from a secondary grinding circuit, which consists of a grinding mill working in closed circuit with a certain number of *hydrocyclones*. Hydrocyclones are classification devices that separate small from coarse particles by the effect of hydrodynamic forces in a vortex motion. Besides producing this size separation, hydrocyclones act as a solid liquid separator by concentrating the underflow and diluting the overflow. See Fig. 1.6.

Flotation processes are partially insensitive to slurry concentration, as long as the pulp is below 33–35 % solid. This would indicate that a denser hydrocylone overflow would contribute to produce a denser tailing. The last cells of a flotation circuit produce middlings with a low percentage solid. A *middling* is a piece of ore larger than the liberation size that will not float efficiently. Since this product is diluted, if returned to the regrinding circuit, it will further liberate the particles and will recover additional water.

The grinding-classification-concentration processes, we have been describing, involves the use of a substantial quantity of water. The final concentrate of a flotation plant usually has more water than is allowed in the next process to recover the metal or in concentrate to be sold as such. On the other hand, tailings to be discarded contain a prohibitive amount of water. In both cases, dewatering of the solids must be done to produce a dryer concentrate in the first case, and to recover the most part of the water in the second.

1.3 Dewatering

Dewatering is a process of *solid–liquid separation* achieved by thickening and filtration. *Thickening* uses the force of gravity to separate the particles from the water by sedimentation in large cylindrical tanks called thickeners, while *filtration* uses pressure forces to pass the slurry through a cloth and separate it into a *filter cake* on the cloth and clean water called *filtrate*. Often dewatering is difficult, especially when the treated particles are very fine as required for concentration. In

Fig. 1.8 View of a conventional thickener for copper tailings

Fig. 1.9 Band filter for tailing filtration

such cases, size enlargement may be beneficial. This is accomplished by *coagulation* and *flocculation*, both of which use the bonding of particles in larger agglomerates by electrical or other bonding forces.

Figures 1.8 and 1.9 show two methods of solid–liquid separation, thickening and filtration respectively.

1.4 Tailings Disposal

Safe and environmentally friendly deposition of mining waste is a major concern in the mining industry world-wide. In sulphide concentration plants, almost the total tonnage of treated mineral is deposited as tailings since the recovered product represents a very small percentage of the total tonnage. In the copper industry, for example, 98 % of the treated ore will be disposed as waste, together with a significant amount of water. To compensate for the increasing production costs associated with decreasing mineral content of the ores, successful mining activities today relay on the treatment of high daily tonnage of ores, for example 200,000 tons per day in the copper industry. The transport, deposition and storage of mine tailings must move large volumes of low density slurries, from which a large amount of the water will be lost by evaporation.

Figures 1.10 and 1.11 show the transport, discharge and deposition of copper tailings in a copper concentrator.

For waste minimization in the minerals industry, it is essential to learn how to dewater and handle suspensions at very high concentrations. Such materials are non-Newtonian fluids affected by shear and compression. Figure 1.12 shows the discharge of a material at very high concentration which behaves as a *Non-Newtonian fluid* (Fig. 1.13).

Fig. 1.10 Tailing channel in a Chilean copper mine

Fig. 1.11 Discharge of tailings into a tailing dam

Fig. 1.12 Material behaving as a non-Newtonian tailing

Fig. 1.13 Copper concentrate transport to the world market in one of Chile's copper mines

1.5 Water Management

Chile is a predominantly a mining and agricultural country. In the desert north, the mining, agriculture and fresh water industries compete for water resources. The consumption of water, as an average for the country in 1999, was 1,350 m^3/s with the following distribution by economic activity (Renner 2005):

• Agriculture	84.6 %
• Other industries	6.5 %
• Mining	4.5 %
• Potable water	4.4 %

In the mining industry, consumption of water depends on the metal recovery process used. Froth flotation uses around 0.72 m^3/ton of ore and hydrometallurgy 0.30 m^3/ton of ore. In the end, the part of the water in concentrators is lost at the tailing dam, where only around 60–70 % of the water can be recovered. Losses are related to retention of water as moisture by the deposited solid and evaporation from the clear water pond. The total water consumption by the mining industry in Chile in the year 2010 was 300 million m^3/year and a projection for 2026 is 631 million m^3/year. The scarcity of water made the Chilean mining industry to install, in 2010, its first big copper concentrator Esperanza operated totally with sea water.

The cost of water in the mining industry depends on local conditions. For example, in desert areas the price of fresh water in 2010 was (Concha 2001):

• Fresh water	0.50 US$/$m^3$
• Recovered process water	0.18 US$/$m^3$
• Sea water	1.20 US$/$m^3$
• Desalinized sea water	2.20 US$/$m^3$

In non-desert areas the cost of fresh water can be as low as 0.2 US$/$m^3$.

1.6 Conclusions

Faced with the need to produce concentrates with adequate moisture for a safe transportation to the world market, and the necessity to recover water from the tailings to lower operating costs by diminishing fresh water requirements, to avoid pollution and reduce liability associated with contamination of land and water sources, the mining industry needs technical knowledge and scientific research in the areas of Solid–liquid Separation, especially in Thickening, Filtration, Rheology and Suspension Transport.

The purpose of this textbook is to give the basic knowledge necessary to comprehend the problems of solid–liquid separation in the mineral industry.

References

Barry, W. (1997). *Mineral processing technology* (6th ed.). Oxford: Butterworth-Heinemann.

Concha, F. (2001). *Manual de Filtración y Separación*. Concepción: Cettem Ltd.

Renner, P. (2005). Opportunities for sustainable development in the use of water resources by the Chilean Mining Industry, keynote address. *Addendum to the Proceedings of the International Seminar on Paste and Thickened Tailings*, Santiago, Chile, April 59–66.

Chapter 2
Theory of Mixtures

Abstract This chapter lays out the conceptual basis for the study of processes of solid–liquid separation. For the study of flows in rigid and deformable porous media and of suspension sedimentation and transport, we must consider bodies formed of different materials. The appropriate tool to do this is the *Theory of Mixtures*. A rigorous but limited account of the Theory of Mixtures of continuum mechanics is given that postulates that each point in space of a body is simultaneously occupied by a finite number of particles, one for each component of the mixture. In this way, the mixture can be represented as a superposition of continuous media, each following its own movement with the restriction imposed by the interaction between components. An introduction discusses the conditions that a multi-component body must fulfill to be considered a continuum. The concepts of body, component, mixture, deformation and rate of deformation are introduced and discussed. Mass and momentum balance equations are formulated for each component of the mixture and the need to establish constitutive equations to complete a dynamic process is discussed.

To study the flow in rigid and deformable porous media and for the study of sedimentation and transport of suspensions, it is convenient to consider a body formed of different materials. The appropriate tool to do this is the *Theory of Mixtures*. There is not one but several Theories of Mixtures, and here we will follow the developments of Truesdell and Toupin (1960), Truesdell (1965, 1984).

The Theory of Mixtures postulates that each point in space is simultaneously occupied by a finite number of particles, one for each component of the mixture. In this way, the mixture may be represented as a superposition of continuous media, each following its own movement with the restriction imposed by the interaction between components. This means that each component will obey the laws of conservation of mass and momentum, incorporating terms to account for the interchange of mass and momentum between components. To obtain a rational theory, we must require that the properties of the mixture follow the same laws as a body of a single component, that is, that the mixture behaves as a single component body. Concha and Barrientos (1993), Concha (2001).

F. Concha A, *Solid–Liquid Separation in the Mining Industry*,
Fluid Mechanics and Its Applications 105, DOI: 10.1007/978-3-319-02484-4_2,

Treatment similar or alternative to this treatment may be found in many articles and books, such as Bowen (1976), Atkin and Crain (1976), Bedford and Drumheller (1983), Drew (1983), Truesdell (1984), Ungarish (1993), Rajagopal and Tao (1995), Drew and Passman (1998).

2.1 Kinematics

2.1.1 Body, Configuration and Type of Mixture

The term *mixture* denote a body B formed by n components $B_\alpha \subset B, \alpha = 1, 2, \ldots, n$. The elements of B_α are called *particles* and are denoted by p_α. Each body occupies a determined region of the Euclidian three-dimensional space E_3 called *configuration* of the body. The elements of the configurations are points $X_\alpha \in E_3$, whose positions are given by the position vector $\boldsymbol{r}$. Thus, the position of a particle $p_\alpha \in B_\alpha$ is given by:

$$\boldsymbol{r} = \boldsymbol{\chi}_\alpha(p_\alpha), \quad \alpha = 1, 2, \ldots, n \tag{2.1}$$

To investigate the properties of $\boldsymbol{\chi}_\alpha$ see Bowen (1976). The configuration $\chi(B)$ of the mixture is:

$$\chi(B) = \bigcup\nolimits_\alpha \chi_\alpha(B_\alpha) \tag{2.2}$$

The volume of $\chi(B)$ is called the material volume and is denoted by $V_m := V(\chi(B))$. To every body B_α we can assign a positive, continuous and additive function m_α that measures the amount of matter it contains, such that:

$$m(B) = \sum_{\alpha=1}^{n} m_\alpha(B_\alpha) \tag{2.3}$$

where m_α and $m(\mathrm{B})$ are the *masses* of the α component and of the mixture respectively. Due to the continuous nature of mass, we can define a *mass density* $\bar{\rho}_\alpha(\boldsymbol{r}, t)$ at point $\boldsymbol{r}$ and time t in the form:

$$\bar{\rho}_\alpha(\boldsymbol{r}, t) = \lim_{k \to \infty} \frac{m_\alpha(P_k)}{V_m(P_k)}, \quad \alpha = 1, 2, \ldots, n \tag{2.4}$$

where $P_{k+1} \subset P_k$ are part of the mixture having the position $\boldsymbol{r}$ in common at time t. Due to the hypothesis that mass for a continuum is an absolutely continuous function of volume, the function $\bar{\rho}_\alpha$ exists almost everywhere in B, see Drew and Passman (1999). This mass density is called the *apparent density* of B_α. The total mass of B_α can be written in terms of $\bar{\rho}_\alpha$ by:

$$m_\alpha = \int\limits_{V_m(t)} \bar{\rho}_\alpha(\boldsymbol{r}, t) dV \tag{2.5}$$

For each body $\boldsymbol{B}_\alpha$ we select a reference configuration $\chi_{\alpha\kappa}$, such that in that configuration it is the only component of the mixture (pure state). Let $\rho_{\alpha\kappa}$ be the mass density of the α component in the reference configuration and call it *material density*. Then we can write:

$$m_\alpha = \int_{V_m(t)} \bar{\rho}_\alpha(\boldsymbol{r}, t) dV = \int_{V_\kappa} \rho_{\alpha\kappa}(\boldsymbol{R}) dV \tag{2.6}$$

The material density of $\boldsymbol{B}_\alpha$ in the actual configuration is denoted by $\rho_\alpha(\boldsymbol{r}, t)$ and defines by the function $\varphi_\alpha(\boldsymbol{r}, t)$:

$$\varphi_\alpha(\boldsymbol{r}, t) = \frac{\bar{\rho}_\alpha(\boldsymbol{r}, t)}{\rho_\alpha(\boldsymbol{r}, t)}, \quad \alpha = 1, 2, \ldots, n \tag{2.7}$$

Substituting into Eq. (2.5) yields:

$$m_\alpha = \int_{V_m(t)} \bar{\rho}_\alpha dV = \int_{V_m(t)} \rho_\alpha \varphi_\alpha dV \tag{2.8}$$

The new element of volume $dV_\alpha := \varphi_\alpha dV$ is defined such that:

$$m_\alpha = \int_{V_m(t)} \bar{\rho}_\alpha dV = \int_{V_\alpha} \rho_\alpha dV_\alpha \tag{2.9}$$

The volume $V_\alpha(t)$ is called the partial volume of α and the function $\varphi_\alpha(\boldsymbol{r}, t)$ the volume fraction of $\boldsymbol{B}_\alpha$ in the present configuration. Since the sum of the partial volumes give the total volume, φ_α should obey the restriction:

$$\sum_{\alpha=1}^{n} \varphi_\alpha(\boldsymbol{r}, t) = 1 \tag{2.10}$$

We can distinguish two types of mixtures: homogeneous and heterogeneous. Homogeneous mixtures fulfil completely the condition of continuity for the material because the mixing between components occurs at the molecular level. Those mixtures are frequently called *solutions*. For homogeneous mixtures, $\bar{\rho}_\alpha$ is the concentration of the component B_α. In heterogeneous mixtures, the mixing of the components is at the macroscopic level, and for them to be considered as a continuum, the size of the integration volume V_m in the previous equations must be greater than that of the mixing level. These mixtures are also called *multiphase mixtures* because each component can be identified as a different phase. In these types of mixtures, $\varphi_\alpha(r, t)$ is a measure of the local structure of the mixture, and $\bar{\rho}_\alpha$ is called the *bulk density*.

It is sometimes convenient to define another reference configuration for B_α, such as $\chi_{\alpha c}$, with material volume V_c, that may or may not correspond to a certain instant in the motion of the mixture. The mass density of B_α in this new reference configuration is denoted by $\bar{\rho}_{\alpha c}$, which is related to $\bar{\rho}_{\alpha\kappa}$ in the following way:

$$m_\alpha = \int_{V_c} \bar{\rho}_{\alpha c} dV = \int_{V_\kappa} \rho_{\alpha\kappa} dV_\alpha \tag{2.11}$$

2.1.2 Deformation and Motion

The position of the particle in space is denoted by the *material point* p_α in the reference configuration $\boldsymbol{\chi}_{\alpha\kappa}$:

$$\boldsymbol{R}_\alpha = \boldsymbol{\chi}_{\alpha\kappa}(p_\alpha) \tag{2.12}$$

We assume that (2.12) has an inverse such that

$$p_\alpha = \boldsymbol{\chi}_{\alpha\kappa}^{-1}(\boldsymbol{R}_\alpha) \tag{2.13}$$

The *motion* of $p_\alpha \in B_\alpha$ is a continuous sequence of configurations over time:

$$\boldsymbol{r} = \boldsymbol{\chi}_\alpha(p_\alpha, t), \quad \alpha = 1, 2, \ldots, n \tag{2.14}$$

Substituting (2.13) into (2.14) yields:

$$\boldsymbol{r} = \boldsymbol{f}_\alpha(\boldsymbol{R}_\alpha, t) \tag{2.15}$$

where $\boldsymbol{f}_\alpha$ is the *deformation function* of the α component:

$$\boldsymbol{f}_\alpha = \boldsymbol{\chi}_\alpha \circ \boldsymbol{\chi}_{k\alpha}^{-1} \tag{2.16}$$

We require $\boldsymbol{f}_\alpha$ to be twice differentiable and to have an inverse, such that:

$$R_\alpha = \boldsymbol{f}_\alpha^{-1}(\boldsymbol{r}, t) \tag{2.17}$$

For a given particle $p_\alpha \in B_\alpha$, that is, for a constant $\boldsymbol{R}_\alpha$ and a variable t, the deformation function, Eq. (2.15), represents the trajectory of the particle in time, and for a constant time, the same equation represents the deformation of the body B_α from the reference configuration $\chi_{\alpha\kappa}$ to the current configuration $\chi_{\alpha t}$.

Spatial and material coordinates

The Cartesian components x_i of $\boldsymbol{r}$ and X_i^α of $\boldsymbol{R}_\alpha$ are the *spatial* and *material* coordinates of p_α:

$$\boldsymbol{r} = x_i \boldsymbol{e}_i \quad \text{and} \quad \boldsymbol{R}_\alpha = X_i^\alpha \boldsymbol{e}_i \tag{2.18}$$

Any property G_α of the body B_α can be described in terms of material or spatial coordinates. For $G_\alpha(p_\alpha, t)$ we can write either:

$$G_\alpha = G_\alpha\left(\chi_{\alpha\kappa}^{-1}(\boldsymbol{R}_\alpha, t)\right) \equiv g_{\alpha 1}(\boldsymbol{R}_\alpha, t)\,, \quad \text{or} \tag{2.19}$$

$$G_\alpha = G_\alpha\left(\chi_\alpha^{-1}(\boldsymbol{r}, t)\right) \equiv g_{\alpha 2}(\boldsymbol{r}, t) \tag{2.20}$$

Of course the properties $g_{\alpha 1}(\boldsymbol{R}_\alpha, t)$ and $g_{\alpha 2}(\boldsymbol{r}, t)$ are equivalent. We refer to the first notation as the *material property* and to the second as the *spatial property* G of the body B_α.

Since the property G_α is the function of two variables $(\boldsymbol{R}_\alpha, t)$ or $(\boldsymbol{r}, t)$, it is possible to obtain the *partial derivatives of G with respect to each of these variables* $\boldsymbol{R}_\alpha$ or $\boldsymbol{r}$. The *gradient* of G_α is the partial derivative of G_α with respect to the space variable. Since there are two such variables, there will be two gradients:

$$\text{Material gradient} \quad \frac{\partial G_\alpha}{\partial \boldsymbol{R}_\alpha} = \frac{\partial (g_{\alpha 1}(\boldsymbol{R}_\alpha, t))}{\partial \boldsymbol{R}_\alpha} = \frac{\partial g_{\alpha 1}}{\partial X_i^\alpha} \boldsymbol{e}_i \equiv \text{grad} G_\alpha \tag{2.21}$$

$$\text{Spatial gradient} \quad \frac{\partial G_\alpha}{\partial \boldsymbol{r}} = \frac{\partial (g_{\alpha 2}(\boldsymbol{r}, t))}{\partial \boldsymbol{r}} = \frac{\partial g_{\alpha 2}}{\partial x_i} \boldsymbol{e}_i \equiv \text{grad} G_\alpha \equiv \nabla G_\alpha \tag{2.22}$$

In the same way, we can define material and spatial time derivatives of G_α:

$$\text{Material derivative} \quad \left.\frac{\partial G_\alpha}{\partial t}\right|_{\boldsymbol{R}_\alpha} = \frac{\partial (g_{\alpha 1}(\boldsymbol{R}_\alpha, t))}{\partial t} = \left.\frac{\partial g_{\alpha 2}(\boldsymbol{r}, t)}{\partial t}\right|_{\boldsymbol{R}_\alpha} \equiv \frac{D G_\alpha}{D t} \equiv \dot{G}_\alpha \tag{2.23}$$

$$\text{Spatial derivative} \quad \frac{\partial G_\alpha}{\partial t} = \frac{\partial g_{\alpha 2}(\boldsymbol{r}, t)}{\partial t} = \left.\frac{\partial g_{\alpha 1}(\boldsymbol{R}_\alpha, t)}{\partial t}\right|_{\boldsymbol{r}} \tag{2.24}$$

The material derivative represents the derivative of G_α with respect to time holding the material point $\boldsymbol{R}_\alpha$ fixed, while the spatial derivative is the derivative of G_α with respect to time holding the place $\boldsymbol{r}$ fixed.

The relationship between the material and the spatial derivatives is obtained by applying the chain rule of differentiation to Eq. (2.23):

$$\begin{aligned} \dot{G}_\alpha &= \left.\frac{\partial g_{\alpha 2}(\boldsymbol{r}, t)}{\partial t}\right|_{\boldsymbol{R}_\alpha} = \frac{\partial g_{\alpha 2}(\boldsymbol{r}, t)}{\partial t} + \frac{\partial g_{\alpha 2}(\boldsymbol{r}, t)}{\partial \boldsymbol{r}} \cdot \left.\frac{\partial \boldsymbol{r}}{\partial t}\right|_{\boldsymbol{R}_\alpha} \\ &= \frac{\partial g_{\alpha 2}}{\partial t} + \nabla g_{\alpha 2} \cdot \dot{\boldsymbol{r}} \\ &= \frac{\partial g_{\alpha 2}}{\partial t} + \nabla G_\alpha \cdot \dot{\boldsymbol{r}} \end{aligned} \tag{2.25}$$

where $\boldsymbol{r}$ (with a point above) is the material derivative of the deformation function $\boldsymbol{r} = \boldsymbol{f}_\alpha(\boldsymbol{R}_\alpha, t)$.

Gradient of deformation tensor

Consider two particles $p_\alpha, q_\alpha \in B_\alpha$ having the positions $\boldsymbol{R}_\alpha$ and $\boldsymbol{R}_\alpha + d\boldsymbol{R}_\alpha$ in the reference configuration; see Fig. 2.1. At time t, their positions are:

$$\boldsymbol{r} = \boldsymbol{f}_\alpha(\boldsymbol{R}_\alpha, t) \quad \text{and} \quad \boldsymbol{r} + d\boldsymbol{r} = \boldsymbol{f}_\alpha(\boldsymbol{R}_\alpha + d\boldsymbol{R}_\alpha, t) \tag{2.26}$$

The position of q_α at time t can be approximated in the vicinity of $\boldsymbol{r}$ by a linear function of $d\boldsymbol{R}_\alpha$:

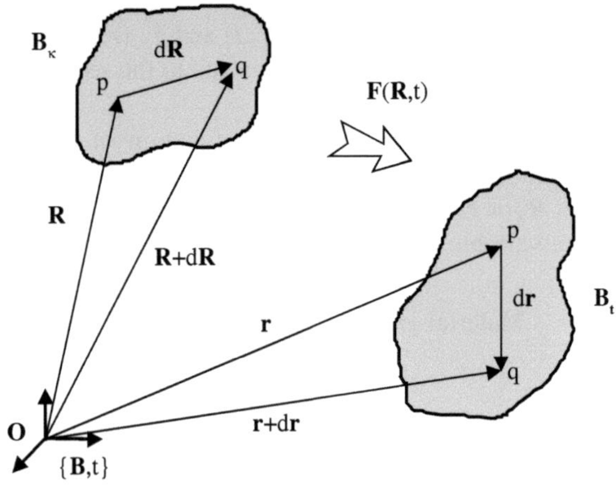

Fig. 2.1 Deformation of a body from the reference configuration $B_{\alpha\kappa}$ to the present configuration $B_{\alpha t}$

$$\boldsymbol{f}_\alpha(\boldsymbol{R}_\alpha + d\boldsymbol{R}_\alpha, t) \approx \boldsymbol{f}_\alpha(\boldsymbol{R}_\alpha, t) + \frac{\partial \boldsymbol{f}_\alpha}{\partial \boldsymbol{R}_\alpha} \cdot d\boldsymbol{R}_\alpha \tag{2.27}$$

Then, from (2.26) and (2.27) we see that:

$$d\boldsymbol{r} = \frac{\partial \boldsymbol{f}_\alpha}{\partial \boldsymbol{R}_\alpha} \cdot d\boldsymbol{R}_\alpha \tag{2.28}$$

The tensor $\partial \boldsymbol{f}_\alpha / \partial \boldsymbol{R}_\alpha$, that approximates the deformation function of $\boldsymbol{f}_\alpha$ in the neighbourhood of $\boldsymbol{r}$ is called the gradient of the *deformation tensor* of the α component, and is denoted by:

$$\boldsymbol{F}_\alpha(\boldsymbol{R}_\alpha, t) = \frac{\partial \boldsymbol{f}_\alpha(\boldsymbol{R}_\alpha, t)}{\partial \boldsymbol{R}_\alpha} = \operatorname{grad} \boldsymbol{r} \tag{2.29}$$

To ensure the existence of an inverse, det $\boldsymbol{F}_\alpha \neq 0$. In Cartesian and matrix notation the deformation tensor can be written in the form:

$$\boldsymbol{F}_\alpha(\boldsymbol{R}_\alpha, t) = \frac{\partial \boldsymbol{r}}{\partial \boldsymbol{R}_\alpha} = \boldsymbol{B}^T \begin{bmatrix} \frac{\partial x_1}{\partial X_1} & \frac{\partial x_1}{\partial X_2} & \frac{\partial x_1}{\partial X_3} \\ \frac{\partial x_2}{\partial X_1} & \frac{\partial x_2}{\partial X_2} & \frac{\partial x_2}{\partial X_3} \\ \frac{\partial x_3}{\partial X_1} & \frac{\partial x_3}{\partial X_2} & \frac{\partial x_3}{\partial X_3} \end{bmatrix} \boldsymbol{B} \tag{2.30}$$

where $\boldsymbol{B}$ is the basis of orthogonal unit vectors.

Equation (2.28) represents the transformation of a line element $d\boldsymbol{R}_\alpha$ from the reference configuration to the present configuration $d\boldsymbol{r}$:

$$d\boldsymbol{r} = \boldsymbol{F}_\alpha(\boldsymbol{R}_\alpha, t) \cdot d\boldsymbol{R}_\alpha \tag{2.31}$$

A deformation is called homogeneous if $\boldsymbol{F}_\alpha$ is independent of $\boldsymbol{R}_\alpha$.

Change of reference configuration

The deformation quantified by $\boldsymbol{F}_\alpha(\boldsymbol{R}_\alpha, t)$ depends on the reference configuration chosen. Since this reference configuration is arbitrary, it is convenient to know how a change of reference configuration affects $\boldsymbol{F}_\alpha$.

Consider two reference configurations $B_{\alpha\kappa}$ and $B_{\alpha c}$, and the actual configuration $B_{\alpha t}$ of the body B_α. Call $\boldsymbol{F}_{\alpha\kappa}$, $\boldsymbol{F}_{\alpha c}$ and $\boldsymbol{P}_\alpha$ the gradient of deformation tensors to go from $B_{\alpha\kappa}$ to $B_{\alpha t}$, from $B_{\alpha c}$ to $B_{\alpha t}$ and $B_{\alpha\kappa}$ to $B_{\alpha c}$ respectively; see Fig. 2.2. We can write:

$$d\boldsymbol{r} = \boldsymbol{F}_{\alpha\kappa} \cdot d\boldsymbol{R}_{\alpha\kappa} = \boldsymbol{F}_{\alpha c} \cdot d\boldsymbol{R}_{\alpha c} \quad \text{and} \quad \mathrm{d}\boldsymbol{R}_{\alpha c} = \boldsymbol{P}_\alpha \cdot d\boldsymbol{R}_{\alpha\kappa}$$

Then $$d\boldsymbol{r} = \boldsymbol{F}_{\alpha\kappa} \cdot d\boldsymbol{R}_{\alpha\kappa} = \boldsymbol{F} \circ (\boldsymbol{P}_\alpha \cdot d\boldsymbol{R}_{\alpha\kappa}),$$

and therefore $$\boldsymbol{F}_{\alpha\kappa} = \boldsymbol{F}_{\alpha c} \circ \boldsymbol{P}_\alpha \tag{2.32}$$

Dilatation

Consider an element of material volume $dV_{\alpha\kappa}$ in the form of a parallelepiped in the reference configuration; see Fig. 2.3, then:

$$dV_{\alpha\kappa} = d\boldsymbol{R}_{\alpha 1} \cdot d\boldsymbol{R}_{\alpha 2} \times d\boldsymbol{R}_{\alpha 3} \equiv [d\boldsymbol{R}_{\alpha 1}, d\boldsymbol{R}_{\alpha 2}, d\boldsymbol{R}_{\alpha 3}]$$

After the deformation, the volume becomes:

$$dV_m = d\boldsymbol{r}_1 \cdot d\boldsymbol{r}_2 \times d\boldsymbol{r}_3 \equiv [d\boldsymbol{r}_1, d\boldsymbol{r}_2, d\boldsymbol{r}_3]$$

Using (2.31):

$$\begin{aligned} dV_m &= [\boldsymbol{F}_{\alpha 1} \cdot d\boldsymbol{R}_{\alpha 1}, \boldsymbol{F}_{\alpha 2} \cdot d\boldsymbol{R}_{\alpha 2}, \boldsymbol{F}_{\alpha 3} \cdot d\boldsymbol{R}_{\alpha 3}] \\ &= \det \boldsymbol{F}_\alpha [d\boldsymbol{R}_{\alpha 1}, d\boldsymbol{R}_{\alpha 2}, d\boldsymbol{R}_{\alpha 3}] \\ &= \det \boldsymbol{F}_\alpha dV_{\alpha\kappa} \end{aligned} \tag{2.33}$$

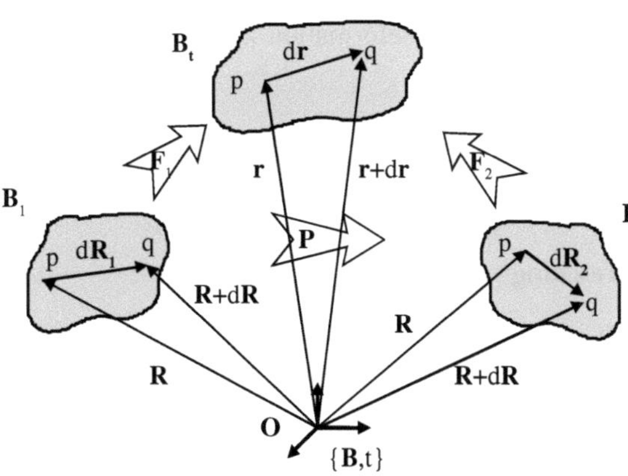

Fig. 2.2 Change of reference configuration

Fig. 2.3 Dilatation

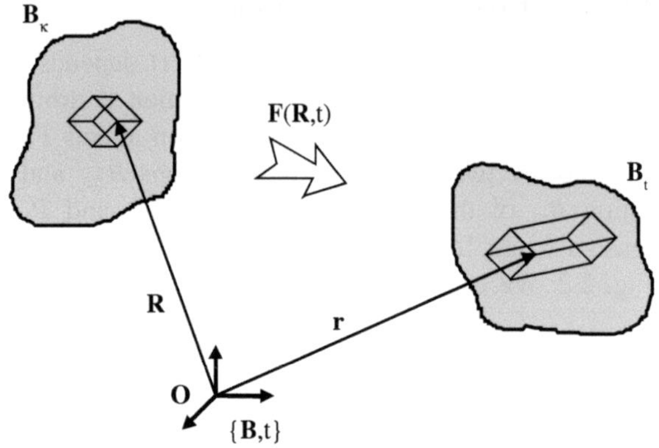

The quotient between the elements of volume before and after the deformation is called the *dilatation* of the body and is denoted by J_α, then:

$$J_\alpha = \det \boldsymbol{F}_\alpha \tag{2.34}$$

The physical meaning of dilatation, expressed by (2.33), shows that $\det \boldsymbol{F}_\alpha > 0$.

Rigid deformation

A special type of deformation is the *rigid deformation* defined as a deformation in which the distances between the particles in a body do not change. Consider two particles $p_\alpha, q_\alpha \in B_\alpha$ which, during the deformation, maintain their distance. Call $d\boldsymbol{R}_\alpha = ds_{\alpha\kappa}\boldsymbol{e}_\kappa$ and $\mathrm{d}\boldsymbol{r} = ds\boldsymbol{e}$, refer to Fig. 2.1 and write:

$$\begin{aligned} d_{\alpha s}^2 - ds_{\alpha\kappa}^2 &= 0 \\ d\boldsymbol{r}\cdot d\boldsymbol{r} - d\boldsymbol{R}_\alpha \cdot d\boldsymbol{R}_\alpha &= 0 \\ (\boldsymbol{F}_\alpha \cdot d\boldsymbol{R}_\alpha)(\boldsymbol{F}_\alpha \cdot d\boldsymbol{R}_\alpha) - d\boldsymbol{R}_\alpha \cdot \boldsymbol{I} \cdot d\boldsymbol{R}_\alpha &= 0 \\ d\boldsymbol{R}_\alpha \cdot \left(\boldsymbol{F}_\alpha^T \cdot \boldsymbol{F}_\alpha - \boldsymbol{I}\right) \cdot d\boldsymbol{R}_\alpha &= 0 \end{aligned}$$

Therefore, a rigid deformation should obey:

$$\boldsymbol{F}_\alpha^T \cdot \boldsymbol{F}_\alpha = \boldsymbol{I} \tag{2.35}$$

There are two cases for which (2.35) is valid: if $\boldsymbol{F}_\alpha = \boldsymbol{I}$, which represents a *translation,* and $\boldsymbol{F}_\alpha = \boldsymbol{Q}_\alpha$, which is a *rotation.*

Stretching

Since for the deformation function $\boldsymbol{f}_\alpha(\boldsymbol{R}_\alpha, t)$, $\det \boldsymbol{F}_\alpha > 0$, the polar decomposition (Gurtin 1981) may be applied to $\boldsymbol{f}_\alpha$:

$$\boldsymbol{F}_\alpha = \boldsymbol{Q}_\alpha \cdot \boldsymbol{U}_\alpha = \boldsymbol{V}_\alpha \cdot \boldsymbol{Q}_\alpha \tag{2.36}$$

Then,

$$U_\alpha^2 = F_\alpha^T \cdot F_\alpha = C_\alpha \quad \text{and} \quad V_\alpha^2 = F_\alpha \cdot F_\alpha^2 = B_\alpha \tag{2.37}$$

Since U_α and V_α are symmetric and positive definite tensors and Q_α is an orthogonal tensor, it can be shown that the characteristic values of U_α and V_α are the same:

$$U_\alpha = \sum_{\alpha=1}^{n} \lambda_k u_k u_k \quad \text{and} \quad V_\alpha = \sum_{\alpha=1}^{n} \lambda_k v_k v_k \tag{2.38}$$

and that the characteristic vectors are related by de rotation:

$$v_k = Q_\alpha \cdot u_k \tag{2.39}$$

Velocity and acceleration

The velocity and acceleration of a particle $p_\alpha \in B_\alpha$ are the first and second material derivatives of the deformation function $r = f_\alpha(R_\alpha, t)$:

$$v_\alpha = \frac{Df_\alpha(R_\alpha, t)}{Dt} = \frac{D_\alpha r}{Dt} = \dot{r}(R_\alpha, t) \tag{2.40}$$

$$a_\alpha = \dot{v}_\alpha = \frac{D_\alpha v_\alpha}{Dt} = \frac{D_\alpha^2 r}{Dt^2} \tag{2.41}$$

If the flow field is expressed in spatial coordinates, the acceleration is:

$$a_\alpha = \frac{D_\alpha v_\alpha}{Dt} = \left.\frac{\partial v_\alpha(R_\alpha, t)}{\partial t}\right|_r + \frac{\partial v_\alpha(R_\alpha, t)}{\partial r} \cdot \frac{\partial r(R_\alpha, t)}{\partial t} = \frac{\partial v_\alpha}{\partial t} + \nabla v_\alpha \cdot v_\alpha \tag{2.42}$$

Velocity gradient: rate of dilatation, stretching and spin

Consider two particles $p_\alpha, q_\alpha \in B_\alpha$. If p_α has a velocity of $v_\alpha(r, t)$, the velocity $v_\alpha(r + dr, t)$ of q_α, can be approximated by:

$$v_\alpha(r + dr, t) = v_\alpha(r, t) + \frac{\partial v_\alpha(r, t)}{\partial r} \cdot dr \tag{2.43}$$

Since $v_\alpha(r + dr, t) = v_\alpha(r, t) + dv_\alpha$, in analogy to 2.26–2.28,

$$dv_\alpha = \frac{\partial v_\alpha(r, t)}{\partial r} \cdot dr \equiv L_\alpha(r, t) \cdot dr \tag{2.44}$$

The linear function $L_\alpha := \partial v_\alpha / \partial r$ is called the *velocity gradient tensor*. In Cartesian tensor and matrix notations L_α can be written in the form:

$$\boldsymbol{L}_\alpha = \nabla \boldsymbol{v}_\alpha = \frac{\partial v_{\alpha i}}{\partial x_j}\boldsymbol{e}_i\boldsymbol{e}_j = \boldsymbol{B}^T \begin{bmatrix} \frac{\partial v_{\alpha 1}}{\partial x_1} & \frac{\partial v_{\alpha 1}}{\partial x_2} & \frac{\partial v_{\alpha 1}}{\partial x_3} \\ \frac{\partial v_{\alpha 2}}{\partial x_1} & \frac{\partial v_{\alpha 2}}{\partial x_2} & \frac{\partial v_{\alpha 2}}{\partial x_3} \\ \frac{\partial v_{\alpha 3}}{\partial x_1} & \frac{\partial v_{\alpha 3}}{\partial x_2} & \frac{\partial v_{\alpha 3}}{\partial x_3} \end{bmatrix} \boldsymbol{B} \tag{2.45}$$

The relationship between $\boldsymbol{L}_\alpha$ and $\boldsymbol{F}_\alpha$ can be obtained by calculating $\dot{\boldsymbol{v}}_\alpha = d\dot{\boldsymbol{r}}$ from Eq. (2.31):

$$\begin{aligned} d\dot{\boldsymbol{v}} = d\dot{\boldsymbol{r}} &= \frac{D}{Dt}(\boldsymbol{F}_\alpha \cdot d\boldsymbol{R}_\alpha) \\ &= \dot{\boldsymbol{F}}_\alpha \cdot d\boldsymbol{R}_\alpha = \dot{\boldsymbol{F}}_\alpha \cdot \boldsymbol{F}_\alpha^{-1} d\boldsymbol{r} \\ &= \boldsymbol{L}_\alpha \cdot d\boldsymbol{r} \end{aligned}$$

Therefore:

$$\boldsymbol{L}_\alpha = \dot{\boldsymbol{F}}_\alpha \boldsymbol{F}_\alpha^{-1} \tag{2.46}$$

The velocity gradient can be separated into three irreducible parts, which are mutually orthogonal:

$$\boldsymbol{L}_\alpha = \underbrace{\frac{1}{3}(\operatorname{tr}\boldsymbol{L}_\alpha)\boldsymbol{I}}_{\substack{\text{Rateofexpansion} \\ \text{orrateofdilatation} \\ \text{tensor.}}} + \underbrace{\left\{\frac{1}{2}\left(\boldsymbol{L}_\alpha + \boldsymbol{L}_\alpha^T\right) - \frac{1}{3}(\operatorname{tr}\boldsymbol{L}_\alpha)\boldsymbol{I}\right\}}_{\text{Rateofsheartensororstrechingtensor.}} + \underbrace{\frac{1}{2}\left(\boldsymbol{L}_\alpha - \boldsymbol{L}_\alpha^T\right)}_{\substack{\text{Rateofrotationtensor} \\ \text{orspintensor.}}} \tag{2.47}$$

To show that $\operatorname{tr}\boldsymbol{L}_\alpha$ represents the rate of dilatation, calculate the following:

$$\begin{aligned} \operatorname{tr}\boldsymbol{L}_\alpha = \operatorname{tr}\nabla\boldsymbol{v}_\alpha &= \frac{\partial v_{\alpha i}}{\partial x_i} \\ &= \nabla \cdot \boldsymbol{v}_\alpha \end{aligned}$$

On the other hand, take de derivative of the dilatation $J(\boldsymbol{r}, t)$:

$$\begin{aligned} \dot{J}_\alpha &= \frac{D}{Dt}(\det \boldsymbol{F}_\alpha) \\ &= \det \boldsymbol{F}_\alpha \operatorname{tr}\left(\dot{\boldsymbol{F}}_\alpha \boldsymbol{F}_\alpha^{-1}\right) \\ &= \det \boldsymbol{F}_\alpha \operatorname{tr}\boldsymbol{L}_\alpha \end{aligned} \tag{2.48}$$

From (2.48) we can write:

$$\operatorname{tr}\boldsymbol{L}_\alpha = \nabla \cdot \boldsymbol{v}_\alpha = \frac{\dot{J}}{J} \tag{2.49}$$

Equation (2.49) shows that $\mathrm{tr}\,\boldsymbol{L}_\alpha$ and $\nabla \cdot \boldsymbol{v}_\alpha$ have the meaning of *rate of dilatation per unit of dilatation.*

Defining the following terms:

$$\text{Rate of expansion tensor:} \quad \boldsymbol{L}_{\alpha E} = \frac{1}{3}(\mathrm{tr}\,\boldsymbol{L}_\alpha)\boldsymbol{I} \tag{2.50}$$

$$\text{Rate of shear(stretching:)} \quad \boldsymbol{D}_\alpha = \left\{ \frac{1}{2}\left(\boldsymbol{L}_\alpha + \boldsymbol{L}_\alpha^T\right) - \frac{1}{3}(\mathrm{tr}\,\boldsymbol{L}_\alpha)\boldsymbol{I} \right\} \tag{2.51}$$

$$\text{Rate of rotation(spin:)} \quad \boldsymbol{W}_\alpha = \frac{1}{2}\left(\boldsymbol{L}_\alpha - \boldsymbol{L}_\alpha^T\right)$$

Equation (2.47) may be written in the form:

$$\underset{\substack{\text{Velocitygradient}\\ \text{tensor}}}{\boldsymbol{L}_\alpha} = \underset{\substack{\text{Rateofexpansion}\\ \text{orrateofdilatation}\\ \text{tensor.}}}{\boldsymbol{L}_{\alpha E}} + \underset{\substack{\text{Rateofshear tensor}\\ \text{orstretchingtensor.}}}{\boldsymbol{D}_\alpha} + \underset{\substack{\text{Rateofrotationtensor}\\ \text{orspintensor.}}}{\boldsymbol{W}_\alpha} \tag{2.52}$$

2.1.3 Mass Balance

Let the rate of mass transfer, *per unit volume*, from all other components to B_α be denoted by $\bar{g}_\alpha(\boldsymbol{r}, t)$. This term $\bar{g}_\alpha(\boldsymbol{r}, t)$ receives the name of *mass growth rate* of the α component. The following balance must be obeyed:

$$\underset{\substack{\text{Mass rate of change of}\\ \text{the }\alpha\text{ component in }V_m}}{\frac{d}{dt}\int_{V_m} \bar{\rho}_\alpha dV} = \underset{\substack{\text{Net rate of generation of}\\ \text{the }\alpha\text{ componente in }V_m}}{\int_{V_m} \bar{g}_\alpha dV} \tag{2.53}$$

where dV is an element of material volume V_m of B_α. Taking the left side of (2.53) to reference configuration yields:

$$
\begin{aligned}
\frac{d}{dt}\int_{V_m} \bar{\rho}_\alpha dV &= \int_{V_\kappa} \frac{D}{Dt}(\bar{\rho}_\alpha J_\alpha) dV \\
&= \int_{V_\kappa} (\dot{\bar{\rho}}_\alpha J_\alpha + \bar{\rho}_\alpha \dot{J}_\alpha) dV \\
&= \int_{V_\kappa} (\dot{\bar{\rho}}_\alpha + \bar{\rho}_\alpha \nabla \cdot \boldsymbol{v}_\alpha) J_\alpha dV \\
&= \int_{V_m} \left(\frac{\partial \bar{\rho}_\alpha}{\partial t} + \nabla \cdot \bar{\rho}_\alpha \boldsymbol{v}_\alpha \right) dV \\
&= \int_{V_m} \frac{\partial \bar{\rho}_\alpha}{\partial t} dV + \int_{V_m} \nabla \cdot \bar{\rho}_\alpha \boldsymbol{v}_\alpha dV \\
&= \int_{V_m} \frac{\partial \bar{\rho}_\alpha}{\partial t} dV + \oint_{S_m} \bar{\rho}_\alpha \boldsymbol{v}_\alpha \cdot \boldsymbol{n} dV
\end{aligned}
\tag{2.54}
$$

Substituting in (2.53) gives a new form of the mass balance of B_α:

$$
\int_{V_m} \frac{\partial \bar{\rho}_\alpha}{\partial t} dV + \oint_{S_m} \bar{\rho}_\alpha (\boldsymbol{v}_\alpha \cdot \boldsymbol{n}) dV = \int_{V_m} \bar{g}_\alpha dV \tag{2.55}
$$

On the other hand, both volume integrals in (2.53) maybe taken to the reference configuration to obtain:

$$
\int_{V_\kappa} \left(\frac{D}{Dt}(\bar{\rho}_\alpha J_\alpha) - \bar{g}_\alpha J_\alpha \right) dV = 0 \tag{2.56}
$$

Performing the material derivative:

$$
\begin{aligned}
&\int_{V_\kappa} (\dot{\bar{\rho}}_\alpha J_\alpha + \bar{\rho}_\alpha \dot{J}_\alpha - \bar{g}_\alpha J_\alpha) dV = 0 \\
&\int_{V_\kappa} (\dot{\bar{\rho}}_\alpha J_\alpha + \bar{\rho}_\alpha \nabla \cdot \boldsymbol{v}_\alpha J_\alpha - \bar{g}_\alpha J_\alpha) dV = 0 \\
&\int_{V_\kappa} (\dot{\bar{\rho}}_\alpha + \bar{\rho}_\alpha \nabla \cdot \boldsymbol{v}_\alpha - \bar{g}_\alpha) J_\alpha dV = 0 \\
&\int_{V_m} (\dot{\bar{\rho}}_\alpha + \bar{\rho}_\alpha \nabla \cdot \boldsymbol{v}_\alpha - \bar{g}_\alpha) dV = 0
\end{aligned}
$$

Using the localization theorem (Gurtin 1981) yields:

$$
\dot{\bar{\rho}}_\alpha + \bar{\rho}_\alpha \nabla \cdot \boldsymbol{v}_\alpha = \bar{g}_\alpha \tag{2.57}
$$

Writing the material derivative in terms of the spatial derivative and combining the result with the second term of Eq. (2.57) gives:

$$
\frac{\partial \bar{\rho}_\alpha}{\partial t} + \nabla \cdot \bar{\rho}_\alpha \boldsymbol{v}_\alpha = \bar{g}_\alpha \tag{2.58}
$$

Equations (2.57) or (2.58) receive the name of *continuity equation*. The last one has a *conservation form*.

Going back to Eq. (2.56), the localization theorem is used directly on this equation to give:

$$\frac{D}{Dt}(\bar{\rho}_\alpha J_\alpha) = \bar{g}_\alpha J_\alpha \tag{2.59}$$

This expression divided by $\bar{\rho}_\alpha J_\alpha$ represent the rate *by unit mass* of growth of the mass of the α component, and is denoted by $\widehat{g}_\alpha = \bar{g}_\alpha / \bar{\rho}_\alpha$. Integrate with boundary condition $\bar{\rho}_\alpha(\boldsymbol{R}_\alpha) = \bar{\rho}_{\alpha K}$ to give:

$$\bar{\rho}_\alpha J_\alpha = \bar{\rho}_{\alpha K} \exp\left(\int_{t_\kappa}^{t} \widehat{g}_\alpha(\tau) d\tau\right) \tag{2.60}$$

In those cases in which there is no mass transfer between components, $\widehat{g}_\alpha = 0$, Eq. (2.60) reduces to:

$$\bar{\rho}_\alpha J_\alpha = \bar{\rho}_{\alpha K} \tag{2.61}$$

Equation (2.61) is the local mass balance for a body B_α that deforms from the reference to the actual configuration.

Taking the material derivative of (2.59) we can obtain the continuity equation:

$$\begin{aligned} \dot{\bar{\rho}}_\alpha J_\alpha + \bar{\rho}_\alpha \dot{J}_\alpha &= J_\alpha \bar{g}_\alpha \\ \dot{\bar{\rho}}_\alpha J_\alpha + \bar{\rho}_\alpha \nabla \cdot \boldsymbol{v}_\alpha J_\alpha &= J_\alpha \bar{g}_\alpha \\ \dot{\bar{\rho}}_\alpha + \bar{\rho}_\alpha \nabla \cdot \boldsymbol{v}_\alpha &= \bar{g}_\alpha \end{aligned} \tag{2.62}$$

Check this equation with (2.57).

Mass balance in a discontinuity

For bodies having discontinuities, the local mass balance equations are not valid. In these cases, it is necessary to analyze the macroscopic mass balance further. Consider a body $B_\alpha \in B$ having a surface of discontinuity S_I that separates the body into two parts B_α^+ and B_α^- in the actual configuration; see Fig. 2.4. The following conditions hold:

$$V_m = V^+ + V^-, \quad S_m = S^+ + S^-, \quad S_I = B_\alpha^+ \cap B_\alpha^- \tag{2.63}$$

Applying the macroscopic balance (2.55) to each side of the body, and noting that the surface of discontinuity is not a material surface, yields:

$$\begin{aligned} &\int_{V^+} \frac{\partial \bar{\rho}_\alpha}{\partial t} dV + \int_{V^-} \frac{\partial \bar{\rho}_\alpha}{\partial t} dV + \int_{S^+} \bar{\rho}_\alpha (\boldsymbol{v}_\alpha \cdot \boldsymbol{n}) dS + \int_{S_I} \bar{\rho}_\alpha^+ \left(v_\alpha^+ - \boldsymbol{v}_I\right) \cdot (-\boldsymbol{e}_I) dS \\ &\quad + \int_{S^-} \bar{\rho}_\alpha (\boldsymbol{v}_\alpha \cdot \boldsymbol{n}) dS + \int_{S_I} \bar{\rho}_\alpha^- \left(\boldsymbol{v}_\alpha^- - \boldsymbol{v}_I\right) \cdot \boldsymbol{e}_I dS = \int_{V^+} \bar{g}_\alpha dV + \int_{V^-} \bar{g}_\alpha dV \end{aligned}$$

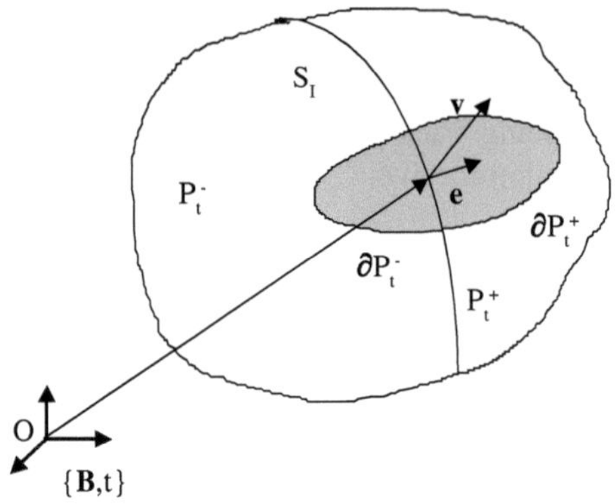

Fig. 2.4 Body B_α with a surface of discontinuity

Adding the volume and surface integrals and using (2.63), we have:

$$\int_{V_m} \frac{\partial \bar{\rho}_\alpha}{\partial t} dV + \int_{S_m} \bar{\rho}_\alpha (\boldsymbol{v}_\alpha \cdot \boldsymbol{n}) dS - \int_{S_I} \left(\bar{\rho}_\alpha^+ \left(\boldsymbol{v}_\alpha^+ - \boldsymbol{v}_I \right) - \bar{\rho}_\alpha^- \left(\boldsymbol{v}_\alpha^- - \boldsymbol{v}_I \right) \right) \cdot \boldsymbol{e}_I dS$$
$$= \int_{V^+} \bar{g}_\alpha dV + \int_{V^-} \bar{g}_\alpha dV$$

Using Eq. (2.55), the previous equation reduces to:

$$\int_{S_I} \left(\bar{\rho}_\alpha^+ \left(\boldsymbol{v}_\alpha^+ - \boldsymbol{v}_I \right) - \bar{\rho}_\alpha^- \left(\boldsymbol{v}_\alpha^- - \boldsymbol{v}_I \right) \right) \cdot \boldsymbol{e}_I dS = 0$$
$$\int_{S_I} \left[\bar{\rho}_\alpha \left(\boldsymbol{v}_\alpha - \boldsymbol{v}_I \right) \cdot \boldsymbol{e}_I \right] dS = 0$$

where the jump of a property G is defined as $[G] = G^+ - G^-$. This equation is called the *macroscopic mass jump balance* at a discontinuity. Using the localization theorem in the previous equation, we obtain the *local mass jump balance* at a discontinuity:

$$\left[\bar{\rho}_\alpha \left(\boldsymbol{v}_\alpha - \boldsymbol{v}_I \right) \cdot \boldsymbol{e}_I \right] = 0 \tag{2.64}$$

This equation can also be written in the following form called the *Rankin-Hugoniot* jump condition (Bustos et al. 1999):

$$\sigma = \frac{\left[\bar{\rho}_\alpha \boldsymbol{v}_\alpha \cdot \boldsymbol{e}_I \right]}{\left[\bar{\rho}_\alpha \right]} \tag{2.65}$$

where $\sigma = \boldsymbol{v}_I \cdot \boldsymbol{e}_I$ is the displacement velocity of the discontinuity.

Average properties of the mixture

Adding the continuity Eq. (2.58) and the mass jump balance (2.65) for all components, those properties for the mixture may be obtained:

$$\frac{\partial}{\partial t}\left(\sum_{\alpha=1}^{n}\bar{\rho}_\alpha\right)+\nabla\cdot\sum_{\alpha=1}^{n}\bar{\rho}_\alpha \boldsymbol{v}_\alpha=\sum_{\alpha=1}^{n}\bar{g}_\alpha \tag{2.66}$$

$$\sigma=\frac{\left[\sum_{\alpha=1}^{n}\bar{\rho}_\alpha \boldsymbol{v}_\alpha\cdot \boldsymbol{e}_I\right]}{\left[\sum_{\alpha=1}^{n}\bar{\rho}_\alpha\right]} \tag{2.67}$$

According to the initial postulates, the mixture should follow the laws of a pure material; therefore, the continuity equation and the mass jump condition for the mixture should be:

$$\frac{\partial \rho}{\partial t}+\nabla\cdot\rho\boldsymbol{v}=0 \tag{2.68}$$

$$\sigma=\frac{[\rho\boldsymbol{v}\cdot\boldsymbol{e}_I]}{[\rho]} \tag{2.69}$$

where ρ and $\boldsymbol{v}$ are the *mass density* and *mass average* or *convective velocity* of the mixture. Comparing Eqs. (2.66) and (2.67) with (2.68) and (2.69) respectively, we deduce the following definitions for the mixture properties:

$$\text{Mass density}\quad \rho=\sum_{\alpha=}^{n}\bar{\rho}_\alpha \tag{2.70}$$

$$\text{Mass average velocity}\quad \boldsymbol{v}=\frac{\sum_{\alpha=1}^{n}\bar{\rho}_\alpha\boldsymbol{v}_\alpha}{\sum_{\alpha=1}^{n}\bar{\rho}_\alpha}=\frac{\sum_{\alpha=1}^{n}\bar{\rho}_\alpha\boldsymbol{v}_\alpha}{\rho} \tag{2.71}$$

$$\text{Mass growth rate}\quad \sum_{\alpha=1}^{n}\bar{g}_\alpha=0 \tag{2.72}$$

This last equation indicates that no net production of mass occurs.

Convective diffusion equation

Sometimes it is convenient to express the continuity equation of each component in terms of the *convective mass flux density* $\boldsymbol{j}_{\alpha C}=\bar{\rho}_\alpha\boldsymbol{v}$ of that component. Adding and subtracting the convective flux per unit volume $\nabla\cdot\bar{\rho}_\alpha\boldsymbol{v}$, yields:

$$\frac{\partial\bar{\rho}_\alpha}{\partial t}+\nabla\cdot\bar{\rho}_\alpha\boldsymbol{v}=-\nabla\cdot\bar{\rho}_\alpha(\boldsymbol{v}_\alpha-\boldsymbol{v})+\bar{g}_\alpha$$

Defining the diffusive flux density by $\boldsymbol{j}_{\alpha D}:=\bar{\rho}_\alpha(\boldsymbol{v}_\alpha-\boldsymbol{v})=\bar{\rho}_\alpha\boldsymbol{u}_\alpha$, where $\boldsymbol{u}_\alpha=\boldsymbol{v}_\alpha-\boldsymbol{v}$ is the diffusion velocity, we can write:

$$\frac{\partial \bar{\rho}_\alpha}{\partial t} + \nabla \cdot \bar{\rho}_\alpha \boldsymbol{v} = -\nabla \cdot \bar{\rho}_\alpha (\boldsymbol{v}_\alpha - \boldsymbol{u}_\alpha) + \bar{g}_\alpha \tag{2.73}$$

Summing this equation over α yields.

$$\frac{\partial}{\partial t}\left(\sum_{\alpha=1}^{n} \bar{\rho}_\alpha\right) + \left(\nabla \cdot \sum_{\alpha=1}^{n} \bar{\rho}_\alpha \boldsymbol{v}\right) = -\nabla \cdot (\bar{\rho}_\alpha \boldsymbol{u}_\alpha) + \sum_{\alpha=1}^{n} \bar{g}_\alpha$$

Using the definitions (2.70)–(2.72), gives:

$$\sum_{\alpha=1}^{n} j_{\alpha D} = \sum_{\alpha=1}^{n} \bar{\rho}_\alpha \boldsymbol{u}_\alpha = 0 \tag{2.74}$$

Mass balance for incompressible mixtures

Incompressible mixtures are those having incompressible components, realizing that the mixture itself can be compressible, because the volume fraction of the components may be changing.

Using the concept of volume fraction given in (2.7), the continuity equation can be written in the form:

$$\frac{\partial}{\partial t}(\bar{\rho}_\alpha \varphi_\alpha) + \nabla \cdot (\bar{\rho}_\alpha \varphi_\alpha \boldsymbol{v}_\alpha) = \bar{g}_\alpha \tag{2.75}$$

$$\sigma = \frac{[\bar{\rho}_\alpha \varphi_\alpha \boldsymbol{v}_\alpha \cdot \boldsymbol{e}_I]}{[\bar{\rho}_\alpha \varphi_\alpha]} \tag{2.76}$$

Since, for an incompressible component ρ_α is constant, dividing by ρ_α the mass balances become volume balances:

$$\frac{\partial \varphi_\alpha}{\partial t} + \nabla \cdot (\varphi_\alpha \boldsymbol{v}_\alpha) = \hat{g}_\alpha \varphi_\alpha, \quad \sigma = \frac{[\varphi_\alpha \boldsymbol{v}_\alpha \cdot \boldsymbol{e}_I]}{[\varphi_\alpha]} \tag{2.77}$$

Summing all components yields:

$$\frac{\partial}{\partial t}\left(\sum_{\alpha=1}^{n} \varphi_\alpha\right) + \nabla \cdot \left(\sum_{\alpha=1}^{n} \varphi_\alpha \boldsymbol{v}_\alpha\right) = \sum_{\alpha=1}^{n} \hat{g}_\alpha \varphi_\alpha \quad \sigma = \frac{\left[\sum_{\alpha=1}^{n} \varphi_\alpha \boldsymbol{v}_\alpha \cdot \boldsymbol{e}_I\right]}{\left[\sum_{\alpha=1}^{n} \varphi_\alpha\right]} \tag{2.78}$$

Using the restriction $\sum_{\alpha=1}^{n} \varphi_\alpha = 1$, and defining the volume average velocity $\boldsymbol{q}$ by :

$$\boldsymbol{q} = \sum_{\alpha=1}^{n} \varphi_\alpha \boldsymbol{v}_\alpha, \tag{2.79}$$

the mass balance equation and the mass jump condition for the mixture become:

$$\nabla \cdot \boldsymbol{q} = \sum_{\alpha=1}^{n} \hat{g}_\alpha \varphi_\alpha \quad [\boldsymbol{q} \cdot \boldsymbol{e}_I] = 0 \tag{2.80}$$

The last equation shows that the volume average velocity suffers no jump across a surface of discontinuity.

2.2 Dynamical Processes

2.2.1 Linear Momentum Balance

Applying the axiom of linear momentum and the Cauchy stress principle to each body B_α, we arrive to the *macroscopic balance of linear momentum*:

$$\underbrace{\frac{d}{dt}\int_{V_m} \bar{\rho}_\alpha \boldsymbol{v}_\alpha dV}_{\substack{\text{Rateofchangeoflinear}\\ \text{momentumof}B_\alpha.}} = \underbrace{\int_{S_m} \boldsymbol{T}_\alpha \cdot \boldsymbol{n} dS}_{\substack{\text{Diffusivefluxof}\\ \text{linermomentumin}B_\alpha.}} + \underbrace{\int_{V_m} (\boldsymbol{b}_\alpha + \boldsymbol{m}_\alpha + \bar{g}_\alpha \boldsymbol{v}_\alpha) dV}_{\substack{\text{Sourceoflinearmomentumduetobody}\\ \text{forces,interactionforces and massgeneration.}}} \tag{2.81}$$

where $\boldsymbol{T}_\alpha$ is the stress tensor field, called *partial stress* and $\boldsymbol{b}_\alpha$ is the body force on B_α, $\boldsymbol{m}_\alpha$ is the interaction force between components, that is, the force by unit volume exerted on B_α by all other components and $\bar{g}_\alpha$ is the rate of mass growth, as defined earlier.

Using the Green-Gauss-Ostrogradsky (GGO) theorem on the surface integral yields:

$$\frac{d}{dt}\int_{V_m} \bar{\rho}_\alpha \boldsymbol{v}_\alpha dV = \int_{V_m} (\nabla \cdot \boldsymbol{T}_\alpha + \boldsymbol{b}_\alpha + \boldsymbol{m}_\alpha + \bar{g}_\alpha \boldsymbol{v}_\alpha) dV$$

Making a change of reference configuration on the left-hand side and taking the material derivative:

$$\begin{aligned}
\int_{V_{\alpha\kappa}} \left(J_\alpha \frac{D}{Dt}(\bar{\rho}_\alpha \boldsymbol{v}_\alpha) + \bar{\rho}_\alpha \boldsymbol{v}_\alpha \frac{DJ_\alpha}{Dt} \right) dV &= \int_{V_m} (\nabla \cdot \boldsymbol{T}_\alpha + \boldsymbol{b}_\alpha + \boldsymbol{m}_\alpha + \bar{g}_\alpha \boldsymbol{v}_\alpha) dV \\
\int_{V_{\alpha\kappa}} \left(\frac{D}{Dt}(\bar{\rho}_\alpha \boldsymbol{v}_\alpha) + \bar{\rho}_\alpha \boldsymbol{v}_\alpha \nabla \cdot \boldsymbol{v}_\alpha \right) J_\alpha dV &= \int_{V_m} (\nabla \cdot \boldsymbol{T}_\alpha + \boldsymbol{b}_\alpha + \boldsymbol{m}_\alpha + \bar{g}_\alpha \boldsymbol{v}_\alpha) dV \\
\int_{V_m} \left(\frac{D}{Dt}(\bar{\rho}_\alpha \boldsymbol{v}_\alpha) + \bar{\rho}_\alpha \boldsymbol{v}_\alpha \nabla \cdot \boldsymbol{v}_\alpha \right) dV &= \int_{V_m} (\nabla \cdot \boldsymbol{T}_\alpha + \boldsymbol{b}_\alpha + \boldsymbol{m}_\alpha + \bar{g}_\alpha \boldsymbol{v}_\alpha) dV
\end{aligned} \tag{2.82}$$

Changing the material to spatial derivative yields:

$$\begin{aligned}
&\int_{V_m} \left(\frac{\partial}{\partial t}(\bar{\rho}_\alpha \boldsymbol{v}_\alpha) + \bar{\rho}_\alpha \boldsymbol{v}_\alpha \cdot \nabla \boldsymbol{v}_\alpha + \bar{\rho}_\alpha \boldsymbol{v}_\alpha \nabla \cdot \boldsymbol{v}_\alpha \right) dV = \int_{V_m} (\nabla \cdot \boldsymbol{T}_\alpha + \boldsymbol{b}_\alpha + \boldsymbol{m}_\alpha + \bar{g}_\alpha \boldsymbol{v}_\alpha) dV \\
&\int_{V_m} \left(\frac{\partial}{\partial t}(\bar{\rho}_\alpha \boldsymbol{v}_\alpha) + \nabla \cdot (\bar{\rho}_\alpha \boldsymbol{v}_\alpha \boldsymbol{v}_\alpha) \right) dV = \int_{V_m} (\nabla \cdot \boldsymbol{T}_\alpha + \boldsymbol{b}_\alpha + \boldsymbol{m}_\alpha + \bar{g}_\alpha \boldsymbol{v}_\alpha) dV \\
&\int_{V_m} \left(\frac{\partial}{\partial t}(\bar{\rho}_\alpha \boldsymbol{v}_\alpha) + \nabla \cdot (\bar{\rho}_\alpha \boldsymbol{v}_\alpha \boldsymbol{v}_\alpha) - \nabla \cdot \boldsymbol{T}_\alpha - \boldsymbol{b}_\alpha - \boldsymbol{m}_\alpha - \bar{g}_\alpha \boldsymbol{v}_\alpha \right) dV = 0
\end{aligned} \tag{2.83}$$

Using the localization theorem (Gurtin 1981) leads to the *linear momentum balance* in the *conservation form*:

$$\frac{\partial}{\partial t}(\bar{\rho}_\alpha \boldsymbol{v}_\alpha) + \nabla \cdot (\bar{\rho}_\alpha \boldsymbol{v}_\alpha \boldsymbol{v}_\alpha) = \nabla \cdot \boldsymbol{T}_\alpha + \boldsymbol{b}_\alpha + \boldsymbol{m}_\alpha + \bar{g}_\alpha \boldsymbol{v}_\alpha \tag{2.84}$$

If instead we take the derivative of (2.82) in the following form:

$$\begin{aligned}
&\int_{V_m} \left(\frac{D}{Dt}(\bar{\rho}_\alpha \boldsymbol{v}_\alpha) + \bar{\rho}_\alpha \boldsymbol{v}_\alpha \nabla \cdot \boldsymbol{v}_\alpha \right) dV = \int_{V_m} (\nabla \cdot \boldsymbol{T}_\alpha + \boldsymbol{b}_\alpha + \boldsymbol{m}_\alpha + \bar{g}_\alpha \boldsymbol{v}_\alpha) dV \\
&\int_{V_m} (\dot{\bar{\rho}}_\alpha \boldsymbol{v}_\alpha + \bar{\rho}_\alpha \dot{\boldsymbol{v}}_\alpha + \bar{\rho}_\alpha \boldsymbol{v}_\alpha \nabla \cdot \boldsymbol{v}_\alpha - \nabla \cdot \boldsymbol{T}_\alpha - \boldsymbol{b}_\alpha - \boldsymbol{m}_\alpha - \bar{g}_\alpha \boldsymbol{v}_\alpha) dV = 0 \\
&\int_{V_m} \left(\underbrace{(\dot{\bar{\rho}}_\alpha + \bar{\rho}_\alpha \nabla \cdot \boldsymbol{v}_\alpha - \bar{g}_\alpha)}_{\text{by continuity equation} = 0} \boldsymbol{v}_\alpha + \bar{\rho}_\alpha \dot{\boldsymbol{v}}_\alpha + -\nabla \cdot \boldsymbol{T}_\alpha - \boldsymbol{b}_\alpha - \boldsymbol{m}_\alpha \right) dV = 0 \\
&\int_{V_m} (\bar{\rho}_\alpha \dot{\boldsymbol{v}}_\alpha + -\nabla \cdot \boldsymbol{T}_\alpha - \boldsymbol{b}_\alpha - \boldsymbol{m}_\alpha) dV = 0\,,
\end{aligned}$$

and using the localization theorem:

$$\bar{\rho}_\alpha \dot{\boldsymbol{v}}_\alpha = \nabla \cdot \boldsymbol{T}_\alpha + \boldsymbol{b}_\alpha + \boldsymbol{m}_\alpha \tag{2.85}$$

Linear momentum jump balance

In regions having discontinuities, Eqs. (2.84) and (2.85) are still valid on each side of the discontinuity, but they are not valid at the discontinuity. Following a procedure similar to that used previously for the mass jump balance, we write the last equation of (2.83) in the form:

$$\int_{V_m} \left(\frac{\partial}{\partial t}(\bar{\rho}_\alpha \boldsymbol{v}_\alpha) - \boldsymbol{b}_\alpha - \boldsymbol{m}_\alpha - \bar{g}_\alpha \boldsymbol{v}_\alpha\right) dV = -\int_{V_m} (\nabla \cdot (\bar{\rho}_\alpha \boldsymbol{v}_\alpha \boldsymbol{v}_\alpha) - \nabla \cdot \boldsymbol{T}_\alpha) dV$$

$$\int_{V_m} \left(\frac{\partial}{\partial t}(\bar{\rho}_\alpha \boldsymbol{v}_\alpha) - \boldsymbol{b}_\alpha - \boldsymbol{m}_\alpha - \bar{g}_\alpha \boldsymbol{v}_\alpha\right) dV = -\oint_{S_m} \bar{\rho}_\alpha \boldsymbol{v}_\alpha (\boldsymbol{v}_\alpha \cdot \boldsymbol{n}) dS - \oint_{S_m} \boldsymbol{T}_\alpha \cdot \boldsymbol{n} dS$$

Applying this equation to each side of the discontinuity yields:

$$\int_{V^+} \left(\frac{\partial}{\partial t}(\bar{\rho}_\alpha \boldsymbol{v}_\alpha) - \boldsymbol{b}_\alpha - \boldsymbol{m}_\alpha - \bar{g}_\alpha \boldsymbol{v}_\alpha\right) dV + \int_{V^-} \left(\frac{\partial}{-}(\bar{\rho}_\alpha \boldsymbol{v}_\alpha) - \boldsymbol{b}_\alpha - \boldsymbol{m}_\alpha - \bar{g}_\alpha \boldsymbol{v}_\alpha\right) dV$$

$$-x = -\oint_{S^+} \bar{\rho}_\alpha \boldsymbol{v}_\alpha (\boldsymbol{v}_\alpha \cdot \boldsymbol{n}) dS - \int_{S_I} \bar{\rho}_\alpha^+ \boldsymbol{v}_\alpha^+ \left(\boldsymbol{v}_\alpha^+ - \boldsymbol{v}_I\right) \cdot (-\boldsymbol{e}_I) dS - \oint_{S^-} \bar{\rho}_\alpha \boldsymbol{v}_\alpha (\boldsymbol{v}_\alpha \cdot \boldsymbol{n}) dS$$

$$-\oint_{S_I} \bar{\rho}_\alpha^- \boldsymbol{v}_\alpha^- \left(\boldsymbol{v}_\alpha^- - \boldsymbol{v}_I\right) \cdot \boldsymbol{e}_I dS - \oint_{S^+} \boldsymbol{T}_\alpha \cdot \boldsymbol{n} dS - \int_{S_I} \boldsymbol{T}_\alpha^+ \cdot (-\boldsymbol{e}_I) dS - \oint_{S^-} \boldsymbol{T}_\alpha \cdot \boldsymbol{n} dS$$

$$-\int_{S_I} \boldsymbol{T}_\alpha^- \cdot \boldsymbol{e}_I dS$$

Adding integrals with (+) and (–) and defining the jump of a property G by $[G] = G^+ - G^-$, yields:

$$\begin{aligned} &\int_{V_m} \left(\frac{\partial}{\partial t}(\bar{\rho}_\alpha \boldsymbol{v}_\alpha) - \boldsymbol{b}_\alpha - \boldsymbol{m}_\alpha - \bar{g}_\alpha \boldsymbol{v}_\alpha\right) dV - \oint_{S_m} \bar{\rho}_\alpha \boldsymbol{v}_\alpha (\boldsymbol{v}_\alpha \cdot \boldsymbol{n}) dS - \oint_{S_m} \boldsymbol{T}_\alpha \cdot \boldsymbol{n} dS \\ &\quad = -\int_{S_I} [\bar{\rho}_\alpha \boldsymbol{v}_\alpha (\boldsymbol{v}_\alpha - \boldsymbol{v}_I) \cdot \boldsymbol{e}_I] dS - \int_{S_I} [\boldsymbol{T}_\alpha \cdot \boldsymbol{e}_I] dS \end{aligned} \tag{2.86}$$

The left hand-side of (2.86) is zero by the macroscopic linear momentum balance, so that:

$$\int_{S_I} [\bar{\rho}_\alpha \boldsymbol{v}_\alpha (\boldsymbol{v}_\alpha - \boldsymbol{v}_I) \cdot \boldsymbol{e}_I] dS + \int_{S_I} [\boldsymbol{T}_\alpha \cdot \boldsymbol{e}_I] dS = \int_{S_I} ([\bar{\rho}_\alpha \boldsymbol{v}_\alpha (\boldsymbol{v}_\alpha - \boldsymbol{v}_I) \cdot \boldsymbol{e}_I] + [\boldsymbol{T}_\alpha \cdot \boldsymbol{e}_I]) dS = 0$$

Applying the localization theorem (Gurtin 1981) yields the *linear momentum jump balance* for the α component:

$$[\bar{\rho}_\alpha \boldsymbol{v}_\alpha (\boldsymbol{v}_\alpha - \boldsymbol{v}_I) \cdot \boldsymbol{e}_I] - [\boldsymbol{T}_\alpha \cdot \boldsymbol{e}_I] = 0, \text{ or } \sigma[\bar{\rho}_\alpha \boldsymbol{v}_\alpha] = [\bar{\rho}_\alpha \boldsymbol{v}_\alpha (\boldsymbol{v}_I \cdot \boldsymbol{e}_I)] - [\boldsymbol{T}_\alpha \cdot \boldsymbol{e}_I] \tag{2.87}$$

where $\sigma = [\boldsymbol{v}_I \cdot \boldsymbol{e}_I]$ is the displacement velocity of the discontinuity.

Linear momentum balance for a mixture
Summing Eq. (2.84) for all component results in:

$$\frac{\partial}{\partial t}\sum_{\alpha=1}^{n}(\bar{\rho}_\alpha \boldsymbol{v}_\alpha) + \nabla\cdot\sum_{\alpha=1}^{n}(\bar{\rho}_\alpha \boldsymbol{v}_\alpha \boldsymbol{v}_\alpha) = \nabla\cdot\sum_{\alpha=1}^{n}\boldsymbol{T}_\alpha + \sum_{\alpha=1}^{n}(\boldsymbol{b}_\alpha + \boldsymbol{m}_\alpha + \bar{g}_\alpha \boldsymbol{v}_\alpha)$$

Substituting the component velocity by the diffusion velocity by means of equation $\boldsymbol{u}_\alpha = \boldsymbol{v}_\alpha - \boldsymbol{v}$ in the second term of the left-hands side yields:

$$\frac{\partial}{\partial t}\sum_{\alpha=1}^{n}(\bar{\rho}_\alpha \boldsymbol{v}_\alpha) + \nabla\cdot\sum_{\alpha=1}^{n}(\bar{\rho}_\alpha(\boldsymbol{u}_\alpha + \boldsymbol{v})(\boldsymbol{u}_\alpha + \boldsymbol{v})) = \nabla\cdot\sum_{\alpha=1}^{n}\boldsymbol{T}_\alpha + \sum_{\alpha=1}^{n}(\boldsymbol{b}_\alpha + \boldsymbol{m}_\alpha + \bar{g}_\alpha \boldsymbol{v}_\alpha)$$

$$\frac{\partial}{\partial t}\sum_{\alpha=1}^{n}(\bar{\rho}_\alpha \boldsymbol{v}_\alpha) + \nabla\cdot\sum_{\alpha=1}^{n}(\bar{\rho}_\alpha \boldsymbol{u}_\alpha \boldsymbol{v}) + \nabla\cdot\sum_{\alpha=1}^{n}(\bar{\rho}_\alpha \boldsymbol{v}\boldsymbol{u}_\alpha) + \nabla\cdot\sum_{\alpha=1}^{n}(\bar{\rho}_\alpha \boldsymbol{v}\boldsymbol{v})$$
$$= \nabla\cdot\sum_{\alpha=1}^{n}\boldsymbol{T}_\alpha - \nabla\cdot\sum_{\alpha=1}^{n}(\bar{\rho}_\alpha \boldsymbol{u}_\alpha \boldsymbol{u}_\alpha) + \sum_{\alpha=1}^{n}(\boldsymbol{b}_\alpha + \boldsymbol{m}_\alpha + \bar{g}_\alpha \boldsymbol{v}_\alpha)$$

Using the definitions (2.70)–(2.72) we get:

$$\frac{\partial \rho}{\partial t} + \nabla\cdot(\rho \boldsymbol{v}\boldsymbol{v}) = \nabla\cdot\left(\sum_{\alpha=1}^{n}\boldsymbol{T}_\alpha - \sum_{\alpha=1}^{n}(\bar{\rho}_\alpha \boldsymbol{u}_\alpha \boldsymbol{u}_\alpha)\right) + \sum_{\alpha=1}^{n}(\boldsymbol{b}_\alpha + \boldsymbol{m}_\alpha + \bar{g}_\alpha \boldsymbol{v}_\alpha)$$

For the mixture the linear momentum of a single component should be valid, then:

$$\frac{\partial \rho}{\partial t} + \nabla\cdot\rho \boldsymbol{v}\boldsymbol{v} = \nabla\cdot\boldsymbol{T} + \boldsymbol{b}$$

Comparing the last two equations we conclude that it is necessary that:

$$\boldsymbol{T} = \boldsymbol{T}_I - \sum_{\alpha=1}^{n}\bar{\rho}_\alpha \boldsymbol{u}_\alpha \boldsymbol{u}_\alpha, \quad \boldsymbol{b} = \sum_{\alpha=1}^{n}\boldsymbol{b}_\alpha, \quad \sum_{\alpha=1}^{n}(\boldsymbol{m}_\alpha + \bar{g}_\alpha \boldsymbol{v}) = 0 \tag{2.88}$$

$$\text{with} \quad \boldsymbol{T}_I = \sum_{\alpha=1}^{n}\boldsymbol{T}_\alpha \tag{2.89}$$

The term $\boldsymbol{T}_I$ receives the name of the internal part of the stress tensor (Truesdell 1984). The last term in (2.88) indicates that no net production of linear momentum exists, and that the growth in one component is done at the expense of the linear momentum of the other components.

2.2.2 Angular Momentum Balance

The application of Euler's second law for the angular momentum and Cauchy's stress principle to the α component of the body gives the *macroscopic angular momentum balance*:

$$\frac{d}{dt}\int_{V_m} \left((\boldsymbol{r}-\boldsymbol{r}_q)\times \bar{\rho}_\alpha \boldsymbol{v}_\alpha\right) dV = \int_{S_m} \left((\boldsymbol{r}-\boldsymbol{r}_q)\times \boldsymbol{T}_\alpha \cdot \boldsymbol{n}\right) dS + \int_{V_m} \left((\boldsymbol{r}-\boldsymbol{r}_q)\times (\boldsymbol{b}_\alpha + \boldsymbol{m}_\alpha + \bar{g}_\alpha \boldsymbol{v}_\alpha)\right) dS + \int_{V_m} \boldsymbol{a}_{\alpha q} dV \tag{2.90}$$

where $\boldsymbol{r}_q$ is the position of a fixed point Q with respect to which the torques and angular momentum are calculated.

When the field variables are smooth and continuous, a procedure similar to that used in the previous section leads to the *local angular momentum balance*:

$$\boldsymbol{T}_\alpha - \boldsymbol{T}_\alpha^T = \boldsymbol{A}_{\alpha q} \tag{2.91}$$

where $\boldsymbol{A}_{\alpha q}$ is the skew tensor corresponding to the axial vector $\bar{\boldsymbol{a}}_{\alpha q}$. If we assume that there is no interchange of angular momentum between components, $\bar{\boldsymbol{a}}_{\alpha q} = 0$ and the stress tensors for the components are symmetric:

$$\boldsymbol{T}_\alpha = \boldsymbol{T}_\alpha^T. \tag{2.92}$$

2.2.3 Dynamic Process

Consider a mixture B formed by component $B_\alpha \subset B$, with $\alpha = 1, 2, \ldots, n$. We say that the following field variables $\boldsymbol{r} = \boldsymbol{f}_\alpha(\boldsymbol{R}_\alpha, t)$, $\bar{\rho}_\alpha = \bar{\rho}_\alpha(\boldsymbol{r}, t)$, $\boldsymbol{T}_\alpha = \boldsymbol{T}_\alpha(\boldsymbol{r}, t)$, $\boldsymbol{b}_\alpha = \boldsymbol{b}_\alpha(\boldsymbol{r}, t)$, $\bar{g}_\alpha = \bar{g}_\alpha(\boldsymbol{r}, t)$ and $\boldsymbol{m}_\alpha = \boldsymbol{m}_\alpha(\boldsymbol{r}, t)$, constitute a dynamic process if they obey the following field equations in regions where they are smooth and continuous:

$$\frac{\partial \bar{\rho}_\alpha}{\partial t} + \nabla(\bar{\rho}_\alpha \boldsymbol{v}_\alpha) = \bar{g}_\alpha \tag{2.93}$$

$$\frac{\partial}{\partial t}(\bar{\rho}_\alpha \boldsymbol{v}_\alpha) + \nabla \cdot (\bar{\rho}_\alpha \boldsymbol{v}_\alpha \boldsymbol{v}_\alpha) = \nabla \cdot \boldsymbol{T}_\alpha + \boldsymbol{m}_\alpha + \bar{\rho}_\alpha \boldsymbol{v}_\alpha \tag{2.94}$$

and the following jump balance at discontinuities:

$$\sigma[\bar{\rho}_\alpha] = [\bar{\rho}_\alpha \boldsymbol{v}_\alpha \cdot \boldsymbol{e}_I], \quad \sigma[\bar{\rho}_\alpha \boldsymbol{v}_\alpha \cdot \boldsymbol{e}_I] = [\bar{\rho}_\alpha \boldsymbol{v}_\alpha \boldsymbol{v}_\alpha \cdot \boldsymbol{e}_I] - [\boldsymbol{T}_\alpha \cdot \boldsymbol{e}_I] \tag{2.95}$$

For this dynamic process to be complete, *constitutive equations* relating the kinematical with the dynamical variables must be postulated: $(\boldsymbol{T}_\alpha, \boldsymbol{r})$, $(\boldsymbol{b}_\alpha, \boldsymbol{r})$, $(\boldsymbol{m}_\alpha, \boldsymbol{r})$ and $(\bar{g}_\alpha, \boldsymbol{r})$. A dynamic process for these six field variables $\boldsymbol{r}$, $\bar{\rho}_\alpha$, $\boldsymbol{T}_\alpha$, $\boldsymbol{b}_\alpha$ and $\bar{g}_\alpha$ is admissible when the six equations are satisfied.

References

Atkin, R. J., & Crain, R. E. (1976). Continuum theories of mixtures: Basic theory and historical development. *The Quarterly Journal of Mechanics and Applied Mathematics, 29*, 209–244.

Bedford, A., & Drumheller, D. S. (1983). Theories of immiscible and structured mixtures. *International Journal of Engineering Science, 21*(8), 863–960.

Bowen, R. M. (1976). Theory of mixtures. In A. C. Eringen (Ed.), *Continuum physics* (Vol. III). Waltham: Academic Press.

Bustos, M. C., Concha, F., Bürger, R., & Tory, E. M. (1999). *Sedimentation and thickening, phenomenological foundation and mathematical theory*. Dordrecht: Kluwer Academic Publications.

Concha, F., & Barrientos, A. (1993). *Mecánica Racional Moderna, Termomecánica del Medio Continuo* (Vol. 2, pp. 248–266). Dirección de Docencia, Universidad de Concepción.

Concha, F., (2001). Manual de Filtración y Separación, CIC-Red Cettec, CETTEM Ltda., Edmundo Larenas 270 Concepción, Chile.

Drew, D. A. (1983). Mathematical modeling of two-phase flow. *Annual Review of Fluid Mechanics, 15*, 261–291.

Drew, A. D., & Passman, S.L. (1998). *Theory of multicomponent fluids*. Berlin: Springer.

Gurtin, M. E. (1981). An Introduction. Academic Press, New York to Continuum Mechanic.

Rajagopal, K. R., & Tao, L. (1995). Mechanics of mixtures, Worlds Scientific, Singapore.

Truesdell, C., & Toupin, R. A. (1960). The classical field theories of mechanics. In S. Flügge (Ed.), *Handbook of physics* (Vol. III–1). New York: Springer.

Truesdell, C. (1965). Sulle basi de la termomecanica, Rend. Acad. Lincei, *22*, 33–88, 1957. Traducción al inglés en: Rational Mechanics of Materials, Int. Sci. Rev. Ser. 292–305, Gordon & Breach, New York.

Truesdell, C. (1984). *Rational thermodynamics* (2nd ed.). New York: Springer.

Ungsrish, M. (1993). Hydrodynamics of suspensions. Berlin: Springer.

Chapter 3
Particulate Systems

Abstract Mixtures of finely divided solid particles in water are the subject of this chapter. Here the equations derived in Chap. 2 are applied to particulate systems. First, the requisite for considering a particulate system as a continuum is laid out. In such a system the motion of the components of the mixture can be described through local mass and linear momentum balances in all regions where the field variables are continuous. At discontinuities the field equations must be replaced by the corresponding jump conditions. Next, the result is applied to a two-component solid–fluid system. For fluids, constitutive equations are proposed for the stresses, defining pressure and viscous stress. The properties of the solid components in particulate systems depend strongly on their concentration. At concentrations of less than those where particles are in permanent contact with each other, the mixture is called a suspension and all interaction forces are transmitted from particle to particle through the fluid, defining the pore pressure. At greater concentrations the mixture is called a porous medium, a porous bed or sediment, where the stresses are transmitted through the fluid and from particle to particle by direct contact through the effective solid stress. A dynamic process for a solid–liquid system is defined when the set of field variables in regions where the variables are continuous obey the field equations and at discontinuities obey the jump conditions.

3.1 Dynamic Process for a Particulate System

Consider a set of solid particles thoroughly mixed with a fluid forming a suspension under the following conditions:

1. All particles are small, with respect to the container, and of the same size form and density.
2. The individual particles and the fluid are incompressible.
3. There is no mass transfer between the solid and the fluid.
4. The only body force is gravity.

F. Concha A, *Solid–Liquid Separation in the Mining Industry*,
Fluid Mechanics and Its Applications 105, DOI: 10.1007/978-3-319-02484-4_3,

In a system such as that described, the motion of each of the components of the mixture may be described through the local mass and linear momentum balances in all regions where the field variables are continuous:

$$\text{Mass balance of the component} \quad \frac{\partial}{\partial t}(\rho_\alpha \varphi_\alpha) + \nabla \cdot (\rho_\alpha \varphi_\alpha \mathbf{v}_\alpha) = \bar{g}_\alpha \tag{3.1}$$

$$\text{Mass balance of the mixture} \quad \nabla \cdot q = \sum_{\alpha=1}^{2} \frac{\bar{g}_\alpha}{\rho_\alpha} \tag{3.2}$$

$$\text{Momentum balance of the component} \quad \rho_\alpha \dot{\boldsymbol{v}}_\alpha = \nabla \cdot \boldsymbol{T}_\alpha + \boldsymbol{b}_\alpha + \boldsymbol{m}_\alpha \tag{3.3}$$

$$\text{Restriction} \quad \sum_{\alpha=1}^{2} (\boldsymbol{m}_\alpha + \bar{g}_\alpha \boldsymbol{v}_\alpha) = 0 \tag{3.4}$$

In these expressions the field variables $\boldsymbol{q} = \sum_\alpha \varphi_\alpha v_\alpha$ is the volume average velocity, $\varphi_\alpha(\boldsymbol{r}, t)$, $\boldsymbol{v}_\alpha(\boldsymbol{r}, t)$ and $\boldsymbol{T}_\alpha(\boldsymbol{r}, t)$ represent the volume fraction, the velocity and the stresses of the α component, $\boldsymbol{b}_\alpha(\boldsymbol{r}, t)$ and $\boldsymbol{m}_\alpha(\boldsymbol{r}, t)$ are the body force and the interaction force between components respectively and $\bar{g}_\alpha(\boldsymbol{r}, t)$ measures the rate at which the other components deliver mass per unit volume to the component α. The operator $\nabla \cdot (\cdot)$ represents the divergence of the function $(\cdot)$.

At discontinuities, Eqs. (3.1)–(3.3) must be replaced by the corresponding jump conditions:

$$\sigma[\varphi_\alpha] = [\varphi_\alpha \boldsymbol{v}_\alpha \cdot \boldsymbol{e}_I] \tag{3.5}$$

$$\sigma[\varphi_\alpha \boldsymbol{v}_\alpha] = [\varphi_\alpha \boldsymbol{v}_\alpha (\boldsymbol{v}_\alpha \cdot \boldsymbol{e}_I)] - [\boldsymbol{T}_\alpha \cdot \boldsymbol{e}_I] \tag{3.6}$$

where $\sigma = \boldsymbol{v}_\alpha \cdot \boldsymbol{e}_I$ is the displacement velocity of the discontinuity.

For a two-component system of small solid particles and a fluid, designate the solid component by $\alpha = 1 = s$ and by $\alpha = 2 = f$ the fluid component. Since, due to assumptions 1 and 2 the material density of each component is constant, all terms may be divided by the corresponding density. On the other hand, there is no mass transfer between components, assumption 3 and the only body force is gravity. Then Eqs. (3.1)–(3.3) reduce to:

$$\text{Volume balance for the solid} \quad \frac{\partial \varphi}{\partial t} + \nabla \cdot (\varphi v_s) = 0 \tag{3.7}$$

$$\text{Volume balance for the mixture} \quad \nabla \cdot \boldsymbol{q} = 0\,, \text{ with } \quad \boldsymbol{q} = \varphi \boldsymbol{v}_s + (1 - \varphi) \boldsymbol{v}_f \tag{3.8}$$

$$\text{Momentum balance for the solid:} \quad \rho_s \varphi \dot{\boldsymbol{v}}_s = \nabla \cdot \boldsymbol{T}_s + \rho_s \varphi \boldsymbol{g} + \boldsymbol{m} \tag{3.9}$$

$$\text{Momentum balance for the fluid:} \quad \rho_f (1 - \varphi) \dot{\boldsymbol{v}}_f = \nabla \cdot \boldsymbol{T}_f + \rho_f (1 - \varphi) \boldsymbol{g} - \boldsymbol{m} \tag{3.10}$$

where $\varphi_s(\boldsymbol{r}, t) = \varphi(\boldsymbol{r}, t)$ and $\varphi_f(\boldsymbol{r}, t) = 1 - \varphi(\boldsymbol{r}, t)$ are the volume fraction of solids and fluid at the position $\boldsymbol{r}$ and time t and $\boldsymbol{v}_s(\boldsymbol{r}, t)$, $\boldsymbol{v}_f(\boldsymbol{r}, t), \boldsymbol{T}_s(\boldsymbol{r}, t)$,

$\mathbf{T}_f(\boldsymbol{r},t), \boldsymbol{m}(\boldsymbol{r},t)$ and $\mathbf{g}$ are the solid and fluid velocities, the solid and fluid partial stress tensors, the solid–fluid interaction force and the gravitational constant.

In most cases of practical interest, the acceleration terms are negligible (see Bustos et al. 1999) and can be disregarded in Eqs. (3.9) and (3.10). The interaction force $\boldsymbol{m}$ may be separated into a static force $\boldsymbol{m}_E$ and a dynamic force $\boldsymbol{m}_D$ that depends on the motion. With these definitions, Eqs. (3.9) and (3.10) become:

$$\text{Momentum balance for the solid: } 0 = \nabla \cdot \boldsymbol{T}_s + \rho_s \varphi \boldsymbol{g} + \boldsymbol{m}_E + \boldsymbol{m}_D \tag{3.11}$$

$$\text{Momentum balance for the fluid: } 0 = \nabla \cdot \boldsymbol{T}_f + \rho_f (1-\varphi) \boldsymbol{g} - \boldsymbol{m}_E - \boldsymbol{m}_D \tag{3.12}$$

3.1.1 Fluid Component

For all types of fluids, the stresses may be divided into en equilibrium isotropic stress $p_f\boldsymbol{I}$ and an extra stress $\boldsymbol{T}_f^E$ depending on the motion:

$$\boldsymbol{T}_f = -p_f \boldsymbol{I} + \boldsymbol{T}_f^E \tag{3.13}$$

where p_f is the *fluid partial pressure*, or simply the *pressure* and $\boldsymbol{T}_f^E$ is the viscous stress, or extra stress, of the fluid. Replacing Eq. (3.13) into the momentum balance of the fluid, we obtain:

$$\nabla p_f = \nabla \cdot \boldsymbol{T}_f + \rho_f (1-\varphi) \boldsymbol{g} - \boldsymbol{m}_E - \boldsymbol{m}_D \tag{3.14}$$

In flows of solid–fluid mixtures, there are two variables related to friction. One is the viscous stress tensor $\boldsymbol{T}_f^E$ representing the friction within the fluid and the other is the interaction force $\boldsymbol{m}_D$ corresponding to the friction between the solid and the fluid. Experience has shown (Marle 1967; Whitaker 1986) that the friction within the fluid is much smaller than the friction between solids and fluids, and therefore may be neglected in Eq. (3.14) leading to:

$$\nabla p_f = \rho_f (1-\varphi) \boldsymbol{g} - \boldsymbol{m}_E - \boldsymbol{m}_D \tag{3.15}$$

Neglecting the viscous term in Eq. (3.13) is equivalent to consider the fluid as an *elastic fluid*.

3.1.2 Pore Pressure

The fluid partial pressure p_f is a variable defined over the entire surface of the mixture (remember the solids and fluids are superimposed continuous media), and therefore it is not experimentally measurable. The experimental pressure associate with the flow in a porous media is the *pore pressure*.

It has been shown in fluid mechanics that for a solid–fluid mixture the fluid pressure is continuous across semi-permeable surfaces, that is, surfaces that let the fluid pass but retain the solids. For this reason, the pore pressure $p(z,t)$ may be measured with a manometer as $p(z,t) = \rho_f g(h - z(t))$. See Fig. 3.1.

Dynamic processes involving particulate systems generally depend on the pressure in excess to the hydrostatic pressure, since at the hydrostatic pressure they come to end. Therefore, the concept of excess pore pressure $p_e(z,t)$ is introduced:

$$p_e(z,t) = p(z,t) - \rho_f g_z(L - z(t)). \tag{3.16}$$

3.1.3 Solid Component

The properties of the solid component in particulate systems depend strongly on its concentration. In diluted systems particles have greater liberty of movement as opposed to more concentrated suspensions, where the motion of one particle is hindered by the presence of other particles. The characteristic concentration at which this change occurs is known as *critical concentration* φ_c, defined as that concentration at which the particles become in contact with another, forming a network capable of transmitting compression forces. At concentration less than the critical, the mixture is called suspension and all interaction forces are transmitted from particle to particle through the fluid. At concentration greater than the critical, the mixture is called a *porous medium*, a *porous bed* or a *sediment* and the stresses are transmitted through the fluid and from particle to particle by directly contact.

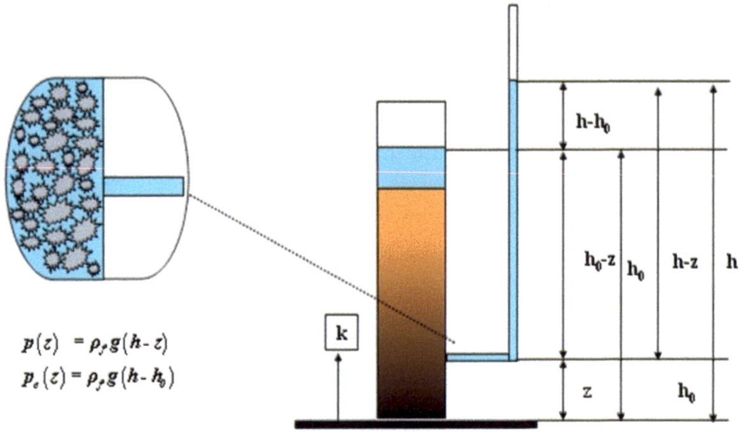

Fig. 3.1 Measurement of pore pressure

For sake of simplicity we will assume that the solid component also behaves as an elastic fluids, so that the tress tensor $\boldsymbol{T}_s$ is given by:

$$\boldsymbol{T}_s = \begin{cases} -\beta \boldsymbol{I} & \text{for} \quad \varphi \leq \varphi_c \\ -(\beta + p_s(\varphi, \varphi_c))\boldsymbol{I} & \text{for} \quad \varphi > \varphi_c \end{cases} \tag{3.17}$$

where φ_c is the critical concentration, β is a constant and $\boldsymbol{I}$ is the second order unit tensor. For concentrations less then the critical the solid stress is constant and it is a function of the concentration and of the critical concentration for values above the critical. This result (but not the basic principle) coincides with more elaborate constitutive equation of the solid stress tensor, for example for isotropic elastic solids (Bustos et al. 1999, p. 43) or a plastic solid (Buscall and White 1987; Green et al. 1996).

Introducing the constitutive Eq. (3.17) into the local momentum balance (3.15) yields:

$$\nabla p_s = \rho_f \varphi \boldsymbol{g} + \boldsymbol{m}_E + \boldsymbol{m}_D \tag{3.18}$$

3.1.4 Solid Effective Stress

Similarly to the case of the fluid component, the "solid pressure p_s" is not an experimentally measurable variable. When a compressive force is applied to a porous medium, the fluid filling the pores between the particles immediately supports the total force exerted, increasing its pore pressure. The excess pore pressure initiates the fluid flow out of the pores, which is accompanied by a reduction of the pore pressure and a progressive transfer of the stress to the solid structure. The rate at which the porous medium deforms depends on the constitutive equation of the solid stress and on the solid–fluid interaction force opposing the fluid flow.

The transient process of stress transfer from the fluid filling the pores to the solid skeleton is called *consolidation*. Fig. 3.2 shows a mechanical analog to consolidation. In the figure, the length of a given spring is shown when it is subjected to different weights, 5, 10, 15, and 20 kg. In the bottom figure, the same spring is submerged in water contained in a frictionless column with a piston and a valve.

- Figure 1: Initially the system is in equilibrium with the valve open and without any weight.
- Figure 2: The valve is closed and a 20 kg weight is put on the piston. Since water is incompressible, it supports the whole weight increasing its "pore pressure". The spring conserves its total length.
- Figure 3: Represents the instant when the valve is open and before the water starts to flow out.

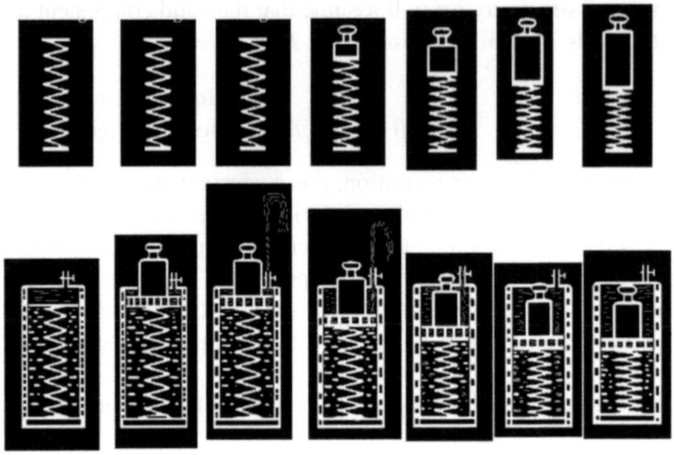

Peso total	0	20	20	20	20	20	20
Presión poros	0	20	20	15	10	5	0
Presión sólido	0	0	0	5	10	15	20
Figura	1	2	3	4	5	6	7

Fig. 3.2 Mechanical analog of consolidation

- Figure 4: With the valve open, the water under pressure starts to flow out through the valve. The rate of outflow is controlled by the amount of excess pore *pressure* and by the friction through the valve (*permeability*). As water leaves, the piston diminishes its level inside the column, the spring compresses until it supports 5 kg. At this moment, the water supports the difference of 15 kg.
- Figure 5: More water flows out and the piston continue dropping. When the spring has compressed to the size shown in the figure, the spring and the water support 10 kg each.
- Figure 6: The water continues flowing out and the spring supports 15 kg while the water supports 5 kg.
- Figure 7: Finally, the spring supports the 20 kg and the water has hydrostatic pressure (zero excess pore pressure).The complete process consists in transferring the 20 kg load from the water (*pore pressure*) to the spring (*solid effective stress*).

3.1.5 Total Pressure

In a particulate system, the measurable total pressure p_t is the sum of the pore pressure p and the effective solid stress σ_e. Then, the relationship between the theoretical and the experimental variables are:

$$p_t = p_f + p_s \equiv p + \sigma_e \tag{3.19}$$

The force exerted by the fluid on a surface S is:

$$\int_S p_f dS = \int_{S_f} p dS_f = \int_S p\left(\varepsilon_f dS\right) \tag{3.20}$$

where S is the cross section of the mixture, S_f is the cross section in the fluid of the porous medium and ε_f is the surface porosity (fraction of the total surface occupied by the fluid). If the surface porosity ε_f is assumed to be equal to the volume porosity ε (fraction of the total volume of the porous medium occupied by the fluid), $\varepsilon_f \equiv \varepsilon = 1 - \varphi$, Eq. (3.20) may be written in the form:

$$\int_S p_f dS = \int_S p(1-\varphi)dS \tag{3.21}$$

from which:

$$p_f = (1-\varphi)p \tag{3.22}$$

Using Eq. (3.19) the following is obtained for the solid pressure:

$$p_s = \varphi p + \sigma_e \tag{3.23}$$

Substituting the theoretical pressures p_f and p_s by its experimental equivalents p *and* σ_e from Eqs. (3.22) and (3.23) into the local balances of momentum (3.18) and (3.15), results in:

$$\nabla(\varphi p) + \nabla\sigma_e = \rho_f \varphi \boldsymbol{g} + \boldsymbol{m}_E + \boldsymbol{m}_D \tag{3.24}$$

$$\nabla((1-\varphi)p) = \rho_f(1-\varphi)\boldsymbol{g} - \boldsymbol{m}_E - \boldsymbol{m}_D. \tag{3.25}$$

3.1.6 Interaction Force at Equilibrium

Consider the local momentum balance of the fluid, Eq. (3.25), at equilibrium, where the dynamic interaction force $\boldsymbol{m}_D = 0$ and the hydrostatic pore pressure is given by $p(z) = \rho_f g(L-z)$, where L is the height of the water level. Substituting into (3.25) yields:

$$(1-\varphi)\rho_f \boldsymbol{g} - p\nabla\varphi|_{equilibrium} = \rho_f(1-\varphi)\boldsymbol{g} - \boldsymbol{m}_E$$

$$\boldsymbol{m}_E = p\nabla\varphi|_{equilibrium} \tag{3.26}$$

Assuming that the functional form of Eq. (3.26) is always valid, we can write for the static interaction force:

$$\boldsymbol{m}_E(\boldsymbol{r}, t) = p\nabla\varphi \tag{3.27}$$

Substituting into the local momentum balances (3.25) and (3.24) results in:

$$\nabla p = \rho_f \boldsymbol{g} - \frac{\boldsymbol{m}_D}{1-\varphi} \tag{3.28}$$

$$\nabla \sigma_e = \Delta\rho\varphi \boldsymbol{g} + \frac{\boldsymbol{m}_D}{1-\varphi} \tag{3.29}$$

In terms of the excess pore pressure p_e, the local momentum balance for the fluid is:

$$\nabla p_e = -\frac{\boldsymbol{m}_D}{1-\varphi} \tag{3.30}$$

Combining (3.29) and (3.30) we can substitute (3.29) by:

$$\nabla p_e + \nabla \sigma_e = \Delta\rho\varphi \boldsymbol{g}. \tag{3.31}$$

3.1.7 *Discontinuities*

It is well known that suspensions develop discontinuities. Therefore, the local field equations must be substituted at discontinuities by jump conditions.

$$\sigma[\varphi] = [\varphi \boldsymbol{v}_s \cdot \boldsymbol{e}_I] \tag{3.32}$$

$$\sigma[\varphi \boldsymbol{v}_s] = [\varphi \boldsymbol{v}_s(\boldsymbol{v}_s \cdot \boldsymbol{e}_I)] - \left[\left(\varphi\left\{p_e + \rho_f g(L-z)\right\} + \sigma_e\right)\boldsymbol{e}_I\right]. \tag{3.33}$$

3.2 Dynamical Process

Collecting the previous results, we can say that the flow of a particulate systems may be represented by the following field variables: the volume fraction of solids $\varphi(\boldsymbol{r},t)$, the solid and fluid velocities $v_s(\boldsymbol{r},t)$ and $v_f(\boldsymbol{r},t)$, the excess pore pressure $p_e(\boldsymbol{r},t)$, the solid effective stress $\sigma_e(\boldsymbol{r},t)$ and the solid–fluid dynamic interaction force $m_D(\boldsymbol{r},t)$. These variables constitute a *dynamical process if,*

1. For regions where the variables are continuous, the obey the field equations:

$$\frac{\partial \varphi}{\partial t} + \nabla \cdot (\varphi \boldsymbol{v}_s) = 0 \tag{3.34}$$

$$\nabla \cdot \boldsymbol{q} = 0\,, \quad con \quad \boldsymbol{q} = \varphi \boldsymbol{v}_s + (1-\varphi)\boldsymbol{v}_f \tag{3.35}$$

$$\nabla \sigma_e = \Delta\rho\varphi \boldsymbol{g} + \frac{\boldsymbol{m}_d}{1-\varphi} \tag{3.36}$$

$$\nabla p_e = -\frac{\boldsymbol{m}_d}{1-\varphi} \tag{3.37}$$

2. At discontinuities they obey the jump conditions:

$$\sigma[\varphi] = [\varphi \boldsymbol{v}_s \cdot \boldsymbol{e}_I] \tag{3.38}$$

$$\sigma[\varphi \boldsymbol{v}_s] = [\varphi \boldsymbol{v}_s(\boldsymbol{v}_s \cdot \boldsymbol{e}_I)] - \left[\left(\varphi\left\{p_e + \rho_f g(L-z)\right\} + \sigma_e\right)\boldsymbol{e}_I\right] \tag{3.39}$$

3. The constitutive equations for the solid–fluid interaction force and for solid effective stress:

$$\boldsymbol{m}_d = \boldsymbol{m}_d(\varphi, \boldsymbol{v}_s, \boldsymbol{q}) \tag{3.40}$$

$$\sigma_e = \sigma_e(\varphi, \boldsymbol{v}_s, \boldsymbol{q}). \tag{3.41}$$

References

Atkin, R. J., & Crain, R. E. (1976). Continuum theories of mixtures, basic theory and historical development. *Quarterly Journal of Mechanics and Applied Mathematics, 29*, 209–244.

Bedford, A., & Drumheller, D. S. (1983). Theories of immiscible and structured mixtures. *International Journal of Engineering Science, 21*(8), 863–960.

Bowen, R.M., (1976). Theory of mixtures. In A. C. Eringen (Ed.), *Continuum physics*. New York: Academic Press.

Buscall, R., & White, L.R. (1987). The consolidation of concentrated suspensions. *Journal of the Chemical Society, Faraday Transactions 1, 83*(3), 873–891.

Bustos, M. C., Concha, F., Bürger, R., & Tory, E. M. (1999). *Sedimentation and thickening, phenomenological foundation and mathematical theory* (pp. 46–47). Dodrecht: Kluwer Academic Publishers.

Concha, F., & Barrientos, A. (1996). *Mecánica racional moderna* (Vol. II, pp. 248–266). Termomecánica Del Medio Continuo: Dirección De Docencia, Universidad De Concepción.

Concha, F., Bustos, M. C., & Barrientos, A. (1996). Phenomenological theory of sedimentation. In E. Tory (Ed.), *Sedimentation of small particles in viscous fluids* (p. 60). Southampton: Computational Mechanics Publications Southampton.

Drew, D. A. (1983). Mathematical modeling of two-phase flow. *Annual Review of Fluid Mechanics, 15*, 261–291.

Green, M. D., Eberl, M., & Landman, K. A. (1996). Compressive yield stress of flocculated suspensions: Determination via experiment. *AIChE Journal, 42*(8), 2308–2318.

Marle, C. M. (1967). Ecoulements monophasiques em milieu poreux. *Review Institute Francaise du Petrole, 22*, 1471–1509.

Truesdell, C. (1984). *Rational themodynamics* (2nd ed.). New York: Springer.

Whitaker, S. (1986). Flow in porous media I: A theoretical derivation of darcy's law. *Transport in Porous Media, 1*, 1–30.

Chapter 4
Sedimentation of Particulate Systems

Abstract This chapter deals with sedimentation of particulate systems considered as discrete media. Sedimentation is the settling of a particle or suspension of particles in a fluid due to the effect of an external force such as gravity, centrifugal force or any other body force. Discrete sedimentation has been successful in establishing constitutive equations for continuous sedimentation processes. The foundation of the motion of particles in fluids is discussed in different flow regimes, Euler's flow, Stokes flow and flows with a boundary layer. Starting from the sedimentation of a sphere in an unbounded fluid, a complete analysis is made of the settling of individual spherical particles and suspensions. The results are extended to isometric particles and to arbitrarily shaped particles. Sphericity as a shape factor is used to describe the form of isometric particles. A hydrodynamic sphericity must be defined for particles with arbitrary shapes by performing sedimentation or fluidization experiments, calculating the drag coefficient for the particles using the volume equivalent diameter and obtaining a sphericity defined for isometric particles that fits experimental values. A modified drag coefficient and sedimentation velocities permits grouping all sedimentation results in one single equation for particles of any shape.

Sedimentation is the settling of a particle, or suspension of particles, in a fluid due to the effect of an external force such as gravity, centrifugal force or any other body force. For many years, workers in the field of Particle Technology have been looking for a simple equation relating the settling velocity of particles to their size, shape and concentration. Such a simple objective has required a formidable effort and it has been solved, only in part, through the work of Newton (1687) and Stokes (1844) on flow around a particle, and the more recent research of Lapple (1940), Heywood (1962), Batchelor (1967), Zenz (1966), Barnea and Mitzrahi (1973) and many others, to those Turton and Levenspiel (1986) and Haider and Levenspiel (1998). Concha and collaborators established in 1979 an heuristic theory of sedimentation, that is, a theory based on the fundamental principles of mechanics, but to a greater or lesser degree to intuition and empirism. These works, (Concha and Almendra 1979a, b; Concha and Barrientos 1982, 1986; Concha and Christiansen 1986), first solve the settling of one particle in a fluid, then, they introduce

F. Concha A, *Solid–Liquid Separation in the Mining Industry*,

Fluid Mechanics and Its Applications 105, DOI: 10.1007/978-3-319-02484-4_4,

corrections for the interaction between particles, through which the settling velocity of a suspension is drastically reduced. Finally, the settling of isometric and non-spherical particles was treated. This approach, which uses principles of particle mechanics, receives the name of the *discrete approach* to sedimentation, or *discrete sedimentation.*

Discrete sedimentation has been successful in establishing constitutive equations in processes using sedimentation in order to analyze the sedimentation properties of a certain particulate material in a given fluid. Nevertheless, to analyze a sedimentation process and to obtain behavioral pattern permitting the prediction of capacities and equipment design procedures, another approach is required, the so-called *continuum approach.* In the present chapter the discrete approach will be analyzed, leaving the continuum approach for later chapters.

4.1 Discrete Sedimentation

The physics underlying sedimentation, that is, the settling of a particle in a fluid has long been known. Stokes showed the equation describing the sedimentation of a sphere in 1851 and that can be considered as the starting point of all discussions of the sedimentation process. Stokes showed that the settling velocity of a sphere in a fluid is directly proportional to the square of the particle radius, to the gravitational force and to the density difference between solid and fluid, and inversely proportional to fluid viscosity. This equation is based on a force balance around the sphere. Nevertheless, the proposed equation is valid only for slow motions, so that in other cases expressions that are more elaborate should be used. The problem is related to the hydrodynamic force between the particle and the fluid.

Consider the incompressible flow of a fluid around a solid sphere. The equations describing the phenomena are the continuity equation and Navier–Stokes equation:

$$\begin{gathered}\nabla\cdot\boldsymbol{v}=0\\ \rho\left(\frac{\partial\boldsymbol{v}}{\partial t}+\underbrace{\nabla\boldsymbol{v}\cdot\boldsymbol{v}}_{\text{Convective force}}\right)=-\nabla p+\underbrace{\mu\nabla^2\boldsymbol{v}}_{\text{Diffusive force}}+\rho\boldsymbol{g}\end{gathered}\tag{4.1}$$

where $\boldsymbol{v}$ and p are the fluid velocity and pressure field, ρ and μ are the fluid density and viscosity and g is the gravity force vector.

Unfortunately, Navier–Stokes equation are non-linear and it is impossible to be solved explicitly in a general form. Therefore, methods have been used to solve it in special cases. It is known that the Reynolds number $Re=\rho_f du/\mu$, where ρ_f, d and u are the fluid density and the particle diameter and velocity respectively, is an important parameter that characterizes the flow. It is a dimensionless number representing the ratio of convective to diffusive forces in Navier–Stokes equation. In dimensionless form, Navier–Stokes equation becomes:

$$\frac{1}{St}\frac{\partial \boldsymbol{v}^*}{\partial t^*} + \nabla^* \boldsymbol{v}^* \cdot \boldsymbol{v}^* = -\frac{1}{Ru}\nabla^* p^* + \frac{1}{\boldsymbol{Re}}\nabla^{*2}\boldsymbol{v}^* - \frac{1}{Fr}\boldsymbol{e}_k \tag{4.2}$$

where the starred terms represent dimensionless variables defined by: $\boldsymbol{v}^* = \boldsymbol{v}/u_0$; $p^* = p/p_0$; $t^* = t/t_0$; $\nabla^* = L\nabla$ and u_0, p_0, t_0 and L are characteristic velocity, pressure and time and length in the problem, and St, Ru, Re and Fr are the Struhal, Ruark, Reynolds and Froud numbers and $\boldsymbol{e}_k$ is the vertical unit vector:

$$\text{Strouhal} \quad St = \frac{t_0 u_0}{L} \tag{4.3}$$

$$\text{Ruark} \quad Ru = \frac{\rho u_0^2}{p_0} \tag{4.4}$$

$$\text{Reynolds} \quad \text{Re} = \frac{\rho u_0 L}{\mu} \tag{4.5}$$

$$\text{Froude} \quad Fr = \frac{u_0^2}{Lg} \tag{4.6}$$

When the Reynolds number is small ($Re \to 0$), for example $Re < 10^{-3}$, convective forces may be neglected in Navier–Stokes equation, obtaining the so called *Stokes Flow*. In dimensional form Stokes Flow is represented by:

$$\begin{aligned} &\nabla \cdot \boldsymbol{v} = 0 \\ &\rho \frac{\partial \boldsymbol{v}}{\partial t} = -\nabla p + \mu \nabla^2 \boldsymbol{v} + \rho \boldsymbol{g} \end{aligned} \tag{4.7}$$

4.1.1 Hydrodynamic Force on a Sphere in Stokes Flow

Due to the linearity of the differential equation in Stokes Flow, the velocity, the pressure and the *hydrodynamic force* in a steady flow are linear functions of the relative solid–fluid velocity. For the hydrodynamic force, the linear function, depends on the size and shape of the particle ($6\pi R$ for the sphere) and on fluid viscosity (μ). Solving the boundary value problem, and neglecting the Basset term of added mass, yields (Happel and Brenner 1965) for a sphere:

$$F_D = -6\pi\mu R u \tag{4.8}$$

It is common to write the hydrodynamic force in its dimensionless form known as *drag coefficient* C_D:

$$C_D = \frac{F_D}{(1/2\rho_f u^2)(\pi R^2)} \tag{4.9}$$

where ρ_f is the fluid density. Substituting (4.9) into (4.8), the drag coefficient on a sphere in Stokes flow is:

$$C_D = \frac{24}{Re} \tag{4.10}$$

4.1.2 Macroscopic Balance on a Sphere in Stokes Flow

Consider a small solid sphere submerged in a viscous fluid and suspended with a string. If the sphere, with density greater than that of the fluid, is in equilibrium, the balance of forces around it is zero. The forces acting on the particles are: (1) gravity F_g, that pulls the sphere down, (2) buoyancy F_b, that is, the pressure forces of the fluid acting on the particle that pushes the sphere upwards and (3) the string resistance F_{string}, that supports the particle from falling, see Fig. 4.1. The force balance gives:

$$0 = F_{string} + \underbrace{F_g}_{-\rho_p V_p g} + \underbrace{F_b}_{+\rho_f V_p g} \tag{4.11}$$

$$0 = F_{string} - \rho_p V_p g + \rho_f V_p g \tag{4.12}$$

$$F_{string} = \underbrace{\left(\rho_p - \rho_f\right)}_{\Delta\rho} V_p g \equiv \Delta\rho V_p g \tag{4.13}$$

If the string is cut, forces become unbalanced and, according to Newton's second Law, the particle will accelerate. The initial acceleration can be obtained from the new force balance, where the string resistance is absent. Figure 4.2 shows this new force balance before the motion begins.

The initial acceleration is:

$$\begin{aligned} \rho_p V_p a(\mathrm{t} = 0) &= \Delta\rho V_p g \\ a(\mathrm{t} = 0) &= \frac{\Delta\rho}{\rho_p} g \end{aligned} \tag{4.14}$$

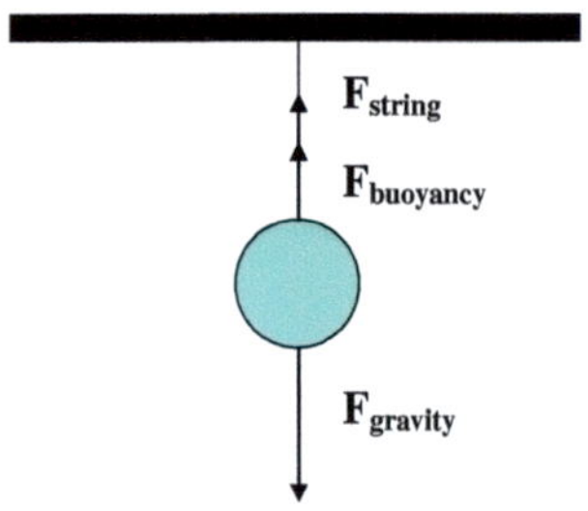

Fig. 4.1 Equilibrium of a sphere submerged in a viscous fluid

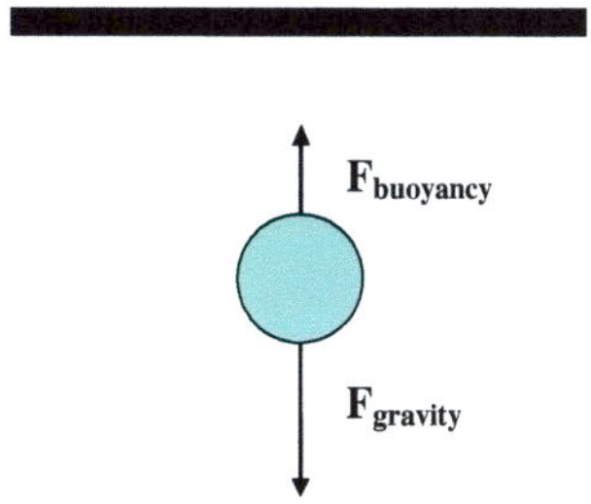

Fig. 4.2 State before the motion initiates

Once the particle is in motion, a new force, the drag, appears representing the resistance opposed by the fluid to the particle motion. This force F_D is proportional to the relative solid–fluid velocity and to the relative particle acceleration. Since the fluid is at rest, it corresponds to the sphere velocity and acceleration. Once the motion starts, the drag force is added and the balance of forces becomes, Fig. 4.3:

$$\rho_p V_p a(t) = \underbrace{\Delta\rho V_p g}_{\text{Netweight}} \underbrace{- 6\pi\mu R u(t) - (1/2)\rho_p V_p a(t)}_{\text{Dragforce}} \tag{4.15}$$

$$\frac{3}{2}\rho_p V_p \mathrm{a}(t) = \Delta\rho g V_p - 6\pi\mu R u(t) \tag{4.16}$$

The term $(1/2)\rho_p V_p$ that was added to the mass $\rho_p V_p$ in the first term of Eq. (4.16) is called *added mass* induced by the acceleration.

Due to the increase in the velocity $u(t)$ with time, the sum of the first and last term of Eq. (4.16) diminishes, becoming zero at a given time. At that time, the velocity becomes a constant called *terminal velocity* $u = u_\infty$, which is a characteristic of the solid–fluid system. From (4.16) with $a(\mathrm{t}) = 0$,

$$\mathrm{u}_\infty = \frac{2}{9}\frac{\Delta\rho R^2 g}{\mu} = \frac{1}{18}\frac{\Delta\rho d^2 g}{\mu} \tag{4.17}$$

This expression receives the name of *Stokes Equation* and is valid for small Reynolds numbers.

Problem 4.1 Calculate the terminal sedimentation velocity of a quartz sphere with a diameter of 10 μ m and 2.65 g/cm^3 in density in water at 20 °C.

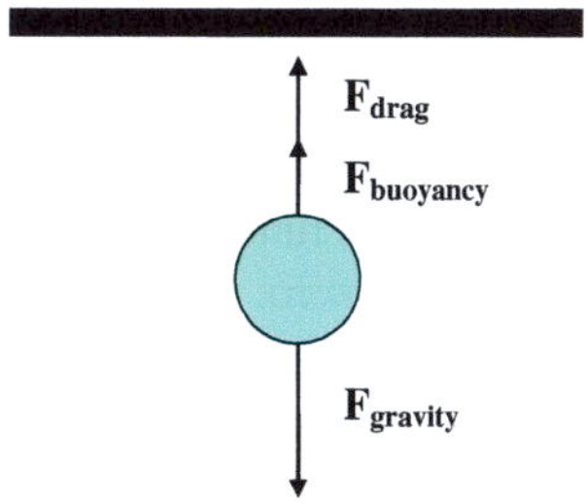

Fig. 4.3 State after the motion starts

The water viscosity at 20 °C is 0.01 g/cm s, then, applying Eq. (4.17) results in:

$$u_\infty = \frac{1}{18}\frac{(2{,}650 - 1{,}000) \times (10/1{,}000{,}000)^2 \times 9.81}{0.001} = 5.00 \times 10^{-6}\ \text{m/s}.$$

Sedimentation dynamics

Equation (4.16) is the differential equation for the settling velocity of a sphere in a gravity field. It can be written in the form:

$$\dot{u}(\mathrm{t}) + \frac{2}{3}\frac{\mu}{18\rho_p d^2}u(\mathrm{t}) - \frac{2\Delta\rho}{3\rho_p}g = 0 \tag{4.18}$$

Its solution is: $$u(\mathrm{t}) = \frac{1}{18}\frac{\Delta\rho d^2 g}{\mu}\left(1 - \exp\left(-\frac{2}{3}\frac{\mu}{18\rho_p d^2}t\right)\right) \tag{4.19}$$

The term inside the exponential term multiplying the time t is called the *Stokes Number* and the term outside the parenthesis is the *terminal velocity*, as we already saw in (4.17).

Problem 4.2 Calculate the terminal settling velocity and the time to reach it for quartz particles, 10, 50 and 100 μm in size and 2.65 g/cm^3 in density, in water at 20 °C.

Applying Eqs. (4.19) and (4.17), the values of terminal velocities of 0.899, 0.225 and 0.00899 cm/s and times of 0.03140, 0.00740 and 0.00030 are obtained for particles with diameters of d = 100, 50 and 10 μm respectively. Figure 4.4 shows the evolution of the particle velocities.

4.1.3 Hydrodynamic Force on a Sphere in Euler's Flow

When the Reynolds number tends to infinity ($Re \to \infty$), viscous forces disappear and Navier–Stokes equation becomes *Euler's Equation for Inviscid Flow.*

$$\begin{aligned} &\nabla \cdot \boldsymbol{v} = 0 \\ &\rho\frac{\partial \boldsymbol{v}}{\partial t}\nabla\boldsymbol{v} \cdot \boldsymbol{v} = -\nabla p + \rho\boldsymbol{g} \end{aligned} \tag{4.20}$$

In this case, the tangential component of the velocity at the particle surface is also a linear function of the relative solid–fluid velocity, but the radial component is equal to zero:

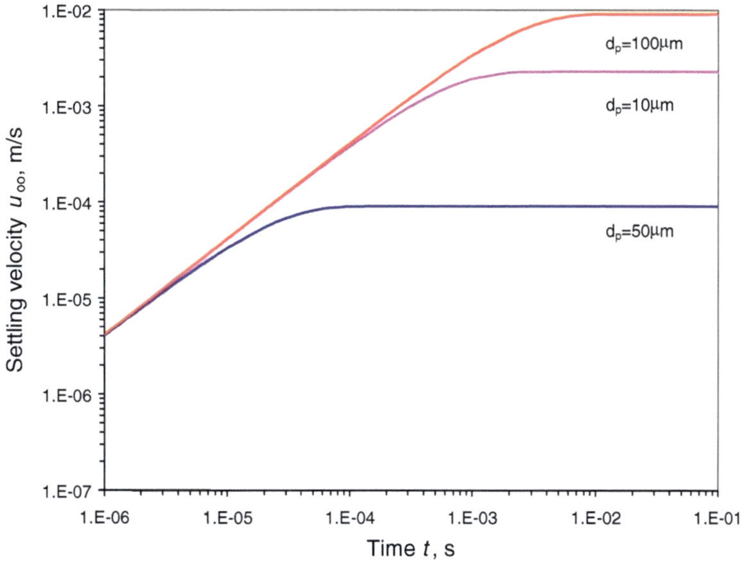

Fig. 4.4 Settling velocity versus time for Problem 4.2

$$u_\theta(\theta) = \left(\frac{3}{2}\text{sen}\theta\right)u \quad \text{and} \quad u_r = 0 \tag{4.21}$$

Now, the pressure is given by a non-linear function called the *Bernoulli equation* (Batchelor 1967):

$$\begin{aligned} p(\theta) + 1/2\rho_f u_\theta^2 &= p + 1/2\rho_f u^2 = \text{constant} \\ p(\theta) - p &= \frac{1}{2}\rho_f u^2\left(1 - \left(\frac{u_\theta}{u}\right)^2\right) \end{aligned} \tag{4.22}$$

Substituting (4.21) into (4.22), the dimensionless pressure, called *pressure coefficient,* defined by $C_p = (p(\theta) - p)/(1/2)\rho_f u^2$, may be expressed in Euler's flow by:

$$C_p = 1 - \frac{9}{2}\text{sen}^2\theta \tag{4.23}$$

Figure 4.5 shows Eq. (4.23) in graphic form, where $p(\theta)$ and u_θ are respectively the pressure and tangential velocity at an angle θ on the surface of the sphere, and p and u are the values of the same variables at the bulk of the flow.

For an inviscid stationary flow, the hydrodynamic force is zero. This result is due to the fact that the friction drag is zero in the absence of viscosity and that the form drag depends on the pressure distribution over the surface of the sphere and this distribution is symmetric, leading to a zero net force.

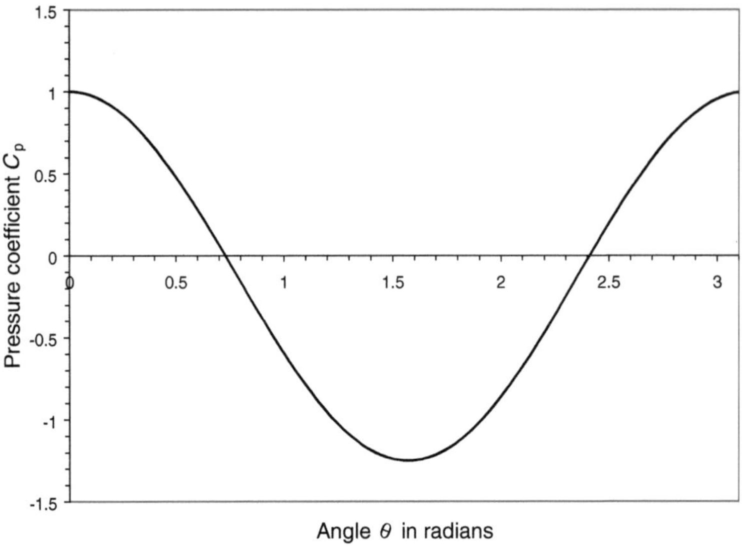

Fig. 4.5 Pressure coefficient in terms of the distance over the sphere in an inviscid flow (Schlichting 1968, p. 21)

4.1.4 Hydrodynamic Force on a Sphere in Prandtl's Flow

For intermediate values of the Reynolds Number, inertial and viscous forces in the fluid are of the same order of magnitude. In this case, the flow may be divided into two parts, an external inviscid flow far from the particle and an internal flow near the particle, where viscosity plays an important role. This picture forms the basis of the *Boundary-Layer Theory* (Meksyn 1961; Rosenhead 1963; Golstein 1965; Schlichting 1968).

In the external inviscid flow, Euler's equations are applicable and the velocity and pressure distribution may be obtained from Eqs. (4.21) and (4.23). The region of viscous flow near the particle is known as the *boundary layer* and it is there where a steep velocity gradient permits the non-slip condition at the solid surface to be satisfied. The energy dissipation produced by the viscous flow within the boundary layer retards the flow and, at a certain point, aided by the adverse pressure gradient, the flow reverses its direction. These phenomena force the fluid particles outwards and away from the solid producing the phenomena called *boundary layer separation*, which occurs at an *angle of separation* given by Lee and Barrow (1968):

$$\theta_s = 214\,Re^{-0.1} \quad \text{for} \quad 24 < Re < 10{,}000 \tag{4.24}$$

For $Re = 24$ the value of the angle of separation is $\theta_s = 155.7$ diminishing to $\theta_s = 85.2$ for Re $= 10{,}000$. For Reynolds numbers exceeding 10,000, the angle of separation diminishes slowly from $\theta_s = 85.2°$ to 84° and then maintains this value

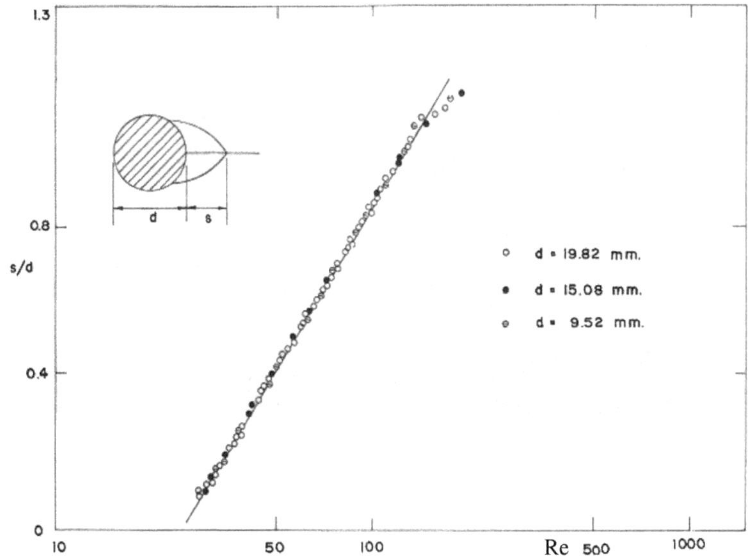

Fig. 4.6 Length of the region of closed streamlines behind a sphere (from Taneda 1956; Batchelor 1967, p. 262)

up to Re ≈ 150,000 (Tomotika 1936; Fage 1937; Amai 1938; Cabtree et al. 1963). Due to the separation of the boundary layer, the region of closed streamlines behind the sphere contains a standing ring-vortex, which first appears at Reynolds number of Re ≈ 24. See Fig. 4.6.

Taneda (1956) determined that beyond $Re \approx 130$ the ring-vortex began to oscillate and that at higher Reynolds numbers the fluid in the region of closed streamlines broke away and was carried downstream forming a wake. Figure 4.7 shows a similar case for the flow around a cylinder.

The thickness "δ" of the boundary layer, defined as the distance from the solid surface to the region where the tangential velocity v_θ reaches 99 % of the value of the external inviscid flow, is proportional to $Re^{-0.5}$ and, at the point of separation, may be written in the form:

$$\frac{\delta}{R} = \frac{\delta_0}{Re^{1/2}} \tag{4.25}$$

McDonald (1954) gives a value of $\delta_0 = 9.06$.

The separation of the boundary layer prevents the recovery of the pressure at the rear of the sphere, resulting in an asymmetrical pressure distribution with a higher pressure at the front of the sphere. Figure 4.8 shows the pressure coefficient of a sphere in terms of the distance from the front stagnation point over the surface of the sphere in an inviscid flow and in boundary layer flow. The figure shows that the pressure has an approximate constant value behind the separation point at Reynolds numbers around $Re \approx 150{,}000$. This dimensionless pressure is called

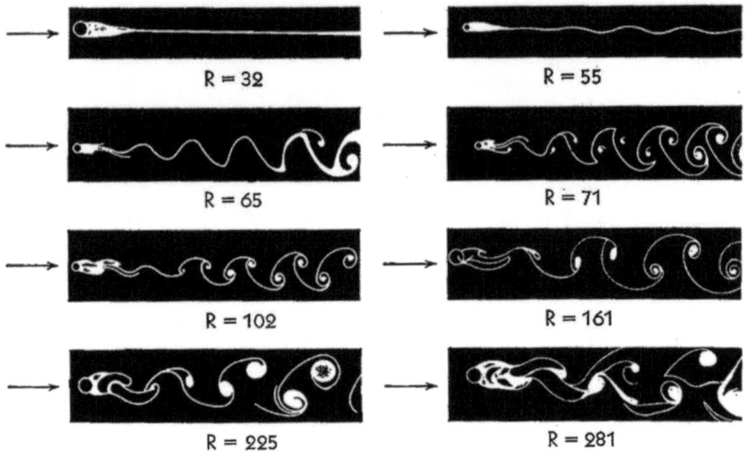

Fig. 4.7 Flow around a cylinder for several values of the Reynolds number

base pressure and has a value of $p_b^* \approx -0.4$ (Fage 1937; Lighthill 1963, p. 108; Goldstein 1965, pp. 15 and 497; Schlichting 1968, p. 21).

The asymmetry of the pressure distribution explains the origin of the form-drag, the magnitude of which is closely related to the position of the point of separation. The farther the separation points from the front stagnation point, the smaller the form-drag. For a sphere at high Reynolds numbers, from $Re = 10{,}000$ up to $Re = 150{,}000$, the position of the separation point does not change very much, except with the change of flow from laminar to turbulent. Therefore, the *form-drag* will remain approximately constant. At the same time, the *friction-drag*, also called *skin friction*, falls proportionally to $Re^{-1/2}$. From these observations, we can conclude that, for Reynolds numbers of about $Re = 1{,}000$, the viscous interaction force has diminished sufficiently for its contribution to the total interaction force to

Fig. 4.8 Pressure coefficient as a function of the distance from the front stagnation point over the surface of a sphere in an inviscid and a boundary layer flow

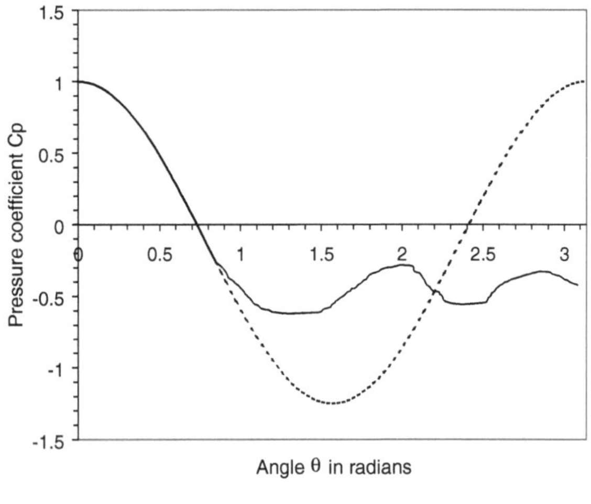

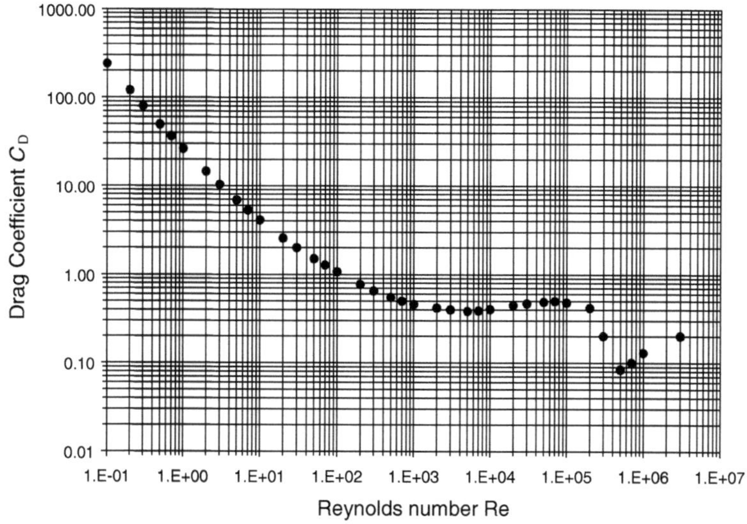

Fig. 4.9 Drag coefficient C_D versus Reynolds number, according to standard data from Lapple and Shepherd (1940). See also Perry (1963, p. 561)

be negligible. Therefore, between $Re \approx 10{,}000$ to $Re \approx 150{,}000$, the drag coefficient is approximately constant at $C_D \approx 0.44$.

For Reynolds numbers greater than $Re \approx 150{,}000$, the flow changes in character and the boundary layer becomes turbulent. The increase in kinetic energy of the external region permits the flow in the boundary layer to reach further to the back of the sphere, shifting the separation point to values close to $\theta_S \approx 110°$ and permitting also the base pressure to rise. The effect of these changes on the drag coefficient is a sudden drop and after that, a sharp increase with the Reynolds number. Figure 4.9 shows a plot of standard experimental values of the drag coefficient versus the Reynolds number (Lapple and Shepherd 1940), where this effect is shown.

4.1.5 Drag Coefficient for a Sphere in the Range 0 < Re < 150,000

Figure 4.9 shows the variation of the drag coefficient of a sphere for different values of the Reynolds number, and confirms that for $Re \to 0$, $C_D \propto Re^{-1/2}$ and that for Re $> 1{,}000$, $C_D \to 0.43$. To obtain a general equation relating de drag coefficient to the Reynolds number, we will use the concept of the *boundary layer* and the knowledge that, for a given position at the surface of the sphere, the pressure inside the boundary layer is equal to the pressure in the inviscid region just outside the boundary layer before the separation point, and is a constant beyond it. We should also remember that the point of separation and the base pressure are constant for Reynolds numbers greater than $Re \approx 1{,}000$.

Consider a solid sphere of radius R with an attached boundary layer of thickness δ submerged in a flow at high Reynolds number (Abraham 1970). Assume that the system of sphere and boundary layer has a spherical shape with a radius equal to a, which can be approximated by $a = R + \delta$, as shown in Fig. 4.10 (Abraham 1970; Concha and Almendra 1979a, b).

Outside the spherical shell of radius a, and up to the point of separation $\theta = \theta_s$, the flow is inviscid and therefore the fluid velocity and the pressure distributions are given by:

$$u_\theta(\theta) = \frac{3}{2} u \sin\theta, \quad \text{for } 0 \le \theta \le \theta_s \tag{4.26}$$

$$p(\theta) = \frac{1}{2}\rho_f u^2 \left(1 - \left(\frac{u_\theta}{u}\right)^2\right), \quad \text{for } 0 \le \theta \le \theta_s \tag{4.27}$$

Beyond the separation point, a region exists where the pressure is constant and equal to the base pressure p_b:

$$p(\theta) = p_b, \quad \text{for } \theta_s \le \theta \le \pi \tag{4.28}$$

Since the effect of viscosity is confined to the interior of the sphere of radius a, the total drag exerted by the fluid on a, consists of a form drag only, Then:

$$F_D = \oint_{S_a} p(\theta) \cos\theta \, dS \tag{4.29}$$

The element of surface of the sphere of radius a is:

$$dS = a^2 \text{sen}\, \theta d\theta d\phi \tag{4.30}$$

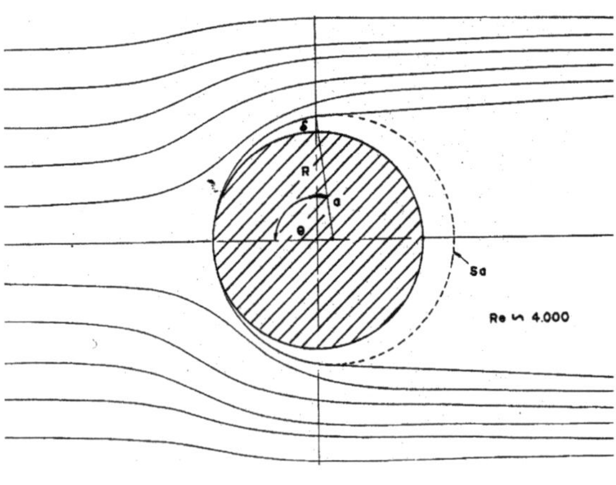

Fig. 4.10 Physical model for the flow in the boundary layer around a sphere (Concha and Almendra 1979a, b)

where ϕ is the azimuthally coordinate. Replacing (4.30) into (4.29) results in:

$$F_D = \int_0^{2\pi}\int_0^{\pi} p(\theta)\text{sen }\theta \cos \theta d\theta d\phi = 2\pi a^2 \int_0^{\pi} p(\theta)\text{sen }\theta \cos \theta d\theta$$

Since the values of $p(\theta)$ are different before and after the separation point, the integrals are separated into two parts, from 0 to θ_s and from θ_s to π:

$$F_D = 2\pi a^2 \left(\int_0^{\theta_s} p(\theta)\text{sen }\theta \, d(\text{sen }\theta) + \int_{\theta_s}^{\pi} p(\theta)\text{sen }\theta \, d(\text{sen }\theta) \right)$$

Substituting the values of $p(\theta)$ from (4.27) and (4.28), and integrating the previous equation we obtain:

$$F_D = \pi a^2 \rho_f u^2 \left(\frac{1}{2}\text{sen}^2 \theta_s - \frac{9}{16}\text{sen}^4 \theta_s - \frac{1}{2} p_b^* \text{sen}^2 \theta_s \right) \tag{4.31}$$

Substituting $a = R + \delta$ and defining the function $f(\theta_s, p_b^*)$ in the form:

$$f(\theta_s, p_b^*) = \frac{1}{2}\text{sen}^2 \theta_s - \frac{9}{16}\text{sen}^4 \theta_s - \frac{1}{2} p_b^* \text{sen}^2 \theta_s \tag{4.32}$$

we can into (4.31) we can write:

$$F_D = \rho_f u^2 \pi R^2 \left(1 + \frac{\delta}{R} \right)^2 f(\theta_s, p_b^*) \tag{4.33}$$

In terms of the drag coefficient we have:

$$C_D = 2f(\theta_s, p_b^*) \left(1 + \frac{\delta}{R} \right)^2 \tag{4.34}$$

and defining the new parameter C_0 in the form:

$$C_0 = 2f(\theta_s, p_b^*) \tag{4.35}$$

Using Eq. (4.32) we obtain:

$$C_D = C_0(\theta_s, p_b^*) \left(1 + \frac{\delta_0}{Re^{1/2}} \right)^2 \tag{4.36}$$

Calculating the value of $f(\theta_s, p_b^*)$ for $\theta_s = 84°$ and $p_b^* \approx -0.4$, we obtain $f(84, -0.4) = 0.142$ and from (4.35) $C_0 = 0.284$. Using the value of $\delta_0 = 9.06$, we finally obtain:

$$C_D = 0.284 \left(1 + \frac{9.06}{Re^{1/2}} \right)^2 \tag{4.37}$$

Expression (4.37) represents the drag coefficient for a sphere in boundary layer flow (Concha and Almendra 1979a, b). A comparison with experimental data from Lapple and Shepherd (1940) is shown in Fig. 4.11.

Several alternative empirical equations have been proposed for the drag coefficient of spherical particles. See earlier articles reviewed by Concha and Almendra (1979a, b), Zigrang and Sylvester (1981), Turton and Levenspiel (1986), Turton and Clark (1987), Haider and Levenspiel (1998), and the more recent work of Ganguly (1990), Thomson and Clark (1991), Ganser (1993), Flemmer et al. (1993), Darby (1996), Nguyen et al. (1997), Chabra et al. (1999) and Tsakalakis and Stamboltzis (2001).

It is worthwhile to mention the work of Brauer and Sucker (1976):

$$C_D = 0.49 + \frac{24}{Re} + \frac{3.73}{Re^{1/2}} - \frac{4.83 \times 10^{-3} Re^{1/2}}{1 + 3.0 \times 10^{-6} Re^{3/2}} \tag{4.38}$$

and that of Haider and Levenspiel (1998):

$$C_D = \frac{24}{Re}\left(1 + 0.1806 Re^{0.6459}\right) + \frac{0.4251}{1 + 6{,}880.95/Re} \tag{4.39}$$

who presented an alternative empirical equation for the drag coefficient of spherical particles in the range of Reynolds numbers less than 260,000. Both Eqs.

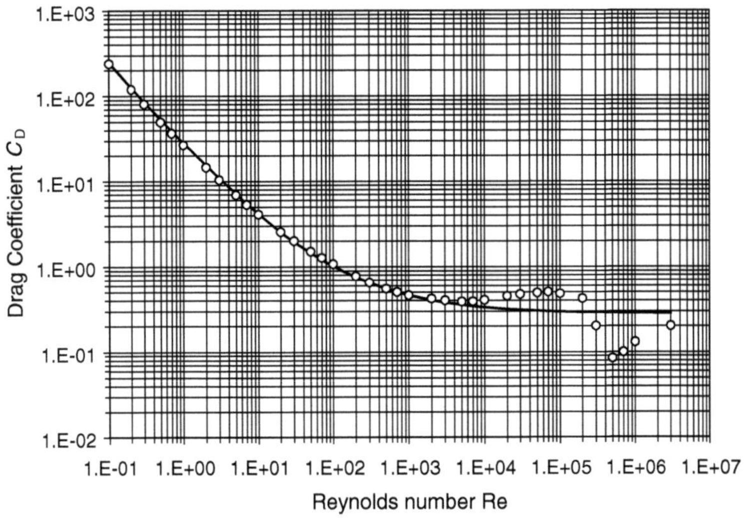

Fig. 4.11 Drag coefficient versus Reynolds number for a sphere. The continuous line is a simulation of Eq. (4.37). Circles are standard data from Lapple and Shepherd (1940) see Table 4.1

(4.38) and (4.39) give better approximations than Concha and Almendra's equation (4.37) for Reynolds numbers in the range of $5 \times 10^3 < Re < 5 \times 10^5$.

Table 4.1 gives the standard drag coefficient of Lappel and Shepherd (1940), L&Sh in table and results of Concha and Almendra (1979a, b), Heider and Levenspiel (1998), H&L in Table 4.1 and Brauer and Sucker (1976), B&Z in Table 4.1 (Fig. 4.12).

4.1.6 Sedimentation Velocity of a Sphere

We have seen that when a particle settles at terminal velocity u_∞, a balance is established between drag force, gravity and buoyancy:

$$\begin{aligned} F_{drag} + F_{gravity} + F_{buoyancy} &= 0 \\ F_{drag} = -\left(F_{gravity} + F_{buoancy}\right) &\equiv \text{net weight of the particle} \\ F_D = -\left(\rho_p V_p(-g) + \rho_f V_p g\right) &\equiv \Delta\rho V_p g \end{aligned} \tag{4.40}$$

In (4.40) $\Delta\rho = \rho_s - \rho_f$ is the solid–fluid density difference. This equation shows that the drag force for a particle in sedimentation is known beforehand once the volume of particle and its density difference to the fluid are known. For a spherical particle, $V_p = 4/3\pi R^3$, so that:

$$F_D = \frac{4}{3}\pi R^3 \Delta\rho g \tag{4.41}$$

and the drag coefficient:

$$C_D = \frac{F_D}{1/2\rho_f u_\infty^2 \pi R^2} \equiv \frac{4}{3}\frac{\Delta\rho d g}{\rho_f u_\infty^2} \tag{4.42}$$

where the sphere diameter is $d = 2R$ and u_∞ is the terminal settling velocity of a sphere in an infinite fluid.

Since the Reynolds number for the motion of one particle in an infinite fluid is defined by:

$$Re_\infty = \frac{d u_\infty \rho_f}{\mu_f}, \tag{4.43}$$

combining it with the drag coefficient yields two dimensionless numbers (Heywood 1962):

$$C_D Re_\infty^2 = \left(\frac{4}{3}\frac{\Delta\rho\rho_f g}{\mu_f^2}\right) d^3 \quad \frac{Re_\infty}{C_D} = \left(\frac{3}{4}\frac{\rho_f^2}{\Delta\rho\mu_f g}\right) u_\infty^3 \tag{4.44}$$

Table 4.1 Drag coefficient versus Reynolds number and dimensionless velocity versus dimensionless diameter

L&Sh						Concha and Almendra			H&L	B&Z
Re	C_D	C_DRe^2	Re/C_D	$d^* = (C_DRe^2)^{1/3}$	$u^* = (Re/C_D)^{1/3}$	C_D sim	d* sim	u* sim	C_D sim	C_D sim
1.000E−01	2.400E+02	2.400E+00	4.167E−04	1.339E+00	7.469E−02	2.462E+02	1.000E−01	4.345E−04	2.498E+02	2.523E+02
2.000E−01	1.200E+02	4.800E+00	1.667E−03	1.687E+00	1.186E−01	1.265E+02	2.000E−01	1.733E−03	1.277E+02	1.288E+02
3.000E−01	8.000E+01	7.200E+00	3.750E−03	1.931E+00	1.554E−01	8.615E+01	3.000E−01	3.887E−03	8.664E+01	8.730E+01
5.000E−01	4.950E+01	1.238E+01	1.010E−02	2.313E+00	2.162E−01	5.342E+01	4.000E−01	6.882E−03	5.354E+01	5.376E+01
7.000E−01	3.650E+01	1.789E+01	1.918E−02	2.615E+00	2.677E−01	3.918E+01	5.000E−01	1.071E−02	3.920E+01	3.923E+01
1.000E+00	2.650E+01	2.650E+01	3.774E−02	2.981E+00	3.354E−01	2.834E+01	6.000E−01	1.534E−02	2.833E+01	2.822E+01
2.000E+00	1.460E+01	5.840E+01	1.370E−01	3.880E+00	5.155E−01	1.536E+01	7.000E−01	2.077E−02	1.539E+01	1.512E+01
3.000E+00	1.040E+01	9.360E+01	2.885E−01	4.540E+00	6.607E−01	1.087E+01	8.000E−01	2.697E−02	1.094E+01	1.064E+01
5.000E+00	6.900E+00	1.725E+02	7.246E−01	5.567E+00	8.982E−01	7.146E+00	9.000E−01	3.393E−02	7.252E+00	6.947E+00
7.000E+00	5.300E+00	2.597E+02	1.321E+00	6.380E+00	1.097E+00	5.481E+00	1.000E+00	4.162E−02	5.605E+00	5.316E+00
1.000E+01	4.100E+00	4.100E+02	2.439E+00	7.429E+00	1.346E+00	4.183E+00	1.100E+00	5.003E−02	4.319E+00	4.054E+00
2.000E+01	2.550E+00	1.020E+03	7.843E+00	1.007E+01	1.987E+00	2.564E+00	1.200E+00	5.913E−02	2.702E+00	2.502E+00
3.000E+01	2.000E+00	1.800E+03	1.500E+01	1.216E+01	2.466E+00	1.972E+00	1.300E+00	6.891E−02	2.102E+00	1.945E+00
5.000E+01	1.500E+00	3.750E+03	3.333E+01	1.554E+01	3.218E+00	1.457E+00	1.400E+00	7.934E−02	1.568E+00	1.463E+00
7.000E+01	1.270E+00	6.223E+03	5.512E+01	1.839E+01	3.806E+00	1.215E+00	1.500E+00	9.041E−02	1.310E+00	1.238E+00
1.000E+02	1.070E+00	1.070E+04	9.346E+01	2.204E+01	4.538E+00	1.017E+00	1.600E+00	1.021E−01	1.095E+00	1.055E+00
2.000E+02	7.700E−01	3.080E+04	2.597E+02	3.135E+01	6.380E+00	7.537E−01	1.700E+00	1.144E−01	7.960E−01	8.060E−01
3.000E+02	6.500E−01	5.850E+04	4.615E+02	3.882E+01	7.728E+00	6.495E−01	1.800E+00	1.272E−01	6.729E−01	7.030E−01
5.000E+02	5.500E−01	1.375E+05	9.091 E+02	5.161E+01	9.687E+00	5.529E−01	1.900E+00	1.406E−01	5.568E−01	6.003E−01
7.000E+02	5.000E−01	2.450E+05	1.400E+03	6.257E+01	1.119E+01	5.046E−01	2.000E+00	1.545E−01	4.996E−01	5.442E−01
1.000E+03	4.600E−01	4.600E+05	2.174E+03	7.719E+01	1.295E+01	4.634E−01	2.100E+00	1.690E−01	4.535E−01	4.924E−01
2.000E+03	4.200E−01	1.680E+06	4.762E+03	1.189E+02	1.682E+01	4.049E−01	2.200E+00	1.839E−01	4.015E−01	4.151E−01
3.000E+03	4.000E−01	3.600E+06	7.500E+03	1.533E+02	1.957E+01	3.803E−01	2.300E+00	1.993E−01	3.916E−01	3.889E−01
5.000E+03	3.850E−01	9.625E+06	1.299E+04	2.127E+02	2.351E+01	3.563E−01	2.400E+00	2.152E−01	3.961E−01	3.818E−01
7.000E+03	3.900E−01	1.911E+07	1.795E+04	2.674E+02	2.618E+01	3.439E−01	2.500E+00	2.315E−01	4.063E−01	3.914E−01

(continued)

Table 4.1 (continued)

L&Sh						Concha and Almendra			H&L	B&Z
Re	C_D	C_DRe^2	Re/C_D	$d^* = (C_DRe^2)^{1/3}$	$u^* = (Re/C_D)^{1/3}$	C_D sim	d* sim	u* sim	C_D sim	C_D sim
1.000E+04	4.050E−01	4.050E+07	2.469E+04	3.434E+02	2.912E+01	3.330E−01	2.600E+00	2.482E−01	4.204E−01	4.090E−01
2.000E+04	4.500E−01	1.800E+08	4.444E+04	5.646E+02	3.542E+01	3.170E−01	2.700E+00	2.653E−01	4.475E−01	4.456E−01
3.000E+04	4.700E−01	4.230E+08	6.383E+04	7.507E+02	3.996E+01	3.101E−01	2.800E+00	2.829E−01	4.592E−01	4.619E−01
5.000E+04	4.900E−01	1.225E+09	1.020E+05	1.070E+03	4.673E+01	3.031E−01	2.900E+00	3.008E−01	4.681 E–01	4.759E−01
7.000E+04	5.000E−01	2.450E+09	1.400E+05	1.348E+03	5.192E+01	2.995E−01	3.000E+00	3.190E−01	4.708E−01	4.818E−01
1.000E+05	4.800E−01	4.800E+09	2.083E+05	1.687E+03	5.928E+01	2.963E−01	3.100E+00	3.376E−01	4.715E−01	4.861E–01
2.000E+05	4.200E−01	1.680E+10	4.762E+05	2.561 E+03	7.809E+01	2.915E−01	3.200E+00	3.566E−01	4.686E−01	4.904E−01
3.000E+05	2.000E−01	1.800E+10	1.500E+06	2.621 E+03	1.145E+02	2.893E−01	3.300E+00	3.758E−01	4.655E−01	4.915E−01
5.000E+05	8.400E−02	2.100E+10	5.952E+06	2.759E+03	1.812E+02	2.872E−01	3.400E+00	3.954E−01	4.610E−01	4.921E–01
7.000E+05	1.000E−01	4.900E+10	7.000E+06	3.659E+03	1.913E+02	2.861E−01	3.500E+00	4.152E−01	4.579E−01	4.922E−01
1.000E+06	1.300E−01	1.300E+11	7.692E+06	5.066E+03	1.974E+02	2.851E−01	3.600E+00	4.353E−01	4.548E−01	4.921E−01
3.000E+06	2.000E−01	1.800E+12	1.500E+07	1.216E+04	2.466E+02	2.829E−01	3.700E+00	4.557E−01	4.462E−01	4.916E−01

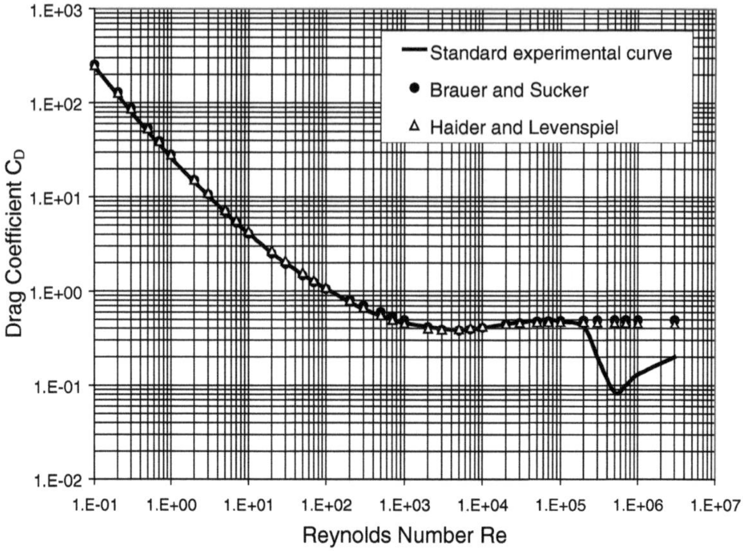

Fig. 4.12 Comparison of the drag coefficient for a sphere, simulated by Brauer and Zucker (1976) and by Haider and Levenspiel (1998), and standard experimental points from Laple and Shepherd (1940)

Concha and Almendra (1979a) defined the characteristic parameters P and Q of the solid–liquid system:

$$P = \left(\frac{3}{4}\frac{\mu_f^2}{\Delta\rho\rho_f g}\right)^{1/3} \quad Q = \left(\frac{4}{3}\frac{\Delta\rho\mu_f g}{\rho_f^2}\right)^{1/3} \tag{4.45}$$

so that Eq. (4.44) may be written in the form:

$$C_D Re_\infty^2 = \left(\frac{d}{P}\right)^3 = d^{*^3} \quad \frac{Re_\infty}{C_D} = \left(\frac{u_\infty}{Q}\right)^3 \equiv u_\infty^{*3} \tag{4.46}$$

Expressions (4.46) define a dimensionless size d^* and a dimensionless velocity u^*, which are characteristics of a solid–liquid system:

$$d^* = \left(\frac{d}{P}\right) \quad u_\infty^* = \left(\frac{u_\infty}{Q}\right) \tag{4.47}$$

Since there is a direct relationship between the Drag Coefficient and the Reynolds Number, see for example Eqs. (4.37)–(4.39), there must be a relationship between the dimensionless groups $C_D Re^2$ and Re/C_D. Table 4.1 gives that relationship.

Multiplying the two terms in Eq. (4.47), we can observe that the Reynolds number may be written in terms of the dimensionless size and dimensionless velocity:

$$Re_\infty = d^* u^*_\infty \tag{4.48}$$

Replacing (4.37) and (4.48) into (4.46) we obtain:

$$d^{*3} = C_0\left(1+\frac{\delta_0}{\left(u^*_\infty d^*\right)^{1/2}}\right)^2\left(u^*_\infty d^*\right)^2$$

$$u^*_\infty d^* + \delta_0\left(u^*_\infty d^*\right)^{1/2} - \frac{d^{*3/2}}{C_0^{1/2}} = 0$$

Solving these algebraic equations, explicit expressions are obtained for the dimensionless settling velocity u^*_∞ of a sphere of dimensionless size d^* and for the dimensionless sphere of diameter d^* settling a dimensionless velocity u^*_∞ (Concha and Almendra 1979a):

$$u^*_\infty = \frac{1}{4}\frac{\delta_0^2}{d^*}\left(\left(1+\frac{4}{C_0^{1/2}\delta_0^2}d^{*3/2}\right)^{1/2}-1\right)^2 \tag{4.49}$$

$$d^* = \frac{1}{4}C_0 u^{*2}_\infty\left(1+\left(1+\frac{4\delta_0}{C_0^{1/2}}u^{*-3/2}_\infty\right)^{1/2}\right)^2 \tag{4.50}$$

Equations (4.49) and (4.50) are known as *Concha and Almendra's equations* for a sphere. These two equations are general for spheres settling in a fluid at any Reynolds number. Introducing the values of $C_0 = 0.284$ and $\delta_0 = 9.06$, the following final equations are obtained [In Concha and Amendra's paper (1979a), $C_0 = 0.28$ was used].

$$u^* = \frac{20.52}{d^*}\left(\left(1+0.0914 d^{*3/2}\right)^{1/2}-1\right)^2 \tag{4.51}$$

$$d^* = 0.071 u^{*2}\left(1+\left(1+68.0 u^{*-3/2}\right)^{1/2}\right)^2 \tag{4.52}$$

Problem 4.3 To calculate sedimentation velocities, it is necessary to know the sedimentation parameters P and Q that depend on the solid and fluid properties. Construct a table for these parameters as functions of the solid density. Assume that the fluid density is 1 g/cm^3 and its viscosity 0.01 g/cm s. Using the definition of P and Q from Eq. (4.45), Table 4.2 is obtained.

Table 4.2 Parameters P and Q for several values of the solid density

ρ_s (g/cm^3)	ρ_f (g/cm^3)	μ_f (g/cms)	P (cm)	Q (cm/s)	ρ_s (g/cm^3)	ρ_f (g/cm^3)	μ_f (g/cm s)	P (cm)	Q (cm/s)
1.10	1	0.0100	9.1470E−03	1.0933E+00	3.35	1	0.0100	3.1934E−03	3.1314E+00
1.15	1	0.0100	7.9906E−03	1.2515E+00	3.40	1	0.0100	3.1711E−03	3.1535E+00
1.20	1	0.0100	7.2600E−03	1.3774E+00	3.45	1	0.0100	3.1494E−03	3.1752E+00
1.25	1	0.0100	6.7396E−03	1.4838E+00	3.50	1	0.0100	3.1282E−03	3.1967E+00
1.30	1	0.0100	6.3422E−03	1.5767E+00	3.55	1	0.0100	3.1076E−03	3.2179E+00
1.35	1	0.0100	6.0245E−03	1.6599E+00	3.60	1	0.0100	3.0876E−03	3.2388E+00
1.40	1	0.0100	5.7622E−03	1.7354E+00	3.65	1	0.0100	3.0681E−03	3.2594E+00
1.45	1	0.0100	5.5404E−03	1.8049E+00	3.70	1	0.0100	3.0490E−03	3.2798E+00
1.50	1	0.0100	5.3492E−03	1.8694E+00	3.75	1	0.0100	3.0304E−03	3.2999E+00
1.55	1	0.0100	5.1819E−03	1.9298E+00	3.80	1	0.0100	3.0123E−03	3.3198E+00
1.60	1	0.0100	5.0338E−03	1.9866E+00	3.85	1	0.0100	2.9945E−03	3.3394E+00
1.65	1	0.0100	4.9013E−03	2.0403E+00	3.90	1	0.0100	2.9772E−03	3.3588E+00
1.70	1	0.0100	4.7817E−03	2.0913E+00	3.95	1	0.0100	2.9603E−03	3.3780E+00
1.75	1	0.0100	4.6730E−03	2.1400E+00	4.00	1	0.0100	2.9438E−03	3.3970E+00
1.80	1	0.0100	4.5735E−03	2.1865E+00	4.05	1	0.0100	2.9276E−03	3.4158E+00
1.85	1	0.0100	4.4820E−03	2.2311E+00	4.10	1	0.0100	2.9118E−03	3.4343E+00
1.90	1	0.0100	4.3974E−03	2.2741E+00	4.15	1	0.0100	2.8963E−03	3.4527E+00
1.95	1	0.0100	4.3189E−03	2.3154E+00	4.20	1	0.0100	2.8811E−03	3.4709E+00
2.00	1	0.0100	4.2457E−03	2.3553E+00	4.25	1	0.0100	2.8663E−03	3.4889E+00
2.05	1	0.0100	4.1772E−03	2.3940E+00	4.30	1	0.0100	2.8517E−03	3.5067E+00
2.10	1	0.0100	4.1129E−03	2.4314E+00	4.35	1	0.0100	2.8375E−03	3.5243E+00
2.15	1	0.0100	4.0524E−03	2.4677E+00	4.40	1	0.0100	2.8235E−03	3.5417E+00
2.20	1	0.0100	3.9953E−03	2.5029E+00	4.45	1	0.0100	2.8098E−03	3.5590E+00
2.25	1	0.0100	3.9413E−03	2.5372E+00	4.50	1	0.0100	2.7963E−03	3.5761E+00
2.30	1	0.0100	3.8901E−03	2.5706E+00	4.55	1	0.0100	2.7831E−03	3.5931E+00
2.35	1	0.0100	3.8415E−03	2.6032E+00	4.60	1	0.0100	2.7702E−03	3.6098E+00

(continued)

Table 4.2 (continued)

ρ_s (g/cm^3)	ρ_f (g/cm^3)	μ_f (g/cms)	P (cm)	Q (cm/s)	ρ_s (g/cm^3)	ρ_f (g/cm^3)	μ_f (g/cm s)	P (cm)	Q (cm/s)
2.40	1	0.0100	3.7952E−03	2.6349E+00	4.65	1	0.0100	2.7575E−03	3.6265E+00
2.45	1	0.0100	3.7511E−03	2.6659E+00	4.70	1	0.0100	2.7450E−03	3.6430E+00
2.50	1	0.0100	3.7089E−03	2.6962E+00	4.75	1	0.0100	2.7328E−03	3.6593E+00
2.55	1	0.0100	3.6686E−03	2.7258E+00	4.80	1	0.0100	2.7207E−03	3.6755E+00
2.60	1	0.0100	3.6300E−03	2.7548E+00	4.85	1	0.0100	2.7089E−03	3.6915E+00
2.65	1	0.0100	3.5929E−03	2.7832E+00	4.90	1	0.0100	2.6973E−03	3.7075E+00
2.70	1	0.0100	3.5574E−03	2.8111E+00	4.95	1	0.0100	2.6858E−03	3.7232E+00
2.75	1	0.0100	3.5232E−03	2.8384E+00	5.00	1	0.0100	2.6746E−03	3.7389E+00
2.80	1	0.0100	3.4902E−03	2.8651E+00	5.05	1	0.0100	2.6635E−03	3.7544E+00
2.85	1	0.0100	3.4585E−03	2.8914E+00	5.10	1	0.0100	2.6527E−03	3.7698E+00
2.90	1	0.0100	3.4279E−03	2.9172E+00	5.15	1	0.0100	2.6420E−03	3.7850E+00
2.95	1	0.0100	3.3983E−03	2.9426E+00	5.20	1	0.0100	2.6315E−03	3.8002E+00
3.00	1	0.0100	3.3698E−03	2.9676E+00	5.25	1	0.0100	2.6211E−03	3.8152E+00
3.05	1	0.0100	3.3422E−03	2.9921E+00	5.30	1	0.0100	2.6109E−03	3.8301E+00
3.10	1	0.0100	3.3154E−03	3.0162E+00	5.35	1	0.0100	2.6009E−03	3.8449E+00
3.15	1	0.0100	3.2895E−03	3.0400E+00	5.40	1	0.0100	2.5910E−03	3.8596E+00
3.20	1	0.0100	3.2644E−03	3.0633E+00	5.45	1	0.0100	2.5812E−03	3.8741E+00
3.25	1	0.0100	3.2400E−03	3.0864E+00	5.50	1	0.0100	2.5716E−03	3.8886E+00
3.30	1	0.0100	3.2164E−03	3.1091E+00	5.55	1	0.0100	2.5622E−03	3.9029E+00

Problem 4.4 Calculate the settling velocity of quartz spheres with density of 2.65 g/cm^3 of the following diameters: 1, 10, 50, 100 and 500 μm, and 0.1, 0.5 and 1.0 cm in water at 20 °C. Use Eq. (4.51):

$$u_\infty^* = \frac{20.52}{d^*}\left(\left(1+0.0914 d^{*3/2}\right)^{1/2} - 1\right)^2, \quad \text{with} \quad d^* = d/P \quad y \quad u_\infty = Q \times u_\infty^*$$

Data								
ρ_s (g/cm^3)=	2.65	2.65	2.65	2.65	2.65	2.65	2.65	2.65
ρ_f (g/cm^3)=	1	1	1	1	1	1	1	1
μ (poises)=	0.01	0.01	0.01	0.01	0.01	0.01	0.01	0.01
d_p (cm)=	0.0001	0.001	0.005	0.01	0.05	0.1	0.5	1
P (cm)=	0.00359	0.00359	0.00359	0.00359	0.00359	0.00359	0.00359	0.00359
Q (cm/s)=	2.78320	2.78320	2.78320	2.78320	2.78320	2.78320	2.78320	2.78320
Results								
d* =	2.78E−02	2.78E−01	1.39E+00	2.78E+00	1.39E+01	2.78E+01	1.39E+02	2.78E+02
u* =	3.32E−05	3.30E−03	7.73E−02	2.76E−01	2.88E+00	5.77E+00	1.88E+01	2.84E+01
u_{oo} (cm/s) =	9.24E−05	9.18E−03	2.15E−01	7.68E−01	8.01E+00	1.61E+01	5.23E+01	7.90E+01
Re =	9.24E−07	9.18E−04	1.08E−01	7.68E−01	4.00E+01	1.61E+02	2.62E+03	7.90E+03

Problem 4.5 Compare the simulated dimensionless velocity versus dimensionless diameter predicted by Concha and Almendra's equation with experimental data from Table 4.1.

Using Eq. (4.51), the values of column d^*sim and u^*sim are obtained. The plot of these data is shown in Fig. 4.13.

Effect of temperatures on the settling velocity

To calculate the settling velocity of spheres at different temperatures, it is necessary to know the density and viscosity of the fluid involved at these temperatures. In the case of water and air, the following correlations can be used for the density and viscosity. Figures 4.14 and 4.15 show the correlations of densities and viscosities for air and water.

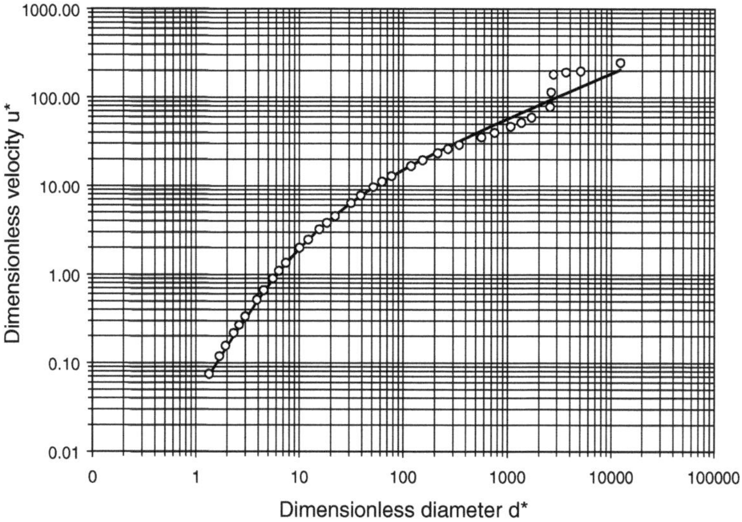

Fig. 4.13 Dimensionless velocity versus dimensionless diameter for the sedimentation of spheres according to Eq. (4.51) of Concha and Almendra. Circles are standard data from Lapple and Shepherd (1940) in Table 4.1

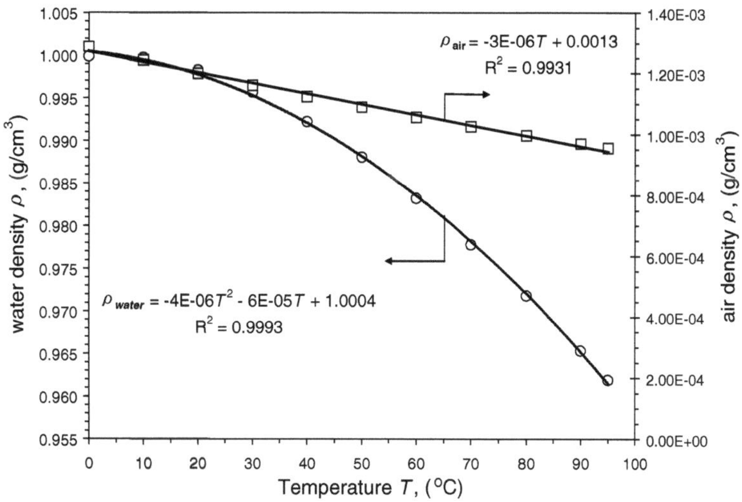

Fig. 4.14 Water and air densities at several temperatures

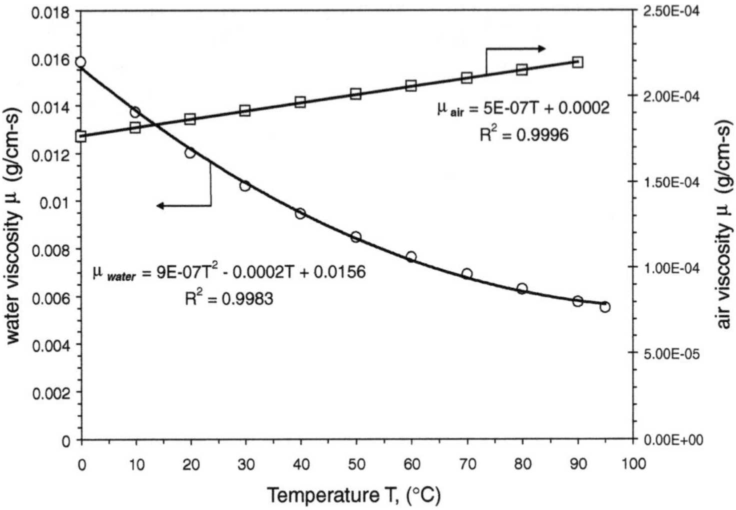

Fig. 4.15 Water and air viscosities at several temperatures

Density:

$$\rho_{water} = -4.0 \times 10^{-6}T^2 - 6.0 \times 10^{-5}T + 1.0004 \text{ g/cm}^3 \tag{4.53}$$

$$\rho_{air} = -3.0 \times 10^{-6}T + 1.3 \times 10^{-3} \text{ g/cm}^3 \tag{4.54}$$

Viscosity:

$$\mu_{water} = 9.0 \times 10^{-7}T^2 - 2.0 \times 10^{-4}T + 1.56 \times 10^{-2} \text{ g/cm s} \tag{4.55}$$

$$\mu_{air} = 5.0 \times 10^{-7}T + 2.0 \times 10^{-4} \text{ g/cm s} \tag{4.56}$$

Problem 4.6 Calculate the settling velocity of a quartz sphere of 300 μm with density 2.65 g/cm^3 in water at 60 °C. From the correlations (4.55), (4.53) and from Eqs. (4.45) we obtain:

$$\rho_{water} = -4.0 \times 10^{-6} \times 20^2 - 6.0 \times 10^{-5} \times 20 + 1.0004 = 0.9976\ \text{g/cm}^3$$

$$\mu_{water} = 9 \times 10^{-7} \times 20^2 - 2 \times 10^{-4} \times 20 + 0.0156 = 0.01196$$

$$P = \left(\frac{3}{4}\frac{\mu_f^2}{\Delta\rho\rho_f g}\right)^{1/3} = 2.550 \times 10^{-3}\ \text{cm}$$

$$Q = \left(\frac{4}{3}\frac{\Delta\rho\mu_f g}{\rho_f^2}\right)^{1/3} = 2.0534\ \text{cm/s}$$

$$d^* = \frac{300}{10{,}000 \times 2.5502 \times 10^{-3}} = 11.764$$

$$u^* = \frac{20.52}{d^*}\left(\left(1 + 0.0921 d^{*3/2}\right)^{1/2} - 1\right)^2 = 1.5017$$

$$u_\infty = u^* \times Q = 1.5017 \times 2.0534 = 3.0826\ \text{m/s}$$

4.1.7 Sedimentation of a Suspension of Spheres

In a suspension, the spheres, surrounding a given sphere, hinder its motion as it settles. This hindrance is due to several effects. In the first place, when the particle changes its position, it can find the new site occupied by another particle, and will collide with it, thus changing its trajectory. The more concentrated the suspension is, the greater the chance of collision. The result is that hindrance is a function of concentration. On the other hand, the settling of each particle of the suspension produces a back flow of the fluid. This back flow will increase the drag on a given particle, retarding its sedimentation. Again, an increase in concentration will increase the hindrance. It is clear that in both cases, the hindrance depends on the fraction of volume occupied by the particles and not on their weight and therefore the appropriate parameter to measure hindrance is the volumetric concentration rather than the percentage by weight.

Several theoretical works have been devoted to study the interaction of particles in a suspension during sedimentation. These types of studies were discussed in Tory (1996). In a recent approach, Quispe et al. used the tools of lattice-gas and cellular automata to study the sedimentation of particles and the fluid flow through an ensemble of settling particles. They were able to obtain some important macroscopic properties of the suspensions. Unfortunately, none of these works has yielded a sufficiently general and simple relationship between the variables of the suspension and its settling velocity to be used for practical purposes.

Concha and Almendra (1979b) assumed that the same equations used for the drag coefficient and for the settling velocity of individual spherical particles are valid for a suspension of particles if parameters P and Q are replaced by

$P(\varphi)$ and $Q(\varphi)$ depending on the volume fraction of solids. Write Eqs. (4.36) and (4.49) in the form:

$$C_D(\varphi) = C_0(\varphi)\left(1 + \frac{\delta_0(\varphi)}{\mathrm{Re}^{1/2}}\right)^2 \tag{4.57}$$

$$U^*(\varphi) = \frac{\delta_0^2}{D^*(\varphi)}\left(\left(1 + \frac{1}{C_0^{1/2}\delta_0^2}D^{*3/2}(\varphi)\right)^{1/2} - 1\right)^2 \tag{4.58}$$

where

$$D^* = \frac{d}{P(\varphi)} \quad \text{and} \quad U^* = \frac{u}{Q(\varphi)} \tag{4.59}$$

It is convenient to express the properties of a suspension, such as viscosity, as the product of the same property of the fluid with a function of concentration. Assume that $P(\varphi)$ and $Q(\varphi)$ can be related with P and Q in that form, then:

$$P(\varphi) = Pf_p(\varphi) \quad \text{and} \quad Q(\varphi) = Qf_q(\varphi) \tag{4.60}$$

Replacing (4.60) into (4.59), and using the definition (4.47) of d^* and u^*, results in:

$$D^* = \frac{d^*}{f_p(\varphi)} \quad \text{and} \quad U^*(\varphi) = \frac{u^*(\varphi)}{f_q(\varphi)} \tag{4.61}$$

With these definitions, Eq. (4.58) may be written in the form:

$$u^*(\varphi) = \frac{\delta_0^2}{d^*}f_p(\varphi)f_q(\varphi)\left(\left(1 + \frac{1}{C_0^{1/2}\delta_0^2}f_p^{-3/2}d^{*3/2}\right)^{1/2} - 1\right)^2 \tag{4.62}$$

Substituting the values of $\delta_0 = 9.06$ and $C_0 = 0.284$, yields:

$$u^*(\varphi) = \frac{20.52}{d^*}f_p(\varphi)f_q(\varphi)\left(1 + 0.0914f_p^{-3/2}(\varphi)d^{*3/2^{1/2}} - 1\right)^2 \tag{4.63}$$

This expression, known as *Concha and Almendra's equation for a suspension of spheres*, permits the calculation of the settling velocity of a sphere of any size and density when it forms part of a suspension with volume fraction φ. To perform the calculations, it is necessary to know the parameters $f_p(\varphi)$ and $f_q(\varphi)$.

Asymptotic expressions for the sedimentation velocity.

For small values of the Reynolds number, $Re \rightarrow 0$, the following expressions may be derived from (4.51) and (4.63), which reduce the settling equation to the expression indicated:

$$0.0921 d^{*3/2} f_p^{-3/2} \ll 1 \Rightarrow u^*(\varphi) = 20.52\left(\frac{0.0914}{2}\right)^2 d^{*2} f_p^{-2}(\varphi) f_q(\varphi)$$
$$0.0921 d^{*3/2} \ll 1 \Rightarrow u^*_\infty = 20.52\left(\frac{0.0914}{2}\right)^2 d^{*2}$$

In these expressions, the symbols u^*_∞ and $u^*(\varphi)$ indicate the settling velocity of a particle in an infinite medium and the velocity of the same particle in a suspension. The quotient between these two terms is:

$$\text{For } Re_\infty \to 0 \quad \frac{u^*(\varphi)}{u^*_\infty} = f_p^{-2}(\varphi) f_q(\varphi) \tag{4.64}$$

With a similar deduction, we can write for high Reynolds numbers, Eq. (4.49) equation reference goes here (4.51) and (4.62) in the form:

$$0.0921 d^{*3/2} f_p^{-3/2} \gg 1 \Rightarrow u^*(\varphi) = 20.52 \times 0.0914 \times d^{*1/2} f_p^{-1/2}(\varphi) f_q(\varphi)$$
$$0.0921 d^{*3/2} \gg 1 \Rightarrow u^*_\infty = 20.52 \times 0.0914 \times d^{*1/2}$$

The quotient between these two equations is:

$$\text{for } Re_\infty \to \infty \quad \frac{u^*(\varphi)}{u^*_\infty} = f_p^{-1/2}(\varphi) f_q(\varphi) \tag{4.65}$$

To find functional forms for the functions $f_p(\varphi)$ and $f_q(\varphi)$ experimental values for the settling velocities u_∞ and $u(\varphi)$ are needed.

Functional form for $f_p(\varphi)$ and $f_q(\varphi)$

Several authors have presented expressions for the velocity ratio u/u_∞. See Concha and Almendra (1979b). Richardson and Zaki (1954), made the most comprehensive and most cited work on the relative particle–fluid flow under gravity. We will use their data in this book.

Richardson and Zaki (1954) performed careful sedimentation and liquid fluidization tests with mono-sized spherical particles in a wide range of particles sizes, fluid densities and viscosities. The authors expressed their result in the form:

$$u(\varphi)/u_\infty = (1-\varphi)^n \text{ for Re} \to 0 \quad \text{and} \quad u(\varphi)/u_\infty = (1-\varphi)^m \text{ for Re} \to \infty \tag{4.66}$$

The characteristics of these particles and fluid are given in Table 4.3. Table 4.4 shows values for $u(\varphi)/u_\infty$ obtained from their experimental results.

Instead of using the values of n presented by Richardson and Zaki (1954), we will optimize the values using all experimental data. From (4.64), (4.65) and (4.66), we can write

Table 4.3 Experimental data of Richardson and Zaki (1954)

N°	d (cm)	ρ_s (g/cm^3)	μ_fx102 (g/cms)	ρ_f (g/cm^3)	v_{st} (cm/s)	Re_{oo}	n + 1	n
P	0.0181	1.058	20.800	1.034	0.00206	0.000185	4.90	3.90
Q	0.0181	1.058	20.800	1.034	0.00206	0.000185	4.79	3.79
K	0.0096	2.923	62.000	1.208	0.01390	0.000216	4.75	3.75
L	0.0096	2.923	62.000	1.208	0.01390	0.000216	4.65	3.65
F	0.0358	1.058	20.800	1.034	0.00807	0.001430	4.92	3.92
G	0.0358	1.058	20.800	1.034	0.00807	0.001430	4.89	3.89
H	0.0096	2.923	20.800	1.034	0.04550	0.002180	4.76	3.76
I	0.0096	2.923	20.800	1.034	0.04550	0.002180	4.72	3.72
J	0.0096	2.923	20.800	1.034	0.04550	0.002180	4.69	3.69
R	0.0230	2.623	62.000	1.208	0.06590	0.002950	4.85	3.85
S	0.0230	2.623	62.000	1.208	0.06590	0.002950	4.80	3.80
T	0.0128	2.960	1.890	2.890	0.03307	0.064700	4.84	3.84
M	0.0128	2.960	1.890	2.890	0.03307	0.064700	4.72	3.72
C	0.0181	1.058	1.530	1.001	0.06400	0.078900	4.76	3.76
A	0.0181	1.058	1.530	1.001	0.06640	0.078900	4.90	3.90
B	0.0181	1.058	1.530	1.001	0.06640	0.078900	4.79	3.79
X	0.1029	2.976	112.900	1.221	0.89100	0.099500	5.30	4.30
Y	0.1029	2.976	112.900	1.221	0.89100	0.099500	5.20	4.20
D	0.0253	1.058	2.910	0.935	0.14750	0.120000	4.94	3.94
E	0.0253	1.058	2.910	0.935	0.14750	0.120000	4.90	3.90
N	0.0096	2.923	1.612	2.170	0.23400	0.030200	4.74	3.74
O	0.0096	2.923	1.612	2.170	0.23400	0.030200	4.65	3.65
2	0.0253	2.78	6.075	1.135	0.82700	0.391000	4.65	3.65
5	0.0253	1.06	1.000	1.000	0.19400	0.490000	4.53	3.53
8	0.0230	2.623	1.890	2.890	0.34900	1.227000	4.450	3.450
9	0.0230	2.623	1.890	2.890	0.34900	1.227000	4.520	3.520
12	0.0230	2.623	1.890	2.890	0.34900	1.227000	4.140	3.140
10	0.0230	2.623	1.612	2.170	0.65250	2.021000	4.300	3.300
11	0.0230	2.623	1.612	2.170	0.65250	2.021000	4.350	3.350
13	0.0230	2.623	1.612	2.170	0.65250	2.021000	4.240	3.240
6	0.0510	2.745	6.075	1.135	2.89000	2.745000	4.22	3.22
14	0.1029	2.976	10.960	1.153	6.03000	6.530000	4.300	3.300
15	0.1029	2.976	10.960	1.153	6.03000	6.530000	4.070	3.070
16	0.1029	2.976	10.960	1.153	6.03000	6.530000	4.000	3.000
4	0.0253	2.78	1.000	1.000	3.55000	8.971000	3.59	2.59
3	0.1029	10.6	15.010	0.875	19.60000	11.750000	3.72	2.72
1	0.1029	2.976	1.890	2.890	1.16000	18.180000	3.800	2.800
	0.1029	2.976	1.890	2.890	1.16000	18.180000	3.640	2.640
	0.1029	2.976	1.890	2.890	1.16000	18.180000	3.860	2.860
	0.1029	2.976	1.839	2.745	2.48000	38.260000	3.340	2.340
	0.1029	2.976	1.839	2.745	2.48000	38.260000	3.560	2.560

(continued)

Table 4.3 (continued)

N°	d (cm)	ρ_s (g/cm^3)	μfx102 (g/cms)	ρ_f (g/cm^3)	v_{st} (cm/s)	Re_{oo}	n + 1	n
	0.1029	2.976	1.839	2.745	2.48000	38.260000	3.500	2.500
	0.0510	2.745	1.000	1.000	8.10000	41.720000	3.11	2.11
	0.1029	2.745	1.000	1.0	7.35000	14.450000	3.78	2.78
	0.1029	2.745	1.000	1.000	7.35000	14.450000	3.78	2.78
	0.4200	2.89	15.010	0.875	31.90000	78.250000	3.34	2.34
	0.1029	10.6	3.810	0.818	36.15000	79.800000	3.08	2.08
	0.2466	11.25	15.010	0.875	58.10000	80.350000	3.39	2.39
	0.3175	7.73	15.010	0.875	54.70000	101.200000	3.17	2.17
	0.4200	2.89	6.075	1.135	34.05000	267.000000	2.58	1.58
	0.1029	10.6	1.000	1.000	47.50000	488.700000	2.43	1.43
	0.4200	2.89	1.000	1.000	48.60000	2,041.000000	2.33	1.33
	0.3175	7.73	1.000	1.000	79.70000	2,530.000000	2.36	1.36
	0.6350	7.74	1.000	1.000	112.70000	7,150.000000	2.38	1.38

$$\begin{aligned} f_p^{-2}(\varphi)f_q(\varphi) &= (1-\varphi)^n \\ f_p^{-1/2}(\varphi)f_q(\varphi) &= (1-\varphi)^m \end{aligned} \tag{4.67}$$

and, by solving this algebraic set, we obtain:

$$f_p = (1-\varphi)^{(2/3)(m-n)} \quad \text{and} \quad f_q = (1-\varphi)^{(1/3)(4m-n)} \tag{4.68}$$

Using Eq. (4.62) and the calculated values in Table 4.5 and Fig. 4.16. The best values for m and n, were $n = 3.90$ and $m = 0.85$.

Then,

$$f_p(\varphi) = (1-\varphi)^{-2.033}, \quad f_q(\varphi) = (1-\varphi)^{-0.167} \tag{4.69}$$

A plot of Eq. (4.69) is given in Fig. 4.16.

Table 4.4 gives the data and Fig. 4.17 shows a plot of the dimensionless settling velocity versus Reynolds number for spheres, according to the experimental data of Richardson and Zaki (1954) and the simulation of Concha and Almendra (1979a, b) with Eqs. (4.62) and (4.69).

If all data of Table 4.4 are plotted in the form U^* versus D^* with the definitions (4.61), Fig. 4.18 is obtained.

On the other hand, Fig. 4.19 shows the settling velocity u^* versus d^* for suspensions of spheres in water at 20 °C for different values of the concentration φ. This figure can be used to visualize the state of flow of particulate systems.

In Chap. 5 of this textbook, Eq. (5.6), we will see that the volume average velocity, also known as *spatial velocity* or *percolation velocity*, is given by:

$$q = v_s - (1-\varphi)v_r \tag{4.70}$$

Table 4.4 Values of u/u_∞ from the experimental results of Richardson and Zaki (1954)

u/u_∞										
Re_{00}	0.000	0.010	0.050	0.100	0.150	0.200	0.300	0.400	0.500	0.585
0.00019	0.0021	0.96156	0.81869	0.66305	0.53056	0.41884	0.24882	0.13639	0.06699	0.03239
0.00019	0.0021	0.96263	0.82333	0.67078	0.54013	0.42925	0.25877	0.14428	0.07229	0.03568
0.00026	0.0139	0.96301	0.82502	0.67361	0.54365	0.43310	0.26249	0.14725	0.07433	0.03696
0.00026	0.0139	0.96398	0.82926	0.68075	0.55256	0.44287	0.27202	0.15497	0.07966	0.04035
0.00144	0.0081	0.96137	0.81786	0.66165	0.52884	0.41698	0.24705	0.13501	0.06606	0.03182
0.00144	0.0081	0.96166	0.81911	0.66375	0.53142	0.41978	0.24971	0.13709	0.06745	0.03267
0.00217	0.0455	0.96292	0.82460	0.67290	0.54277	0.43213	0.26156	0.14650	0.07381	0.03663
0.00217	0.0455	0.96330	0.82629	0.67574	0.54631	0.43601	0.26532	0.14953	0.07589	0.03794
0.00217	0.0455	0.96359	0.82756	0.67788	0.54898	0.43894	0.26817	0.15184	0.07748	0.03896
0.00295	0.0659	0.96205	0.82080	0.66655	0.53489	0.42354	0.25330	0.13992	0.06935	0.03384
0.00295	0.0659	0.96253	0.82290	0.67007	0.53925	0.42829	0.25785	0.14354	0.07179	0.03537
0.06473	0.0331	0.96214	0.82122	0.66725	0.53576	0.42449	0.25420	0.14064	0.06983	0.03414
0.06473	0.0331	0.96330	0.82629	0.67574	0.54631	0.43601	0.26532	0.14953	0.07589	0.03794
0.07579	0.0640	0.96292	0.82460	0.67290	0.54277	0.43213	0.26156	0.14650	0.07381	0.03663
0.07863	0.0664	0.96156	0.81869	0.66305	0.53056	0.41884	0.24882	0.13639	0.06699	0.03239
0.07863	0.0664	0.96263	0.82333	0.67078	0.54013	0.42925	0.25877	0.14428	0.07229	0.03568
0.09916	0.8910	0.95770	0.80207	0.63569	0.49717	0.38308	0.21574	0.11119	0.05077	0.02278
0.09916	0.8910	0.95867	0.80619	0.64242	0.50531	0.39172	0.22357	0.11701	0.05441	0.02488
0.11990	0.1475	0.96118	0.81702	0.66026	0.52712	0.41512	0.24529	0.13363	0.06515	0.03127
0.11990	0.1475	0.96156	0.81869	0.66305	0.53056	0.41884	0.24882	0.13639	0.06699	0.03239
0.30240	0.2340	0.96311	0.82544	0.67432	0.54454	0.43407	0.26343	0.14801	0.07484	0.03728
0.30240	0.2340	0.96398	0.82926	0.68075	0.55256	0.44287	0.27202	0.15497	0.07966	0.04035
0.39091	0.8270	0.96398	0.82926	0.68075	0.55256	0.44287	0.27202	0.15497	0.07966	0.04035
0.49082	0.1940	0.96514	0.83438	0.68941	0.56344	0.45489	0.28392	0.16477	0.08657	0.04484

(continued)

Table 4.4 (continued)

u/u_∞										
Re_{00}	0.000	0.010	0.050	0.100	0.150	0.200	0.300	0.400	0.500	0.585
1.22741	0.3490	0.96592	0.83781	0.69524	0.57081	0.46308	0.29214	0.17164	0.09151	0.04811
1.22741	0.3490	0.96524	0.83481	0.69013	0.56436	0.45591	0.28493	0.16561	0.08717	0.04524
1.22741	0.3490	0.96893	0.85124	0.71833	0.60031	0.49625	0.32629	0.20109	0.11344	0.06319
2.02024	0.6525	0.96738	0.84428	0.70632	0.58490	0.47885	0.30819	0.18531	0.10153	0.05490
2.02024	0.6525	0.96689	0.84212	0.70261	0.58017	0.47353	0.30275	0.18064	0.09807	0.05254
2.02024	0.6525	0.96796	0.84689	0.71080	0.59063	0.48530	0.31486	0.19108	0.10584	0.05787
2.75371	2.8900	0.96816	0.84775	0.71230	0.59256	0.48747	0.31711	0.19304	0.10732	0.05890
6.52757	6.0300	0.96738	0.84428	0.70632	0.58490	0.47885	0.30819	0.18531	0.10153	0.05490
6.52757	6.0300	0.96962	0.85430	0.72364	0.60718	0.50406	0.33454	0.20841	0.11908	0.06721
6.52757	6.0300	0.97030	0.85738	0.72900	0.61413	0.51200	0.34300	0.21600	0.12500	0.07147
8.98150	3.5500	0.97431	0.87560	0.76118	0.65644	0.56105	0.39701	0.26632	0.16609	0.10250
11.7571	1.9600	0.97303	0.86978	0.75083	0.64272	0.54501	0.37902	0.24921	0.15177	0.09143
18.2520	1.1600	0.97225	0.86622	0.74452	0.63441	0.53537	0.36836	0.23923	0.14359	0.08522
18.2520	1.1600	0.97382	0.87335	0.75718	0.65113	0.55483	0.38999	0.25961	0.16043	0.09810
18.2520	1.1600	0.97167	0.86355	0.73983	0.62826	0.52825	0.36056	0.23201	0.13774	0.08084
38.0915	2.4800	0.97676	0.88690	0.78150	0.68366	0.59324	0.43404	0.30260	0.19751	0.12771
38.0915	2.4800	0.97460	0.87695	0.76359	0.65965	0.56482	0.40128	0.27044	0.16958	0.10525
38.0915	2.4800	0.97519	0.87965	0.76843	0.66611	0.57243	0.40996	0.27885	0.17678	0.11095
41.3100	8.1000	0.97902	0.89742	0.80067	0.70970	0.62448	0.47115	0.34033	0.23165	0.15634
75.6315	7.3500	0.97245	0.86710	0.74610	0.63648	0.53776	0.37100	0.24169	0.14559	0.08673
75.6315	7.3500	0.97245	0.86710	0.74610	0.63648	0.53776	0.37100	0.24169	0.14559	0.08673
78.1029	31.900	0.97676	0.88690	0.78150	0.68366	0.59324	0.43404	0.30260	0.19751	0.12771
79.8642	36.150	0.97931	0.89880	0.80320	0.71317	0.62868	0.47622	0.34558	0.23651	0.16052
83.5212	58.100	0.97627	0.88463	0.77739	0.67813	0.58666	0.42637	0.29497	0.19078	0.12222

(continued)

Table 4.4 (continued)

u/u_∞										
Re_{00}	0.000	0.010	0.050	0.100	0.150	0.200	0.300	0.400	0.500	0.585
101.241	54.700	0.97843	0.89466	0.79562	0.70281	0.61618	0.46117	0.33006	0.22221	0.14831
267.187	34.050	0.98425	0.92215	0.84665	0.77354	0.70288	0.56919	0.44615	0.33448	0.24918
488.775	47.500	0.98573	0.92928	0.86014	0.79263	0.72681	0.60047	0.48168	0.37113	0.28432
2,041.20	48.600	0.98672	0.93405	0.86925	0.80561	0.74321	0.62227	0.50692	0.39777	0.31046
2,530.48	79.700	0.98642	0.93262	0.86650	0.80170	0.73825	0.61565	0.49921	0.38958	0.30237
7,156.45	112.70	0.98623	0.93166	0.86468	0.79909	0.73496	0.61127	0.49414	0.38422	0.29710

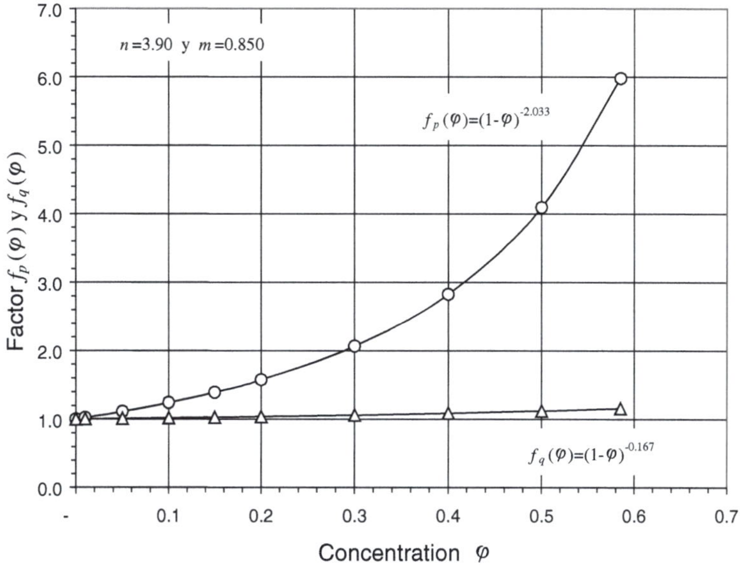

Fig. 4.16 Parameters $f_p(\varphi)$ and $f_q(\varphi)$ from data by Richardson and Zaki

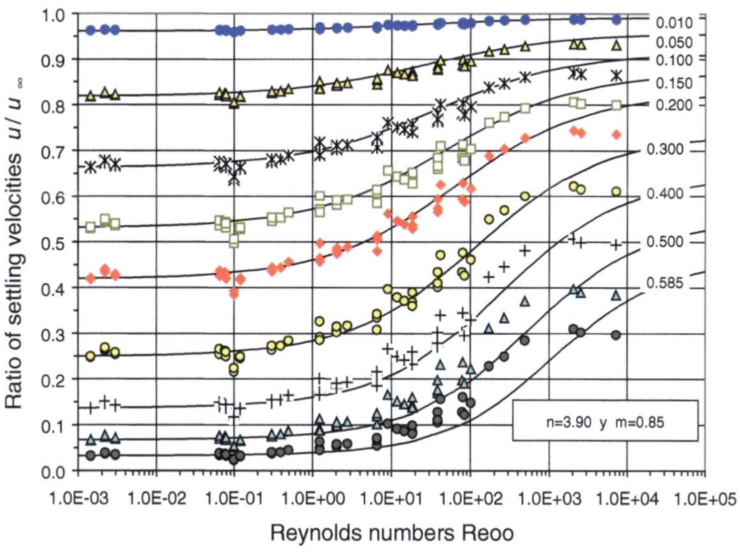

Fig. 4.17 Dimensionless settling velocities versus Reynolds number, together with data obtained from Richardson and Zaki (1954) experimental results

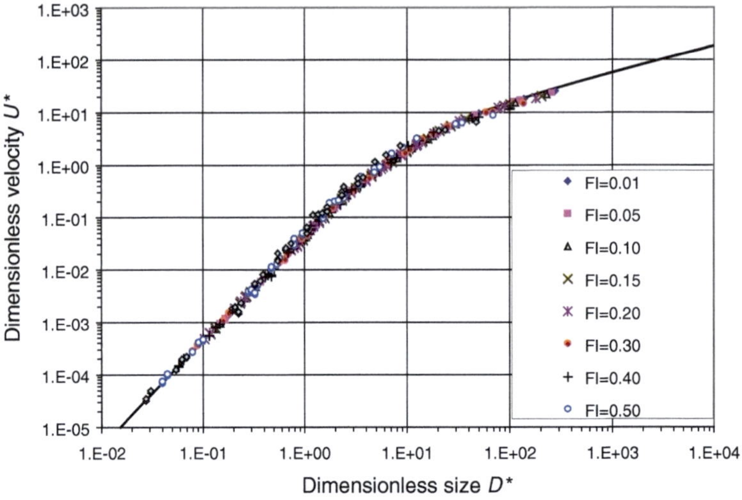

Fig. 4.18 Dimensionless velocity U^* for suspensions of spheres of any size and density versus the dimensionless diameter D^*, with experimental data of Richardson and Zaki (1954)

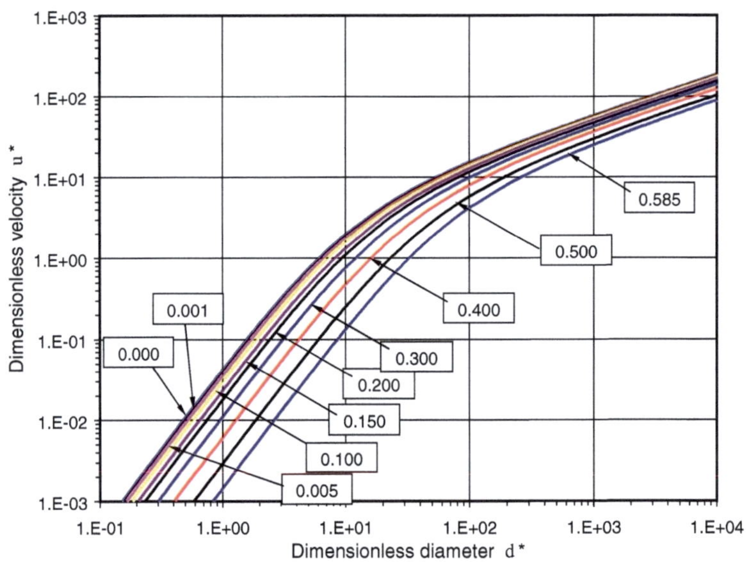

Fig. 4.19 Simulation of the dimensionless velocity u^* versus the dimensionless size d^* for spherical particles settling in a suspension in water at 20 °C with the volume fraction of solids a parameter

where q, v_s and v_r are the spatial velocity, the solid component velocity and the relative solid–fluid velocity. Figure 4.17 divides the $u^* \times d^*$ plane into three regions: a *porous bed*, between the d^* axis and the line of constant concentration $\varphi = 0.585$ [Barnea and Mednick (1975) demonstrated that this concentration corresponds to the minimum fluidization velocity], a second region of a fluidized bed between $0.585 \leq \varphi \leq 0$ and a third region of *hydraulic or pneumatic transport*, for values of velocities above concentration $\varphi = 0$.

Drag Coefficient for a suspension of spheres

From Eqs. (4.49) and (4.62) we deduce that:

$$\tilde{\delta}_0^2 = \delta_0^2 f_p(\varphi) f_q(\varphi)$$
$$\tilde{C}_0^{1/2} \tilde{\delta}_0^2 = C_0^{1/2} \delta_0^2 f_p^{3/2}(\varphi)$$

therefore, the parameters of the Drag Coefficient are:

$$\tilde{C}_0 = C_0 f_p(\varphi) f_q^{-2}(\varphi) \tag{4.71}$$

$$\tilde{\delta}_0 = \delta_0 f_p^{1/2}(\varphi) f_q^{1/2}(\varphi) \tag{4.72}$$

and the Drag Coefficient of the sphere can be written in the form:

$$C_D = C_0 f_p(\varphi) f_q^{-2}(\varphi) \left(1 + \frac{\delta_0 f_p^{1/2}(\varphi) f_q^{1/2}(\varphi)}{\mathrm{Re}^{1/2}}\right)^2 \tag{4.73}$$

Using the values for $f_p(\varphi)$ and $f_q(\varphi)$ from (4.69), we obtain finally:

$$C_D = 0.284(1-\varphi)^{-2.01} \left(1 + \frac{9.08(1-\varphi)^{-1.83}}{\mathrm{Re}^{1/2}}\right)^2 \tag{4.74}$$

Problem 4.7 Consider a porous bed formed by spherical particles of dimensionless diameter $d^* = 10$. A fluid percolates through the bed at a dimensionless velocity $u^* = 0.001$. If the velocity is increased, establish the range of velocities at which the three regimes are present.

Drawing a vertical line in the plot of dimensionless velocity versus dimensionless size, Fig. 4.20, for $d^* = 10$, see lines in red in the next figure, we find that the system of particles forms a porous bed until the dimensionless velocity $u^* \approx 0.130$, which is the dimensionless minimum fluidization velocity. Fluidization exists for the range of dimensionless velocities $0.13 \leq u^* \leq 2.00$. For greater velocities, particles are transported.

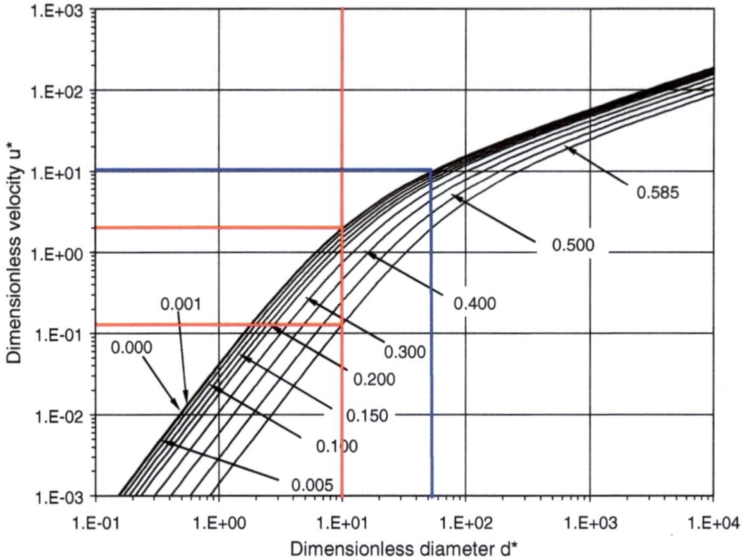

Fig. 4.20 Plot of dimensionless diameter vesus dimensionless size for sphweres of Problem 4.7

Problem 4.8 Determine the sedimentation velocity of a sphere, 150 μm in diameter and 2.65 g/cm^3 in density at 15 °C, forming a suspension of 40 % of solid by weight.

The volume fraction of solids is:

$$\varphi = \frac{0.9989 \times 40}{2.65 \times (100 - 40) + 0.9989 \times 40} = 0.201$$

Parameters for the solid–fluid system:

$$\rho_{water} = 0.9986,\ \mu_{water} = 0.01280,\ P = 2.66 \times 10^{-3},\ Q = 2.099$$

$$f_p(0.201) = 1.5761,\ f_q(0.201) = 1.0381$$

$$d^* = 5.622$$

$$u^* = \frac{20.52}{d^*} f_p(\varphi) f_q(\varphi) \left(\left(1 + 0.0921 f_p^{-3/2} d^{*3/2} \right)^{1/2} - 1 \right)^2 = 1.863$$

$$u = 1.863 \times 2.0987 = 3.910 \text{ cm/s } (\boldsymbol{u} = -3.91\, \boldsymbol{k} \text{ cm/s})$$

Problem 4.9 Determine the fluidization velocity of a 40 % by weight suspension of mono-sized quartz spherical particles, 150 μm in diameter and 2.65 g/cm^3 in density at 15 °C. Calculate at which volume average velocity these particles begin to be transported.

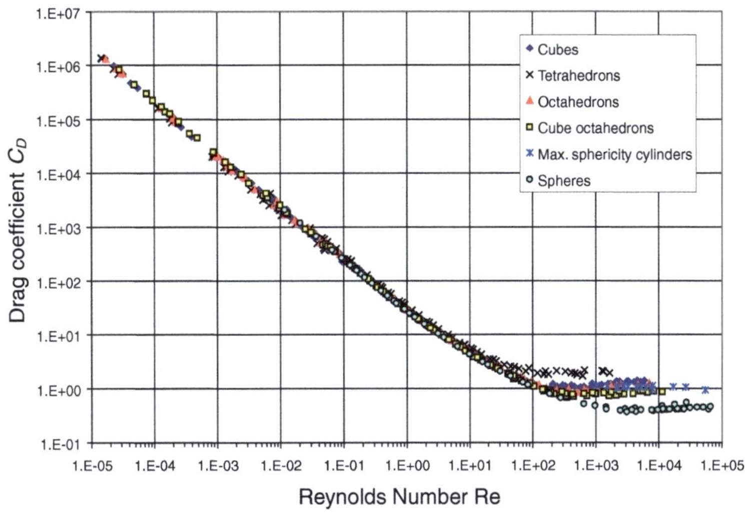

Fig. 4.21 Plot of the drag coefficient versus the Reynolds number for the settling of isometric particles according to Pettyjohn and Christiansen (1948) and Barker (1951)

From the previous problem, we know that the sedimentation velocity for the same suspension is $u = -3.91$ cm/s. The volume average velocity is given by $q = v_s - (1 - \varphi)u$, and since for fluidization $v_s = 0$, $q = -(1 - \varphi)u$

$$q = (1 - 0.201) \times 3.91 = 3.13 \text{ cm/s}.$$

The dimensionless particle size is $d^* = 5.62$. A straight line for this value in blue in Fig. 4.21 gives a transport velocity of $u^* = 10$, which corresponds to a velocity $u = -10 \times 3.91 = -39.1$ cm/s. Then:

$$u = -39.1 \text{ cm/s}, \; q = -(1 - 0) \times -39.1 = 39.1 \text{ cm/s}.$$

4.1.8 Sedimentation of Isometric Particles

The behavior of non-spherical particles is different than that of spherical particles during sedimentation. While spherical particles fall in a vertical trajectory, non-spherical particles rotate, vibrate and follow spiral trajectories. Several authors have studied the sedimentation of *isometric particles*, which have a high degree of symmetry with three equal mutually perpendicular symmetry axes, such as the tetrahedron, octahedron and dodecahedron. Wadell (1932, 1934), Pettyjohn and Christiansen (1948) and Christiansen and Barker (1965) show that isometric particles follow vertical trajectories at low Reynolds numbers, but rotate and vibrate and show helicoidally trajectories for Reynolds numbers between 300 and 150,000. Figure 4.18 shows the drag coefficient versus Reynolds number for these particles.

Pettyjohn and Christiansen (1948) demonstrate that velocities in Stokes flow for isometric particles may be described with the following expression:

$$\frac{u_p}{u_e} = 0.843\log\left(\frac{\psi}{0.065}\right) \quad \text{with} \quad u_e = \frac{\Delta\rho d_e^2 g}{18\mu_f} \tag{4.75}$$

where d_e is the volume equivalent diameter, that is, the diameter of a sphere with the same volume as the particle, and u_e is its settling velocity.

In the range of $2{,}000 \leq Re \leq 17{,}000$, the same authors derived the following equation for the settling velocity:

$$u_e = \frac{4}{3}\frac{\Delta\rho d_e g}{\rho_f C_D}, \tag{4.76}$$

with the drag coefficient C_D given by: $C_D = (5.31 - 4.88\psi)/(1.433 \times 0.43)$. The value of 1.433 in the denominator of this equation is a factor that takes the theoretical value of $C_D = 0.3$ (see Fig. 4.11) to the average experimental value $C_D = 0.43$ (see Fig. 4.19).

As we have already said, for $Re > 300$, the particles begin to rotate and oscillate, which depends on the particle density. To take into account these behavior, Barker (1951) introduced the particle to fluid density ratio as a new variable in the form:

$$C_D(\psi, \lambda) = \lambda^{1/18}\frac{(5.31 - 4.88\psi)}{0.62}, \tag{4.77}$$

where λ is the quotient between the solid and fluid densities $\lambda = \rho_p/\rho_f$.

Data from Pettyjohn and Christiansen (1948) and from Barker (1951) are plotted in Fig. 4.21. Figure 4.22 gives details of the higher end of the Reynolds range.

Drag coefficient and sedimentation velocity

Results obtained for spherical particles (Concha and Almendra 1979a, b), may be used to develop functions for the drag coefficient and sedimentation velocity of isometric particles.

Assume that Eqs. (4.73) and (4.62) valid for isometric particles, with values of C_0 and δ_0 as functions of the sphericity ψ and of the density quotient λ (Concha and Barrientos 1986):

$$C_D(\psi, \lambda) = \tilde{C}_0(\psi, \lambda)\left(1 + \frac{\delta_0(\psi, \lambda)}{\mathrm{Re}^{1/2}}\right)^2 \tag{4.78}$$

$$u_p^* = \frac{1}{4}\frac{\tilde{\delta}_0^2(\psi, \lambda)}{d^*}\left(\left(1 + \frac{4}{\tilde{C}_0^{1/2}(\psi, \lambda)\tilde{\delta}_0^2(\psi, \lambda)}d^{*3/2}\right)^{1/2} - 1\right)^2 \tag{4.79}$$

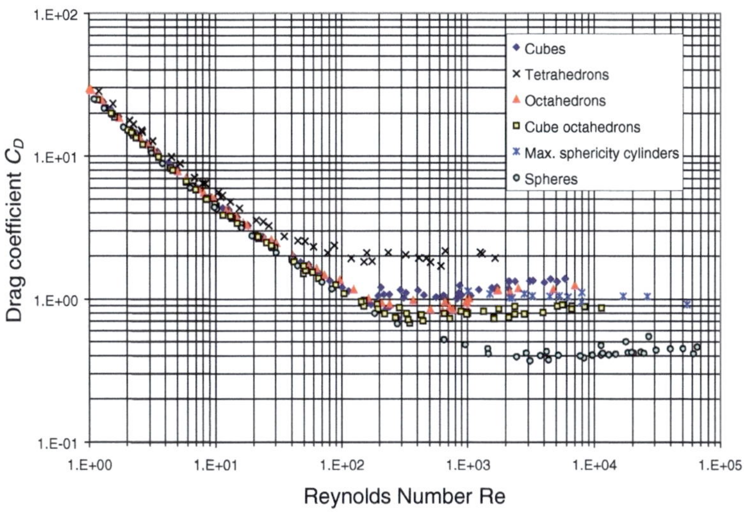

Fig. 4.22 Details for Fig. 4.18 at the high end oh Reynolds range (Pettyjohn and Christiansen 1948; Barker 1951)

where the Reynolds number is defined using the volume equivalent diameter. Assume also that:

$$\tilde{C}_0(\psi,\lambda) = C_0 f_A(\psi) f_C(\lambda) \tag{4.80}$$

$$\tilde{\delta}_0(\psi,\lambda) = \delta_0 f_B(\psi) f_D(\lambda) \tag{4.81}$$

where C_0 and δ_0 are the same parameters of a sphere.

We have already demonstrated that for a sphere (volume equivalent sphere in this case) at a low Reynolds number, $Re \to 0$, the dimensionless velocity can be approximated by Eq. (4.64).

Assume that we can approximate the velocity of isometric particles at low Reynolds numbers, Re → 0, in the same way as for spherical particles. Then:

$$u_e^* = \frac{d_e^{*2}}{C_0\delta_0^2} \quad \text{and} \quad u_p^* = \frac{d_e^{*2}}{\tilde{C}_0(\psi,\lambda)\tilde{\delta}_0^2(\psi,\lambda)} \tag{4.82}$$

Taking the quotient of these terms and substituting (4.80) and (4.81), results in:

$$Re \to 0, \quad \frac{u_e^*}{u_p^*} \equiv \frac{u_e}{u_p} = f_A(\psi) f_B^2(\psi) f_C(\lambda) f_D^2(\lambda) \tag{4.83}$$

On the other hand, for $Re \to \infty$:

$$\frac{C_D(\psi,\lambda)}{C_D} = \frac{\tilde{C}_0(\psi,\lambda)}{C_0} \quad \text{and} \quad \frac{C_D(\psi,\lambda)}{C_D} = f_A(\psi) f_C(\lambda) \tag{4.84}$$

To determine the functions f_A,f_B,f_C and f_D we will use the correlations presented by Pettyjohn and Christiansen (4.75) and (4.77), and by Barker (1951). From (4.83) and (4.75) we can write:

$$f_A(\psi)f_B^2(\psi)f_C(\lambda)f_D^2(\lambda)=\left\{0.843\log\frac{\psi}{0.065}\right\}^{-1} \tag{4.85}$$

$$f_A(\psi)f_C(\lambda)=\lambda^{1/18}\frac{5.31-4.88\psi}{0.62} \tag{4.86}$$

From (4.77) and (4.86) we deduce that:

$$f_A(\psi)=\frac{5.31-4.88\psi}{0.62}\quad f_C(\lambda)=\lambda^{1/18}. \tag{4.87}$$

Since in the Stokes regime the density does not influence the flow, Eq. (4.85) implies that (Fig. 4.23):

$$f_C(\lambda)f_D^2(\lambda)=1\Rightarrow f_D(\lambda)=\lambda^{-1/36} \tag{4.88}$$

Therefore:

$$f_B(\psi)=\left\{\frac{5.31-4.88\psi}{0.62}\times 0.843\log\frac{\psi}{0.065}\right\}^{-1/2} \tag{4.89}$$

Problem 4.10 Using Concha and Barrientos (1986) model for isometric particles, determine the values of the dimensionless velocity versus the dimensionless size for particles with sphericities 0.4, 0.5, 0.6, 0.7, 0.8, 0.9 and 1.0 for values of d^* from 0.01 to 100.000.

The result is shown in Fig. 4.24.

Haider and Levenspiel (1998) give an alternative equation for the Drag Coefficient and the settling velocity of isometric particles based on their equation for spherical particles:

$$\begin{aligned}C_D(\psi)=&\frac{24}{Re}\left(1+(8.1716\times\exp(-4.0655\psi))\mathrm{Re}^{(0.0964+0.5565\psi)}\right)\\&+\frac{73.69\mathrm{Re}\times\exp(-5.0740\psi)}{Re+5.378\times\exp(6.2122\psi)}\end{aligned} \tag{4.90}$$

$$u^*=\left(\frac{18}{d^{*2}}+\frac{3.1131-2.3252\psi}{d^{*0.5}}\right)^{-1},\quad \text{for } 0.5\le\psi\le 1 \tag{4.91}$$

(a)

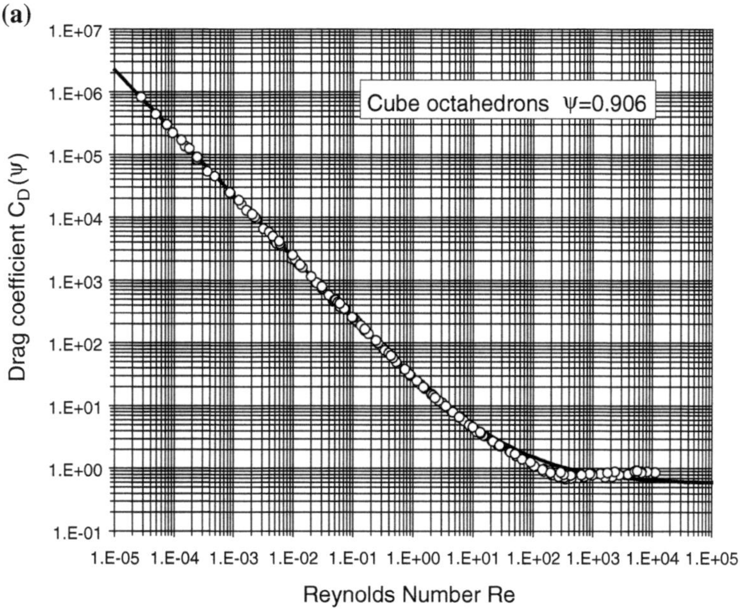

(b)

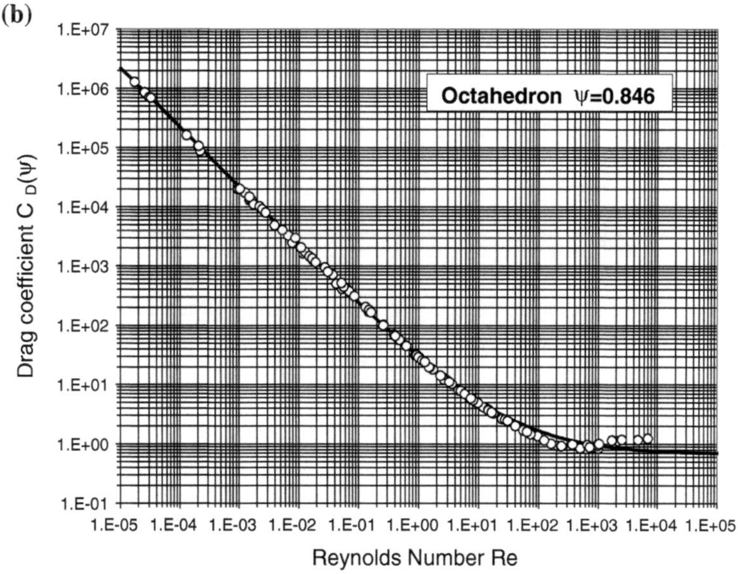

Fig. 4.23 Simulation with Concha and Barrientos' equation (4.78) and experimental values for isometric particles from Pettyjohn and Christiansen (1948) and Barker (1951) for cube octahedrons (**a**), for octahedrons (**b**), for cubes (**c**), for tetrahedrons (**d**)

(c)

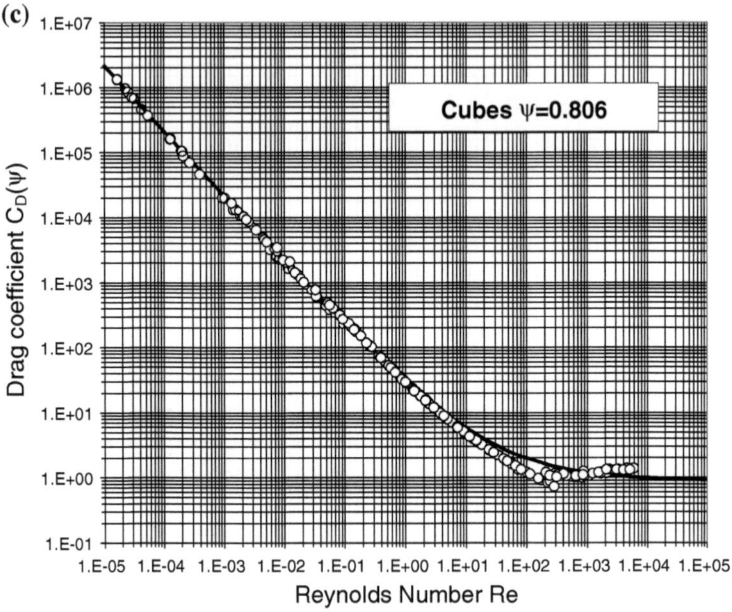

(d)

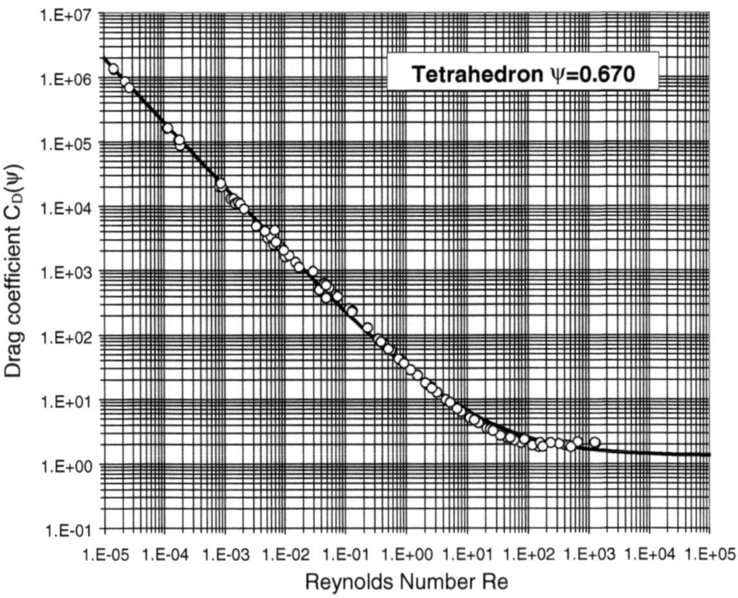

Fig. 4.23 (continued)

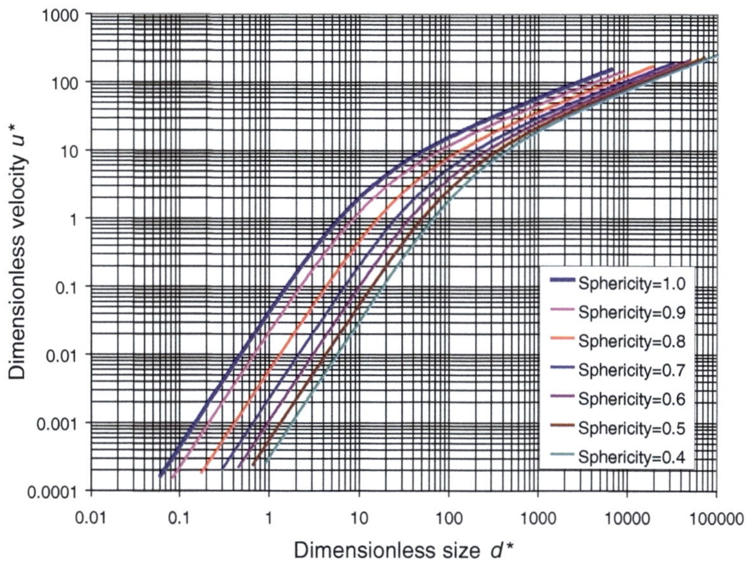

Fig. 4.24 Dimensionles size versus dimensionless velocity for isometric particles according to Concha and Barrientos (1986)

$$C_D = \frac{1}{K_1}\frac{24}{\text{Re}}\left(1 + 0.1118(K_1K_2\text{Re})^{0.6567}\right) + \frac{0.4305K_1K_2^2\text{Re}}{3305 + K_1K_2\text{Re}} \tag{4.92}$$

where

$$K_1 = \left(\frac{1}{3} + \frac{2}{3}\psi^{-1/2}\right)^{-1}, \quad K_2 = 10^{1.8148(-\log\psi)^{0.5743}} \tag{4.93}$$

Modified drag coefficient and sedimentation velocity

Introducing the values for f_A,f_B,f_C and f_D into Eq. (4.78) results in:

$$\frac{C_D(\psi,\lambda)}{f_A(\psi)f_C(\lambda)} = C_0\left(1 + \frac{\delta_0}{\left(\text{Re}/(f_B^2(\psi)f_D^2(\lambda))\right)^{0.5}}\right)^2 \tag{4.94}$$

Defining the modified Drag Coefficient C_{DM} and the modified Reynolds number Re_M by:

$$C_{DM} = \frac{C_D(\psi,\lambda)}{f_A(\psi)f_C(\lambda)} \quad \text{Re}_M = \left(\frac{\text{Re}}{f_B^2(\psi)f_D^2(\lambda)}\right) \tag{4.95}$$

we can write the drag coefficient in the form of Eq. (4.78). Plotting C_{DM} versus Re_M, Fig. 4.25 is obtained.

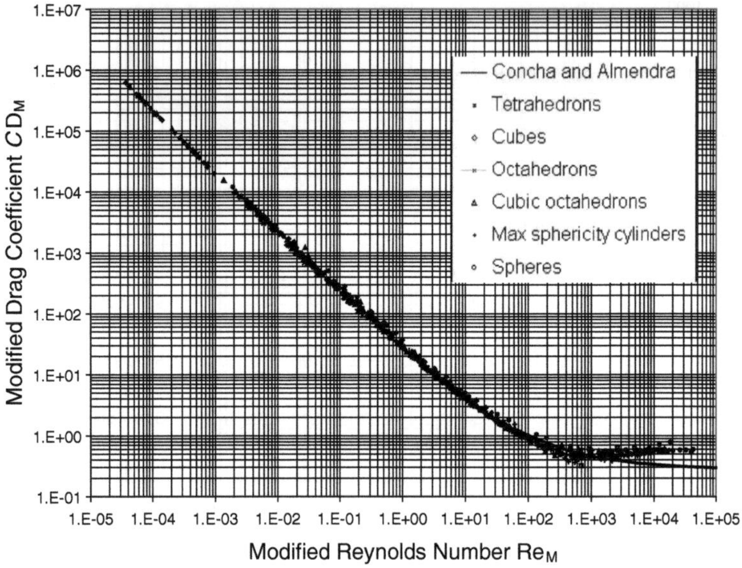

Fig. 4.25 Experimental data of Pettyjohn and Christiansen (1948), and Barker (1951), plotted as C_{DM} versus Re_M for isometric particles

A similar result may be obtained for the sedimentation velocity. Defining modified dimensionless diameter d^*_{eM} and velocity u^*_M:

$$d^*_{eM} = \frac{d^*_e(\psi,\lambda)}{\left(f_A^{1/2}(\psi)f_B^2(\psi)f_C^{1/2}f_D^2(\lambda)\right)^{2/3}} \quad \text{and} \quad u^*_{eM} = \frac{u^*_p(\psi,\lambda)}{f_B(\psi)f_D(\lambda)}. \tag{4.96}$$

The unified u^*_{eM} versus d^*_{eM} curve is given in Fig. 4.26, for Pettyjohn and Christiansen (1948) and Christiansen and Barker (1965) data.

$$u^*_{eM} = \frac{20.52}{d^*_{eM}}\left(\left(1+0.0921 d_{eM}^{*3/2}\right)^{1/2} - 1\right)^2 \tag{4.97}$$

The experimental data used in the previous correlations are 655 points including spheres, cubes-octahedrons, maximum sphericity cylinders, octahedrons and tetrahedrons in the following ranges:

$0.1\ \text{cm} < d_e < 5\ \text{cm}$
$1.7\ \text{g/cm}^3 < \rho_s < 11.2\ \text{g/cm}^3$
$0.67 < \psi < 1$
$0.87\ \text{g/cm}^3 < \rho_f < 1.43\ \text{g/cm}^3$
$9 \times 10^{-3}\ \text{g/cm s} < \mu < 900\ \text{g/cm s}$
$5 \times 10^{-3} < Re < 2 \times 10^4$

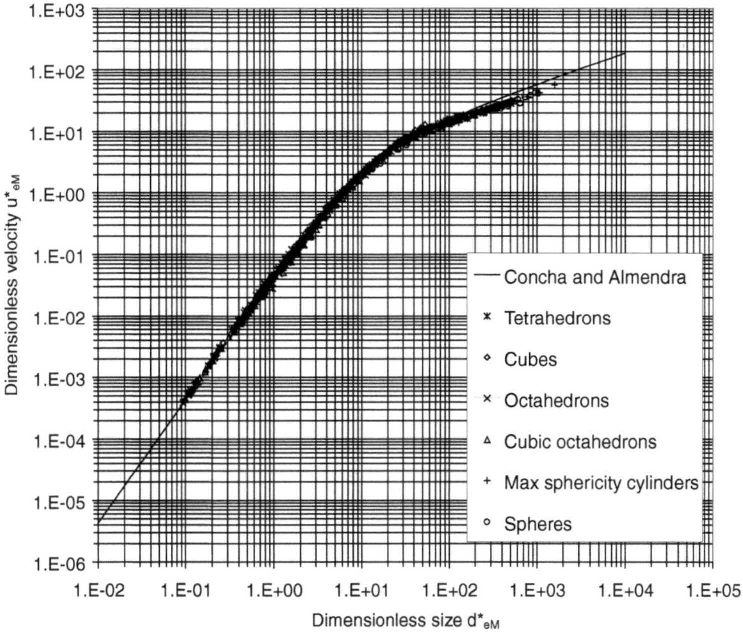

Fig. 4.26 Unified u^*_M versus d^*_{eM} curve for data of Pettyjohn and Christiansen (1948) and Barker (1951)

with the following values for the particles sphericity (Happel and Brenner 1965; Barker 1951) and the parameters of Concha and Barrientos (1986)

	Ψ	$f_A\ (\psi)$	$f_B\ (\psi)$
Sphere	1.000	1.0000	1.0000
Cube octahec	0.906	1.4334	0.8826
Octahedron	0.846	1.9057	1.2904
Cube	0.806	2.2205	1.5468
Tetrahedron	0.670	3.2910	2.3108
Max sph cylin	0.875	1.6774	1.0966

Problem 4.11 Determine the sphericity and the settling velocity of a quartz cube of 1 mm in size and a density of 2.65 g/cm^3 in water at 25 °C. Use the methods of Concha and Barrientos and that of Haider and Levenspiel.

By definition, sphericity is the ratio of the surface of a volume-equivalent sphere and the surface of the particle. For a cube of 1 mm in size, the surface is 6 mm^2 and its volume is 1 mm^3. The volume equivalent sphere has a diameter of:

$$d_e = (6V/\pi)^{1/3} = (6/\pi)^{1/3} = (6/\pi)^{1/3}$$

The sphericity is:

$$\psi = \frac{\pi(6/\pi)^{2/3}}{6} = \left(\frac{\pi}{6}\right)^{1/3} = 0.806$$

Then, from Concha and Barrientos:

$$f_A(\psi) = \frac{5.31 - 4.88\psi}{0.62} = \frac{5.31 - 4.88 \times 0.806}{0.62} = 2.2205$$

$$f_B(\psi) = \left(\frac{5.31 - 4.88 \times 0.806}{0.62} \times 0.843 \log \frac{0.806}{0.065}\right)^{-1/2} = 0.699$$

$$\rho_{water} = -4.0 \times 10^{-6} \times 25^2 - 6.0 \times 10^{-5} \times 25 + 1.0004 = 0.9964\ (\mathrm{g/cm^3})$$

$$f_C(\psi) = \lambda^{-1/18} = \left(\frac{2.65}{0.9964}\right)^{-1/18} = 0.9471$$

$$f_D(\psi) = \lambda^{1/36} = \left(\frac{2.65}{0.9964}\right)^{1/36} = 1.0275$$

$$\tilde{\delta}_0(\psi,\lambda) = 9.08 \times f_B(\psi) f_D(\lambda) = 9.08 \times 0.699 \times 1.0275 = 6.5215$$

$$\tilde{C}_0(\psi,\lambda) = 0.28 \times f_A(\psi) f_C(\lambda) = 0.28 \times 2.2205 \times 0.9471 = 0.5889$$

$$\mu_{water} = 9.0 \times 10^{-7} \times 25^2 - 2.0 \times 10^{-4} \times 25 + 1.56 \times 10^{-2} = 0.0112 (\mathrm{g/cm \times s})$$

$$P = \left(\frac{3}{4}\frac{\mu_f^2}{\Delta\rho\rho_f g}\right)^{1/3} = \left(\frac{3 \times 0.0112^2}{4(2.65 - 0.9964) \times 0.9964 \times 980.1}\right)^{1/3} = 0.00387$$

$$Q = \left(\frac{4}{3}\frac{\Delta\rho\mu_f g}{\rho_f^2}\right)^{1/3} = \left(\frac{4 \times (2.65 - 0.9964) \times 0.0112 \times 980.1}{3 \times 0.9964^2}\right)^{1/3} = 2.896$$

$$d^* = \frac{d_e}{P} = \frac{0.1}{0.00387} = 25.85$$

$$d^*_{eM} = \frac{d^*}{f_A^{0.5}(\psi) f_B^2(\lambda) f_C^{0.5}(\psi) f_D^2(\lambda)} = \frac{25.85}{2.2205^{0.5} \times 0.699^2 \times 0.9471^{0.5} \times 1.0275^2} = 34.56$$

$$u^*_{eM} = \frac{20.52}{d^*_{eM}}\left(\left(1 + 0.0921 d^{*3/2}_{eM}\right)^{1/2} - 1\right)^2 = \frac{20.52}{40.67}\left(\left(1 + 0.0921 \times 40.67^{3/2}\right)^{1/2} - 1\right)^2 = 7.025$$

$$u^* = u^*_{eM} f_B(\psi) f_D(\lambda) = 8.028 \times 0.699 \times 1.0275 = 5.046$$

$$u_p = Q \times u^* = 2.896 \times 5.046 = 14.61\ \mathrm{cm/s}$$

Using the method of Haider and Levenspiel, Eq. (4.91), we get:

$$u_p^* = \left(\frac{18}{25.85^2} + \frac{3.1131 - 2.3252 \times 0.809}{25.85^{0.5}}\right)^{-1} = 3.6954$$

$$u_p = Q \times u_p^* = 2.89631 \times 3.694 = 10.703\ \mathrm{cm/s}$$

4.1.9 Sedimentation of Particles of Arbitrary Shape

Concha and Christiansen (1986) extended the validity of Eqs. (4.78) and (4.79) to suspensions of particles of arbitrary shape.

$$C_D(\psi,\lambda,\varphi) = \tilde{C}_0(\psi,\lambda,\varphi)\left(1+\frac{\delta_0(\psi,\lambda,\varphi)}{Re^{1/2}}\right)^2 \tag{4.98}$$

$$u_p^*(\psi,\lambda,\varphi) = \frac{1}{4}\frac{\tilde{\delta}_0^2(\psi,\lambda,\varphi)}{d^*}\left(\left(1+\frac{4}{\tilde{C}_0^{1/2}(\psi,\lambda,\varphi)\tilde{\delta}_0^2(\psi,\lambda,\varphi)}d^{*3/2}\right)^{1/2}-1\right)^2 \tag{4.99}$$

where ψ, λ and φ are the sphericity of the particles, the density ratio of solid and fluid and the volume fraction of solid in the suspension.

Similarly as in the case of isometric particles, they assumed that the functions $\tilde{C}_0$ and $\tilde{\delta}_0$ may be written in the form:

$$\tilde{C}_0(\psi,\lambda,\varphi) = C_0 f_A(\psi) f_C(\lambda) f_p(\varphi) f_q^{-2}(\varphi) \tag{4.100}$$

$$\tilde{\delta}_0(\psi,\lambda,\varphi) = \delta_0 f_B(\psi) f_D(\lambda) f_F(\varphi) f_p^{1/2}(\varphi) f_q^{1/2}(\varphi) \tag{4.101}$$

with

$$f_A(\psi) = \frac{5.31-4.88\psi}{0.62}$$

$$f_B(\psi) = \left\{\frac{5.31-4.48\psi}{0.62}\times 0.843\log\frac{\psi}{0.065}\right\}^{-1/2}$$

$$f_C(\lambda) = \lambda^{1/18}, \quad f_D(\lambda) = \lambda^{-1/36}$$

$$f_p(\varphi) = (1-\varphi)^{-2.033}, \quad f_q(\varphi) = (1-\varphi)^{-0.167}$$

Hydrodynamic shape factor

Concha and Christiansen (1986) found it necessary to define a *hydrodynamic shape factor* to be used with the above equations, since the usual methods to measure sphericity did not gave good results. They defined the *effective hydrodynamic sphericity* of a particle as the *sphericity of an isometric particle having the same drag (volume) and the same settling velocity as the particle.*

The hydrodynamic sphericity may be obtained by performing sedimentation or fluidization experiments, calculating the drag coefficient for the particles using the volume equivalent diameter and obtaining the sphericity (defined for isometric particles) that fit the experimental value. Figure 4.27 give simulated Drag Coefficients curves for several sphericities.

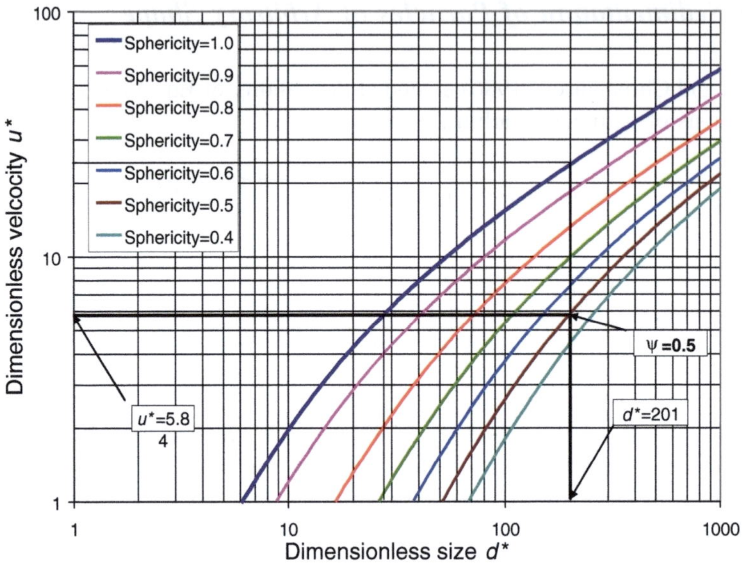

Fig. 4.27 Simulated dimensionless velocity versus dimensionless size for isometric particles and several sphericities

Problem 4.12 Estimate the sphericity of crushed quartz particle of 7.2 mm, with 2.65 g/cm^3 in density, which gives an average settling velocity of 16.3 cm/s in water at 20 °C.

$$d_e = 0.5\ \text{cm}\ ,\ \rho_s = 2.65\,\text{g/cm}^3,\ \rho_f = 1.00\ \text{g/cm}^3,\ \mu = 0.01\ \text{g/cm s}$$
$$P = 0.00359,\ Q = 2.7921$$

Results:

$$d_e^* = \frac{d_e}{P} = \frac{0.72}{0.00359} = 2{,}001,\ u_p^* = \frac{u_p}{Q} = \frac{16.3}{2.7921} = 5.84$$

With these values for the equivalent dimensionless diameter and the dimensionless settling velocity of the particle d_e^* and u_p^*, from a plot of u^* versus d^*, see Fig. 4.27 we obtain a sphericity of $\psi = 0.5$ for the quartz.

Modified drag coefficient and sedimentation velocity

A unified correlation can also be obtained for the drag coefficient and the sedimentation velocity of irregular particles forming a suspension. Defining C_{DM}, Re_M, d_{eM} and u_{pM} in the following form:

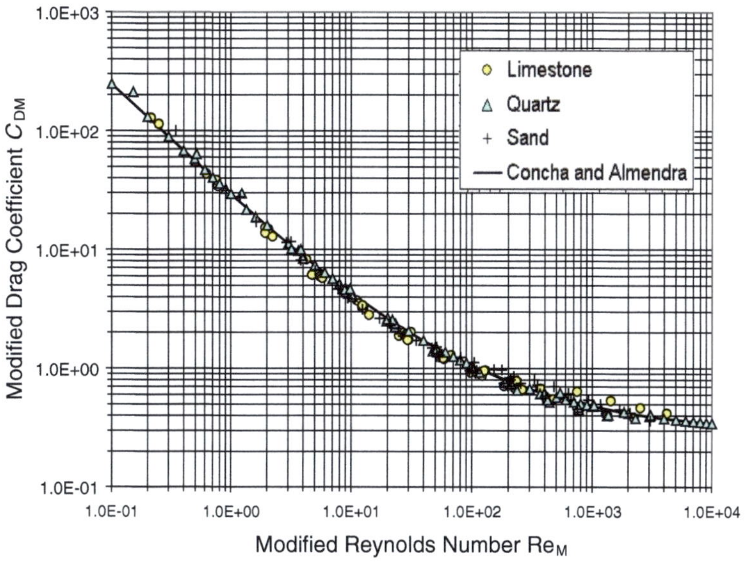

Fig. 4.28 Unified drag coefficient versus Reynolds number for quartz, limestone and sand particles [the same data as in Fig. 4.22 (Concha and Christiansen 1986)]

$$C_{DM} = \frac{C_D(\psi,\lambda)}{f_A(\psi)f_C(\lambda)f_P(\varphi)} \qquad Re_M = \left(\frac{Re}{f_B^2(\psi)f_D^2(\lambda)f_p^2(\varphi)}\right) \tag{4.102}$$

$$d_{eM}^* = \frac{d_e^*(\psi,\lambda)}{f_p(\varphi)\times\left(f_A(\psi)^{1/2}\times f_B^2(\psi)\times f_C(\lambda)^{1/2}\times f_D^2(\lambda)\right)^{2/3}} \tag{4.103}$$

$$u_{pM}^* = \frac{u_p^*(\psi,\lambda)}{f_B(\psi)f_D(\lambda)f_q(\varphi)} \tag{4.104}$$

Figures 4.28 and 4.29 show the unified correlations for the data from Concha and Christiansen (1986).

Problem 4.13 Determine the minimum fluidization velocity of quartz particles with 250 microns in size, density $\rho_s = 2.65\ \text{g}/\text{cm}^3$ and sphericity $\psi = 0.55$, in water and 20 °C.

The minimum fluidization velocity occurs at $\varphi = 0.585$, therefore, we have the following results:

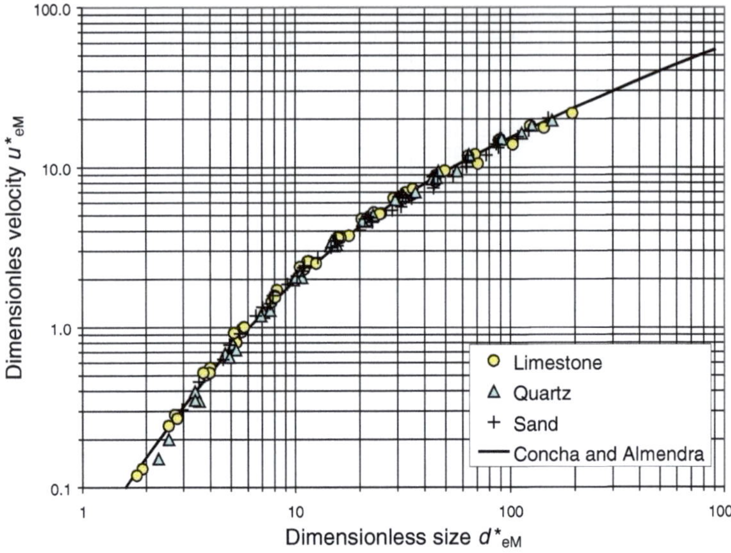

Fig. 4.29 Unified sedimentation velocity versus size for limestone, quartz and sand particles [the same data as in Fig. 4.22 (Concha and Christiansen 1986)]

$$u^* = \frac{20.52}{d^*} f_p \times f_q \times f_B^2 \times f_D^2 \times \left(\left(1 + \frac{0.0921 \times d^{*-3/2} f_p^{-3/2}}{f_A^{0.5} \times f_C^{0.5} \times f_B^2 \times f_D^2} \right)^{0.5} - 1 \right)^2$$

$$q = (1 - \varphi) \times u_p$$
$$\rho_s = 2.65 \text{ g/cm}^3$$
$$\rho_f = 1.0.9976 \text{ g/cm}^3, \quad \mu_f = 0.01196 \text{ g/cm - s}$$
$$P = 0.00405, \quad Q = 2.96151$$
$$f_A(\psi) = 4.2355\ ,\ f_B(\psi) = 0.5495,\ f_C(\lambda) = 1.0558,\ f_D(\lambda) = 0.9732$$
$$\varphi = 0.585, \quad f_p(\varphi) = 5.97734, \quad f_q(\varphi) = 1.1582.$$
$$d^* = d/P = 6.1737\ ,\ u^* = 0.03898\ ,\ u_p = u^* \times Q = 0.1154$$
$$q = (1 - \varphi) \times u_p = 0.04791 \text{ cm/s}.$$

Ganser (1993) proposed an empirical equation for the drag coefficient of non-spherical non-isometric particles, including irregular particles, similar to that given earlier for spherical particles (4.92), but with different values for the parameters K_1.

$$C_D = \frac{1}{K_1} \frac{24}{\text{Re}} \left(1 + 0.1118 (K_1 K_2 \text{Re})^{0.6567} \right) + \frac{0.4305 K_1 K_2^2 \text{Re}}{3305 + K_1 K_2 \text{Re}} \tag{4.105}$$

where

$$K_1 = \left(\frac{1}{3}\frac{d_p}{d_e} + \frac{2}{3}\psi^{-1/2}\right)^{-1}, \quad K_2 = 10^{1.8148(-\log\psi)^{0.5743}} \tag{4.106}$$

In the equation for K_1, d_e and d_p are the *volume equivalent* and the *projected area* equivalent diameters of the irregular particle respectively.

Finally, it is interesting to mention the work of Yin et al. who analyzed the settling of cylindrical particles analytically and obtained, by linear and angular momentum balances, the forces and torques applied to the particle during their fall. Using Ganser's equation for the drag coefficient, they solved the differential equations of motion numerically obtaining results close to those measured experimentally by them.

References

Abraham, F. F. (1970). Functional dependence of the drag coefficient of a sphere on Reynolds number. *Physics of Fluids, 13*, 2194–2195.

Barker, H. (1951). The effect of shape and density on the free settling rates of particles at high Reynolds Numbers, Ph.D. Thesis, University of Utah, Table 7, pp. 124–132; Table 9, pp. 148–153.

Barnea, E., & Mitzrahi, J. (1973). A generalized approach to the fluid dynamics of particulate systems, 1. General correlation for fluidization and sedimentation in solid multiparticle systems. *Chemical Engineering Journal, 5*, 171–189.

Barnea, E., & Mednick, R. L. (1975). Correlation for minimum fluidization velocity. *Transactions on Institute of Chemical Engineers, 53*, 278–281.

Batchelor, G. K. (1967). *An introduction for fluid dynamics* (p. 262). Cambridge: Cambridge University Press.

Brauer, H., & Sucker, D. (1976). Umströmung von Platten, Zylindern un Kugeln. *Chemie Ingenieur Technik, 48*, 665–671.

Cabtree, L. F. (1963). Three dimensional boundary layers. In L. Rosenhead (Ed.), *Laminar boundary layers*, Oxford: Oxford Univesity press, p. 423.

Chabra, R. P., Agarwal, L., & Sinha, N. K. (1999). Drag on non-spherical particles: An evaluation of available methods. *Powder Technology, 101*, 288–295.

Concha, F., & Almendra, E. R. (1979a). Settling velocities of particulate systems, 1. Settling velocities of individual spherical particles. *International Journal of Mineral Processing, 5*, 349–367.

Concha, F., & Almendra, E. R. (1979b). Settling velocities of particulate systems, 2. Settling velocities of suspensions of spherical particles. *International Journal of Mineral Processing, 6*, 31–41.

Concha, F., & Barrientos, A. (1986). Settling velocities of particulate systems, 4. settling of no spherical isometric particles. *International Journal of Mineral Processing, 18*, 297–308.

Concha, F., & Christiansen, A. (1986). Settling velocities of particulate systems, 5. Settling velocities of suspensiones of particles of arbitrary shape. *International Journal of Mineral Processing, 18*, 309–322.

Christiansen, E. B., & Barker, D. H. (1965). The effect of shape and density on the free settling rate of particles at high Reynolds numbers. *AIChE Journal, 11*(1), 145–151.

Darby, R. (1996). Determining settling rates of particles, *Chemical Engineering*, 109–112.

Fage, A. (1937). Experiments on a sphere at critical Reynolds number (pp. 108, 423). London: Aeronautical Research Council, Reports and Memoranda, N°1766.

Flemmer, R. L., Pickett, J., & Clark, N. N. (1993). An experimental study on the effect of particle shape on fluidization behavior. *Powder Technology, 77*, 123–133.
Ganguly, U. P. (1990). On the prediction of terminal settling velocity in solids-liquid systems. *International Journal of Mineral Processing, 29*, 235–247.
Ganser, G. H. (1993). A rational approach to drag prediction of spherical and non-spherical particles, 1993. *Powder Technology, 77*, 143–152.
Goldstein, S. (Ed.). (1965). *Modern development in fluid dynamics* (Vols. 1, 2, p. 702). New York: Dover.
Haider, A., & Levenspiel, O. (1998). Drag coefficient and terminal velocity of spherical and non-spherical particles. *Powder Technology, 58*, 63–70.
Happel, J., & Brenner, H. (1965). Low Reynolds hydrodynamics (p. 220). NJ: Prentice-Hall Inc.
Heywood, H. (1962). *Uniform and non-uniform motion of particles in fluids: Proceeding of the Symposium on the Interaction between Fluid and Particles*, Institute of Chemical Engineering, London, pp. 1–8.
Lapple, C. E., & Shepherd, C. B. (1940). Calculation of particle trajectories. *Industrial and Engineering Chemistry, 32*, 605.
Lee, K., & Barrow, H. (1968). Transport process in flow around a sphere with particular reference to the transfer of mass. *International Journal of Heat and Mass Transfer, 11*, 1020.
Lighthill, M. J. (1963). Boundary layer theory. In L. Rosenhead (Ed.), *Laminar boundary layers*, Oxford: Oxford University, p. 87.
Massarani, G. (1984). Problemas em Sistemas Particulados, Blücher Ltda., Río de Janeiro, Brazil, pp. 102–109.
McDonal, J. E. (1954). *Journal of Applied Meteorology, 11*, p. 478.
Meksyn, D. (1961). New methods in boundary layer theory, New York.
Newton, I. (1687). Filosofiae naturalis principia mathematica, London.
Nguyen, A. V., Stechemesser, H., Zobel, G., & Schulze, H. J. (1997). An improved formula for terminal velocity of rigid spheres. *International Journal of Mineral Processing, 50*, 53–61.
Richardson, J. F., & Zaki, W. N. (1954). Sedimentation and fluidization: Part I. *Transactions on Institute of Chemical Engineers, 32*, 35–53.
Perry 1963, p. 5.61.
Pettyjohn, E. S., & Christiansen, E. B. (1948). Effect of particle shape on free-settling of isometric particles. *Chemical Engineering Progress, 44*(2), 157–172.
Rosenhead, L. (ed) (1963). *Laminar boundary layers* (pp. 87, 423, 687). Oxford: Oxford University Press.
Schlichting, H. (1968). *Boundary layer theory*. New York: McGraw-Hill. 747.
Stokes, G. G. (1844). On the theories of internal friction of fluids in motion and of the equilibrium and motion of elastic solids. *Transactions on Cambridge Philosophical Society, 8*(9), 287–319.
Taneda, S. (1956). *Rep. Res. Inst. Appl. Mech., 4*, 99.
Thomson, T., & Clark, N. N. (1991). A holistic approach to particle drag prediction. *Powder Technology, 67*, 57–66.
Tomotika, A.R.A. (1936). Reports and memoranda N°1678, See also Goldstein 1965, p. 498.
Tory, E. M. (1996). *Sedimentation of small particles in a viscous fluid.* UK: Computational Mechanics Publications Inc.
Tsakalis, K. G., & Stamboltzis, G. A. (2001). Prediction of the settling velocity of irregularly shaped particles. *Minerals Engineering, 14*(3), 349–357.
Tourton, R., & Clark, N. N. (1987). An explicit relationship to predict spherical termninal velocity. *Powder Technology, 53*, 127–129.
Tourton, R., & Levenspiel, O. (1986). A short Note on the drag correlation for spheres. *Powder Technology, 47*, 83–86.
Wadell, J. (1932). Volume shape and roundness of rock particles, *Journal of Geology, 15*, 43–451.

Wadell, J. (1934). The coefficient of resistance s a function of the Reynolds number for solids of various shapes, *Journal of the Franklin Institute 217*, 459–490.
Zens, F. A., & Othme, D. F. (1966). Fluidization and fluid-particle systems. New York: McGraw-Hill.
Zigrang, D. J., & Sylvester, N. D. (1981). An explicit equation for particle settling velocities in solid-liquid systems. *AIChE Journal, 27*, 1043–1044.

Chapter 5
Kynch Theory of Sedimentation

Abstract This chapter studies sedimentation of suspensions treated as continuous media. Sedimentation processes are studied from two perspectives; a discrete approach and a continuum approach, in which dynamic processes are established. This chapter uses the continuum approach and presents the concept of an ideal suspension and an ideal thickener. Suspensions described by solid concentration, solid component velocity and fluid component velocity constitute the sedimentation process provided they obey the mass conservation equations. Sedimentation can be performed in batches or continuously. Batch sedimentation is studied first and the Modes of batch sedimentation are established. These observations are extended to continuous processes. Finally the capacity of an ideal continuous thickener is derived. Kynch sedimentation theory, besides correctly describing the behavior of incompressible suspensions, forms part of the more general theory of compressible materials. The exercise of constructing solutions to Kynch sedimentation processes allows for a better understanding of the sedimentation of compressible pulps. Anyone wanting to understand the phenomenological theory of sedimentation must first master Kynch sedimentation processes.

To have the ability to predict the different modes particle will settle from a suspension under the effect of gravity, sedimentation processes must be studied from a fundamental point of view. We have seen in previous chapters that particulate systems can be viewed from two different approaches. Chapter 4 analyzes sedimentation with a *discrete approach*, in which the laws of mechanics are applied to individual particles in the system. Discrete sedimentation has been successful to establish constitutive equations for the sedimentation properties of a certain particulate material in a given fluid. Nevertheless, to analyze a sedimentation process and to obtain behavioral pattern permitting the prediction of capacities and equipment design procedures, the *continuum approach* must be used. The theory we present in this chapter uses this approach and is based on the works of Kynch (1952) and those of Concha and Bustos (1991) and Bustos et al. (1999).

F. Concha A, *Solid–Liquid Separation in the Mining Industry*,
Fluid Mechanics and Its Applications 105, DOI: 10.1007/978-3-319-02484-4_5,

5.1 Concepts of an Ideal Suspension and an Ideal Thickener

Consider a mixture of solid particles in a fluid contained in a vessel, and assume that the suspension satisfies all the requirements to be considered a super-imposed continuous media. The assumptions are:

1. All solid particles are of the same size, form and density and are small with respect to the vessel in which they are contained.
2. The solid and the fluid are incompressible.
3. There is no mass transport between solid and the fluid.
4. The solid–fluid relative velocity in the mixture is a function of the solid concentration only.
5. The solid concentration is the same in all cross-sections of the vessel.

Assumption 1, together with assumption 3, allows establishing a *unique settling velocity* to all particles; Assumption 2 establishes constant *material densities* for the suspension components; Assumption 4, key to Kynch's theory, is a constitutive assumption for the settling velocity that makes unnecessary the use of the momentum balances.

Mixtures fulfilling assumptions 1–4, receive the name of *ideal suspension* (Shannon and Tory 1966; Bustos and Concha 1988; Concha and Bustos 1992) and may be considered a superposition of two continuous medium. The ideal suspension is a model of great utility. It has, in mechanics, a similar connotation that the ideal gas has in thermodynamics. The ideal suspension does not exist really, but many materials behave as ideal suspensions in certain special cases. The Theory of Mixtures predicts the behavior of an ideal suspension that describes, with good approximation, the settling of a suspension of small glass beads (Shannon and Tory 1966; Davies et al. 1991), the sedimentation of un-flocculated copper concentrates and diluted un-flocculated flotation tailings of many metallic and non-metallic ores (Concha 2001).

In general, the solid concentration of a suspension is function of the three dimensions of space, but assumption 4 allows describing the suspension with one space dimension only, that is, the concentration depends on one space dimension and on time. This assumption defines the concept of and *ideal thickener* as a vessel with no wall effect (Shannon and Tory 1966; Bustos et al. 1990a; Concha and Bustos 1992), where the feed, the underflow and the overflow are represented as surface sources or surface sinks.

All models related to sedimentation use the concept of ideal thickener, irrespectively if they use ideal or real suspensions. The reason for this is that the modeling of thickeners feed and discharge mechanism, including the rakes, are to complicated and, that in spite of that, measurements (Becker 1982) have shown that concentration distribution in an industrial thickener is approximately one-dimensional. See Fig. 5.1.

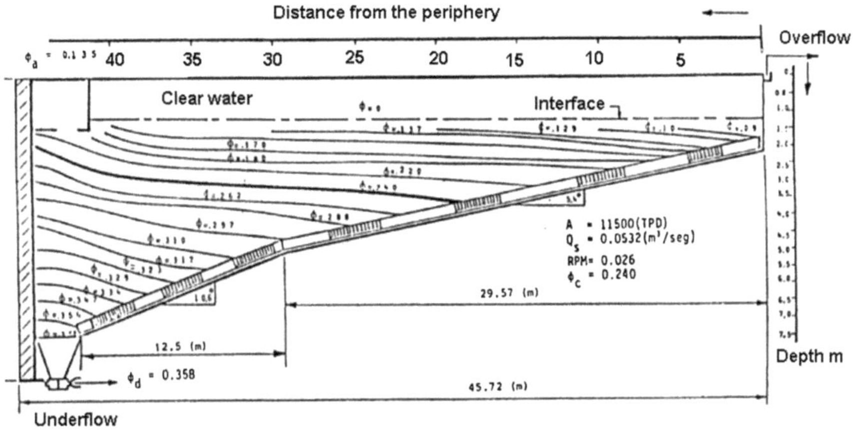

Fig. 5.1 Concentration distribution in an industrial thickener treating copper tailings (Becker 1982)

5.2 Field Equations

Sedimentation of an *ideal suspension* may be described by the following field variables: (1) the solid concentration, as volume fraction of solids, $\varphi(z,t)$, (2) the solid component velocity $\boldsymbol{v}_s(\varphi)$ and (3) the fluid component velocity $\boldsymbol{v}_f(\varphi)$. These field variables must obey the *mass conservation equations* (3.34) and (3.35) :

$$\frac{\partial \varphi}{\partial t} + \nabla \cdot (\varphi \boldsymbol{v}_s) = 0 \tag{5.1}$$

$$\nabla \cdot \boldsymbol{q} = 0, \quad \text{with} \quad \boldsymbol{q} = \varphi \boldsymbol{v}_s + (1-\varphi)\boldsymbol{v}_f \tag{5.2}$$

where $\boldsymbol{q}(\varphi,t)$ is the volume average velocity.

Solutions to these conservation equations are generally discontinuous. This means that discontinuities may appear in the suspension and that Eqs. (5.1) and (5.2) are valid only in those regions where the variables are continuous. At discontinuities they must be replaced by the *mass jump conditions* [Eq. (3.38)]:

$$\sigma[\varphi] = [\varphi \boldsymbol{v}_s \cdot \boldsymbol{e}_I] \quad \text{and} \quad [\boldsymbol{q}] = 0 \tag{5.3}$$

where σ is the rate of propagation of the discontinuity in the direction normal to the discontinuity surface and $[\circ]$ is the difference of value of the variable at each side of the discontinuity.

If the sedimentation vessel is an *ideal thickener*, all equations reduce to one space dimension, then:

$$\frac{\partial \varphi}{\partial t} + \frac{\partial (\varphi v_s)}{\partial z} = 0$$
$$\frac{\partial q}{\partial z} = 0 \quad \text{with} \quad q = \varphi v_s + (1-\varphi) v_f = v_s - (1-\varphi) u$$
$$\sigma = \frac{[\varphi v_s]}{[\varphi]}$$

where $u = v_s - v_f$ is the relative solid fluid velocity.

Define the new variable *solid flux density* by the product of the velocity v_s and concentration φ, that is, $f(\varphi) = \varphi v_s(\varphi)$. In terms of the solid flux density the field equations, for regions where the variables are continuous are:

$$\frac{\partial \varphi}{\partial t} + \frac{\partial f(\varphi)}{\partial z} = 0, \quad \frac{\partial q}{\partial z} = 0 \tag{5.4}$$

and at discontinuities:

$$\sigma = \frac{[f(\varphi)]}{[\varphi]} \quad [q] = 0 \tag{5.5}$$

where $f = q\varphi + \varphi(1-\varphi)u$.

Since discontinuities imply non-uniqueness in the solution, a certain criterion should be used to select the admissible solutions. One of these criteria is *Lax entropy condition* (Bustos and Concha 1988):

$$f_{bk}(\varphi^+) \geq \sigma(\varphi^+, \varphi^-) \geq f_{bk}(\varphi^-), \quad \text{with } \varphi^- \leq \varphi \leq \varphi^+ \tag{5.6}$$

5.2.1 Batch and Continuous Sedimentation

Sedimentation can be performed in a batch or continuous manner. *Batch sedimentation* is usually used in the laboratory. The suspension is introduced in a graduate cylinder with closed bottom, see Fig. 5.2a. The suspension is allowed to settle under the effect of gravity and the water-suspension interface is recorded as a function of time. The *ideal thickener*, for the batch case, is called *settling column* and is depicted in Fig. 5.2b.

If a pulp of solid volume fraction φ_0 is introduced in a column of volume $V = A \times L$, where A is the column cross-section and L the suspension height, the *total solid mass M* and *solid volume per unit area W* in the column are:

$$M = \rho_s A L \varphi_0 \quad \text{and} \quad W = (V/A)\varphi_0 = L\varphi_0 \tag{5.7}$$

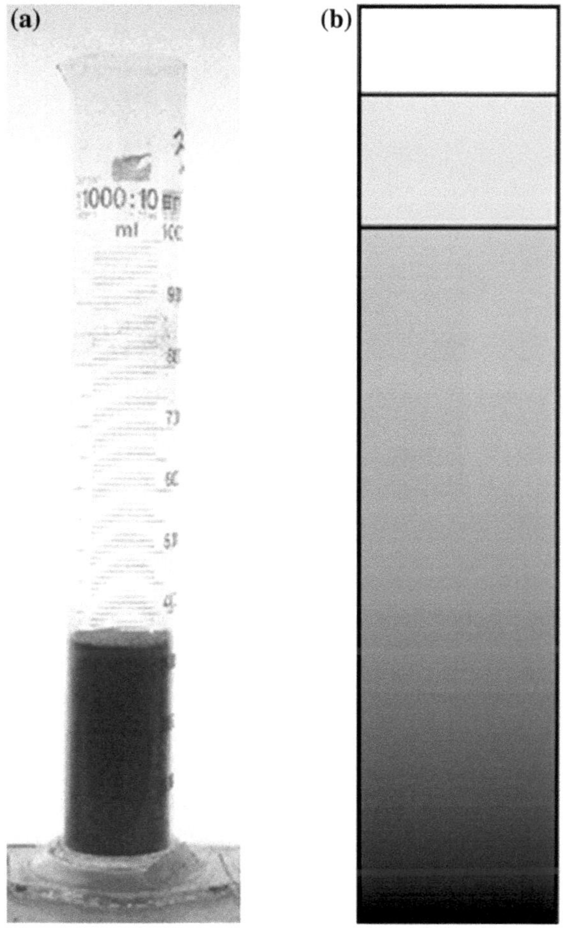

Fig. 5.2 Vessels for batch sedimentation. **a** Laboratory glass graduate cylinder. **b** Settling column

Continuous sedimentation is performed in a cylindrical vessel called *continuous thickener*, with a feedwell at the top and rakes at the bottom of the tank. Figure 5.3 represents a schematic view of the original Dorr thickener.

A volume feed rate $Q_F(t)$ with concentration $\varphi_F(t)$ enters through the feedwell and an underflow rate $Q_D(t)$ with concentration $\varphi_D(t)$ leaves the thickener at the bottom and center of the tank. At the top and periphery of the tank clear water leaves through the overflow launder at volume flowrate of $Q_O(t)$.

The *solid volume flux* in and out of the thickener are given by $F = Q_F\varphi_F$ and $D = Q_D\varphi_D$. The ratio of the solid volume flux to the thickener cross section S is the *solid volume flux density*, or just *solid flux density*, so that the solid feed and underflow flux densities are:

$$f_F = \frac{Q_F\varphi_F}{S} \quad \text{and} \quad f_D = \frac{Q_D\varphi_D}{S} \tag{5.8}$$

Fig. 5.3 Dorr thickener (1905) vessels for continuous sedimentation

The *volume average velocity* q of the pulp is:

$$q = Q_D/S \tag{5.9}$$

A macroscopic balance in the thickener at steady state gives:

Solid mass: $F = D$ (5.10)

Pulp volume: $Q_F = Q_D + Q_O$ (5.11)

Solid volume: $Q_F\varphi_F = Q_D\varphi_D$ (5.12)

Solid flux density: $f_F = q\varphi_D$ (5.13)

5.3 Batch Kynch Sedimentation Process

When an *ideal suspension* settles under the effect of gravity in a *settling column*, the following steps can be distinguished:

(a) Before settling begins, the suspension is homogenized by agitation obtaining a suspension of constant concentration.

(b) When sedimentation starts, all particles settle at the same velocity forming a water-suspension interface that moves at the velocity of the solid particles. This step is called *hindered settling* because the presence of other particles hinders the settling of an individual particle. See Fig. 5.4. There are special cases, noted by Bürger and Tory (2000), where during a short time not all particles settle at the same velocity.

(c) Particles that reach the bottom of the column very rapidly cover the whole column cross sectional area. This material, with concentration φ_∞, is called *sediment*. The particles in the sediment start piling up moving the interface between the sediment and the suspension at a constant characteristic upward velocity. See Fig. 5.4.

(d) At a given time $t = t_c$, called *critical time*, the interface water-suspension meets with the interface sediment-suspension at a *critical height* $z = z_c$. These coordinate (z_c, t_c) define the *critical point* were sedimentation ends.

Based on the description of batch sedimentation, we can add the following assumptions to the 5 general assumptions given at the beginning:

1. There is no inflow or outflow of suspension from the settling column, therefore the volume average velocity $q = 0$.
2. The suspension has an initial constant concentration φ_0.

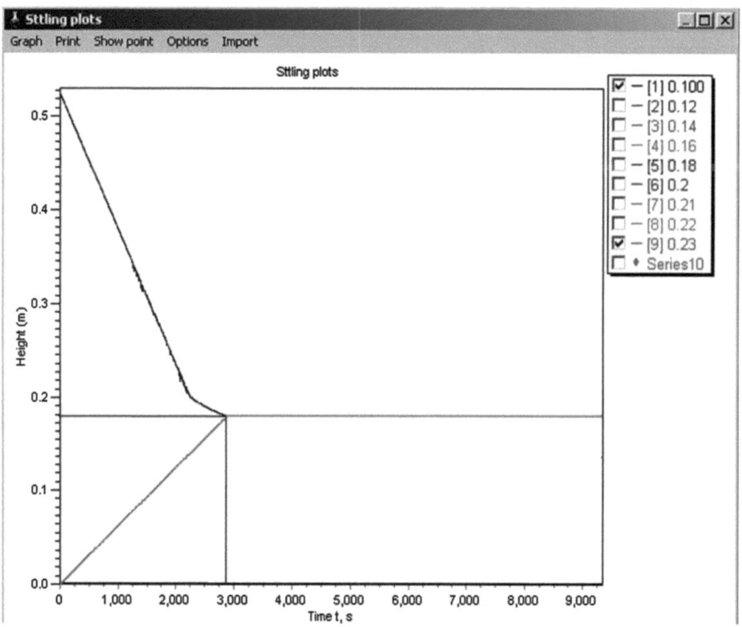

Fig. 5.4 Settling *curve* showing the water-suspension and the sediment-suspension interfaces for the settling of a suspension with initial concentration $\varphi_0 = 0.10$ and sediment concentration $\varphi_\infty = 0.23$

3. The final sediment concentration is φ_∞.

Under these addition assumptions, Eqs. (5.4) and (5.5) may be written, for regions where the variables are continuous, in the form:

$$\frac{\partial \varphi}{\partial t}+\frac{\partial f_{bk}(\varphi)}{\partial z}=0 \tag{5.14}$$

$$f_{bk}(\varphi)=\varphi(1-\varphi)u \tag{5.15}$$

where f_{bk} is called *Kynch batch flux density function* and is a constitutive equation to be determined experimentally.

At discontinuities Eq. (5.14) is not satisfied and is replaced by the Rankin-Hugoniot condition (Bustos and Concha 1988; Concha and Bustos 1991):

$$\sigma=\frac{[f_{bk}(\varphi)]}{[\varphi]},\quad 0\leq z\leq L \tag{5.16}$$

However, discontinuous solutions, satisfying (5.14) at points of continuity and (5.16) at discontinuities, are in general not unique, therefore an additional criterion, or *entropy principle*, is necessary to select the physically relevant discontinuous solution. This solution is called *entropy weak solution.*

One of these entropy criteria is the *Oleinik jump entropy condition,* requiring that:

$$\frac{f_{bk}(\varphi)-f_{bk}(\varphi^-)}{\varphi^+-\varphi^-}\leq\sigma(\varphi^+,\varphi^-)\leq\frac{f_{bk}(\varphi)-f_{bk}(\varphi^+)}{\varphi^--\varphi^+},\quad \text{for all } \varphi^-\leq\varphi\leq\varphi^+ \tag{5.17}$$

An interpretation of this entropy condition indicates that (5.17) is satisfied if, and only if, in a $f_{bk}(\varphi)$ versus φ plot, the chord joining the point $(\varphi^+, f_{bk}(\varphi^+))$ and $(\varphi^-, f_{bk}(\varphi^-))$ remains above the curve $f_{bk}(\varphi)$ for $\varphi^+<\varphi^-$ and below the curve $f_{bk}(\varphi)$ for $\varphi^+>\varphi^-$. See Fig. 5.5.

Discontinuities satisfying (5.16) and (5.17) are called *shocks*. If, in addition,

$$f'_{bk}(\varphi^-)=\sigma(\varphi^+,\varphi^-)\quad\text{or}\quad f'_{bk}(\varphi^+)=\sigma(\varphi^+,\varphi^-) \tag{5.18}$$

are satisfied, the discontinuity is called *contact discontinuity*.

Initial and boundary conditions for the conservation law expressed by Eqs. (5.14) to (5.17) are:

$$\begin{aligned}\varphi(z,0)&=\varphi_0\quad\text{for}\quad 0\leq z\leq L\\ \varphi(L,t)&=\varphi_L\quad\text{for}\quad t>0\end{aligned} \tag{5.19}$$

$$\varphi(0,t)=\varphi_\infty\quad\text{for}\quad t>0 \tag{5.20}$$

where concentrations φ_L, φ_0 y φ_∞ are all constant. Kynch batch flux-density functions should obey the following conditions:

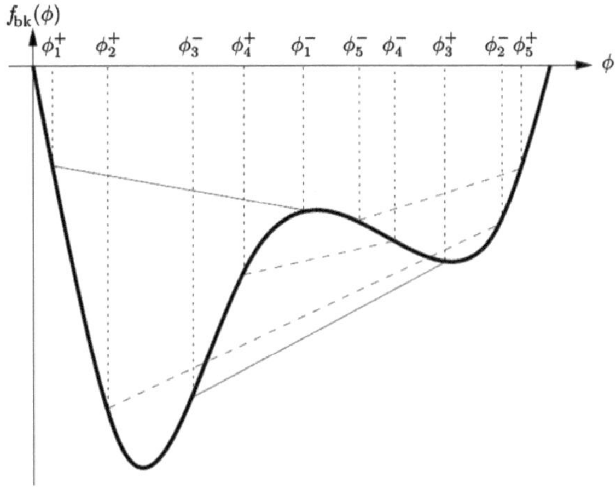

Fig. 5.5 Geometrical interpretation of Oleinik's criterion

$$\begin{aligned} &f_{bk}(\varphi)<0 \quad \text{for} \quad 0\leq\varphi\leq\varphi_\infty, \quad f_{bk}(0)=f_{bk}(\varphi_\infty)=0\\ &f'_{bk}(0)<0 \quad \text{and} \quad f'_{bk}(\varphi_\infty)>0 \end{aligned} \tag{5.21}$$

Equations (5.14)–(5.21) form an initial-boundary value problem called *Batch Kynch Sedimentation Process* (BKSP) (Bustos and Concha 1988).

5.3.1 Solution to the Batch Kynch Sedimentation Process

Equation (5.14) may be written in the form:

$$\frac{\partial\varphi}{\partial t}+f'_{bk}(\varphi)\frac{\partial\varphi}{\partial z}=0, \quad \text{with} \quad f'_{bk}(\varphi)=df_{bk}/d\varphi \tag{5.22}$$

Since $f_{bk}(0)=f_{bk}(\varphi_\infty)=0$, initial and boundary conditions (5.19) and (5.20) may be written as initial conditions only:

$$\varphi(z,0)=\begin{cases} 0 & \text{for} \quad L<z\\ \varphi_0 & \text{for} \quad 0\leq z\leq L\\ \varphi_\infty & \text{for} \quad z<0 \end{cases} \tag{5.23}$$

Summarizing, we can state that sedimentation of ideal suspensions may be represented by the *volume fraction of solids* $\varphi(z,t)$ and the *Kynch batch flux density function* $f_{bk}(\varphi)$. These two functions constitute a BKSP if, in the region of space $0\leq z\leq L$ and time $t>0$, they obey Eq. (5.22) where variables are continuos and Eqs. (5.16) and (5.17) at discontinuities. Additionally they must satisfy the initial conditions (5.23).

(a) **Solution by the method of characteristics**

Written in this form, a BKSP can be treated as an initial value problem and may be solved by the *method of characteristics* for constant initial values. According to Courant and Hilbert (1963), straight lines, called *characteristics* can be drawn in the (z, t) plane, where Eq. (5.22) is valid. These lines obey the following conditions:

$$\frac{dz(\varphi, t)}{dt} = f'_{bk}(\varphi(z(t), t)), \quad \text{for } t > 0,\ z(0) = 0 \tag{5.24}$$

Here $dz(\varphi, t)/dt$ represents the *propagation velocity of concentration waves* of constant value φ in the (z,t) domain. Since φ is constant along these lines, $f'_{bk}(\varphi)$ is also constant and the characteristics are straight lines.

As an example, take the case with initial concentration $\varphi_0 < \varphi_\infty^{**}$, where φ_∞^{**} is the point where a tangent drawn from point $(\varphi_\infty, f_{bk}(\varphi_\infty))$ cuts Kynch flux density curve. See Fig. 5.6. Since the initial values for $z<0$, $0 \le z \le L$ and $z > L$ are constant, characteristic starting from the ordinate axis are parallel straight lines with slope given by $f'_{bk}(0)$, $f'_{bk}(\varphi_0)$ and $f'_{bk}(\varphi_\infty)$ respectively. On the other hand, we can see that $f'_{bk}(0)$, $f'_{bk}(\varphi_0)$ and $f'_{bk}(\varphi_\infty)$ are lines tangent to Kynch flux density curve at $\varphi = 0$, $\varphi = \varphi_0$ and $\varphi = \varphi_\infty$. Where these lines intersect, the solution is no longer unique and discontinuities, with slope $\sigma(0, \varphi_0)$ and $\sigma(0, \varphi_\infty)$, are formed in the form of cords drawn from $(0, f_{bk}(0))$ to $(\varphi_0, f_{bk}(\varphi_0))$ and $(\varphi_0, f_{bk}(\varphi_0))$ to $(\varphi_\infty, f_{bk}(\varphi_\infty))$ respectively. See Fig. 5.6.

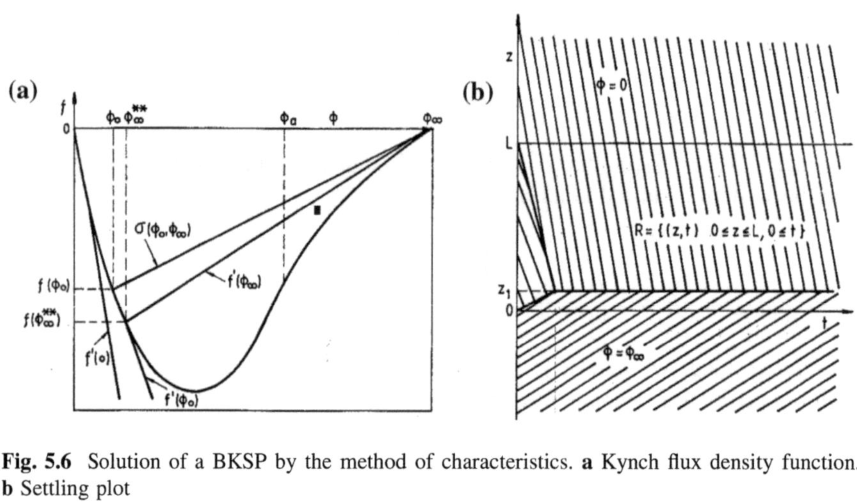

Fig. 5.6 Solution of a BKSP by the method of characteristics. **a** Kynch flux density function. **b** Settling plot

Modes of Sedimentation

In his work Kynch (1952) mentioned modes of sedimentation as manners in which a suspension may settle, but without giving a specific meaning to them. Precise description of this concept was given by Bustos and Concha (1988) and Concha and Bustos (1991) who defined *Modes of Batch Sedimentation* (MBS) as the different possible BKSP, that is, as the possible entropy weak solutions to the batch sedimentation problem that can be constructed for a given initial data and Kynch flux density function. There are 7 MBS for Kynch flux density functions with at most two inflection points (Bustos et al. 1999).

The type of MBS depends on how the zones of constant concentration φ_0 and φ_∞ are separated after sedimentation is complete. Figures 5.7 and 5.8 show the 7 Modes of Batch Sedimentation, including the flux density function, settling plot and concentration profile.

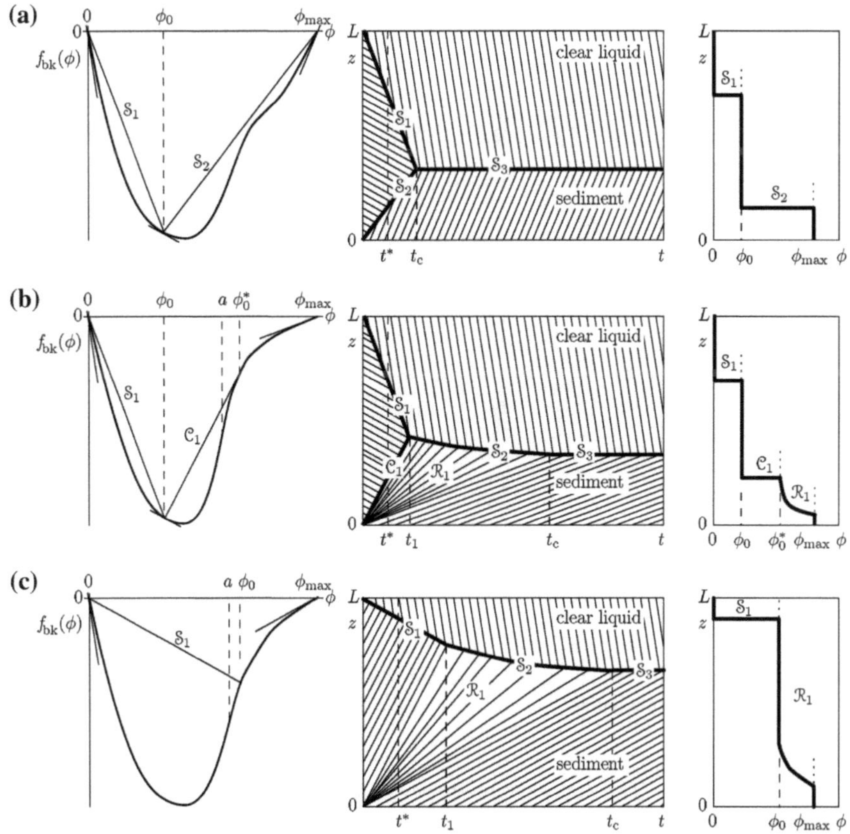

Fig. 5.7 Modes of batch sedimentation processes, *MBS-1–MBS-3*, for batch Kynch flux density function with one and two inflection points. In these figures the shocks are described by $\boldsymbol{\delta}_i$ and the contact discontinuity by C

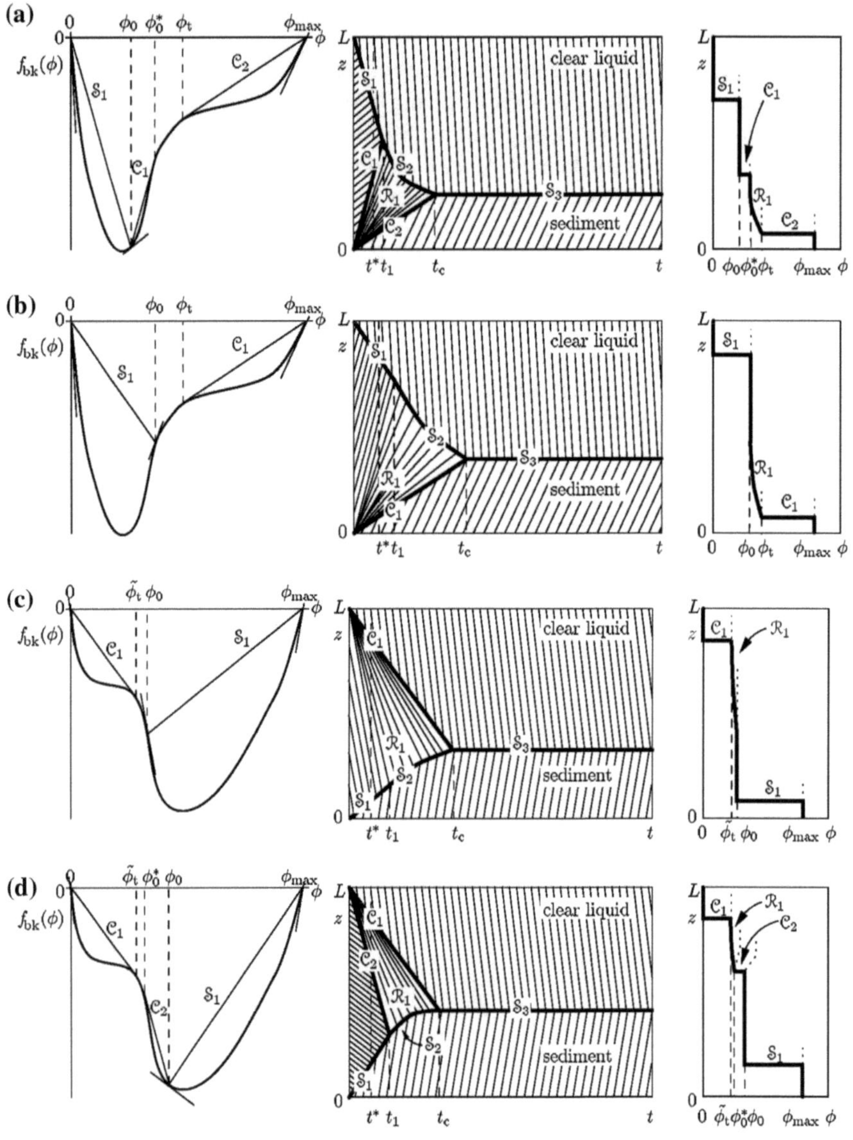

Fig. 5.8 Modes of batch sedimentation processes, *MBS-4–MBS-7*, for batch Kynch flux density function with two inflection points. In these figures the shocks are described by $\boldsymbol{\delta}_i$ and the contact discontinuity by C

- MBS-1 A shock. The supernatant-suspension and the suspension sediment are both shocks meeting at the critical time t_c. See Fig. 5.7a.
- MBS-2 The rising shock is replaced, from bottom to top, by a contact discontinuity followed by a rarefaction wave. This MBS can occur only with a Kynch Batch flux density function with one inflection point. See Fig. 5.7b.

- MBS-3 A rarefaction wave. In this case no cord can be drawn from $\left(\varphi_0, f'_{bk}(\varphi_0)\right)$ to $\left(\varphi_0^*, f'_{bk}(\varphi_0^*)\right)$ and the contact discontinuity becomes a line of continuity. This MBS can occur only with a Kynch Batch flux density function with one inflection point. See Fig. 5.7c.
- MBS-4 Two contact discontinuities separated by a rarefaction wave. See Fig. 5.8a.
- MBS-5 One rarefaction wave followed by a contact discontinuity. See Fig. 5.8b.
- MBS-6 One rarefaction wave followed by a convex shock. This mode occurs if the inflection points a and b are on the left of the minimum in Kynch flux density curve and the tangency point $(\varphi_t, f_{bk}(\varphi_t))$ of a line drawn from $(0, f_{bk}(0))$ to the Kynch flux density curve is in the range $a < \varphi_t < b$, and the initial concentration is in the interval $\varphi_t < \varphi_0 < b$. See Fig. 5.8c.
- MBS-7 One shock followed by a contact discontinuity and a curved shock. This mode occurs if the inflection points a and b are on the left of the minimum in Kynch flux density curve and the tangency point $(\varphi_t, f_{bk}(\varphi_t))$ of a line drawn from $(0, f_{bk}(0))$ to the Kynch flux density curve is in the range $a < \varphi_t < b$, and the initial concentration is $\varphi_0 > b$. See Fig. 5.8d.

In all cases independently of the type of MSB, the final state is a sediment of concentration φ_∞. The height of the sediment is given by a mass balance as:

$$z_\infty = L\varphi_0/\varphi_\infty \tag{5.25}$$

5.4 Continuous Kynch Sedimentation Process

Figure 5.9 shows a continuous thickener showing the feedwell and rakes.

It has been established that four zones exists in a continuous thickener.

Zone I. Zone I correspond to *clear water,* which is located in the region above and outside the feed well

Zone II. Below the clear water zone a region of constant concentration forms. This zone is called *hindered settling zone* and has a concentration called *conjugate concentration.*

Zone III. Under the hindered settling zone, a *transition zone* takes the conjugate concentration to the sediment concentration. This can happen through a shock wave, a rarefaction wave or a combination of them.

Zone IV. Finally, we have the *sediment zone*, a zone of constant and final concentration (Fig. 5.10).

Fig. 5.9 Continuous thickener

Fig. 5.10 Ideal Continuous Thickener (ICT)

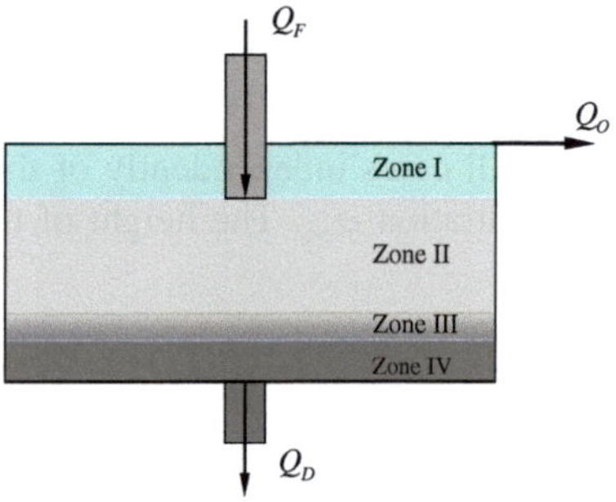

5.4.1 Solution to the Continuous Kynch Sedimentation Process

Based on the description of continuous sedimentation, we can add the following assumptions to the 5 general assumptions given at the beginning:

1. The outflow velocity of suspension from the ideal thickener is $q(t) = Q_D(t)/S$.
2. The suspension has an initialconcentration distribution $\varphi_I(z)$.
3. Solids never enter zone I, that is, the domain of the solution is $0 \leq z \leq L$, where L is the base of the feedwell.

Under these additional assumptions, Eqs. (5.4) and (5.5) may be written, for regions where the variables are continuous, in the form:

$$\frac{\partial \varphi}{\partial t} + f_k'(\varphi)\frac{\partial \varphi}{\partial z} = 0 \quad \text{where } f_k'(\varphi) = df_k'/dz \tag{5.26}$$

$$\frac{\partial q}{\partial z} = 0, \quad \text{with } f_k = q\varphi + f_{bk}(\varphi) \tag{5.27}$$

and (5.5) for discontinuities:

$$\sigma = \frac{[f_k(\varphi)]}{[\varphi]} \qquad [q] = 0 \tag{5.28}$$

The jump condition must satisfy Oleinik jump entropy condition:

$$\frac{f_k(\varphi) - f_k(\varphi^-)}{\varphi^+ - \varphi^-} \leq \sigma(\varphi^+, \varphi^-) \leq \frac{f_k(\varphi) - f_k(\varphi^+)}{\varphi^- - \varphi^+}, \quad \text{for all } \varphi^- \leq \varphi \leq \varphi^+ \tag{5.29}$$

Summarizing, we can state that the *continuous sedimentation of ideal suspensions* in an *ideal thickener* may be represented by the *volume fraction of solids* $\varphi(z,t)$, the *volume average velocity* $q(z,t)$ and the *continuous Kynch flux density function* $f_k(\varphi)$. These functions constitute a *Continuous Kynch Sedimentation Process* (CKSP) if, in the region of space $0 \leq z \leq L$ and time $t > 0$ where the variables are continuous, the obey Eqs. (5.26) and (5.27), and at discontinuities they obey Eqs. (5.28) and (5.29).

Solution by the Method of Characteristics

In the majority of cases, Kynch batch flux density functions are functions with one infection point only at φ_a. If φ_∞ is the concentration at the end of the sedimentation process, the Kynch flux density function must obey the following properties. See Fig. 5.11.

$$\begin{aligned} &f_k(\varphi,t) = q(t)\varphi + f_{bk}(\varphi) \leq 0, \quad 0 \leq \varphi \leq \varphi_\infty \\ &f_k(0,t) = 0 \end{aligned} \tag{5.30}$$

$$f_k'(\varphi,t) < 0, \quad 0 \leq \varphi \leq \varphi_a \tag{5.31}$$

$$f_k'(\varphi,t) > 0, \quad \varphi_a \leq \varphi \leq \varphi_\infty \tag{5.32}$$

$$f_k''(\varphi,t) < 0, \quad \varphi_a \leq \varphi \leq \varphi_\infty \tag{5.33}$$

Equation (5.27) shows that $q = q(t)$ is independent of the z coordinate. In the rest of this work we will assume that q is a constant independent of time (steady state).

Solutions to the CKSP are straight lines, called *characteristics,* drawn in the (z,t) plane, where Eq. (5.26) is valid. These lines obey the following conditions:

$$\frac{dz(\varphi,t)}{dt} = f_k'(\varphi(z(t),t)), \quad \text{for} \quad t > 0 \tag{5.34}$$

Here $dz(\varphi,t)/dt = f_k'(\varphi)$ represents the *propagation velocity of concentration waves* of constant value φ.

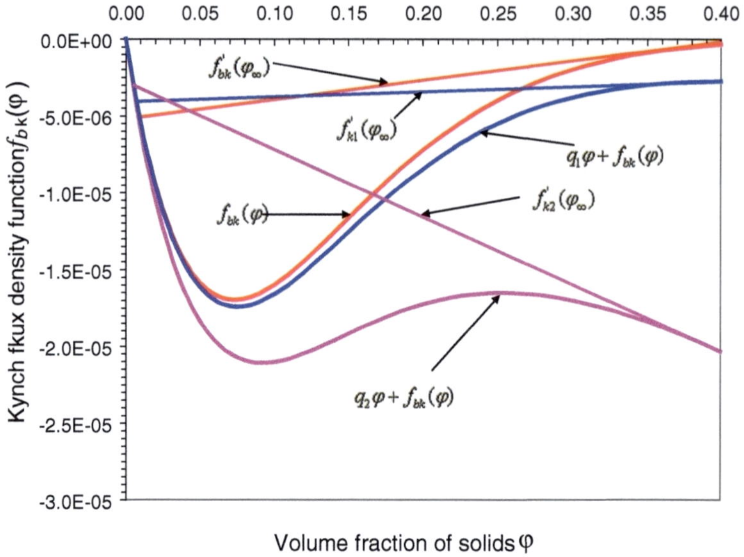

Fig. 5.11 Continuous Kynch flux density function for two values of the volume average velocity $q_1 = 5 \times 10^{-5}$; $q_2 = 6 \times 10^{-6}$ and the Batch Kynch flux density function with one inflection point, $f_{\mathrm{bk}} = 6.05 \times 10^{-4} \times \varphi(1 - \varphi)^{12.59}$

Modes of continuous sedimentation

Mode of Continuous Sedimentation (MCS) are the different possible CKSP, that is, the possible entropy weak solutions to the continuous sedimentation problem that can be constructed for a given initial data and Kynch flux density function (Bustos et al. 1999). Since these MSC depend entirely on the Kynch flux density function and on the initial conditions, it is necessary to choose these material properties and conditions.

We will consider the following initial conditions:

$$\varphi_I(z) = \begin{cases} \varphi_L & \text{for } A \leq z \leq L \\ \varphi_\infty & \text{for } 0 \leq z < A \end{cases} \tag{5.35}$$

From Eq. (5.27), the value of φ_L is obtained by solving the implicit equation:

$$f_F = q\varphi_L + f_b(\varphi_L(t)) \tag{5.36}$$

where f_F is the feed solid flux density function. In case Eq. (5.36) admits more than one solution φ_L, the relevant one is selected by the physical argument that the feed suspension is always diluted on entering the thickener, as shown by Comings et al. (1954).

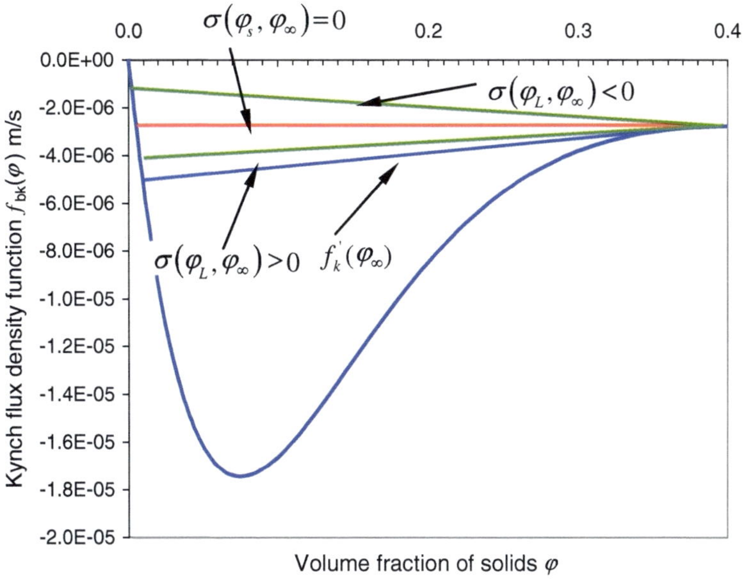

Fig. 5.12 Continuous Kynch flux density functions with one inflection point; $f_k'(\varphi_\infty) > 0$

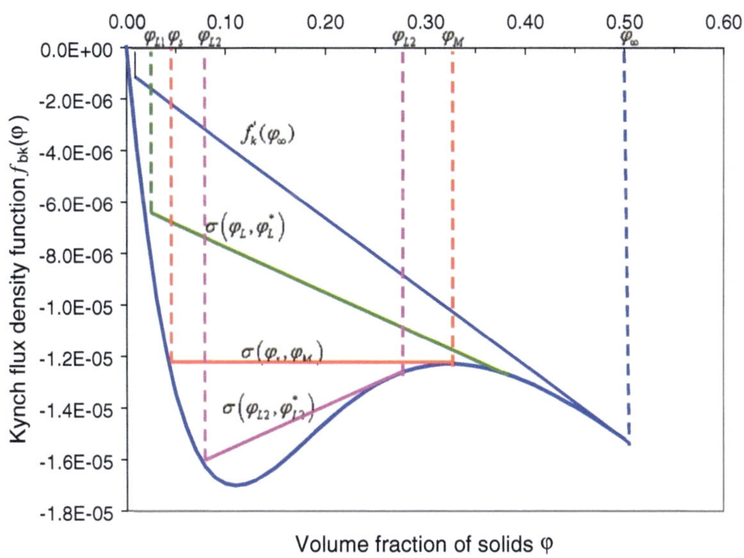

Fig. 5.13 Continuous Kynch flux density functions with one inflection point; $f_k'(\varphi_\infty) < 0$

For a Kynch flux density function with one inflection point, the important parameters are shown in Figs. 5.12 and 5.13 for two values of the volume average velocity q.

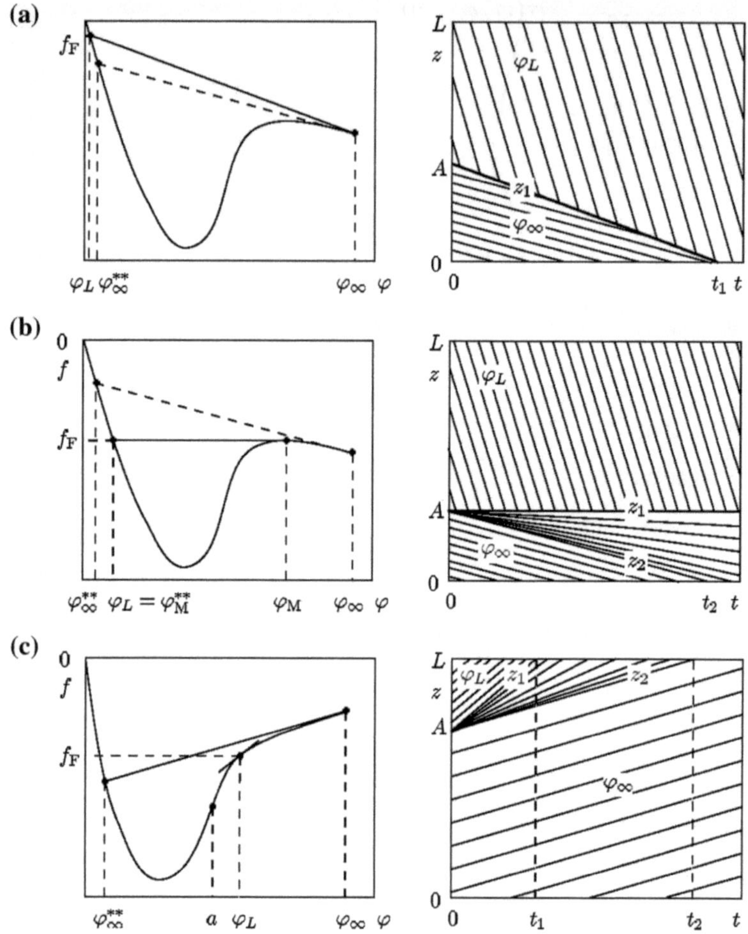

Fig. 5.14 Modes of continuous sedimentation processes. **a** MCS-1. **b** MCS-2. **c** MCS-3

The type of MCS depends on how the zones of constant concentration φ_0 and φ_∞ are separated after sedimentation is complete. Three MCS exist for a flux density function with one infection point (Concha and Bustos 1992).

- MCS-1 A shock separating two zones of continuous sedimentation φ_L and φ_∞. See Fig. 5.14a.
- MCS-2 A contact discontinuity separating two zones of continuous sedimentation φ_L and φ_∞. See Fig. 5.14b.
- MCS-3 A rarefaction wave separating two zones of continuous sedimentation φ_L and φ_∞. See Fig. 5.14c.

The fact that the exact location and propagation speed of the sediment-suspension interface level is always known, permits to formulate a simple control

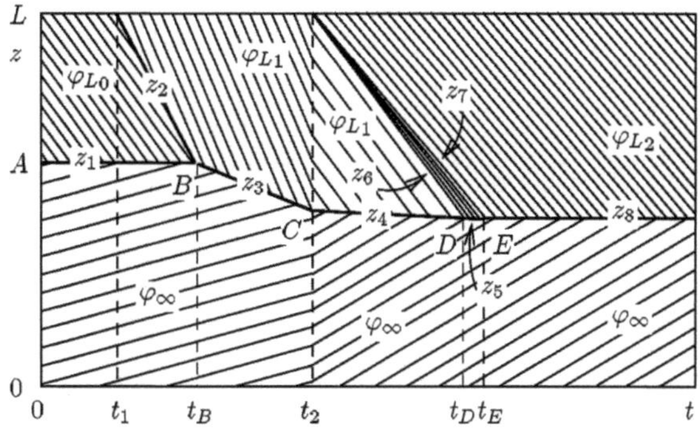

Fig. 5.15 Control of continuous sedimentation after Bustos et al. (1990b)

model for the transient continuous sedimentation. It can be shown (Bustos et al. 1990b) that the steady state corresponding to an MCS-1 can always be recovered after a perturbation of the feed flux density by solving two initial-boundary value problems at known times and with parameters q and φ_L that can be calculate a priori. See Fig. 5.15.

5.4.2 Steady State of an Ideal Continuous Thickener

In MCS-1 and MCS-2 the thickener overflows, empty or attain a steady state, while in MCS-3 no steady state can be attained:

(a) **A MCS-1** can reach a steady state if $f_k'(\infty) > 0$ and $\varphi_L = \varphi_s$, where φ_s is defined in Fig. 5.12. See Fig. 5.16.
(b) **A MCS-2** can reach a steady state if $f_k'(\varphi_\infty) < 0$ and $\varphi_L = \varphi_M^{**}$, so that $\sigma(\varphi_L, \varphi_L^*) = f_k'(\varphi_M) = 0$ and a contact horizontal discontinuity forms. φ_M corresponds to the concentrations of the maximum point in the flux density curve and φ_M^{**} to its conjugate concentration. See Fig. 5.17.

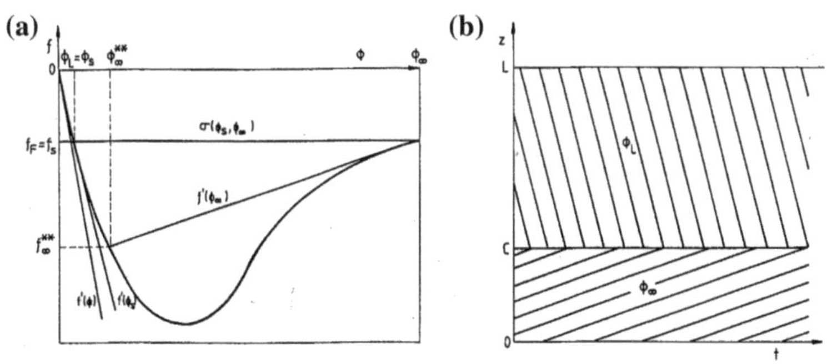

Fig. 5.16 Global weak solution for a MSC-1 leading to a steady state

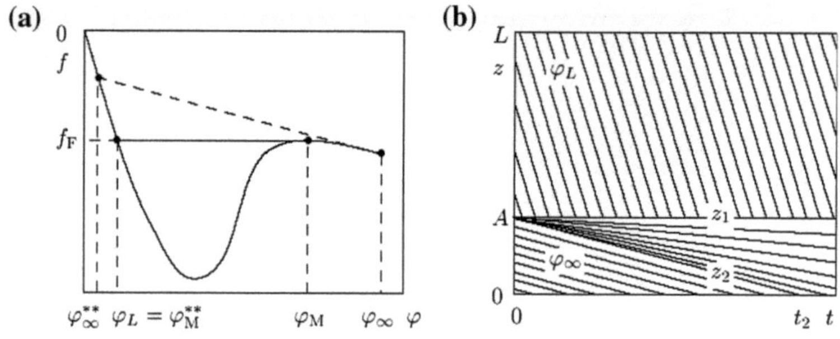

Fig. 5.17 Global weak solution for a MSC-2 leading to a steady state

Capacity of an Ideal Continuos Thickener

At steady state, from Eq. (5.26):

$$\frac{df_k(\varphi)}{dz} = 0, \quad \text{for } 0 \leq z \leq L \tag{5.37}$$

with boundary conditions $z = 0$, $f_k(\varphi(0)) = q\varphi_D$, $z = L$, $f_k(\varphi(L)) = f_F$, where f_F is the feed flux density function. Then,

$$f_k(\varphi(L)) = q\varphi_L + f_{bk}(\varphi_L) = f_F \tag{5.38}$$

$$f_k(\varphi(0)) = f_k(\varphi(0)) = q\varphi_D = f_F \tag{5.39}$$

Substituting (5.39) into (5.40) yields:

$$\begin{aligned} f_F &= \frac{f_F}{\varphi_D}\varphi_L + f_{bk}(\varphi_L) \\ f_F\left(\frac{1}{\varphi_L} - \frac{1}{\varphi_D}\right) &= \frac{f_{bk}(\varphi_L)}{\varphi_L} \\ \frac{1}{f_F} &= \frac{\varphi_L}{f_{bk}(\varphi_L)}\left(\frac{1}{\varphi_L} - \frac{1}{\varphi_D}\right) \end{aligned} \tag{5.40}$$

Since the mass flow to the thickener is $F = \rho_s Q_F \varphi_F$ and the solid flux density is defined by $f_F = -Q_F \varphi_F / S$, we can write:

$$f_F = -\frac{F}{\rho_s S} \tag{5.41}$$

Defining the Unit Area $UA = S/F$, where S is the thickener cross sectional area:

$$UA = -\frac{1}{\rho_s f_F} \tag{5.42}$$

Replacing this expression into Eq. (5.40) yields an equation for the Unit Area.

$$\begin{aligned} UA &= \frac{\varphi_L}{\rho_s f_{bk}(\varphi_L)}\left(\frac{1}{\varphi_D} - \frac{1}{\varphi_L}\right) \\ UA &= \frac{1}{\rho_s f_{bk}(\varphi_L)}\left(\frac{\varphi_L}{\varphi_D} - 1\right) \end{aligned} \tag{5.43}$$

We have seen that for a MCS-1 at steady state, $\varphi_L = \varphi_s$ and $\varphi_D = \varphi_\infty$, then from (5.43):

$$UA = \frac{1}{\rho_s f_{bk}(\varphi_s)}\left(\frac{\varphi_s}{\varphi_\infty} - 1\right) \tag{5.44}$$

and for a MCS-2, $\varphi_L = \varphi_M^{**}$ and $\varphi_D = \varphi_M$, then from (5.43):

$$UA = \frac{1}{\rho_s f_{bk}(\varphi_M^{**})}\left(\frac{\varphi_M^{**}}{\varphi_M} - 1\right) \tag{5.45}$$

Kynch sedimentation theory, besides describing correctly the behavior of incompressible suspensions, forms part of the more general theory of compressible materials. The exercise of constructing *global weak solutions* to Kynch Sedimentation Processes in graphical form, allows a better understanding of the sedimentation of compressible pulps. Every person willing to understand the phenomenological theory of sedimentation must first master Kynch Sedimentation Processes.

References

Becker, R. (1982). Continuous thickening, design and simulation, Eng. Thesis, University of Concepcion, Chile (in Spanish).

Bürger, R., & Tory, E. M. (2000). On upper rarefaction waves in batch settling. *Powder Technology, 108*, 74–87.

Bustos, M. C., & Concha, F. (1988). On the construction of global weak solutions in the Kynch theory of sedimentation. *Mathematical Methods in the Applied Sciences, 10*, 245–264.

Bustos, M. C., Concha, F., & Wendland, W. (1990). Global weak solutions to the problem of continuous sedimentation of an ideal suspension. *Mathematical Methods in the Applied Sciences, 13*, 1–22.

Bustos, M. C., Paiva, F., & Wendland, W. (1990b). Control of continuous sedimentation of ideal suspensions as an initial and boundary value problem. *Mathematical Methods in the Applied Sciences, 12*, 533–548.

Bustos, M. C., Concha, F., Bürger, R., & Tory, E. M. (1999). *Sedimentation and thickening: Phenomenological foundation and mathematical theory* (pp. 111–148). Dordrecht: Kluwer Academic Publishers.

Comings, E. W., Pruiss, C. E., & De Bord, C. (1954). Countinuous settling and thickening. *Industrial and Engineering Chemistry Process Design and Development, 46*, 1164–1172.

Concha, F. (2001). Filtration and separatilon manual, Ed. Red Cettec, Concepci6n, Chile.

Concha, F., & Bustos, M. C. (1991). Settling velocities of particulate systems, 6. Kynch sedimentation processes: Batch settling. *International Journal of Mineral Processing, 32*, 193–212.

Concha, F., & Bustos, M. C. (1992). Settling velocities of particulate systems, 7. Kynch sedimentation processes: Continuous thickening. *International Journal of Mineral Processing, 34*, 33–51.

Courant, R., & Hilbert, D. (1963). *Methods of mathematical physics* (Vol. 2), Interscience. New York, p. 830.

Davies, K. E., Russel, W. B., & Glantschnig, W. J. (1991). Settling suspensions of colloidal silica: Observations and X-ray measurements. *Journal of the Chemical Society, Faraday Transactions, 87*, 411–424.

Kynch, G. J. (1952). A theory of sedimentation. *Transactions of the Faraday Society, 48*, 166–176.

Shannon, P. T., & Tory, E. M. (1966). The analysis of continuous thickening. *Society of Mining Engineers of AIME, 2*, p. 1.

Chapter 6
Flow Through Rigid Porous Media

Abstract The best way to analyze the flow through a porous medium is to divide it into two phenomena, depending on whether the porous matrix is compressible or incompressible. In both cases, the result is useful for the process industry. Rigid porous media form the basis for a simplified filtration theory and compressible porous media form part of thickening theory. In this chapter the fundamental equations for the flow through *rigid porous media* based on the Theory of Mixtures is developed. Consider the flow of a incompressible viscous fluid through a bed of small solid incompressible particles with no mass transfer between the solid and the fluid. Such a mixture of particles is called an *incompressible porous medium* and can be described with the equations for particulate systems presented in Chap. 3. It is convenient in this case to use porosity as a variable instead of the solid volume fraction. Local balances are laid down for mass and momentum and Darcy's and Forchheimer's equations are used as constitutive equations. For a mono-phase flow, permeability is defined and for the case of a two-phase flow, the concepts of relative permeability, saturation and capillary pressure are introduced.

In 1856 Darcy proposed the first experimental correlation for viscous flow through an incompressible porous medium formed by consolidated particles. Darcy established a linear relationship between the volume flow per unit area and the pressure drop through the bed. Massarani (1984) observed that Darcy's correlation was not universally valid, but was restricted to slow flows, and proposed an extension for faster flows. Since then, slow flows through porous media are called Darcy flows and those at higher velocities are called Forcheimer flows. In addition to these pioneering works, Kozeny (1927) and Carman (1937) contributed with classical papers.

The best way to analyze the flow through a porous medium is to divide it into two phenomena, depending whether the porous matrix is compressible or incompressible. In both cases, the result is useful for the process industry. Rigid porous media form the basis for a simplified filtration theory and compressible porous media forms part of the thickening theory. In this chapter we will develop the fundamental equations for the flow through *rigid porous media* based on the Theory of Mixtures.

F. Concha A, *Solid–Liquid Separation in the Mining Industry*,
Fluid Mechanics and Its Applications 105, DOI: 10.1007/978-3-319-02484-4_6,

6.1 Dynamic Process of a Single Phase Flow in a Rigid Porous Media

Consider the flow of a viscous fluid through a bed of solid particles, where the following assumptions are valid:

1. The particles are small with respect to the containing vessel and have the same size and form.
2. The particles and the bed are incompressible.
3. There is no mass transfer between the solid and the fluid.
4. The particles are contained in a vessel with impervious frictionless walls.

Such a mixture of particles is called an *incompressible porous medium* and can be described with the equations for particulate systems described in Chap. 3. Since in this case the only active component is the fluid, due to their incompressibility the solid velocity is zero, it is convenient to use porosity $\varepsilon = 1 - \varphi$ as variable instead of the solid volume fraction $\varphi. v_s = 0$.

6.1.1 Local Balances

The flow through a rigid porous bed can be described by the mass and linear momentum balances. From Chap. 5 and the previous assumptions we can write:

Volume balance for the fluid:

$$\frac{\partial \varepsilon}{\partial t} + \nabla \cdot \left(\varepsilon \boldsymbol{v}_f\right) = 0 \tag{6.1}$$

Volume balance for the mixture:

$$\nabla \cdot \boldsymbol{q} = 0, \quad \text{with} \quad \boldsymbol{q} = \varepsilon \boldsymbol{v}_f \quad \text{and} \quad \boldsymbol{v}_s = 0 \tag{6.2}$$

Linear momentum balance for the fluid:

$$\nabla p_e = -\frac{\boldsymbol{m}_d}{\varepsilon} \tag{6.3}$$

where r is the position vector, $\varepsilon(\boldsymbol{r}, t) = (1 - \varphi(\boldsymbol{r}, t))$ is the porosity of the bed, $\boldsymbol{v}_f(\boldsymbol{r}, t)$ is the fluid velocity, $\boldsymbol{q}(\boldsymbol{r}, t)$ is the volume average velocity, also called *spatial fluid velocity* or *percolation velocity*, $p_e(\boldsymbol{r}, t)$ is the excess pore pressure and $\boldsymbol{m}_d(\boldsymbol{r}, t)$ is the solid–fluid interaction force.

6.1.2 Constitutive Equation for the Solid–Fluid Interaction Force

It is reasonable to assume that the solid–liquid interaction force, originating from the fluid flow through the porous bed, is a function of the bed porosity $\varepsilon(\boldsymbol{r}, t)$ and the solid–fluid relative velocity $\boldsymbol{v}_r(\boldsymbol{r}, t)$:

$$\boldsymbol{m}_d = \boldsymbol{m}_d(\varepsilon, \boldsymbol{v}_r) \tag{6.4}$$

The most general isotropic function for the hydrodynamic force $\boldsymbol{m}_d$:

$$\boldsymbol{m}_d = -\alpha(\varepsilon, |\boldsymbol{v}_r|)\boldsymbol{v}_r \tag{6.5}$$

The coefficient of resistance $\alpha(\varepsilon, |\boldsymbol{v}_r|)$ can be expanded in a Taylor series:

$$\alpha(\varepsilon, |\boldsymbol{v}_r|) = \alpha_0(\varepsilon) + \alpha_1(\varepsilon|\boldsymbol{v}_r|) + \alpha_2(\varepsilon)|\boldsymbol{v}_r|^2 + \cdots \tag{6.6}$$

6.1.3 Darcy's Law

If the relative solid–fluid velocity is low, it is possible to conserve only the first term in the series (6.6), so that:

$$\boldsymbol{m}_d = -\alpha_0(\varepsilon)\boldsymbol{v}_r \tag{6.7}$$

Expression (6.7) is called *Darcy's law,* and corresponds to a linear relationship for the hydrodynamic force. Substituting Darcy's law with the linear momentum equation, we obtain:

$$\nabla p_e = \frac{\alpha(\varepsilon)}{\varepsilon}\boldsymbol{v}_r \tag{6.8}$$

6.1.4 Forcheimer's Law

If the relative solid–fluid velocity is high, it is necessary to use another term for the series (6.6):

$$\boldsymbol{m}_d = -(\alpha_0(\varepsilon) + \alpha_1(\varepsilon)|\boldsymbol{v}_r|)\boldsymbol{v}_r \tag{6.9}$$

Equation (6.9) is called *Forcheimer's law*. Substituting it with the balance of linear momentum, the following result is obtained:

$$\nabla p_e = \left(\frac{\alpha_0(\varepsilon)}{\varepsilon} + \frac{\alpha_1(\varepsilon)}{\varepsilon}|\boldsymbol{v}_r|\right)\boldsymbol{v}_r \tag{6.10}$$

This equation represents a force balance in a porous medium for greater fluid velocities.

Geometrical parameters in a rigid porous bed

To obtain the relationship between the α_i parameters of the geometry of the porous bed, Eq. (6.10) is applied to the well-known case of the flow of a viscous fluid through a horizontal rigid porous bed with constant porosity ε. See Fig. 6.1, where $\boldsymbol{v}_f = v_f\boldsymbol{k}$, $\boldsymbol{q} = q\boldsymbol{k}$ and $p_e = p$, so that $\nabla p_e = (\partial p/\partial z)\boldsymbol{k}$.

For the case of Fig. 6.1, the field equations become:

$$\frac{\partial \varepsilon}{\partial t}+\frac{\partial}{\partial z}(\varepsilon v_f) = 0 \quad \text{and} \quad \frac{\partial q}{\partial z} = 0, \quad \text{with} \quad q = \varepsilon \boldsymbol{v}_f \tag{6.11}$$

$$\frac{\partial p_e}{\partial z} = \left(\frac{\alpha_0(\varepsilon)}{\varepsilon}+\frac{\alpha_1(\varepsilon)}{\varepsilon}|v_r|\right)v_r \tag{6.12}$$

Since the porosity ε is constant, and $v_s = 0$, from (6.11) we see that $q = \varepsilon v_f =$ constant and since $v_r = -v_f$, $q = -\varepsilon v_r$.

The linear momentum equation becomes:

$$\frac{\partial p_e}{\partial z} = -\left(\frac{\alpha_0(\varepsilon)}{\varepsilon^2}+\frac{\alpha_1(\varepsilon)}{\varepsilon^3}|q|\right)q \tag{6.13}$$

Integrating this equation with the following boundary conditions, $p_e(z=0) = p_0$ and $p_e(z=L) = p_L$, where $\Delta p = p_0 - p_L$, we obtain:

$$\int_{p_0}^{p_L} dp_e = -\left(\frac{\alpha_0(\varepsilon)}{\varepsilon^2}+\frac{\alpha_1(\varepsilon)}{\varepsilon^3}|q|\right)q\int_0^L dz$$

$$p_L - p_0 = -\Delta p_e = -\left(\frac{\alpha_0(\varepsilon)}{\varepsilon^2}+\frac{\alpha_1(\varepsilon)}{\varepsilon^3}|q|\right)qL \tag{6.14}$$

Since $q > 0$, we can eliminate the absolute value in Eq. (6.14) and write:

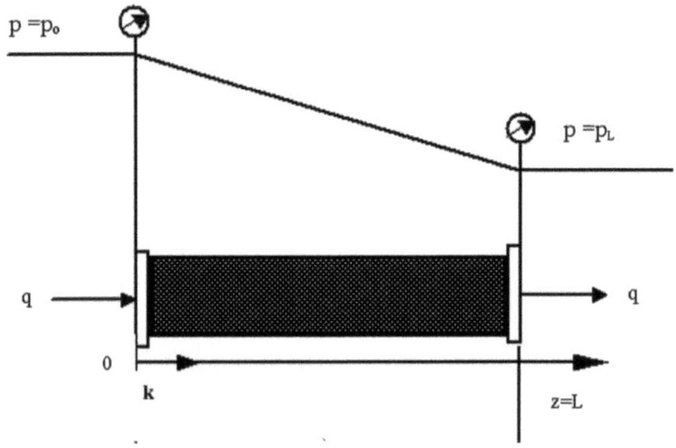

Fig. 6.1 Measurement of parameters in a *horizontal* rigid porous bed

$$\frac{1}{q}\frac{\Delta p_e}{L} = \frac{\alpha_0(\varepsilon)}{\varepsilon^2} + \frac{\alpha_1(\varepsilon)}{\varepsilon^3} q \tag{6.15}$$

Plotting Eq. (6.15) with q as the abscissa and $\Delta p_e / qL$ as ordinate, we get a straight line. See Fig. 6.2.

The extrapolation of the straight line to $q = 0$ and the slope of the straight line are:

$$a(\varepsilon) = \frac{1}{q}\frac{\Delta p}{L}\bigg|_{q \to 0} = \frac{\alpha_0(\varepsilon)}{\varepsilon^2}. \quad b(\varepsilon) = \frac{\alpha_1(\varepsilon)}{\varepsilon^3} \tag{6.16}$$

It is common practice to define the following parameters (Massarani 1984):

$$a(\varepsilon) = \frac{\alpha_0(\varepsilon)}{\varepsilon^2} \equiv \frac{\mu}{k(\varepsilon)} \quad \text{and} \quad b(\varepsilon) = \frac{\alpha_1(\varepsilon)}{\varepsilon^3} \equiv \frac{\rho_f c(\varepsilon)}{\sqrt{k(\varepsilon)}} \tag{6.17}$$

where $k(\varepsilon)$ is the permeability of the porous medium, measured in Darcy's (1 Darcy $= 10^{-3}$ cm^2), and $c(\varepsilon)$ is a dimensionless parameter. See Fig. 6.2.

With definitions (6.17), Eqs. (6.13) and (6.15) become:

$$\frac{\partial p_e}{\partial z} = -\left(\frac{\mu}{k(\varepsilon)} + \frac{\rho_f c(\varepsilon)}{\sqrt{k(\varepsilon)}}|q|\right) q \tag{6.18}$$

$$\frac{1}{q}\frac{\Delta p_e}{L} = \frac{\mu}{k(\varepsilon)} + \frac{\rho_f c(\varepsilon)}{\sqrt{k(\varepsilon)}} q \tag{6.19}$$

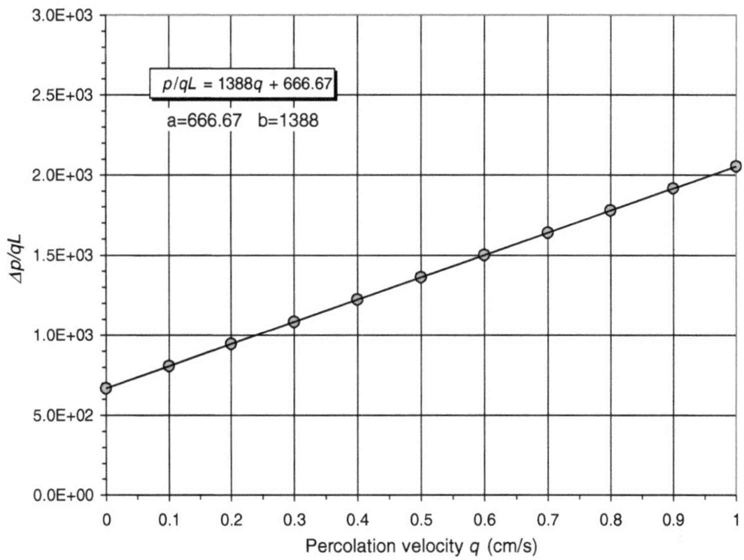

Fig. 6.2 Measurements of the geometrical parameters of a rigid porous bed

Problem 6.1 Determine the parameters of the straight line in Fig. 6.2.

The parameters are $a = 666.67$ and $b = 1{,}388$. With these values the permeability "k" and the parameter "c" are:

$$k = \frac{\mu}{a} = \frac{0.001}{666.67} = 1.5 \times 10^{-6}\ \mathrm{cm}^2$$

$$c = b \times \frac{k^{0.5}}{\rho_f} = \frac{1{,}388 \times (1.5 \times 10^{-6})^{0.5}}{1} = 1.7$$

Problem 6.2 (Massarani 1984) The length and cross-sectional area of a porous media, formed by consolidated sand of 14/20 mesh and sphericity $\psi = 0.6$, with a porosity of $\varepsilon = 0.37$, are $L = 2.1$ cm and 16.8 cm^2 respectively. Determine $k(\varepsilon)$ and $c(\varepsilon)$. Experiments for the flow of water, with 1.00 g/cm^3 in density and 0.0118 (g/cm-s) in viscosity through the porous media give the results shown in the table below:

q (cm/s)	6.33	7.47	10.2	12.7	15.2	17.7	20.3	23.9
Δp_e (cm Hg)	4.69	6.24	10.4	15.2	21.2	28.0	35.9	48.9

The units for the pressure drop must be changed:

$$p_e = \rho_{Hg}\, gh = 13.595 \times 980.7 \times h\,(\mathrm{g/cm\ s^2})$$

where h is measured in cm of Hg. The following table gives the data for the Fig. 6.3.

From the plot: $$\frac{\Delta p_e}{qL} = 1{,}709.8 + 469.6q$$

from which, the parameters can be calculated:

$$k(\varepsilon) = \frac{\mu}{(\Delta p/qL)_{q=0}} = \frac{0.0118}{1{,}697.8} = 6.90 \times 10^{-6}\ \mathrm{cm}^2$$

$$c(\varepsilon) = \sqrt{k(\varepsilon)} \times \left(\frac{\Delta p/q^2L}{\rho}\right) = \frac{6.90 \times 10^{-6} \times 469.69}{1.00} = 1.23$$

Massarani (1989) gives an order of magnitude of the parameters of diverse porous media in Table 6.1.

Several equations have been proposed to define the dependence of the permeability and the c parameter on the porosity. The most commonly used are those of Kozeny (1927), Carman (1937), Ergun (1952) and Massarani (1997) (see following section).

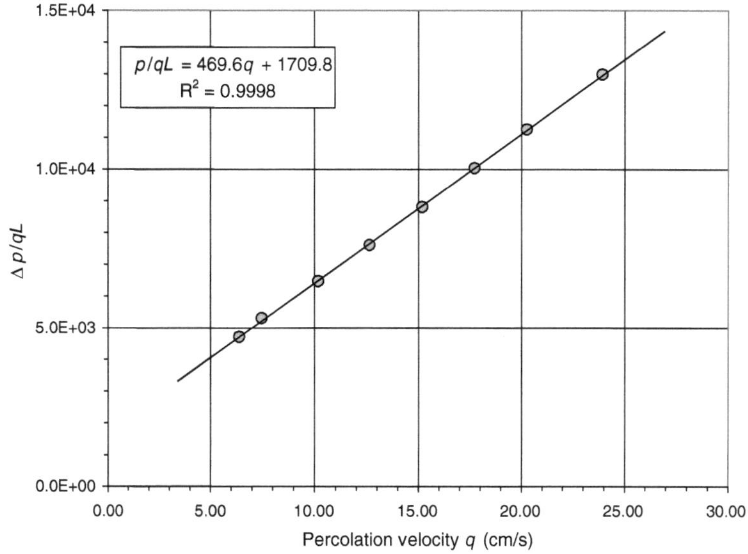

Fig. 6.3 Measurements of the geometrical parameters of a rigid porous bed formed by 14/20 Tyler mesh particles with sphericity $\psi = 0.6$

6.1.5 Forcheimer's and Darcy's Equations

Substituting the values of $\alpha_0(\varepsilon)$ and $\alpha_1(\varepsilon)$ from (6.17) with (6.9) and (6.10) we get:

$$\boldsymbol{m}_d = \frac{\mu\varepsilon}{k(\varepsilon)}\left(1 + \frac{\rho_f\sqrt{k(\varepsilon)}c(\varepsilon)|\boldsymbol{q}|}{\mu}\right)\boldsymbol{q} \tag{6.20}$$

$$\nabla p_e = -\frac{\mu}{k(\varepsilon)}\left(1 + \frac{\rho_f c(\varepsilon)\sqrt{k(\varepsilon)}|\boldsymbol{q}|}{\mu}\right)\boldsymbol{q} \tag{6.21}$$

This last expression is called *Forcheimer's equation.*

The second term in the parenthesis of (6.20) and (6.21) has the form of a Reynolds Number. Defining the *Percolation Reynolds Number* Re^* by:

Table 6.1 Order of magnitude of $k(\varepsilon)$ and $c(\varepsilon)$ (Massarani 1984)

Medium	ε	k (ε) (cm²)	c(ε) (–)
Petroleum sand	0.03	1.0×10^{-12}	3×10^5
Porous metallic plate	0.26	1.0×10^{-7}	15
Copper concentrate	0.43	5.0×10^{-9}	
28/35 mesh quartz sand	0.42	1.5×10^{-6}	1.7
6 mm glass beads	0.40	4.0×10^{-4}	0.49

$$\mathrm{Re}^* = \frac{\rho_f c(\varepsilon)\sqrt{k(\varepsilon)}|\boldsymbol{q}|}{\mu} \tag{6.22}$$

we can write Eqs. (6.20) and (6.21) in the form:

$$\boldsymbol{m}_d = \frac{\mu\varepsilon}{k(\varepsilon)}(1 + Re^*)\boldsymbol{q} \tag{6.23}$$

$$\nabla p_e = -\frac{\mu}{k(\varepsilon)}(1 + Re^*)\boldsymbol{q} \tag{6.24}$$

The Percolation Reynolds Number measures the deviation of the solid–liquid interaction force in Darcy's law. In effect, if $\mathrm{Re}^* \ll 1$, Eq. (6.24) becomes *Darcy's equation*:

$$\nabla p_e = -\frac{\mu}{k(\varepsilon)}\boldsymbol{q} \tag{6.25}$$

Problem 6.3 From the results of Problem 6.1, decide if it corresponds to a Darcy's or Forcheimer's regime. Calculating the Percolation Reynolds Numbers with Eq. (6.22), we obtain the results of Fig. 6.4.

6.1.6 Darcy's and Forcheimer's Equations in Terms of the Piezometric Height

It is sometimes convenient to write Darcy's and Forcheimer's equations in terms of the piezometric height and hydraulic conductivity instead of the excess pore pressure and the percolation velocity. Since $p_e = \rho_f g(h - h_0)$, where $\rho_f g h_0$ is the hydrostatic pressure and h_0 is the hydrostatic head, from (6.24) we have:

$$\nabla h = -\frac{\mu}{\rho_f g k(\varepsilon)}(1 + Re^*)\boldsymbol{q}$$

Defining the *hydraulic conductivity* $i(\varepsilon)$, measured in (cm/s), by:

$$i(\varepsilon) = \frac{\rho_f g k(\varepsilon)}{\mu} \tag{6.26}$$

Darcy's and Forcheimer's equations become:

$$\nabla h = -\frac{1}{i(\varepsilon)}(1 + Re^*)\boldsymbol{q}; \quad \text{with} \quad Re^* = \frac{\rho_f c(\varepsilon)\sqrt{k(\varepsilon)}|\boldsymbol{q}|}{\mu} \tag{6.27}$$

$$\nabla h = -\frac{1}{i(\varepsilon)}\boldsymbol{q} \tag{6.28}$$

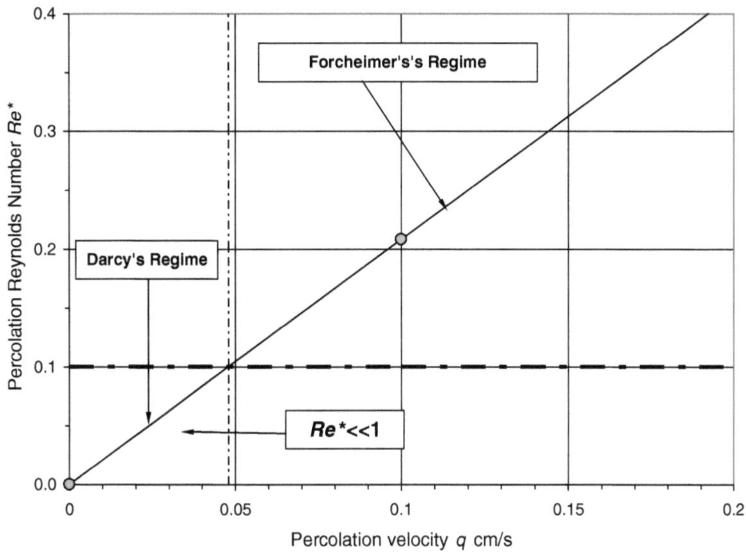

Fig. 6.4 Percolation Reynolds Numbers Re^* versus percolation velocity q for Problem 6.3, showing that Darcy's regime is valid up to around percolation velocity $q = 0.05$ cm/s

6.1.7 Capillary Model of a Rigid Porous Bed

The *capillary model* for a porous bed, also known as the *Kozeny–Carman model*, is based on the assumption that the capillaries consist of a bundle of tubes of arbitrary cross-section. Through these capillaries, a fluid flows following Poiseuille's equation:

$$\nabla p_e = -\frac{\mu\beta}{R_h^2}\bar{\boldsymbol{v}}$$

where R_h is the hydraulic radius, defined as the ratio of the wetted cross-sectional area to the wetted perimeter of the tubes, β is a parameter called *tortuosity*, which is related to the length of the capillaries per unit length of the porous media and $\bar{\boldsymbol{v}}$ is the average velocity in the tubes. This velocity corresponds to the interstitial velocity in de porous medium. In terms of the percolation velocity, $\boldsymbol{q} = \varepsilon\bar{\boldsymbol{v}}$, we have:

$$\nabla p_e = -\frac{\mu\beta}{\varepsilon R_h^2}\boldsymbol{q} \tag{6.29}$$

The hydraulic radius can be written in the form:

$$R_h = \frac{wetted\,cross - sectional\,area}{wetted\,perimeter} \times \frac{L}{L} \times \frac{V}{V} = \frac{(wetted\,cross - sectional\,area \times L)/V}{(wetted\,perimeter \times L)/V}$$
$$= \frac{pore\,volume/bed\,volume}{particle\,surface/bed\,volume}$$
$$= \frac{\varepsilon}{S_p/V} = \frac{\varepsilon}{S_p(1-\varepsilon)/V(1-\varepsilon)} = \frac{\varepsilon}{(1-\varepsilon)S_p/V_p} = \frac{\varepsilon}{(1-\varepsilon)\bar{S}}$$

where $\bar{S}$ is the specific surface per unit volume. Substituting this expression with Eq. (6.29) yields:

$$\nabla p_e = -\frac{\mu(1-\varepsilon)^2\beta\bar{S}^2}{\varepsilon^3}\boldsymbol{q}$$

Comparing this equation to (6.25) gives the following functional form for the permeability:

$$k(\varepsilon) = \frac{\varepsilon^3}{(1-\varepsilon)^2\beta\bar{S}^2} \tag{6.30}$$

If the particles forming the bed are spherical, the specific surface is $\bar{S} = 6/d$ and the permeability is:

$$k(\varepsilon) = \frac{\varepsilon^3 d_p^2}{36(1-\varepsilon)^2\beta} \tag{6.31}$$

where d is the diameter of the sphere.

If the bed is formed with non-spherical particles, the sphericity can be used to characterize the particle shape (see Chap. 4). Sphericity is the ratio of the surface of a sphere to the surface of the particle, both having the same volume $V_e \equiv V_p$. The diameter of this sphere d_e is the *volume-equivalent diameter*:

$$\psi = \frac{S_e}{S_p} = \frac{S_e/V_e}{S_p/V_p} = \frac{6/d_e}{\bar{S}}, \tag{6.32}$$

therefore

$$\bar{S} = \frac{6}{\psi d_e}$$

and the permeability can now be written as:

$$k(\varepsilon) = \frac{\varepsilon^3 d_e^2\psi^2}{36(1-\varepsilon)^2\beta} \tag{6.33}$$

When the particles forming the bed have a size distribution, it is possible to use the average surface-volume diameter $\bar{x}_{12}$ to characterize the size of the particles. According to the definition, the specific surface $\bar{S}$ of the particulate system is:

$$\bar{S} = \frac{\alpha_S \int x^2 f(x)dx}{\alpha_V \int x^3 f(x)dx} = \frac{\alpha_S}{\alpha_V} \frac{\int (1/x) f(x)dx}{\int f(x)dx} = \frac{\alpha_S}{\alpha_V} \int \frac{f(x)}{x} dx = \frac{\alpha_S}{\alpha_V} \frac{1}{\bar{x}_{12}} \tag{6.34}$$

$\bar{x}_{12}$ is also called permeability-size. The parameters α_S and α_V are the particles volume and surface shape factors. For this case, permeability is:

$$k(\varepsilon) = \frac{\varepsilon^3 \bar{x}_{12}^2}{\beta (1-\varepsilon)^2 (\alpha_S / \alpha_V)^2} \tag{6.35}$$

Other models for the permeability and the c parameter are (Table 6.2):

Kozeny–Carman's model (1952) for 0.35 < ε 0.45:

$$\mathrm{k}(\varepsilon) = \frac{\varepsilon^3 (\mathrm{d_e} \psi)^2}{170(1-\varepsilon)^2} \tag{6.36}$$

Ergun's model (1952) for 0.35 < ε 0.45:

$$\text{for } 0.35 < \varepsilon < 0.45: \quad \mathrm{c}(\varepsilon) = \frac{0.143}{\varepsilon^{3/2}} \tag{6.37}$$

Massarani's model (1997):

$$\mathrm{c}(\varepsilon) = \frac{1}{\varepsilon^{3/2}} \left\{ 0.13 \left(\frac{\mathrm{k_0}}{\mathrm{k}} \right)^{0.37} + 0.10 \left(\frac{\mathrm{k_0}}{\mathrm{k}} \right)^{0.01} \right\}^{0.98}, \quad \mathrm{k_0} = 10^{-6} (\mathrm{cm}^2) \tag{6.38}$$

Problem 6.4 For the data of Problem 6.2, calculate the permeability and the c parameter with the equations of Kozeny–Carman, Ergun and Massarani.

The volume-equivalent diameter for particles of $\bar{x}_m$ mesh, is $d_e = 0.986\, \bar{x}_m$. For the 14/20 Tyler mesh range, the average size is $\bar{x}_m = \sqrt{850 \times 1{,}180} = 1{,}002$ μm, therefore, $d_e = 0.986 \times 1{,}002 = 987.5$ μm

Kozeny–Carman:

$$k(\varepsilon) = \frac{\varepsilon^3 d_e^2 \psi^2}{36(1-\varepsilon)^2 \beta} = \frac{(0.37)^3 \times (987.5 \times 10^{-4})^2 \times (0.6)^2}{36 \times (1-0.37)^2 \times 5} = 2.5 \times 10^{-6}\ \mathrm{cm}^2$$

Table 6.2 Values for the β parameter

Type of conduct	β	Reference
Circular	2	Becker (1963)
Elliptical	2.00–2.46	Becker (1963)
Triangular	1.67	Becker (1963)
Bed of particles with porosities in the range of $0.3 \le \varepsilon_0 \le 0.5$	5	Coulson and Richardson (1968)

Ergun:

$$k(\varepsilon) = \frac{\varepsilon^3 (d_e \psi)^2}{150(1-\varepsilon)^2} = \frac{(0.37)^3 \times (987.5 \times 10^{-4})^2}{150 \times (1-0.37)^2} = 8.3 \times 10^{-6}\ \text{cm}^2$$

$$c(\varepsilon) = \frac{0.143}{\varepsilon^{3/2}} = \frac{0.143}{(0.37)^{3/2}} = 0.635$$

Massarani:

$$c(\varepsilon) = \frac{1}{(0.37)^{3/2}} \left\{ 0.13 \left(\frac{10^{-6}}{8.3 \times 10^{-6}} \right)^{0.37} + 0.10 \left(\frac{10^{-6}}{8.3 \times 10^{-6}} \right)^{0.01} \right\}^{0.98} = 2.27$$

6.1.8 Dynamic Process for a Rigid Porous Bed

The stationary flow of a Newtonian fluid through a rigid porous bed is represented by the following field variables: porosity $\varepsilon(\boldsymbol{r}, t)$, percolation velocity $\boldsymbol{q}(\boldsymbol{r}, t)$ and excess pore pressure $p_e(\boldsymbol{r}, t)$. These variables constitute a *dynamic process* if the following field equations are satisfied:

$$\varepsilon(\boldsymbol{r}, t) = \varepsilon_0 \quad \text{and} \quad \boldsymbol{q}(\boldsymbol{r}, t) = \boldsymbol{q}_0 \tag{6.39}$$

$$\nabla p_e = -\frac{\mu}{k(\varepsilon_0)} (1 + Re^*) \boldsymbol{q}_0 \tag{6.40}$$

where $Re^* = \rho_f c(\varepsilon_0) \sqrt{k(\varepsilon_0)} |\boldsymbol{q}_0| / \mu$ is the Percolation Reynolds Number and $k(\varepsilon)$ and $c(\varepsilon)$ are the porous bed characteristic parameters: permeability and resistance parameter, while ρ_f and μ are the fluid density and viscosity respectively.

If the flow is slow, Eq. (6.40) reduces to:

$$\nabla p_e = -\frac{\mu}{k(\varepsilon_0)} \boldsymbol{q}_0. \tag{6.41}$$

6.2 Dynamic Process of a Two-Phase Flow Through a Rigid Porous Bed

The joint flow of a liquid and a gas through a porous bed is of great importance in several industrial processes such as filtration, heap leaching and water seepage through soils. In this section, we will analyze the flow of water and air through a rigid porous bed, such as during the dehumidification step in filtration.

Consider a mixture of water and air flowing through a rigid immobile porous bed. The four assumptions of Sect. 6.1 are valid. Call ε the constant *porosity* of the bed and s the *saturation*, that is, the fraction of pore volume filled with water. Then, the fraction containing air in a pore volume is $(1-s)$. The apparent densities of the solid, water and air are respectively: $\rho_s(1-\varepsilon)$, $\rho_w \varepsilon s$ and $\rho_a \varepsilon(1-s)$, where ρ_s, ρ_w and ρ_a are the solid, water and air material densities. Then, the local mass balances are given by:

For water:

$$\frac{\partial}{\partial t}(\rho_w \varepsilon s) + \nabla \cdot (\rho_w \varepsilon s \boldsymbol{v}_x) = 0 \tag{6.42}$$

For air:

$$\frac{\partial}{\partial t}(\rho_a \varepsilon(1-s)) + \nabla \cdot (\rho_a \varepsilon(1-s)\boldsymbol{v}_a) = 0 \tag{6.43}$$

Since the porous bed is incompressible, $\boldsymbol{v}_s = 0$. Adding Eqs. (6.42) and (6.43) the continuity equation of the fluid mixture is obtained:

$$\frac{\partial \rho}{\partial t} + \nabla \cdot (\rho \boldsymbol{v}) = 0$$

where the fluid mixture density ρ and mass average velocity are given by

$$\rho = \rho_w \varepsilon s + \rho_a \varepsilon(1-s) \tag{6.44}$$

$$\rho \boldsymbol{v} = \rho_w \boldsymbol{q}_w + \rho_a \boldsymbol{q}_a \tag{6.45}$$

where

$$\boldsymbol{q}_w = \varepsilon s \boldsymbol{v}_w \text{ and } \boldsymbol{q}_a = \varepsilon(1-s)\boldsymbol{v}_a \tag{6.46}$$

The local momentum balances for the water and air component are:

For the water:

$$\rho_w \varepsilon s \dot{\boldsymbol{v}}_w = \nabla \cdot \boldsymbol{T}_w + \rho_w \varepsilon s \boldsymbol{g} - \boldsymbol{m}_w \tag{6.47}$$

For the air:

$$\rho_a \varepsilon(1-s)\dot{\boldsymbol{v}}_a = \nabla \cdot \boldsymbol{T}_a + \rho_a \varepsilon(1-s)\boldsymbol{g} - \boldsymbol{m}_a \tag{6.48}$$

where $\dot{\boldsymbol{v}}_\alpha$, $\boldsymbol{T}_\alpha$ and $\boldsymbol{b}_\alpha$ are the rate of change of linear momentum, the stress tensor, and the body forces, respectively, and where $\boldsymbol{m}_w$ and $\boldsymbol{m}_a$ are the interaction forces between components.

Consider slow flows at a steady state, so that the convective terms are negligible. We will also consider that the friction within the fluid is much less than the friction between the fluids and the solid skeleton. With these assumptions, Eqs. (6.47) and (6.48) reduce to:

$$\nabla p_w = \rho_w \varepsilon s \boldsymbol{g} - \boldsymbol{m}_w \tag{6.49}$$

$$\nabla p_a = \rho_a \varepsilon (1 - s) \boldsymbol{g} - \boldsymbol{m}_a \tag{6.50}$$

Adding these to expressions, the linear momentum of the fluid is obtained:

$$\nabla p_t = \rho \boldsymbol{g} - \boldsymbol{m} \tag{6.51}$$

where $\rho = \rho_w \varepsilon s + \rho_a \varepsilon (1 - s)$ is the fluid mixture density, $\boldsymbol{m} = (\boldsymbol{m}_w + \boldsymbol{m}_a)$ is the interaction force that the mixture of fluid exerts on the solid skeleton and $p_f = p_a + p_w$ is the total pressure.

6.2.1 Constitutive Equations for the Pressures

The pressures p_w and p_a are variables associated with the respective components if considered as occupying the whole volume of the mixture of fluids. Consequently, they are not measurable variables. Calling p the pore pressure of the liquid and p_A the measurable air pressure, the following relationship should be valid:

$$p_W = \varepsilon s p \quad y \quad p_a = \varepsilon (1 - s) p_A \tag{6.52}$$

$$p_f = \varepsilon p_A - \varepsilon s (p_A - p) \tag{6.53}$$

When three phases meet in a capillary tube, such as a liquid, that wets the solid and a gas that does not, generates a force called the *capillary force* in the interface among the three phases, which force depends on the interfacial tensions of the components.

The pressure difference between the gaseous and liquid phases is called *capillary pressure,* which depends on the liquid surface tension and the radius of curvature of the meniscus. The capillary pressure is given by the Young–Laplace equation (Dullien 1992):

$$p_c = p_A - p = \gamma_{\ell g} \cos\theta \left(\frac{1}{r_1} + \frac{1}{r_2} \right) \tag{6.54}$$

where $\gamma_{\ell g}$ is the surface tension of the liquid and r_1 and r_2 are the radius of curvature of the meniscus.

As an example, place the bottom of four capillary tubes of different inside diameter in a pool of liquid that wets the solid surface. The liquid will climb through the tubes at different heights, as shown in Fig. 6.5a.

For a capillary, the radius of curvature is $d/2$ and Eq. (6.54) reduces to:

$$p_c = \gamma_{\ell g}(4/d) \cos\theta \tag{6.55}$$

where θ is the contact angle. The magnitude of the capillary force is:

$$f_c = \left(\pi d^2/4\right) \times p_c = \pi d \gamma_{\ell g} \cos\theta \tag{6.56}$$

The height at which the liquid rises in the tube depends on the force balance between the capillary and the gravitational force:

$$\begin{aligned} \rho_\ell \frac{\pi d^2}{4} h &= \pi d \gamma_{\ell g} \cos\theta \\ h &= \frac{4\gamma_{\ell g} \cos\theta}{\rho_\ell d} \end{aligned} \tag{6.57}$$

Therefore, the liquid rises higher in the finer capillaries. See Fig. 6.5a.

If the submerged capillary tubes are taken out of the liquid pool and placed horizontally on an air stream under an air pressure gradient $\Delta p/L$, see Fig. 6.5a, b force $(\pi d^2/4)(\Delta p/L)$ will start displacing the water in the tubes. If the air force is greater than the capillary force:

$$\begin{aligned} \left(\pi d^2/4\right) \times (\Delta p/L) &> \pi d \gamma_{\ell g} \cos\theta \\ \Delta p &\geq \frac{4L\gamma_{\ell g} \cos\theta}{d} \end{aligned} \tag{6.58}$$

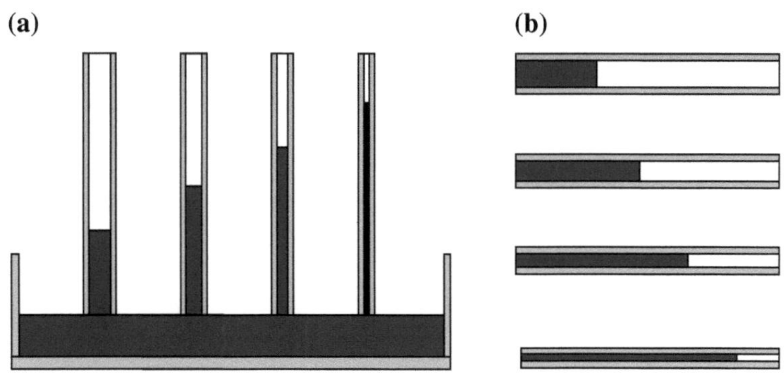

Fig. 6.5 Capillary rise in tubes: **a** *Vertical* tubes; **b** *Horizontal* tubes

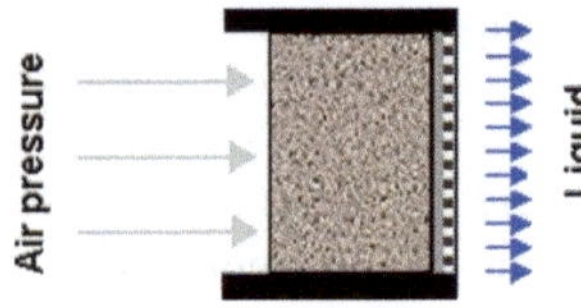

Fig. 6.6 Water displaced from a rigid porous medium

a volume flow rate of water Q will flow out of the tubes according to Poiseuille's law. As long as the inequality holds, the air will be displaced the water from the tubes. This will happen first in the tubes with greater diameter. In tubes where inequality (6.58) becomes an equality, that is where the diameter is $d = 4L\gamma_{\ell g}\cos\theta/\Delta p$, there will be no liquid motion and the tube will remain saturated. If the air pressure is further increased, the diameter of the tubes that remain saturated will decrease.

Going back to a porous medium, consider the capillary model in which the medium is formed by a bundle of randomly distributed capillaries with different diameters. If the porous medium is saturated with water and is subjected to an air pressure gradient, the water will flow out of the capillaries, depending on their diameter and on the liquid properties, according to inequality (6.58). The air will displace the water from the larger capillaries, but the smaller ones will remain saturated. See Fig. 6.6.

For a given air pressure drop, there will be a determined amount of water retained in the medium. Only an increase in air pressure will removed part of it. For the very small capillaries, there is no pressure gradient sufficient to evacuate the water, so that will always be *residual saturation* in a porous medium. The residual saturation is denoted by s_∞.

For a porous medium, the capillary force is a function of saturation, in addition to the variables already mentioned. Figure 6.7 shows such functional dependency.

6.2.2 Constitutive Equations for the Resistance Force

It is well known that the interaction force between a solid and a fluid is much greater than that among particles of the same fluid, which suggests that the constitutive equations for the solid–fluid interaction force are only functions of porosity, saturation and the relative solid–fluids velocities. Since the solid is stationary, the relative solid-fluid velocity is the velocity of the fluid. Thus:

$$\boldsymbol{m}_\ell(|\boldsymbol{q}_\ell|, \varepsilon, s) = \left(\frac{\alpha_0(\varepsilon, s)}{\varepsilon^2} + \frac{\alpha_1(\varepsilon, s)}{\varepsilon^3}|\boldsymbol{q}_\ell|\right)\boldsymbol{q}_\ell \tag{6.59}$$

$$\boldsymbol{m}_a(|\boldsymbol{q}_a|, \varepsilon, s) = \left(\frac{\beta_0(\varepsilon, s)}{\varepsilon^2} + \frac{\beta_1(\varepsilon, s)}{\varepsilon^3}|\boldsymbol{q}_a|\right)\boldsymbol{q}_a \tag{6.60}$$

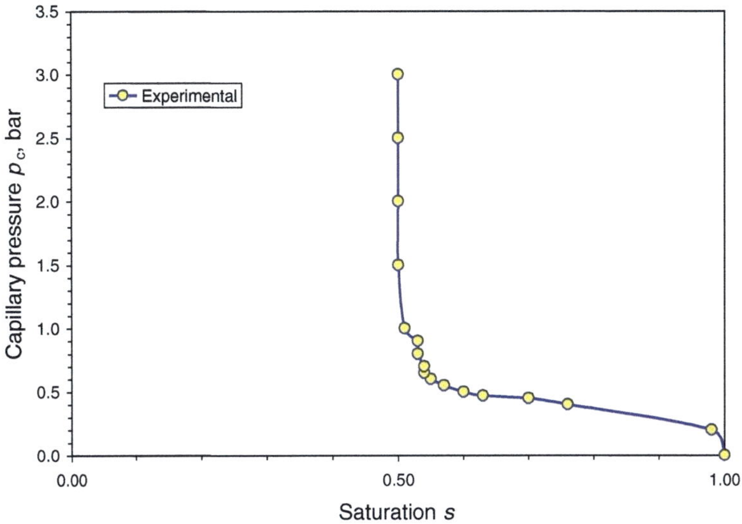

Fig. 6.7 Capillary pressure versus saturation for a copper concentrate

Substituting the expressions (6.52), (6.59) and (6.60) with Eqs. (6.49) and (6.50) results in the water and air pressure gradients:

$$\nabla(\varepsilon s p) = \rho_\ell \varepsilon s \boldsymbol{g} - \left(\frac{\alpha_0(\varepsilon, s)}{\varepsilon^2} + \frac{\alpha_1(\varepsilon, s)}{\varepsilon^3}|\boldsymbol{q}_\ell|\right)\boldsymbol{q}_\ell \tag{6.61}$$

$$\nabla(\varepsilon(1-s)p_a) = \rho_a \varepsilon(1-s)\boldsymbol{g} - \left(\frac{\beta_0(\varepsilon, s)}{\varepsilon^2} + \frac{\beta_1(\varepsilon, s)}{\varepsilon^3}|\boldsymbol{q}_a|\right)\boldsymbol{q}_a \tag{6.62}$$

6.2.3 Percolation in a Non-saturated Porous Medium

Percolation in a porous medium is governed by gravity and is a slow process. Therefore the gradient of the porous pressure and the quadratic terms in q can be neglected in Eq. (6.61):

$$\varepsilon p \nabla s = \rho_\ell \varepsilon s \boldsymbol{g} - \frac{\alpha_0(\varepsilon, s)}{\varepsilon^2}\boldsymbol{q}_\ell \tag{6.63}$$

The liquid volume velocity is then:

$$\boldsymbol{q}_\ell = -\frac{1}{\alpha_0(\varepsilon, s)}\left(\varepsilon^3 p \nabla s - \rho_\ell \varepsilon^3 s \boldsymbol{g}\right) \tag{6.64}$$

Defining the *hydraulic conductivity* $C(\varepsilon, s)$ and the *hydraulic diffusivity* $D(\varepsilon, s)$ in the form:

$$C(\varepsilon, s) = \frac{\rho_\ell \varepsilon^3 sg}{\alpha_0(\varepsilon, s)} \quad \text{y} \quad D(\varepsilon, s) = \frac{\varepsilon^3 p}{\alpha_0(\varepsilon, s)} \tag{6.65}$$

and writing $\boldsymbol{g} = -g\boldsymbol{k}$, the percolation velocity becomes:

$$\boldsymbol{q}_\ell = C(\varepsilon, s)\boldsymbol{k} - D(\varepsilon, s)\nabla s \tag{6.66}$$

This equation is known in the scientific literature as the *Darcy–Buckingham* equation (Massarani 1997).

6.2.4 Pressure Flow Through a Non-saturated Porous Medium

When the pressure gradient is more important than the gradient of saturation, $\partial p/\partial z \gg \partial s/\partial z \, and \, \partial p_a/\partial z \gg \partial s/\partial z$, we can neglect the latter from the momentum Eqs. (6.61) and (6.62). The result is:

$$\nabla p = \rho_\ell \boldsymbol{g} - \left(\frac{\alpha_0(\varepsilon, s)}{s\varepsilon^3} + \frac{\alpha_1(\varepsilon, s)}{s\varepsilon^4}|\boldsymbol{q}_\ell|\right)\boldsymbol{q}_\ell \tag{6.67}$$

$$\nabla p_a = \rho_a \boldsymbol{g} - \left(\frac{\beta_0(\varepsilon, s)}{\varepsilon^3(1 - s)} + \frac{\beta_1(\varepsilon, s)}{\varepsilon^4(1 - s)}|\boldsymbol{q}_a|\right)\boldsymbol{q}_a \tag{6.68}$$

Defining the *relative permeabilities* $k_\ell(\varepsilon, s) \, y \, k_a(\varepsilon, s)$ and the *dimensionless parameters* $c_{r\ell}(\varepsilon, s) \, y \, c_{ra}(\varepsilon, s)$, in each component:

$$k_l(\varepsilon, s) = \frac{\mu_l/k(\varepsilon)}{\alpha_0(\varepsilon, s)/\varepsilon^3 s} \quad \text{and} \quad k_\alpha(\varepsilon, s) = \frac{\mu_l/k(\varepsilon)}{\beta_0(\varepsilon, s)/\varepsilon^3(1 - s)} \tag{6.69}$$

$$c_l(\varepsilon, s) = \frac{\sqrt{k(\varepsilon)}/\rho_l}{\beta_0(\varepsilon, s)/\varepsilon^4 s} \quad \text{and} \quad c_\alpha(\varepsilon, s) = \frac{\sqrt{k(\varepsilon)}/\rho_\alpha}{\beta_1(\varepsilon, s)/\varepsilon^4(1 - s)} \tag{6.70}$$

where $k(\varepsilon)$ is the permeability of the porous medium, which depends exclusively on the structure of the porous matrix and is independent of the fluid and the flow regime. With these definitions and considering that the air pressure is much greater than the weight of the air in the porous medium, we can write:

$$\nabla p = \rho_\ell \boldsymbol{g} - \left(\frac{\mu_\ell}{k(\varepsilon)k_\ell(\varepsilon s)} + \frac{c_\ell(\varepsilon, s)\rho_\ell}{\sqrt{k(\varepsilon)}}|\boldsymbol{q}_\ell|\right)\boldsymbol{q}_\ell \tag{6.71}$$

$$\nabla p_a = \left(\frac{\mu_a}{k(\varepsilon)k_a(\varepsilon s)} - \frac{c_a(\varepsilon, s)\rho_a}{\sqrt{k(\varepsilon)}} |\boldsymbol{q}_a| \right) \boldsymbol{q}_a \tag{6.72}$$

If the flow is slow, $\rho_i c_i(\varepsilon, s)\sqrt{k(\varepsilon)}|\mathbf{q}|/\mu_i \ll 1$, Eqs. (6.71) and (6.72) become:

$$\nabla p = \rho_\ell \boldsymbol{g} - \frac{\mu_\ell}{k(\varepsilon)k_\ell(\varepsilon, s)} \boldsymbol{q}_\ell \quad \nabla p_a = -\frac{\mu_\ell}{k(\varepsilon)k_a(\varepsilon, s)} \boldsymbol{q}_a \tag{6.73}$$

According to their definition, relative permeability can be determined by measuring the respective fluid flow in non-saturated and saturated media. Massarani (1997) measured the permeabilities of water in different types of soils, and found that the relative permeabilities were independent of the soil properties and could be written for all cases in terms of just saturation:

$$k_\ell(s) = 1.36 \times 10^{-6} \times \exp(13.6\,\mathrm{s}) \tag{6.74}$$

Problem 6.5 Laboratory experiments of water flow through an incompressible porous medium, with an area of 6.55 cm^2 and 3.8 cm in thickness, gave a porosity of $\varepsilon = 0.52$ and a permeability of $k(\varepsilon) = 2.906 \times 10^{-10}$cm^2. Air with a viscosity of 2.1×10^{-2} mPas-s was blown at 6 (bars) for 121 (s) to displace water with a density of 1.00 g/cm^3 and 1.2 mPas-s viscosity. Volumes of water and air are given in Table 6.3.

Integrating Eqs. (6.71) and (6.72) for a given time leads to:

$$\frac{\Delta p}{L} = -\frac{\mu_w}{k(\varepsilon)k_w(t)} q_w(t) = -\frac{\mu_a}{k(\varepsilon)k_a(t)} q_a(t) \tag{6.75}$$

and from these equations:

$$k_w(t) = \frac{L}{\Delta p} \frac{\mu_\ell}{k(\varepsilon)S} Q_w(t) \quad k_a(t) = \frac{L}{\Delta p} \frac{\mu_\ell}{k(\varepsilon)S} Q_a(t) \tag{6.76}$$

Using the data from Table 6.3, values of the relative permeabilities were calculated and are plotted in Fig. 6.8.

6.2.5 Reduced and Residual Saturation

The *reduced* saturation s_r is defined by:

$$s_r = \frac{s - s_\infty}{1 - s_\infty} \tag{6.77}$$

The reduced saturation is correlated to the capillary pressure in the form:

Table 6.3 Data of two-phase flows in a porous medium and relative permeabilities

t	s	$V_f(t)$	$Q_w = dV_f/dt$ $(cm)^3/s$	Q_a $(cm)^3/s$	k_w	k_a
0.00	1.000	0.001	0.0000	0.000	1.000	0.000
2.03	0.943	0.523	0.1035	8.168	0.321	0.493
4.06	0.923	0.666	0.0514	10.530	0.159	0.635
6.09	0.913	0.754	0. 0376	11.980	0.117	0.722
8.13	0.895	0.824	0.0313	12.986	0.097	0.783
10.16	0.878	0.883	0.0272	13.715	0.084	0.827
12.19	0.870	0.935	0.0243	14.263	0.075	0.860
14.22	0.863	0.982	0.0220	14.684	0.068	0.886
16.26	0.853	1.025	0.0202	15.013	0.063	0.905
18.29	0.848	1.065	0.0187	15.269	0.058	0.921
20.32	0.841	1.101	0.0175	15.472	0.054	0.933
22.35	0.834	1.136	0.0164	15.633	0.051	0.943
24.38	0.826	1.168	0.0155	15.761	0.048	0.950
26.42	0.827	1.199	0.0147	15.864	0.045	0.957
28.45	0.820	1.228	0.0139	15.947	0.043	0.962
30.48	0.819	1.255	0.0133	16.014	0.041	0.966
32.51	0.819	1.282	0.0127	16.070	0.039	0.969
34.55	0.819	1.307	0.0122	16.117	0.038	0.972
36.58	0.812	1.331	0.0117	16.156	0.036	0.974
38.61	0.812	1.355	0.0113	16.191	0.035	0.976
40.64	0.812	1.377	0.0109	16.223	0.034	0.978
42.67	0.806	1.399	0.0105	16.252	0.033	0.980
44.71	0.806	1.420	0.0102	16.280	0.031	0.982
46.74	0.806	1.440	0.0098	16.308	0.031	0.983
48.77	0.801	1.460	0.0095	16.336	0.030	0.985
50.80	0.795	1.479	0.0093	16.365	0.029	0.987
52.84	0.794	1.498	0.0090	16.395	0.028	0.989
54.87	0.788	1.516	0.0088	16.426	0.027	0.991
56.90	0.788	1.533	0.0086	16.459	0.027	0.993
58.93	0.783	1.551	0.0084	16.493	0.026	0.995
60.96	0.781	1.567	0.0082	16.529	0.025	0.997
63.00	0.783	1.584	0.0080	16.567	0.025	0.999
65.03	0.783	1.600	0.0078	16.606	0.024	1.000
67.06	0.776	1.615	0.0076	16.646	0.024	1.000
69.09	0.776	1.631	0.0075	16.688	0.023	1.000
71.13	0.776	1.646	0.0073	16.731	0.023	1.000
73.16	0.769	1.661	0.0072	16.774	0.022	1.000
75.19	0.769	1.675	0.0071	16.819	0.022	1.000
77.22	0.763	1.689	0.0069	16.863	0.022	1.000
79.26	0.765	1.703	0.0068	16.908	0.021	1.000
81.29	0.758	1.717	0.0067	16.952	0.021	1.000
83.32	0.758	1.730	0.0066	16.995	0.020	1.000

(continued)

Table 6.3 (continued)

t	s	$V_f(t)$	$Q_w = dV_f/dt$ (cm)^3/s	Q_a (cm)^3/s	k_w	k_a
85.35	0.752	1.744	0.0065	17.037	0.020	1.000
87.38	0.752	1.757	0.0064	17.079	0.020	1.000
89.42	0.752	1.770	0.0063	17.118	0.020	1.000
91.45	0.752	1.783	0.0062	17.155	0.019	1.000
93.48	0.747	1.795	0.0061	17.190	0.019	1.000
95.51	0.748	1.807	0.0061	17.222	0.019	1.000
97.55	0.748	1.820	0.0060	17.251	0.019	1.000
99.58	0.748	1.832	0.0059	17.276	0.018	1.000
101.61	0.748	1.844	0.0058	17.297	0.018	1.000
103.64	0.748	1.856	0.0058	17.313	0.018	1.000
105.67	0.743	1.867	0.0057	17.324	0.018	1.000
107.71	0.741	1.879	0.0057	17.330	0.018	1.000
109.74	0.743	1.890	0.0056	17.330	0.017	1.000
111.77	0.743	1.902	0.0055	17.324	0.017	1.000
113.80	0.737	1.913	0.0055	17.312	0.017	1.000
115.84	0.737	1.924	0.0054	17.292	0.017	1.000
117.87	0.737	1.935	0.0054	17.264	0.017	1.000
119.90	0.737	1.946	0.0054	17.229	0.017	1.000
120.56	0.732	1.949	0.0053	17.216	0.017	1.000

$$s_r = 1 - \left\{1 + \left(\frac{p_c}{p_{50}}\right)^{\delta}\right\}^{-1}, \quad \lambda < 0 \tag{6.78}$$

where p_{50} is the value of the capillary pressure for $s_r = 0.50$ and in terms of p_{50} the saturation is written in the form:

$$s = s_\infty + (1 - s_\infty) \times s_r \equiv s_\infty + (1 - s_\infty) \times \left(1 - \left\{1 + \left(\frac{p_c}{p_{50}}\right)^{\delta}\right\}^{-1}\right) \tag{6.79}$$

where $\delta = \ln(0.25/0.75)/\ln(p_{25}/p_{50})$.

Problem 6.6 Increasing pressure was incrementally applied to a saturated porous medium containing 34.3 cm^3 of water while the percolated water was collected. The results are shown in Table 6.4 and in Fig. 6.9.

From the Fig. 6.9, $s_\infty = 0.466$, and from the Table 6.4, $p_{25} = 0.965$ and $p_{50} = 0.740$ $\delta = \ln(0.25/0.75)/\ln(0.965/0.740) = -4.148$.

$$s_r = 1 - \left(1 + \left(\frac{p_c}{0.74}\right)^{-4.138}\right) \tag{6.80}$$

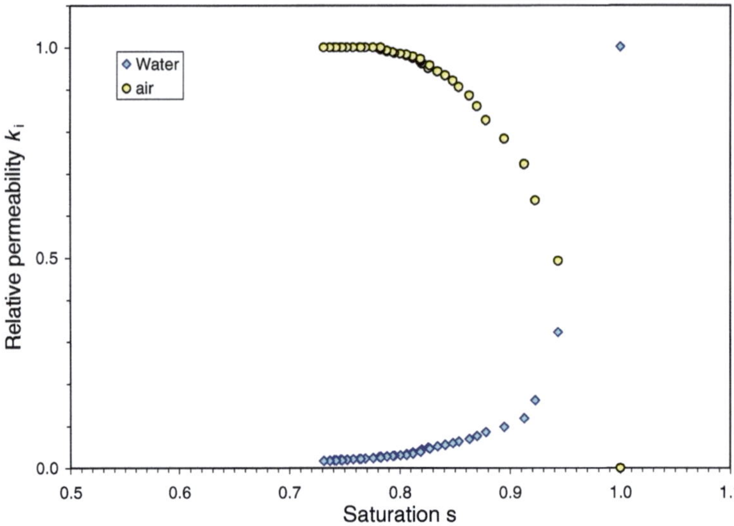

Fig. 6.8 Relative permeabilities of water and air

Table 6.4 Percolation in a porous medium

pc (bars)	$V_{wpercolated}$ (cm^3)	V_w (cm^3)	s	$s_r = (s - s_{oo})/(1 - S_{oo})$
0.0	0.00	34.30	1.000	1.00000
0.2	0.50	33.80	0.985	0.97272
0.3	0.29	33.51	0.977	0.95690
0.4	0.30	33.21	0.968	0.94053
0.5	3.67	29.54	0.861	0.74032
0.6	0.32	29.22	0.852	0.72286
1.0	6.86	22.36	0.652	0.34861
2.0	5.70	16.66	0.486	0.03764
4.0	0.69	15.97	0.466	0.00000

$$s = 0.466 + 0.534 \times \left\{ 1 - \left(1 + \left(\frac{p_c}{0.74} \right)^{-4.148} \right) \right\} \tag{6.81}$$

Figure 6.10 shows a plot of Eq. (6.81).

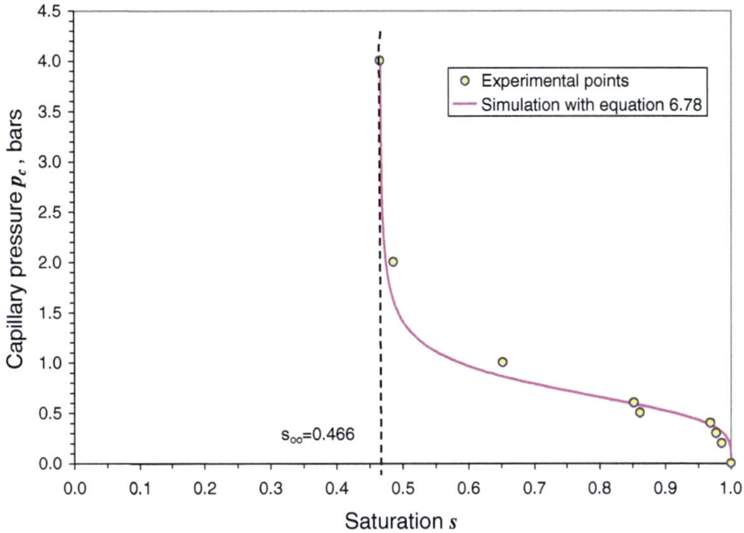

Fig. 6.9 Capillary pressure versus saturation. Experimental points from Table 6.4 and the simulation from Eq. (6.77)

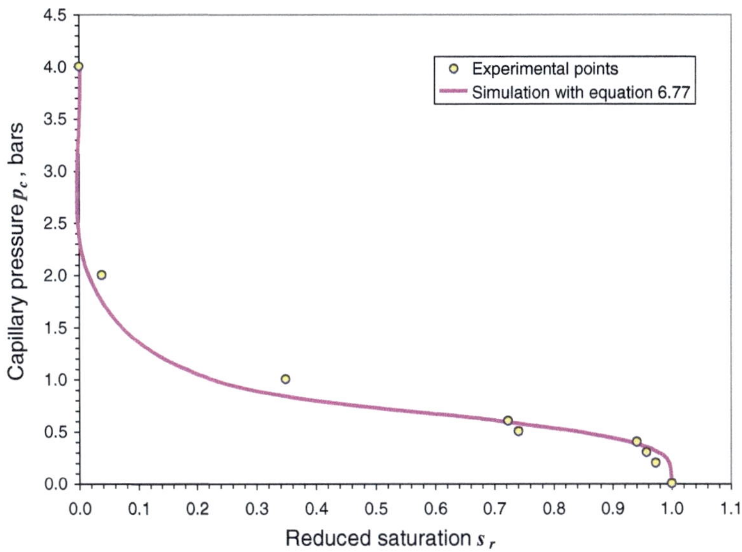

Fig. 6.10 Capillary pressure versus reduced saturation. Experimental point from Table 6.4 and simulation from Eq. (6.78).

References

Becker, R. (1963). Integration des equations du mouvement d'un fluide viscqueaux incompressible, Hanbuch der Physik VIII, Springer Verlag.

Carman, P. C. (1937). *Transactions Institute of Chemical Engineers, London*, *15*, 150.

Coulson, J. M. & Richardson, J. F. (1964). Chemical engineering, (Vol. I), A Pergamon Press Book, New York: MacMillan.

Darcy, H. (1856). *Les Fontaines Publiques de la Ville de Dijon*. París: Dalmont.

Dullien, F. A. L. (1992). *Porous media: Fluid transport and pores structure* (2nd ed, p. 121). New York: Academic Press, INC.

Ergun, S. (1952). Fluid flow through packed columns. *Chemical Engineering Progress, 48*, 89.

Kozeny, J. (1927). *Sitzber. Abad. Wiss. Wien, Math. Naturw. Klasse*, *136* (Abt.Iia), 271.

Massarani, G. (1984). *Problemas em sistemas particulados*. Río de Janeiro: Ed. Edgard Blucher Ltda.

Massarani, G. (1989). Aspectos da fluidodinámica em meios porosos. *Revista brasileira de Engenharía, Número Especial*, 96 pp.

Massarani, G. (1997). Fuid dynamics of particulate systems, Ed. UFR, Rio de Janeiro, Brazi. (in Portuguese)

Chapter 7
Particle Aggregation by Coagulation and Flocculation

Abstract This chapter considers particle aggregation. When agglomerated particles in a suspension increase in size they acquire greater sedimentation velocity essential to obtain a good separation by sedimentation. Two methods for increasing the size of solid particles are studied in this chapter, coagulation by reducing inter-particle electrostatic repulsion and flocculation by bridging particles with polymeric agents. Most mineral particles suspended in water in the neutral pH range have negative surface charges. Positive ions in solution are attracted and adsorbed at the negatively charged surface forming the so-called double layer with its Stern plane and the diffuse layer. The Zeta potential measures the difference in the electrical potential of the charged surface and the bulk of the solution in commercial instruments. If particle surfaces come close together, they attract each other by van der Waals force. If there is no counteracting force, the particles will coagulate and settle out of the suspension. The study of orthokinetic coagulation follows. It is generally accepted that polymers used as flocculants in mineral processing plants aggregate fine particle suspensions by bridging mechanisms. Such bridging links the particles into loose flocs and incomplete surface coverage, which ensures that there is sufficient unoccupied surface available on each particle for adsorption during collisions of chain segments attached to the particles. The description of flocs as fractal objects permits a better understanding of their behavior. Flocculation kinetics shows that a short and highly intensive mixing gives the best results for particle aggregation.

7.1 Introduction

It is common knowledge that particle size in a suspension plays an important role in mineral processing. Concentration results are strongly dependent on the size of the treated particles and there is an optimum size range where recovery is optimal. Recovery is low with particles outside this optimal range. Because it is necessary to reduce particle to achieve proper liberation, over-grinding is unavoidable.

F. Concha A, *Solid–Liquid Separation in the Mining Industry*,
Fluid Mechanics and Its Applications 105, DOI: 10.1007/978-3-319-02484-4_7,

Unfortunately, concentrates or tailings dewatering by thickening and filtration is difficult when treating particles that are very fine. In those cases, size enlargement by particle aggregation is beneficial.

Agglomerated particles in a suspension increase in size and acquire a greater sedimentation velocity essential to obtain good separation by sedimentation. The agglomerates also form more permeable cakes accelerating filtration processes.

Different methods are available to increase the size of solid particles. The most common are *coagulation*, by reduction of inter-particle electrostatic repulsion, *flocculation* by bridging particles with water soluble polymeric agents called flocculants and *oil agglomeration*, in which an oily hydrocarbon is used to agglomerate hydrophobic particles suspended in water, hence producing hydrophobic agglomerates.

7.2 Coagulation

Most mineral particles suspended in water in the range of neutral pH possess negative surface charge, as shown schematically in Fig. 7.1. Positive ions in solution are attracted and adsorb at the negatively charged surface. The adsorbed layer remains rigidly attached and forms what is known as the *Stern plane* or *Stern layer.* Outside the Stern layer, there is a *diffuse layer* in which positive ions outnumber negative ones and balance the excess charge in the Stern layer. The electrostatic potential at the near particle surface decreases linearly through the Stern layer and then exponentially through the diffuse layer reaching zero at the bulk of the solution. The Stern and diffuse layers are called the *double layer* and are important in determining surface forces acting among particles. When the solution flows along the particle, it becomes stagnant at a certain distance from the particle surface. This distance is called the *shear plane*. The difference of electrical potential between the shear plane and the bulk of the solution is called *Zeta Potential*, which can be measured experimentally with several commercial instruments.

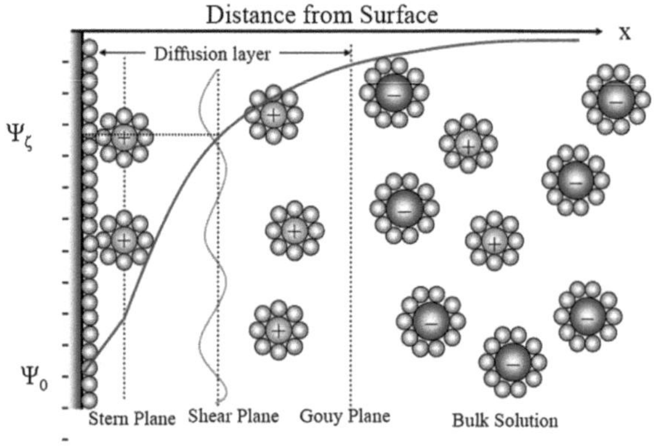

Fig. 7.1 Electrical potential distribution near a particle surface submerged in an electrolyte

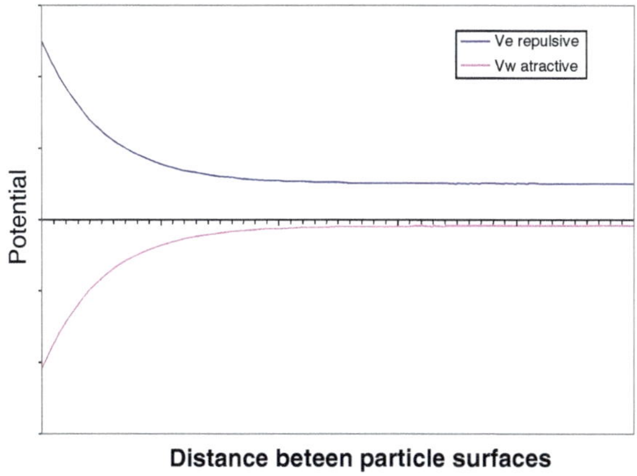

Fig. 7.2 Schematic representation of potential particle interaction energy. Ve represents ion-electrostatic and Vw represents London-Van der Waals molecular or dispersion components

Particle interactions in a solvent depend on surface forces. The most universal components of them are long-range ionic electrostatic repulsive forces and short-range London-van der Waals attractive forces. Figure 7.2 shows the distribution of the potential energy of particle interaction.

The combination of molecular dispersive and ion-electrostatic forces occurring when the diffuse parts of the electrical double layers of two particles overlap, defines the stability of the suspension. Electrostatic repulsive forces are responsible for particle stability by establishing a barrier to prevent particle surfaces from coming sufficiently close to each other for van der Waals attractive forces to act.

If particle surface come close together, they attract each other by van der Waals force. If there is no counteracting force, the particles will *coagulate*, that is, they will aggregate and settle out of the suspension. The stability of fine hydrophobic particles in a suspension is described by the DLVO theory (Derjagin-Landau-Verwey-Overbeek). Figure 7.3 gives the energy of particle interactions versus distance between two particle surfaces and the different cases of attraction and repulsion that can occur.

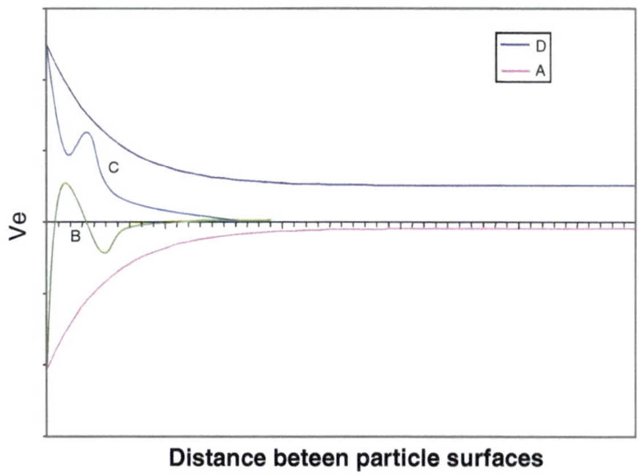

Fig. 7.3 Energy versus inter-particle surface distance, showing four types of interaction depending on the ratio of magnitude of repulsive forces: *A* Rapid coagulation, *B* Slow primary and secondary coagulation. *C* Coagulation with energy barrier and *D* Stable dispersion

In case A, particles only experience attractive forces and fall directly in the primary minimum potential; the system is totally unstable and particles coagulate rapidly and irreversibly. In case B, repulsion produces sufficiently a high potential barrier V_{max} to prevent particles from coagulating spontaneously, but permits a slow coagulation in the secondary minimum potential if $V_{min} < kT$. These *coagula* can be easily re-dispersed by introducing some agitation or increasing the temperature of the system. Curve C represents a sufficiently high potential to prevent coagulation. Curve D represents a system with such high level of repulsive forces that coagulation cannot occur under any circumstance, resulting in a stable dispersed system. V_{max} and V_{min} barriers can be lowered by adding electrolytes, which increase the ionic strength of the solution by diminishing the thickness of the electrical double layer or diminishing the surface charge by adding counter-ions.

In general, mineral particles suspended in water acquire negative charge. Therefore, adding multivalent ions, such as Al^{3+}, Fe^{3+}, Fe^{2+}, Ca^{2+} o Mg^{2+} form coagula by neutralizing the surface charge or compressing the double layer, diminishing its thickness, permitting particles to come enough close together for attractive forces to act. Coagulation conditions can be quantified through the *Zeta potential* ζ, see Fig. 7.1. This parameter measures the magnitude of the electrical potential at the shear plane of the electrical double layer (measured, for example, by electrophoresis), which is proportional to the particle surface potential. The potential at which particles begin to coagulate is known as *critical Zeta potential.* It has been empirically found that the ionic concentration at which particles flocculate is proportional to z^{-6}, where z is the ion valence. This relationship is known as the Schultze-Hardy rule.

Particle suspensions in the various stages of mineral processing are much larger than those of a colloidal system. The particle range for what are termed fines in these processes is from 1 to 10 microns. Therefore, the dominant forces are inertial and convective, which are produced by an agitator, a pump or the action of gravity. These flow conditions can impose enormous energy on the particles, permitting them to overcome the energy barriers of electrostatic repulsive forces without the necessity of neutralizing the charge. An excessively violent agitation can re-disperse the particles and coagulation never occurs.

The results of these two types of coagulation, by charge neutralization or by hydrodynamics overcoming the energy barrier, are different, and in the first case a neutral coagula is obtained while in the second the aggregate has an electric charge.

7.2.1 Coagulation Kinetics

Orthokinetic coagulation of colloidal systems is described by the Smoluchowki-Müller theory (Laskowski and Pugh 1992; Gregory 1986a, b). If the collision efficiency is represented by E_c, the rate of disappearance of particles n by coagulation in a suspension is given by:

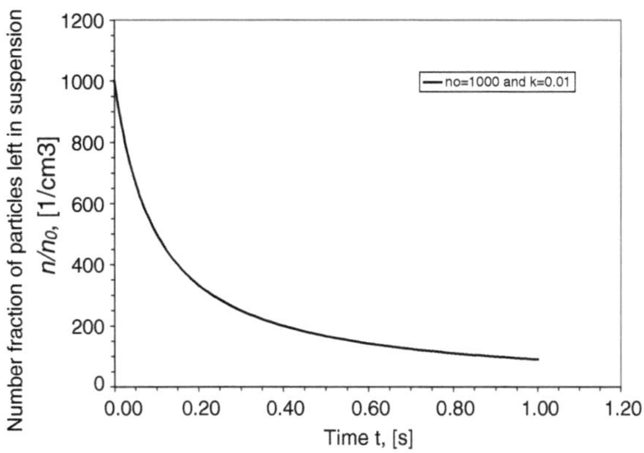

Fig. 7.4 Fraction of remaining particles in the dispersion after a time t

$$\frac{dn}{dt} = -k_c n^2 \quad \text{where} \quad k_c = \frac{4E_c kT}{3\mu} \tag{7.1}$$

where k is the Boltzman constant, μ is the liquid viscosity and T is the absolute temperature. The collision efficiency, with values between 0 and 1, depends on suspension stability, where $E_c = 1$ corresponds to a completely destabilized system. The fines in mineral processing are relatively large compared to colloidal particles. Therefore, the suspension can be considered completely destabilized with $E_c = 1$. Integrating Eq. (7.1) results in:

$$n = n_0 \frac{1}{1 + n_0 k_c t} \tag{7.2}$$

where n_0 is the initial particle concentration.

Problem 7.1 Calculate the fraction of particles remaining in suspension in a coagulation process if the initial particle concentration is $n_0 = 10{,}000\ \left(\text{particles/cm}^3\right)$ and the kinetic constant $k_c = 0.01\ \text{cm}^3/\text{s}$.

From Eq. (7.2) we get the results of Fig. 7.4.

7.3 Flocculation

The aggregation of particles from dispersions by adsorption of large polymer chains of several particles is called *flocculation*. There are several types of flocculation which depend on the way the polymers act on the particle surface. The most important type is *bridging flocculation* in which small quantities of large chain flocculant are adsorbed over many particles simultaneously, producing strong flocs. At higher flocculant concentrations than needed, they adsorb completely in one particle, leaving less possibility of simultaneous adsorption in other

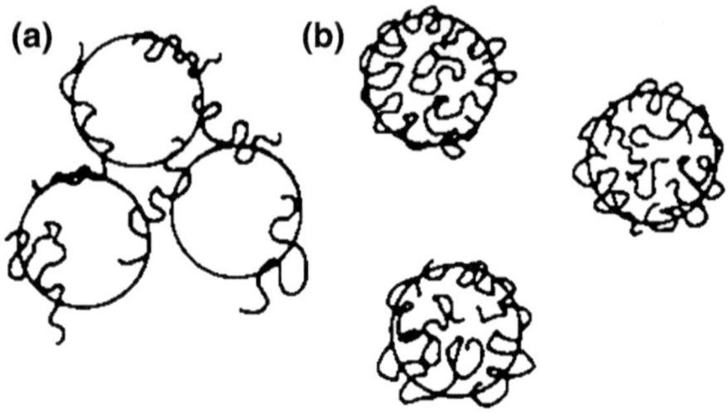

Fig. 7.5 Flocculation by molecular flocculant bridges. **a** Flocculation of several particles. **b** Re-stabilization by excess flocculant

particles, which re-stabilizes the suspension. Consequently, there are optimal doses of polymeric flocculants. If one or more linear polymers adsorbs on particles, a gel forms with a three-dimensional network. In this case, the aggregation is termed *network flocculation*. Strict control of linear or ramified flocculants can produce very compact flocs known as pellets. This type of flocculation is called *pellet flocculation*. See Figs. 7.5 and 7.6.

These three types of flocculation can occur with all types of flocculants, neutral, cationic and anionic. In the case of neutral flocculants, pellet flocculation is the only mechanism possible.

The action of cationic polymers on negatively charged particles is similar to coagulation, where surface charge neutralization predominates over formation of polymer bridges. This aggregation is called *electrostatic flocculation* and should be distinguished from bridging flocculation. The mining industry uses anionic flocculants more than cationic flocculants with negatively charged particles. In this case, the adsorption mechanism is covalent bonds or chemical reaction on the particle surface. This aggregation is called *flocculation by salt bond*. Bridging flocculation is the most important of these aggregation mechanisms.

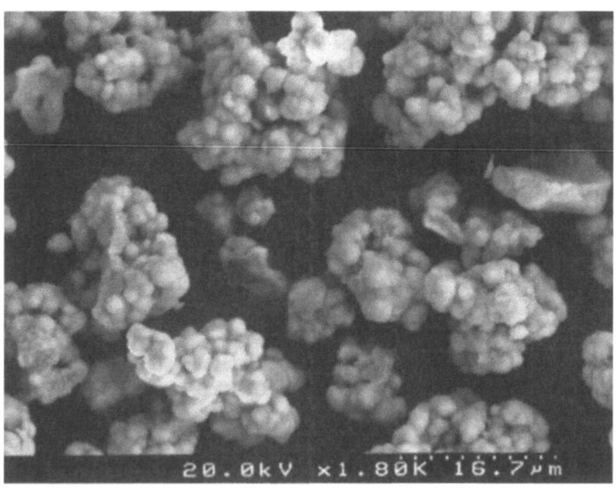

Fig. 7.6 Pellet flocculation (Glover et al. 2000)

It is generally accepted that polymers used as flocculants in mineral processing plants aggregate fine particle suspensions by bridging mechanisms. Bridging is considered the consequence of adsorption of segments of flocculant macromolecules on the surface of more than one particle. Such bridging links the particles into loose flocs and incomplete surface coverage, which ensures that there is sufficient unoccupied surface available on each particle for adsorption during collisions of chain segments attached to the particles. However, it is important to note that adsorption and flocculation are not separate sequential processes, but rather occur simultaneously. The merit of modern polymeric flocculants is their ability to produce larger and stronger flocs than those obtained by coagulation.

Theoretically, the flocculants can be applied either after destabilizing the suspension via coagulation or without prior destabilization. The flocculants are known to be not very effective in treating stable suspensions and so the first option, which involves prior destabilization by coagulation, is always better.

The type of solid, its surface charge and the electrolytes present determine the appropriate type of flocculant and flocculation.

7.3.1 Flocs as Fractal Objects

A fractal is an object that displays self-similarity on all scales, that is, the same type of structure appears at every scale. A plot of a property of the object versus its scale on a log–log graph gives a straight line, the slope of which is called its *fractal dimension.*

A floc is a fractal object because its basic form repeats itself as it grows. Figure 7.7 shows a computer generated floc formed by the successive aggregation of two primary particles.

Based on this basic form, Biggs et al. (2000) constructed a larger floc shown in Fig. 7.8.

It is customary in mineral processing to describe the size of mineral particles with the *volume equivalent diameter* d_e, that is, with the diameter of a sphere having the same volume as the particle. Flocs formed by aggregation of particles can be measured in a similar manner. The mass of a floc is proportional to its diameter d_F, called *mass fractal size,* elevated to the exponent D_m, called mass *fractal dimension,* or just *fractal size* (Fig. 7.9).

$$m(d_F) \propto d_F^{D_m} \tag{7.3}$$

Fig. 7.7 Floc formed by the successive aggregation of two particles

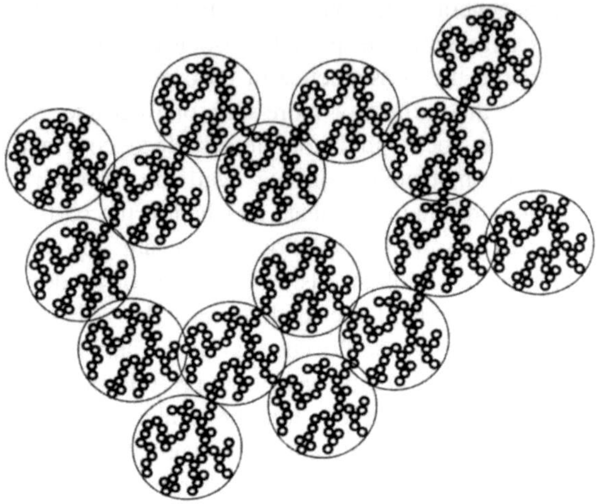

Fig. 7.8 Schematic representation of the fractal structure of flocs (Biggs et al. 2000)

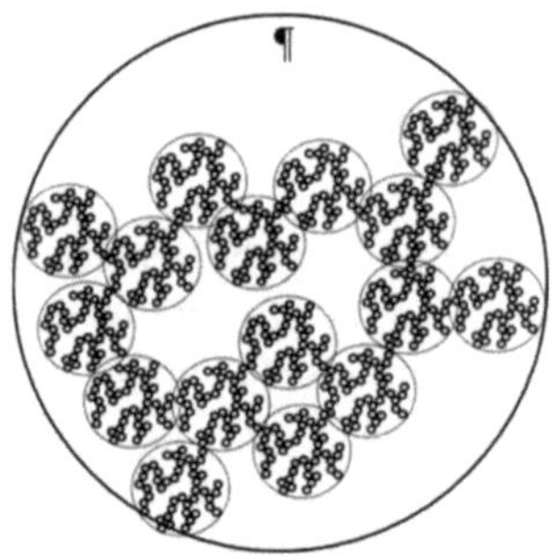

Fig. 7.9 Fractal diameter d_F of a floc

Figure 7.10 show, on the left, a computer simulation by Bushell (2005) of several flocs with different fractal dimensions D_m and, on the right, a photograph of flocs of copper flotation tailings.

The fractal dimension D_m of a floc varies in the range $1 < D_m < 3$ and gives a good indication of the compactness of the floccules. Small values of D_m indicate loose floccules while higher values indicate more compact floccules. The majority of floccules have sizes in the range of $1.7 < D_m < 2.5$. The flocs in Fig. 7.10b have an approximate fractal dimension of $D_m = 2$.

The fractal dimensions of a floc have a major effect on its properties. For example, the relationship between the number of primary particles n_{pF} in one floc, called the *aggregate number*, the single particle equivalent diameter d_p, the floc fractal diameter d_F and the fractal dimension D_m can be expressed by the empirical equation (Gregory 2009):

$$n_{pF} = (1.33 \pm 0.10) \times \left(\frac{d_F}{d_p}\right)^{D_m}, \tag{7.4}$$

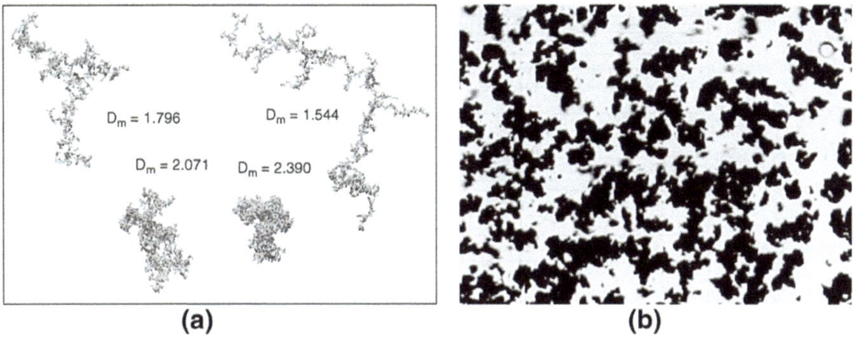

Fig. 7.10 Floc structure. **a** Structure with different mass fractal dimensions simulated by Bushell (2005). **b** Actual flocs of a copper flotation tailing (Concha and Segovia 2012)

From this expression, a floc containing a number of primary particles n_{pF}, has a dimensionless size d_F/d_p given by:

$$\frac{d_F}{d_p}=\left(\frac{n_{pF}}{1.33}\right)^{1/D_m} \tag{7.5}$$

Problem 7.2 Calculate the dimensionless floc size, for flocs with mass fractal dimensions of $1.5<D_m<3$ containing particles between $2\le n_{pF}\le 10.000$.

Figures 7.11 and 7.12 show plots of Eq. (7.5) with the mass fractal dimension as parameter and with the number of primary particles per floc as the parameter.

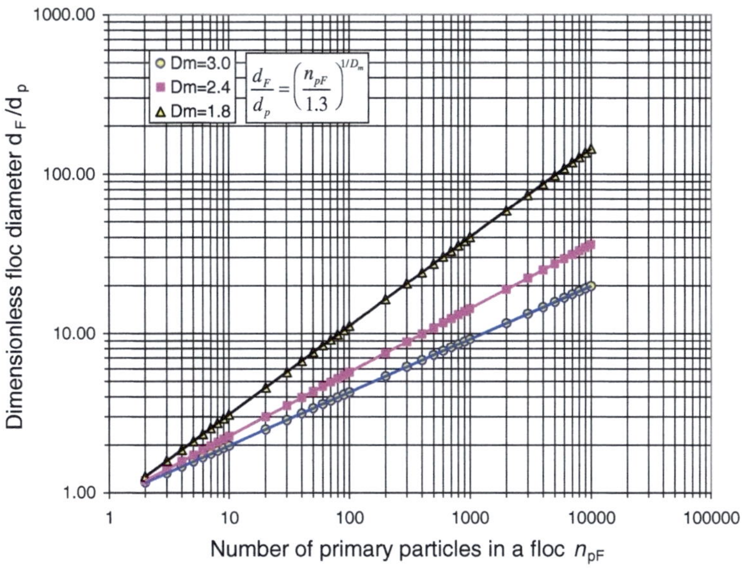

Fig. 7.11 Dimensionless floc sizes versus the number of primary particles per floc with the mass fractal dimension as a parameter

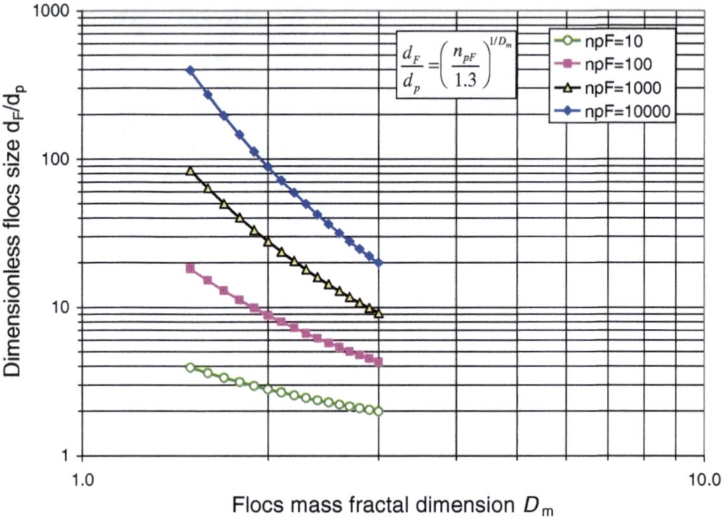

Fig. 7.12 Dimensionless floccules sizes versus mass fractal dimension for several primary particles per floc

7.3.2 Floc Size Measurement

The most commonly used experimental method for mass fractal dimension measurement uses the slope of the straight line obtained by plotting the mass of the aggregates versus the mass fractal size (Fig. 7.13).

In Gregory (2009) proposed light dispersion to determine the fractal mass dimension of an aggregate. If a light ray of magnitude Q is directed toward a

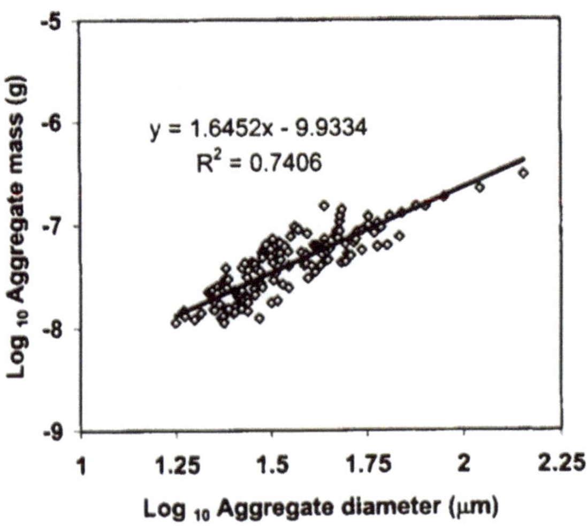

Fig. 7.13 Mass versus fractal diameter of a floc according to Glover et al. (2000), where $D_m = 1.6452$

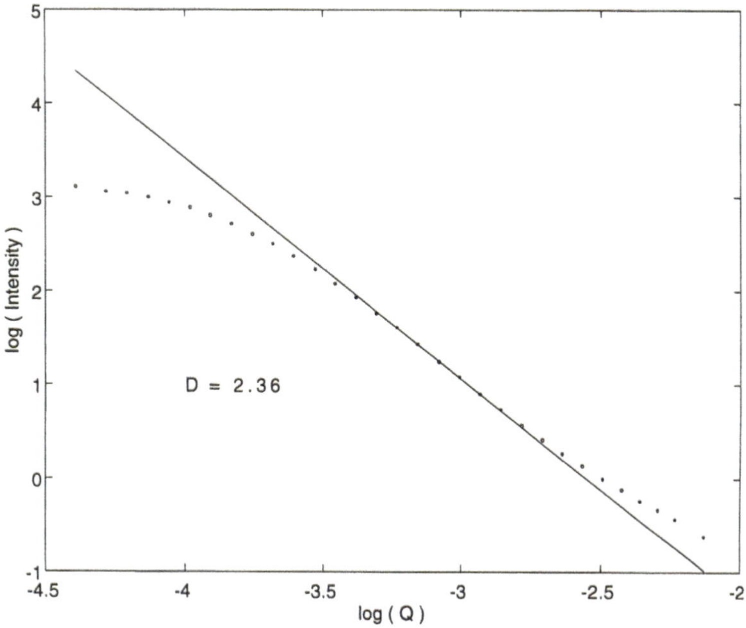

Fig. 7.14 Light intensity versus the magnitude q of the dispersion vector in logarithmic scales (Bushell et al. 1998)

sample of flocs, the dispersed light $I(Q)$ is proportional to the magnitude Q to the power of D_m:

$$I(Q) \propto q^{-D_m} \quad \text{with} \quad q = (1/\lambda_0) 4\pi v_0 \sin(\theta/2) \tag{7.6}$$

where v_0 is the dispersion medium refractive index λ_0 is the wavelength of the incident light and θ is the dispersion angle. Figure 7.14 gives an example where $D_m = 2.12$ is obtained from the slope of the straight-line region of the curve.

A more direct method of floc measurement is visualization with a video camera during the flocculation process (Wu et al. 1999). The images are digitalized and the particle size is measured by applying a statistical technique. Concha and Segovia (2012) installed a floc window, shown in Figs. 7.15 and 7.16, in an on-line

Fig. 7.15 Floc cell, AMIRA project P996

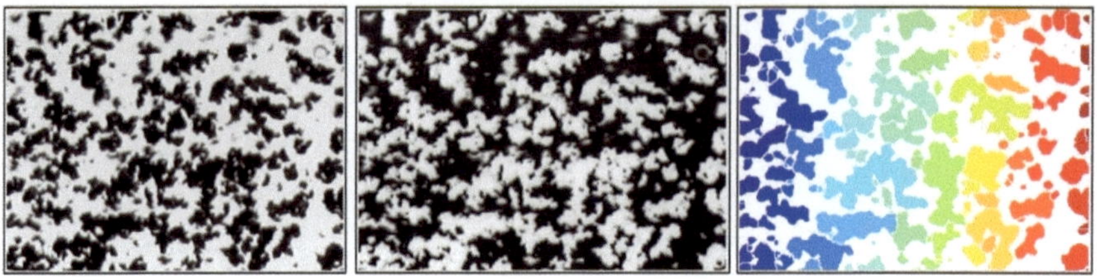

Fig. 7.16 Normal and digitalized images of flocculated copper tailings

instrument to determine thickening parameters in an industrial thickener. They obtained the following size analysis: $x_{10} = 10.30$; $x_{25} = 18.70$; $x_{50} = 31.45$; $x_{75} = 45.40$ and $x_{90} = 66.70\mu$, with an average of $\bar{x} = 41.00\mu$ and a standard deviation of $\sigma = 25.4$.

7.3.3 Orthokinetic Theory of Flocculation

The optimal flocculation condition for a given concentrate or tailing refers to the agitation time to obtain the highest settling velocity of the aggregates, that is, the greatest floc size and density. To obtain this information we must resort to flocculation kinetics.

Consider flocculation as the phenomena of the formation of aggregates of two or more primary particles in a suspension. These aggregates are called flocs. The aggregation occurs by collision of the primary particles in a hydrodynamic field and aggregation by adsorption of long chains of polymer molecules. The fractal form of the flocs suggests that all flocs grow in a similar way, although there is some evidence of re-arrangements among flocs (Selomulya et al. 2001). This fact permits analyzing the formation of flocs by studying the formation of one floc.

Since collision among particles is produced by mechanical energy in a shear field, it is essential that the method of producing the agitation permit the calculation of the shear rate of the flow. The magnitude of shear rate and the application time is important since, on the one hand, it improves the contact among particles to produce agglomeration and, on the other; the strong agitation destroys the flocs. The balance of these two effects is fundamental to obtain larger and denser flocs.

Traditionally it was thought that the shear rates $\dot{\gamma}$ during flocculation should be less than $\dot{\gamma} \approx 100\,(\mathrm{s}^{-1})$ to avoid rupturing the formed flocs (Hogg 2000), but Rulyov (2004) showed that values much greater than these, $\dot{\gamma}$ up to $2,000\,(\mathrm{s}^{-1})$, could be used successfully for very short time, less than 10 (s). This method is termed *ultra-flocculation*.

Degree of flocculation

The rate at which flocculation occurs is principally governed by the frequency of collision of particles but also by the fraction of surface covered by the flocculant. When collision is the result of mechanical or hydro-dynamical agitation it is called *orthokinetic flocculation*.

According to Gregory (1986a, b) orthokinetic flocculation is produced for mono-sized particles at the rate:

$$V_T \frac{dn_p}{dt} = -k_1(\theta) \times k_2(E_c, \dot{\gamma}) n_p^2(t) \tag{7.7}$$

where n_p is the number of primary particles per unit volume remaining in a volume V_T of suspension at time t, d_p is the particle size and $k_1(\theta)$ and $k_2(E_c, \dot{\gamma})$ are the kinetic constants with values:

$$k_1(\theta) = \theta(1 - \theta) \quad \text{and} \quad \mathrm{k}_2(E_c, \dot{\gamma}) = \frac{8}{3} E_c \dot{\gamma} d_p^3 \tag{7.8}$$

where θ is the surface coverage with flocculant, E_c is the collision efficiency and $\dot{\gamma}$ the shear rate.

The maximum value of $k_1 = 0.25$ is obtained when $\theta = 0.5$ therefore, for this flocculant coverage, Eq. (7.8) becomes:

$$V_T \frac{dn_p(t)}{dt} = -\frac{2}{3} \dot{\gamma} E_c d_p^3 n_p(t)^2 \tag{7.9}$$

This equation is the starting point of the majority of flocculation kinetic studies and was first developed by Smoluchowski in 1917 (Elimelech et al. 1998).

In many works, assumptions associated with using Eq. (7.9) are not clearly established and some of them interpret $n_p(t)$ as the number of primary particles remaining at time t and others as the number of flocs produced at time t. To solve Eq. (7.9) many authors assume that the composition of the suspension is constant over time, which can only be true at the beginning of the process. This assumption permits establishing a relationship between the number of particles and the initial solid volume fraction of the suspension, transforming, in this way, Eq. (7.9) into a linear equation to obtain an exponential solution for the number of particles.

Here, we will consider that a primary particle of size d_p will join other primary particles to form an aggregate. Therefore, while the number of primary particles decreases (primary particles in the suspension diminishes) the number and concentration of flocs increase in the system. We will interpret Eq. (7.9) as the rate of disappearance of primary particles with time, where $n_p(t)$ is the number of primary particles remaining in the suspension at time t and n_{p0} is the number of original primary particles present in the suspension.

For constant values of $\dot{\gamma}$, d_p and E_c, the fraction of primary particles remaining in suspension at time t is:

$$\frac{n_p(t)}{n_{p0}} = \frac{1}{1 + (2/3)n_{p0}\dot{\gamma}E_c\, d_p^3 t/V_T} \tag{7.10}$$

If φ_0 is the volume concentration of primary particles at $t = 0$, the initial total volume of primary particles is $V_T\varphi_0 = (\pi/6)d_p^3 n_{p0}$, then the Eq. (7.10) can be written in the form:

$$\frac{n_p(t)}{n_{p0}} = \frac{1}{\pi + 4\varphi_0\dot{\gamma}Ed_p^3 t_c} \tag{7.11}$$

The number of primary particles contained in all flocs at time *t*, is equal to the number of primary particles that have disappeared at time *t*, that is, $(n_{p0} - n_p(t))$. On the other hand, the number of flocs n_f present in the suspension at time *t* is equal to the number of primary particles contained in all flocs divided by the number of primary particles contained in one floc n_{pf}. Then, the number of flocs is $n_f(t) = n_0 - n_p(t)/n_{pf}$. From (7.11) the number of flocs in the suspension at time *t*:

$$\frac{n_f(t)}{n_{p0}} n_{pf} = \frac{4\varphi_0\dot{\gamma}Ed_p^3 t_c}{\pi + 4\varphi_0\dot{\gamma}Ed_p^3 t_c} \tag{7.12}$$

Problem 7.3 Determine the relative number of particles n/n_{p0} left at time *t* and the number of flocs n_F present at time *t* for a suspension with $\varphi_0 = 0.04$, $E_c = 1$ and $\dot{\gamma} = 1;\ 100$ and $500\,1/\text{s}$.

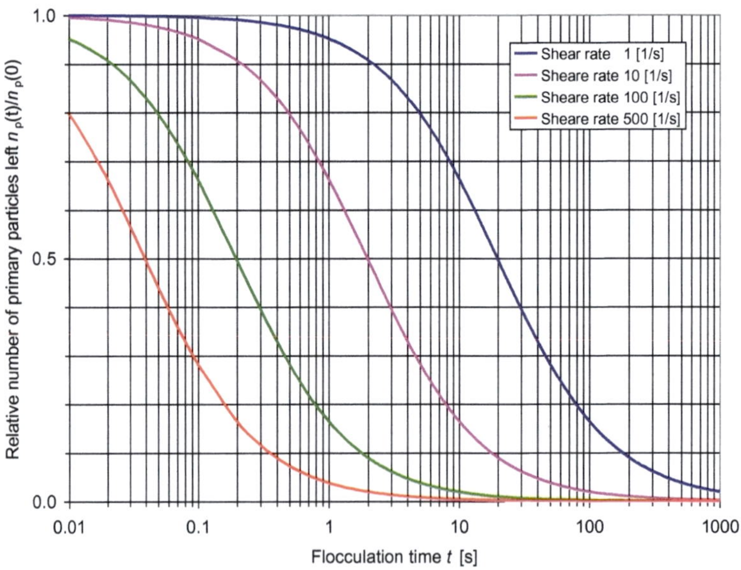

Fig. 7.17 Number of primary particles *left* in the suspension at time *t*, with the shear rate as a parameter

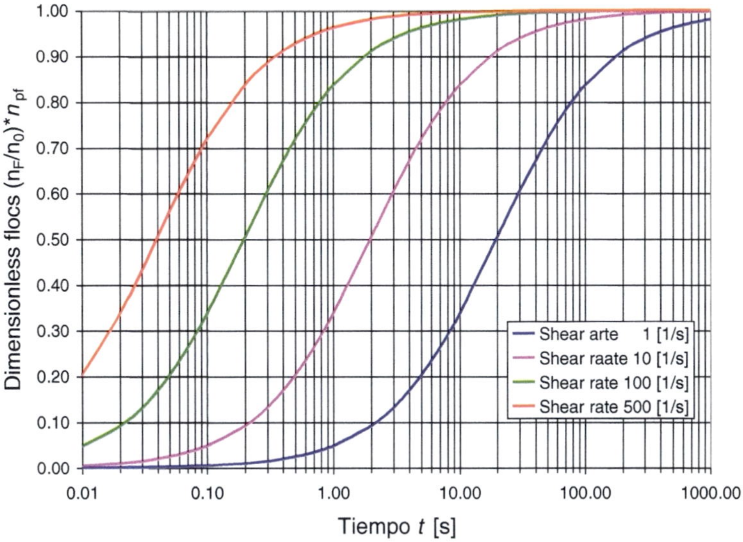

Fig. 7.18 Number of flocs at time t in a suspension of one liter with $\varphi_0 = 0.04$, 1,000 primary particles per floc and the shear rate as a parameter

Figures 7.17 and 7.18 show that, with high shear rates, a few seconds are sufficient for complete flocculation.

Considering that the relationship between the number of primary particles $n_p(t)$ and n_{p0}, and their volume fraction is $n_p(t) = 6V_T\varphi_p(t)/\pi d_p^3$ and $n_0 = 6V_T\varphi_0/\pi d_p^3$, from (7.11) we have the relationship between $\varphi_p(t)$ and φ_0 as:

$$\varphi_{p(t)} = \frac{\pi\varphi_0}{\pi + 4\varphi_0\dot{\gamma}E_c t} \tag{7.13}$$

If we define the *degree of flocculation* $G_f = 100\big((\varphi_0 - \varphi_p(t))/\varphi_p(t)\big)$ as the percentage of particles that have disappeared from the suspension (incorporated into the flocs) at time *t*, the degree of flocculation at time *t*, and the time *t* to obtain a given degree of flocculation are:

$$G_f = 100\frac{4\varphi_0\dot{\gamma}E_c t}{\pi + 4\varphi_0\dot{\gamma}E_c t} \qquad t = \frac{\pi G_f}{\big(100 - G_f\big) \times 4\varphi_0\dot{\gamma}E_c} \tag{7.14}$$

Problem 7.4 Assuming a suspension with an initial volume fraction of $\varphi_0 = 0.04$ (10 % solid) and a collision efficiency of $E_c = 1.0$, calculate: (a) the degree of flocculation as a function of time and (b) the time necessary to obtain a certain degree of flocculation for shear rates $\dot{\gamma} = 1;\ 10;\ 100$ and $1{,}000\,\text{s}^{-1}$. Figures 7.19, 7.20 show the results.

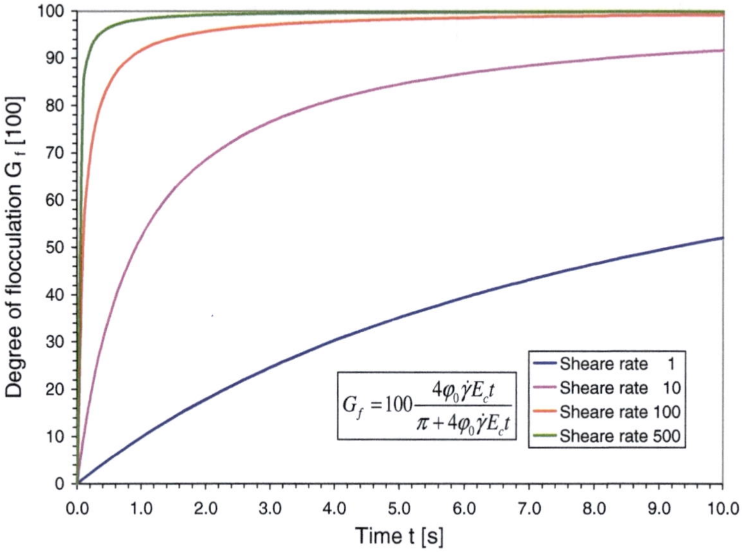

Fig. 7.19 Degree of flocculation versus time for several shear rates with primary particles with an initial concentration of $\varphi_0 = 0.04$ and a collision efficiency of $E_c = 1.0$

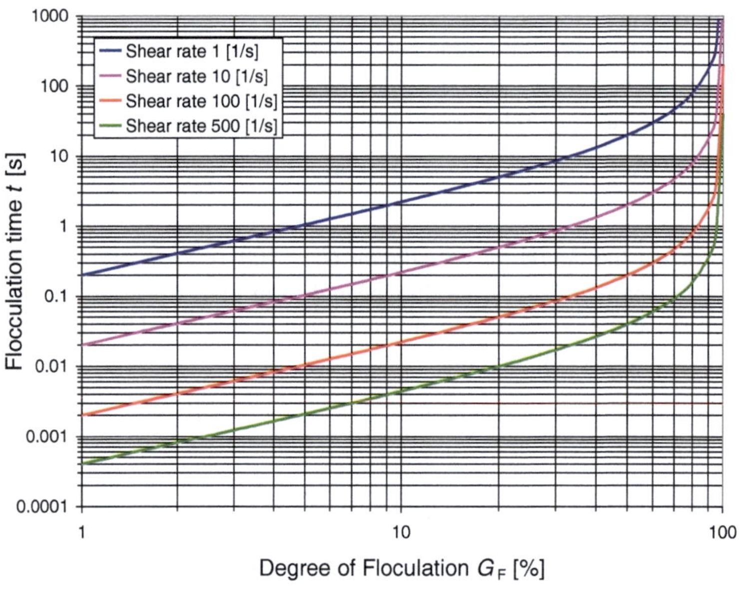

Fig. 7.20 Time to obtain a given degree of flocculation G_r % from a volume fraction of solid particles $\varphi_0 = 0.04$ and an efficiency of collision $E_c = 1.0$

These plots show the immense effect of the shear rate on the degree of flocculation. To obtain a $G_f = 90\,(\%)$ for different values of $\dot{\gamma}$ the following times are needed: 200 (s) for $\dot{\gamma} = 1\,\mathrm{s}^{-1}$, for 200 s; $\dot{\gamma} = 10\,\mathrm{s}^{-1}$, for 20 s; $\dot{\gamma} = 100\,\mathrm{s}^{-1}$; for 2 s, $\dot{\gamma} = 500\,\mathrm{s}^{-1}$ for 0.2 s.

7.3.4 Flocculation Hydrodynamics

In the kinetic model of flocculation the term $\dot{\gamma}$ depends on the hydrodynamic conditions under which flocculation occurs. The value of the efficiency of collision E_c depends on the stability of the suspension of primary particles. A totally unstable suspension has a value $E_c = 1.0$. This is the case for the majority of mineral suspensions since they have sizes much greater then colloidal particles and tend to settle in gravity fields.

We have seen the effect of shear rate $\dot{\gamma}$ on the degree of flocculation; therefore it is important to investigate the shear rate in the several ways a thickener can be fed. These shear rates can be calculated from the hydrodynamics of the flow. As examples, we will analyze the cases of flows in a pipe, a feedwell, a shear vessel and an agitated tank.

(a) **Flocculation in a pipe**

Pulp circulating in a circular tube with a diameter D at a flow rate of Q and an average velocity of $\bar{v}_z$ has an average shear rate $\bar{\dot{\gamma}}$ of (see Chap. 11):

$$\bar{\dot{\gamma}} = \frac{16}{3D}\bar{v}_z\,; \quad \bar{\dot{\gamma}} = \frac{64Q}{3\pi D^3} \tag{7.15}$$

Problem 7.5 Calculate the average shear rate in a 48-inch diameter pipe feeding pulp to a thickener at 2.0 m/s.

•

Data	
D (in)	48
V (m/s)	2
Results	
D (m)	1.2
Q (m3/s)	2.33
$\bar{\dot{\gamma}}$	8.75

This problem shows that flocculation in a cylindrical tube requires a much smaller tube diameter to obtain higher shear rates. The AJ Parker CRC presented the pipe flocculation reactor shown in Fig. 7.21. Figure 7.22 shows the effect of reaction time and shear rate on the size of the flocs. For all shear rates, the floc

Fig. 7.21 Tube reactor in an operation of the Bayer process according to the AJ Parker CRC

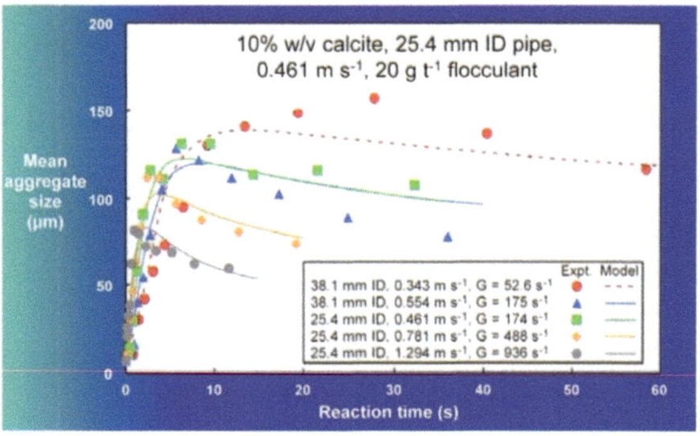

Fig. 7.22 Floc size as a function of time for several pipe diameters and flow rates

diameters increase, reaching a maximum at a time which increases as the shear rate diminishes. This experiment shows that optimum flocculation is obtained for less than 10 s with a shear rate of 175 s^{-1}, with the time increasing as the shear rate decreases.

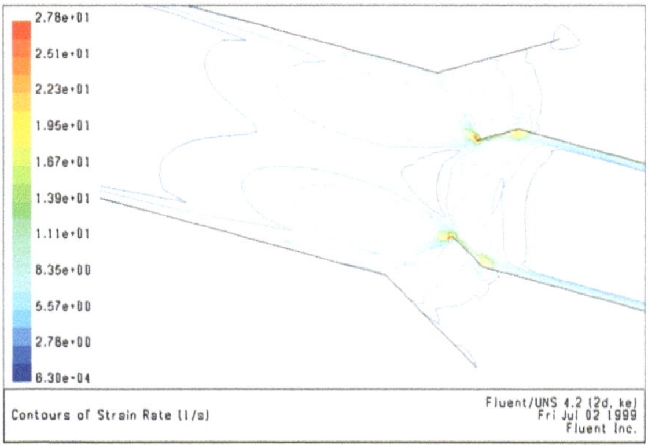

Fig. 7.23 Shear rate mapping in the interior of an E-Duc feeding a tailing thickener (Köck and Concha 1999)

Köck and Concha (1999) simulated the flow in the E-Duc feeding system of a copper tailing thickener. The simulation showed that in the main part of the system the shear rates were on the order of $\dot{\gamma} = \sim 7\,\mathrm{s}^{-1}$, with small regions with shear rates $\dot{\gamma} = \sim 18\,\mathrm{s}^{-1}$ and a maximum shear rate of $\dot{\gamma} = \sim 28\,\mathrm{s}^{-1}$ (Fig. 7.23).

(b) **Flocculation in a feedwell**

Obtaining the shear rate in a feedwell is difficult and the best tool to do it is CFD. Köck and Concha (1999) used Fluent to model a simple feedwell with a tangent entry. They found that the average shear rate was on the order of 1 s^{-1}. For this shear rate, a copper tailing thickener with a volume feed rate of $Q = 2{,}450\ \mathrm{m}^3/\mathrm{h}$ at 25 % solids, assuming an efficiency of collision of $E_c = 85\ \%$, would have only a 44 % degree of flocculation.

A major and recent improvement in thickener design is the feedwell. For example, Outotec describes the Vane feedwell (Outotec web page): "the upper zone, into which feed, dilution water and flocculants are added, provides enhanced mixing and energy dissipation. This maximizes flocculant adsorption, eliminates the possibility of coarse/fine segregation and ensures that all particles are aggregated by the flocculant. Efficient operation is maintained in this upper zone at varying feed rates. The lower zone promotes gentle mixing for continued aggregate growth, with the option of secondary flocculant dosing. This zone also enables aggregates to uniformly discharge under low shear conditions". The idea of this feedwell is flocculation at two different shear rates (Fig. 7.24).

(c) **Flocculation in a shear chamber**

A shear chamber is a cylindrical vessel with a second co-axial cylinder that is a couple of millimeters smaller in diameter than the vessel. The internal cylinder R_{int} rotates at a constant speed Ω rad/s concentrically to the inner cylinder R_{ext}, thus producing a Couette flow. This device is the essential part of a viscometer. The advantage of a shear vessel is that the shear rate can be determined precisely (Schramm 1998) for any rotation speed in a laminar flow (Fig. 7.25):

Fig. 7.24 Outotec Vane feedwell (Outotec web page)

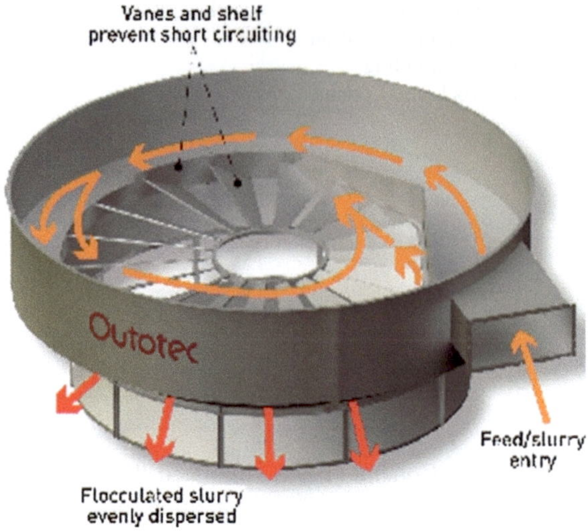

Fig. 7.25 Couette flow (Schramm 1998)

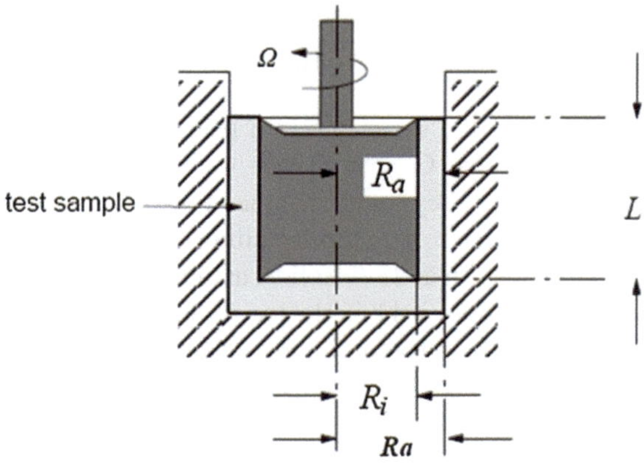

$$\dot{\gamma} = 2\Omega \times \frac{R_{ext}^2}{R_{ext}^2 - R_{int}^2} \ \mathrm{s}^{-1} \tag{7.16}$$

where $\Omega = 2\pi N/60$ rad/s is the angular velocity, N is the rotational speed in (rpm) and R_{ext} and R_{int} are the radius of the exterior and interior cylinder. Rulyov (2010) indicates that in turbulent flow the shear rate is 4.5 times higher than that of laminar flow.

Problem 7.5 Determine the shear rate for a Couette viscometer with exterior/interior cylinder diameters of 102/100 mm, and rotational speed of 100 rpm.

Data	
D_{ext} (mm)	102
D_{int} (mm)	100
N (rpm)	100
Results	
Ω (s^{-1})	10.5
γ (s − 1)	539

The continuous shear chamber proposed by Farrow and Swift (Farrow and Swift 1996a, b, c) is based on Couette geometry. The flocculant is introduced between the concentric cylinders where the suspension flow is circulated. The shear rate is selected by changing the rotational speed and the reaction time is obtained by varying the suspension flow rate, or changing the addition port from 1 to 4 in Fig. 7.26. Figure 7.27 shows the shear chamber of the AJ Parker CRC, Perth, Australia.

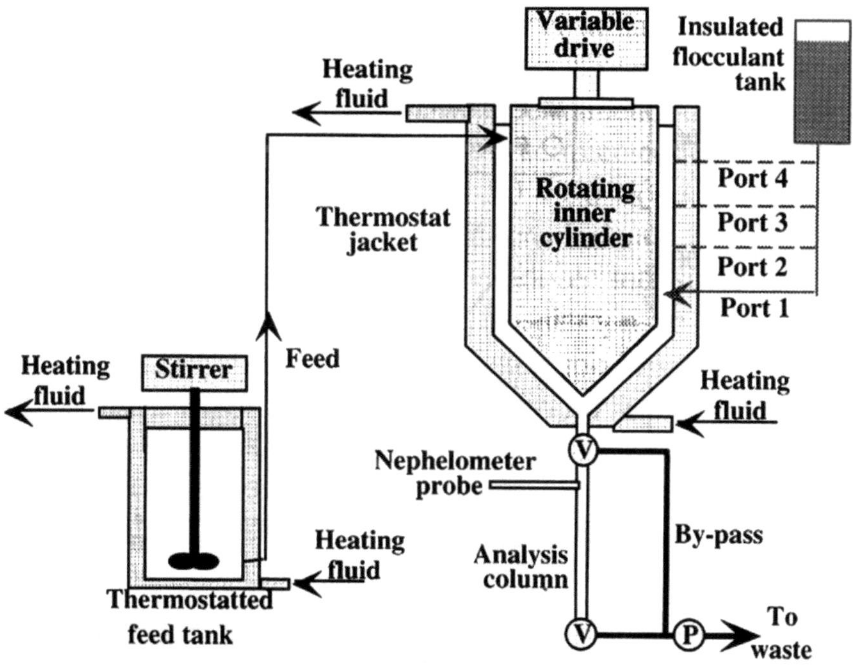

Fig. 7.26 Shear chamber according to Farrow and Swift (1996a, b, c)

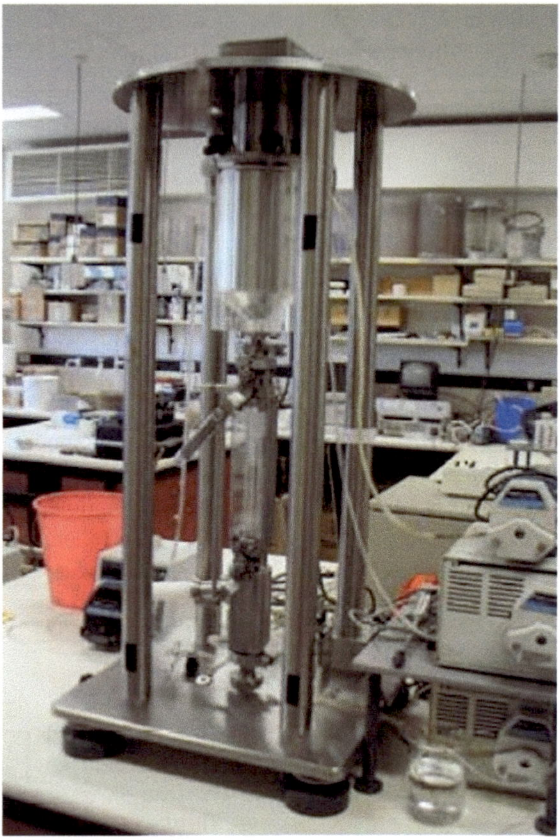

Fig. 7.27 Continuous shear vessel at AJ Parker CRC, Perth, vessel

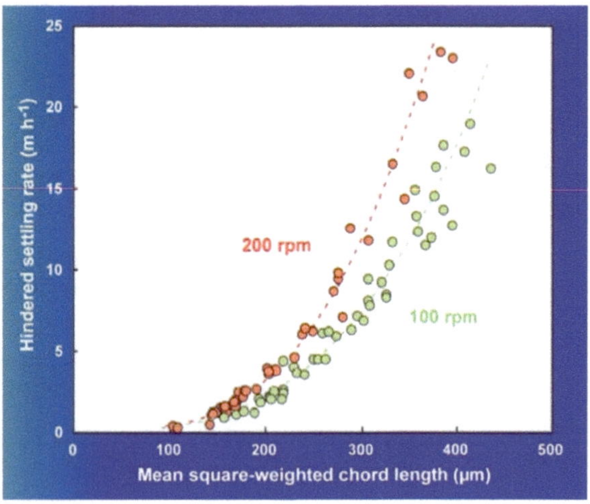

Fig. 7.28 Effect of floc sizes on their settling velocity for two rotational speeds of the shear Farrow and Swift (1996a, b, c)

Figure 7.28 shows the effect of the rotational speed of the shear chamber on the efficiency of flocculation, measured by the settling velocity of the flocs, where a change from 100 to 200 rpm results in a 50 % increase of the settling velocity. The result was interpreted as a change from open flocs to flocs of the pellet type.

(d) **Ultrafloculation**

In general flocculation is performed at shear rates lower than $\dot{\gamma} < 200\,s^{-1}$ to avoid the rupture of flocs. Lower values of the shear rate imply longer reaction time. Rulyov (2004) and Ruylov and Korolov (2008) demonstrated that very high shear rates could be applied to avoid rupturing flocs if very short reaction times are considered. Longer times are needed for the flocs to grow. For example, a suspension of fine coal is completely flocculated in only one or two seconds at $1{,}000 < \dot{\gamma} < 2{,}000\,s^{-1}$.

Recently Rulyov (2004) designed the *Ultrafloc Tester,* a laboratory instrument to characterize flocculation. The instrument is based on the suggestion by Gregory (Elimelech et al. 1998; Glover et al. 2000; Gregory 2009) of using light dispersion to obtain the fractal dimension of the flocs. Figure 7.29 shows the instrument.

Based on his Laboratory Ultrafloc Tester, Rulyov developed industrial Ultraflocculators, such as depicted in Fig. 7.30. The version shown in the figure has a capacity of 200 m^3/h and was applied for coal and quartz suspensions. Figure 7.31 shows results for fine quartz suspensions.

(e) **Flocculation in a stirred tank**

Hogg (1987) recommended stirred tanks for flocculation in the laboratory. These devices consist of cylindrical tanks with four baffles agitated by a six-blade turbine impeller. Figure 7.32 shows the standardized version. Figure 7.32 shows a turbulent stirred tank used by Spicer et al. (1996) to obtain intimate contact between the solid–liquid phases.

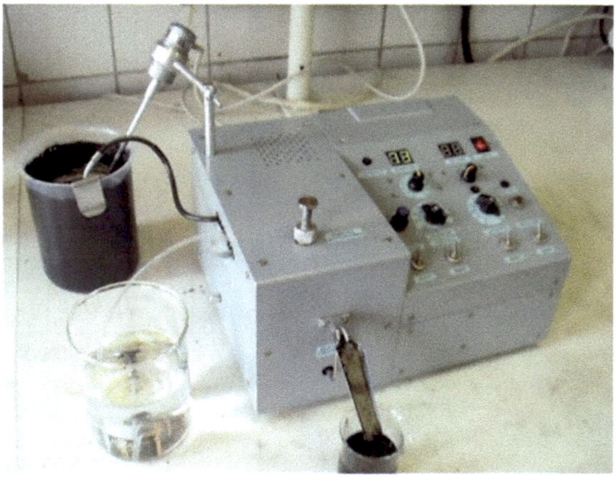

"UltraflocTester-2007"

Fig. 7.29 Laboratory ultra-flocculator with shear rates in the range of $100 < \dot{\gamma} < 40{,}000\,s^{-1}$ (Rulyov et al. 2011)

Fig. 7.30 Ultraflocculator for the industrial flocculation of coal flotation tailings (Rulyov et al. 2009)

Industrial Ultra-flocculator of the upright design (capacity 200 m^3/h)

Fig. 7.31 Degree of flocculation versus shear rate for quartz suspensions with sizes $<5\,\mu$ treated for 1, 6 and 12 s

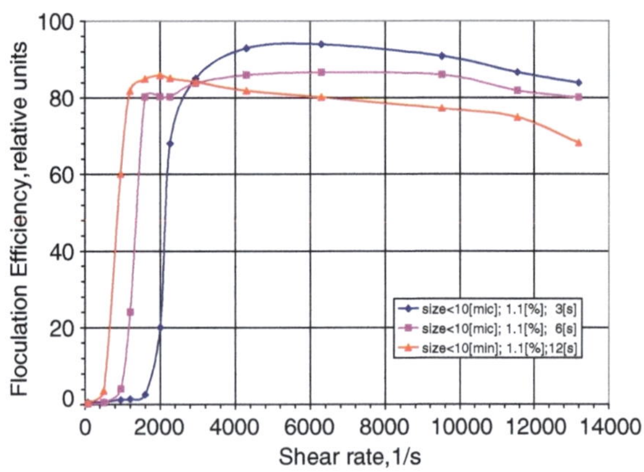

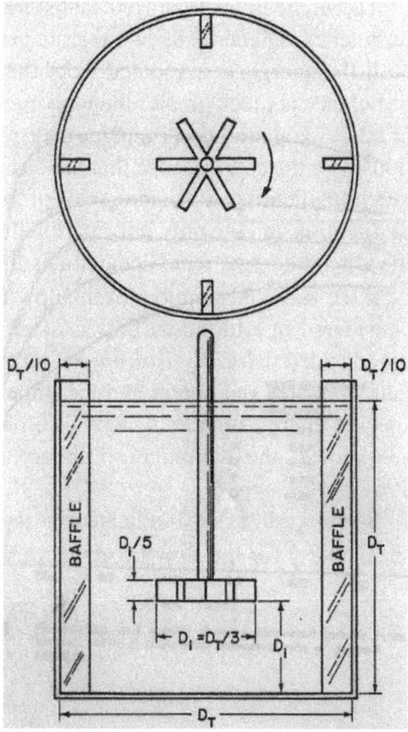

Fig. 7.32 Standard flocculation vessel according to Hogg (1987)

The dimensions of Hogg's flocculation vessels are given in the table.

Suspension volume V cm^3	Tank diameter D_T cm	Impeller speed range rpm
300	7.6	500–1,500
900	10.3	300–1,000
2,700	15.5	200–800

In these stirred tanks the turbulence is described by the *macro* and *micro scale of turbulence*. The macro scale ℓ refers to the size of the eddies where the turbulence originates, and the micro scale λ_0 to the size of the eddies where the turbulence is dissipated. If the tank is properly designed, it is possible to establish the average shear rate in terms of the properties of the system (Concha 1985).

The rotation of the impeller in a stirred tank produces eddies with magnitude ℓ on the order of the tank diameter $\ell \sim D_T$. If the speed u of the liquid at the top of the impeller is on the order of $\bar{u} \sim ND_T$, where N is the impeller speed in (rpm), then the Reynolds number can be written in the form $Re = D_T N^2/v$, where v is kinematic viscosity.

Turbulent flows in stirred tanks are anisotropic. The largest eddies, where the turbulence is generated, disintegrate gradually producing successive smaller eddies until the energy is dissipated from the smallest eddies. If the macros-scale of the turbulence is much greater than the micro-scale, there will be an intermediate size λ of eddies that slightly dissipates energy. These eddies transport energy in all directions with the consequence that the directional information is lost. Therefore, eddies much smaller than the macro-scale of the turbulence $\lambda \sim \ell$ are statistical independent of direction and the only information that they transmit is the *rate of dissipation* ε of the kinetic energy, which constitutes the principal property of the system.

If the Reynolds number of the flow $Re = D_T N^2/\nu$ is high, and the volume being considered in a turbulent flow is smaller than the macro-scale, the turbulence can be considered *locally isotropic*. Since the variables in stirred tanks, such as the tank diameter and height and the impeller diameter, affect the process through the specific energy function ε and since only average values of the dissipation energy are known, the Kolmogoroff theory will be valid for similar systems (Levich 1962).

Spicer et al. (1996) indicate that the specific energy function $\hat{\varepsilon}$ and the average shear rate $\bar{\dot{\gamma}}$ for the flow in stirred tank are given by:

$$\hat{\varepsilon} = \left(\frac{P_N N^3 D_T^2}{V}\right) \quad \text{and} \quad \dot{\gamma} = \left(\frac{\hat{\varepsilon}}{\nu}\right)^{1/2} \tag{7.17}$$

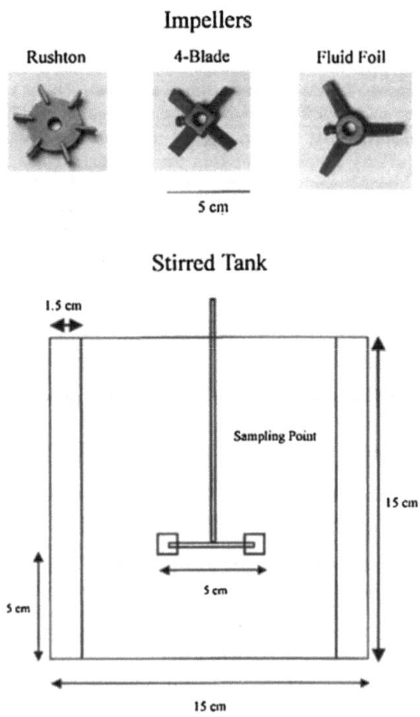

Fig. 7.33 A stirred tank with Rushton impellers of 1.2 cm (Spicer et al. 1996)

where N is the rotational speed of the impeller in rps, D_T is the diameter of the stirred tank in m and V is the liquid volume in m^3. If we let $A = \lambda_A D_T$ and $H = \lambda_H D_T$, the volume of the tank is $V = (\pi/4)\lambda_D^2 \lambda_H D_T^3$:

The shear rate is:

$$\dot{\gamma} = \left(\frac{4P_n}{\pi \nu \lambda_A^2 \lambda_H} \right) N^{3/2} D_T \tag{7.18}$$

Equation (7.18) can be written in the form:

$$\dot{\gamma} = kN^{3/2} D_T \quad \text{where} \quad k = \left(\frac{4P_n}{\pi \nu \lambda_A^2 \lambda_H} \right) \tag{7.19}$$

For small Reynolds numbers, the power number P_N is a function of Re^{-1}, and becomes a constant for $Re > 10^4$.

Problem 7.6 Calculate the shear rate in the stirred tank in Fig. 7.33 with a diameter of $A = 15$ cm, height of fluid $H = 15$ cm and an impeller diameter of $D = 5$ cm at the rotational speeds of 300, 500 and 700 rpm.

D (m)	0.05	0.05	0.05
T (m)	0.15	15	15
N (rpm)	200	500	700
N (rps)	3.333	8.333	11.667
μ[Pa-s] = (kg/ms^2)	1.00E−03	1.00E−03	1.00E−03
ρ (kg/m^3)	1,000	1,000	1,000
V (m^2/s)	1.00E−06	1.00E−06	1.00E−06
λa = A/D(−)	3	3	3
λh = H/D(−)	3.00E+00	3.00E+00	3.00E+00
Re (−)	5.56E+05	3.48E+06	6.81 E+06
V (m^3)	2.649E−03	2.649E+03	2.649E+03
Pn (−)	1.200E+00	4.333E−01	4.333E−01
k (1/ms$^{5/2}$)	247.7	247.7	247.7
γ (1/s)	75.4	297.9	493.5

7.4 Flocculant Properties

Polymeric flocculant are organic reagents with long molecular chains and molecular weights in the range of 10^6 (g-mol) or more. They can be natural polysaccharides, such as gelatins, alginates and agar–agar, or synthetically based acrylic materials derived from petroleum, such as high molecular weight polyacrylamide. The advantage of synthetic flocculants is that their molecular structure, in terms of molecular weight and the degree of hydrolysis or ionicity, can be

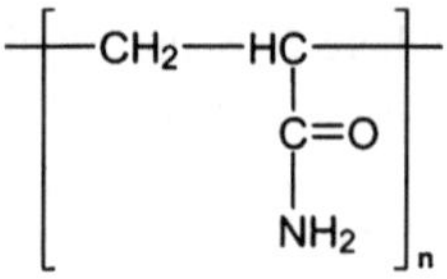

Fig. 7.34 Polyacrylamide

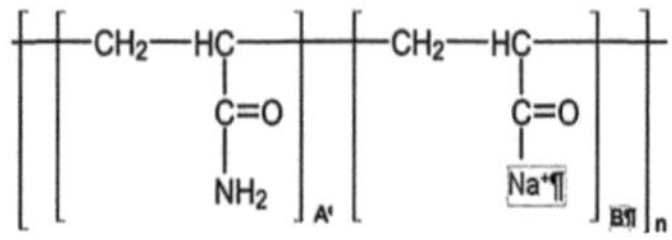

Fig. 7.35 Hydrolized polyacrylamide

designed at will according to industrial requirements. Since they are obtained from chemical reagents, their quality is more consistent than that of natural flocculants. Furthermore, they can be designed with higher molecular weights, which is a technical and economical advantage.

Polyacrylamide is often listed as a nonionic flocculant. However, the vast majority of commercial flocculants are hydrolyzed polyacrylamides (or co-polymers of acrylamide and acrylic acid). Consequently, they contain some carboxylic groups that are expressed as the *degree of anionicity*. Figure 7.34 shows a pure polyacrylamide and Fig. 7.35 a hydrolyzed polyacrylamide.

Other examples of synthetic polymers are:

- Non-ionic: polyethylene oxide and polyvinyl alcohol
- Anionic: Sodium polyestyrene sulphonate
- Cationic: Polyethylamine, dialidimetyl ammonium chloride.

Polymers in general act through hydrogen bridges, but ionic polymers also react by electrostatic interaction. For this reason, the election of the proper flocculant for a given task depends on the surface charge of the particle and on the physico-chemical conditions of the solution, such as pH and ionic force.

The most important characteristics of polymeric flocculants are their molecular weight, the nature of their functional groups and, especially, their charge density in the case of ionic groups. The charge density depends on the ionizable group and on its degree of ionization. The charge is established during reagent synthesis and the degree of ionization depends on the nature of the ionic group and the conditions of the solution. Strongly ionized groups like sulphonates or quaternary amines are totally charged in most cases, while weakly ionized groups like carboxyl and tertiary amines are affected by the pH of the solution. Anionic polymers with carboxylic groups are not completely charged until pH is over 6. In more acidic solutions, the ionic degree decreases. Cationic polymers, based on tertiary amine groups, have reduced charges when pH is lower than 8.

When negatively charged particles come in contact with non-ionic or anionic polymers, flocculation does not occur unless a cation, such as Ca^{2+} is added to promote the adsorption of the polymer on the particle surface by bonding simultaneously to the ionic groups of the polymer and to the negative sites of the mineral surface. An example is flocculation of kaolin with anionic polymers. Kaolin is not flocculated by NaCl but flocculates in the presence of small quantities (1-2 mmolar) of Ca^{2+} ions.

An optimal polyelectrolyte charge density is necessary for bridge-flocculation to occur. For example, flocculation with polyacrylamide improves as the polymer molecular weight increases and at an optimum value of hydrolysis of about 30 %. An increase in hydrolysis produces the mutual repulsion of the different polymer segments; the hydrocarbon chain straightens and expands. Simultaneously, the increase in ionicity decreases polymer adsorption on the negative particles. The optimum hydrolysis value lies between these two tendencies.

Polymers with opposite charges to the particles adsorb strongly due to ionic attraction. Cationic polymers are widely used to flocculate negative particles and charge neutralization fully explains the flocculation mechanism. The optimum flocculation condition in this case is charge neutralization. Excess flocculants can re-stabilize the suspension by giving the particles a positive charge. The best flocculant for these tasks are polymers with strong electrical charges, their molecular weight being a secondary factor.

References

Bushell, G., Amal, R., & Raper, J. (1998). The effect of polydispersity in primary particle size on measurement of the fractal dimension of aggregates. *Particle and Particle Systems Characterization, 15*, 3–8.

Bushell, G. (2005). Forward Light scattering to characteristic structure of blocs composed of large particles. *Chemical Engineering Journal, 11*, 145–149.

Concha, F. (1985). Hydrodynamic characteristics of stirred tanks, Class Notes, Graduate curse. Department of Metallurgical Engineering, University of Concepción (in Spanish).

Concha, F., & Segovia, J. P. (2012). Report to AMIRA of project P996.

Elimelech, M., Gregory, J., Jia, X., & Williams, R. A. (1998). *Particle deposition and aggregation* (p. 167). Oxford: Butterworth-Heinemann.

Farrow, J. B., & Swift, J. D. (1996a). A new process for assessing the performance of flocculants. *International Journal of Mineral Processing, 46*, 263–275.

Farrow, J. B., & Swift, J. D. (1996b). *International Journal of Mineral Processing*, *46*, 263–275.

Farrow, J. B., & Swift, J. D. (1996c). *4th international alumina quality workshop* (pp. 355–363). Darwin, Australia, 2–7 June.

Glover, S. M., Yanh, Y., Jamson, G. J., & Biggs, S. (2000). Bridging flocculation studies by light scattering and settling, *Journal of Chemical Engineering*, *80*,1–3, p 3–12.

Gregory, J. (1986a). Flocculation, in progress in filtration and separation 4. In R. J. Wakeman (Ed.), (pp. 55–99). New York: Elsevier.

Gregory, J. (1986b). Flocculation, in progress in filtration and separation 4. R. J. Wakeman (ed.), Elsevier, pp. 83–90.

Gregory, J. (2009a). Monitoring particle aggregation processes. *Advances in Colloid and Interface Science, 147*, 112.

Gregory, J. (2009b). Monitoring particle aggregation processes. *Advances in Colloid and Interface Science, 147–148*, 109–123.

Healy & La Mer (1964). Energetics of flocculation and redispersion by poloymersm. *Journal Colloid Sciences, 19*, 323–332.

Hogg, R. (1987). Coal preparation wastewater and fine refuse treatment. In S. K. Mishra & R. R. Klimpel (Eds.), *Fine coal processing* (Chap. 11). New Jersey: Noyes Publications.

Hogg, R. (2000). Flocculation and dewatering. *International Journal of Mineral Processing, 58*, 223–236.

Köck, R., & Concha, F. (1999). CFD as tool for the design of thickeners, III Woekshop on Filtration and Separation, Santiago, Chile (in Spanish).

Laskowski, J. S., & Pugh, R. J. (1992). Dispersion stability and dispersing agents. In J. S. Laskowski & J. Ralston (Eds.), *Colloid chemistry in mineral processing* (pp. 115–171). New York: Elsevier.

Levich, V. G. (1962). *Physicochemical Hydrodynamics, Prentice-Hall international series in the physical and a chemical engineering sciences*. Englewood Cliffs: Prentice Hall Inc.

Rulyov, N. N. (2004). Ultra-flocculation: Theory, experiment, applications. In J. S. Laskowski (Ed.), *5th UBC-McGill Biennial International Symposium, 43rd Annual Conference of Metallurgist of CIM* (pp. 197–214), Hamilton, Ontario, Canada.

Ruylov, N. N., Koolyov, V. J., & Kovalchuck, N. M., Ultra-flocculation of quartz suspensions: Effect of shear rate, particle size distribution and solids content. Institute of Biocolloid Chemistry, Ukrainian National, Academy of Sciences, Kiev, Ukraine.

Rulyov, N. N., & Korolyov, V. J. (2008). Ultraflocculatiohn in Wastewater Treatment, In *Proceedings 11 International Mineral Processing Symposium*, Belek-Antalaya, Turkey.

Ruylov, N. N. (2010). Personal communication.

Rulyov N. N., Laskowski, J., & Concha, F. (2011). The use of ultra-flocculation in optimization of experimental flocculation procedures, *Physicochemical Problems of Mineral Processing, 47*, 5–16.

Schramm, G. (1998). *A practical approach to rheology and rheometry*. Karlsruhe: Thermo Haake Gmbh.

Selomulya, C., Amal, R., Bushell, G., & Waite, T. D. (2001). Evidence of shear rate dependence of restructuring and breakup of latex aggregates. *Journal of Colloid and Interface Science, 236*, 67–77.

Spicer, P. Y., Keller, W., & Pratsinis, S. E. (1996). The effect of impeller type on floc size and structure during shear-induced flocculation. *Journal of Colloid and Interface science, 184*, 112–122.

Wu, R. M., Tsou, G. W., & Lee, D. J. (1999). *Estimate of sludge floc permeability*. Taipei: National Taiwan University.

Chapter 8
Thickening

Abstract The chapter analyzes thickening in-depth. In an extensive introduction, the history of thickening is laid out from the Stone Age to the present, emphasizing people and institutions that have been important actors. The chapter then reviews the thickeners used in the mining-mineral industry. During sedimentation, particles settle individually except for collision among them, exerting interaction between them solely through the fluid. At a certain concentration, particles begin to touch each other permanently transforming the suspension into a network of solid particles called sediment. At that point forces among particles are transmitted directly from particle to particle. If settling particles that reach the bottom of the vessel and lie one on top of the other are incompressible, such as glass beads, the whole process ends, but if they are compressible, as in the case of flocculated copper flotation tailings, the weight of the sediment compresses the flocs lying underneath expelling the water from the pores. This phenomenon of extracting water by compression is called consolidation. The theory of sedimentation-consolidation is deduced from the equations for a particulate system and constitutive equations for the solid-fluid interaction force and sediment compressibility are postulated. Batch and continuous sedimentation are analyzed and simulations are compared to data from the literature. Experimental determination of thickening parameters and instruments for their determination are presented. Old and new methods for thickening design are reviewed and software for the design and simulation of batch and continuous thickening are presented. Finally, strategies for the operation and control of industrial thickening are discussed.

8.1 Introduction

8.1.1 From the Stone Age to the Middle Ages

Thickening is not a modern technique and was certainly not invented in the Americas. Whenever people have tried to obtain concentrates from ore, two processes have been used inseparably, *crushing and washing*. There is evidence

F. Concha A, *Solid–Liquid Separation in the Mining Industry*,
Fluid Mechanics and Its Applications 105, DOI: 10.1007/978-3-319-02484-4_8,

that in the 4th Egyptian dynasty, around 2,500 BC, the ancient Egyptians dug for and washed gold. There is also evidence of washing gold in Sudan in the 12th dynasty. Nevertheless, the earliest written reference to crushing and washing in Egypt is Agatharchides, a Greek geographer who lived 200 years before Christ. Ardillon in his book "Les mines du Laurion dans l'antiquité" described the process used in the extensive installation for crushing and washing ores in Greece between the 5th and 3rd centuries BC. In his book "The living rock", Wilson (1994) described mining gold and copper in the Mediterranean from the fall of the Egyptian Dynasties to the Middle Ages and the Renaissance (Concha and Bürger 2002a).

The development of mineral processing from unskilled labor to craftsmanship and eventually to an industry governed by scientific discipline is largely due to the Saxons in Germany and Cornishmen in England, beginning in the 16th century. An international exchange of technology began between these two countries and continued for an extended period. But it was in Saxony where Agricola (1950) wrote his book "De Re Metallica 1950", the first major contribution to the development and understanding of the mining industry, published in Latin in 1556, and shortly after translated to German and Italian. Agricola's book had a tremendous impact, not only on the mineral industry but also on society in general, and continued to be the leading textbook for miners and metallurgists outside the English-speaking world for at least another 300 years. Apart from its immense practical value as a manual, the greatest influence of *De Re Metallica* was in preparing the ground for the introduction of a system of mining education, which, with various modifications to suit local conditions, was later to be adopted internationally.

Agricola's book describes several methods for washing gold, silver, tin and other metallic ores. He described settling tanks used as classifiers, jigs and thickeners, and settling ponds used as thickeners or clarifiers. These devices operated in batches or semi-continuously. A typical description is as follows; see Figs. 8.1 and 8.2:

> To concentrate copper at the Neusohol in the Carpathians, the ore is crushed and washed and passed through three consecutive washer-sifters. The fine particles are washed through a sieve in a tub full of water, where the undersize settle to the bottom of the tub. At a certain stage of filling tub with sediment the plug is drawn to let the water run out. Then, the mud is removed with a shovel and taken to a second tub and then to a third tub where the whole process is repeated. The copper concentrate that has settled in the last tub is taken out and smelted.

It is evident from these references that by using washing and sifting processes, from the ancient Egyptians and Greeks to the mediaeval Germans and Cornishmen knew the practical effect specific gravity of the various components of an ore and used sedimentation in operations that can now be identified as classification, clarification and thickening. There is also evidence that in the early years no distinction was made among these three operations.

Fig. 8.1 Settling tanks described by Agricola

Fig. 8.2 Washing and settling according to Agricola

8.1.2 The Invention of the Dorr Thickener

Clarification and thickening involve the settling of one substance in solid particulate form on a second substance in liquid form. While clarification deals with very dilute suspensions, thickening produces more concentrated pulps. Perhaps this is why clarification was the first of these operations to be described mathematically. The work of Hazen in 1904 was the first to analyze factors affecting particle settling from dilute suspensions in water. He showed that the detention time is not a factor of the design of settling tanks, but rather that the solid removed is proportional to the surface area of the tank, to the settling properties of the solid matter, and inversely proportional to the flow rate through the tank.

The invention of the Dorr thickener in 1905 can be considered a starting point of modern thickening. It made continuous dewatering of dilute pulp possible, whereby a regular discharge of a thick pulp of uniform density takes place concurrently with the overflow of a clarified solution. Scraper blades or rakes driven by a suitable mechanism and rotating slowly over the bottom of the tank, which usually sloped gently toward the center, moved the material as fast as it settled without enough agitation to interfere with the settling (Dorr 1915).

The first reference to variables affecting sedimentation was in 1908. Authors such as Ashley (1909), Nichols (1908a, b), Forbes (1912), Clark (1915), Free (1912) and Ralston (1916) studied the effect of solid and electrolyte concentrations, the degree of flocculation and temperature on the process.

Mishler (1912, 1916), an engineer and superintendent at the Tigre Mining Company concentrator in the Sonora desert in Mexico, was the first to show by experiment that the rate of settling slimes is different for diluted and concentrated suspensions. While the settling speed of diluted slimes is usually independent of the depth of the settling column, thick slime sedimentation rates increase with the depth of the settling column. He devised a formula by which laboratory results could be used in continuous thickeners.

Based on Clark's results (Clark 1915) and their own experiments, Coe and Clevenger (1916) recognized that the settling of an initially homogenous flocculent suspension gives rise to four settling zones. From top to bottom, they distinguished (see Fig. 8.3) a clear water zone A, a zone of constant initial concentration B, a transition zone of variable concentration C and a compression zone D.

Coe and Clevenger (1916) argued that the *solid handling capacity*, today called *solid flux*, has a maximum value in the thickener at a certain dilution, between the

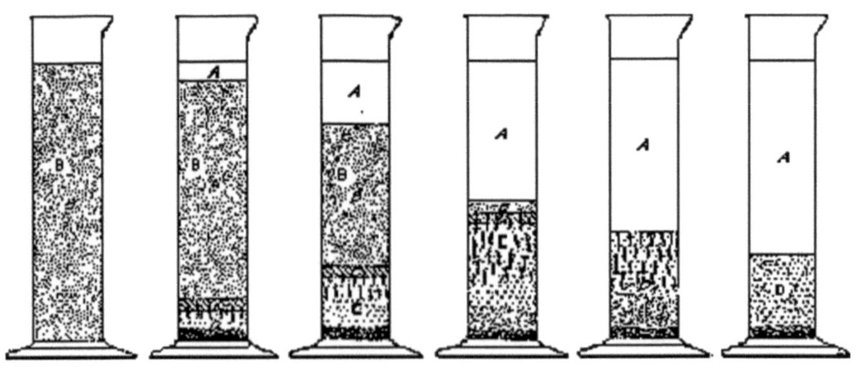

Fig. 8.3 Batch settling of a compressible pulp according to Coe and Clevenger (1916)

feed and the discharge concentration. Independently from Mishler (1912) they developed a similar equation, which with certain corrections, continues to be the most reliable method of thickener design to date.

Several authors described the settling of suspensions by extending Stokes' equation, or by empirical modeling (Egolf and McCabe 1937; Ward and Kammermeyer 1949; Work and Kohler 1940; Kammermeyer 1941), but no further important contributions were made on thickening technology until the 1940s. Stewart and Roberts (1933), reviewing the state-of-the-art of thickening, wrote:

> The basic theory is old but limitations and modifications are still but partially developed. Especially in the realm of flocculent suspensions is the underlying theory incomplete. Practical testing methods for determining the size of machines to be used are available, but the invention and development of new machines will no doubt be greatly stimulated by further investigation of the many interesting phenomena observed in practice and as fresh problems are uncovered.

8.1.3 Operating Variables in a Continuous Thickener

In the decade after Comings published his paper "Thickening of calcium carbonate slurries" (Comings 1940) at least nine engineering theses under his guidance at the University of Illinois considered the effect of operating variables in continuous thickening. Comings and his co-workers described these findings in an important paper (Comings et al. 1954). They show four zones in a continuous thickener: the clarification zone at the top, the settling zone underneath, the upper compression zone further down and the rake action zone at the bottom. Two of the most important features in the operation of a thickener were expressed for the first time, firstly that the concentration in the settling zone is nearly constant for a thickener at a steady state, with the concentration depending on the rate at which the solid is fed to the thickener and not on the concentration of the feed. It was verified that in most cases the feed was diluted to an unknown concentration on entering the thickener. The second finding was that for the same feed rate, increasing or decreasing the sediment depth could adjust the underflow concentration.

Roberts (1949) advanced the empirical hypothesis that the rate at which water is eliminated from a pulp in compression is always proportional to the amount of water left in the sediment.

8.1.4 Kinematical Theory of Sedimentation

Coe and Clevenger's design procedure (1916), which was the only quantitative work on sedimentation during the first half of the 20th century, was based on a macroscopic balance of the solid and fluid in a sedimentation vessel and on the observation of the different concentrations established in the thickener. No underlying sedimentation theory existed.

In his celebrated paper "A theory of sedimentation" in 1952 G. J. Kynch, a mathematician at the University of Birmingham in Great Britain, presented a kinematical theory of sedimentation based on the propagation of concentration waves in the suspension (Kynch 1952). The suspension was considered a continuum and the sedimentation process was represented by the continuity equation of the solid phase (See Chap. 5 of his book).

The basic assumption of Kynch's theory is that at any point in the flow field the settling velocity is a function solely of the concentration of local solids. Kynch showed that if the form of the flux-density function and the initial concentration of a suspension are known, the solution of the continuity equation can be constructed by the method of characteristics, and that this procedure describes the complete sedimentation process.

This chapter had a major influence on the development of thickening thereafter. When Comings moved from Illinois to Purdue University, research on thickening continued there for another 10 years. Three theses by Tory (1961), Stroupe (1962) and De Haas (1963) analyzed Kynch's theory and proved its validity by experiments with glass beads. Their results were published in a series of joint papers (Shannon et al. 1963; Shannon and Tory 1965, 1966). Batch sedimentation was regarded as the process of propagating concentration waves upwards from the bottom of the settling vessel. The concept of an *ideal suspension* and *ideal thickener* was presented for the first time by Shannon and Tory (1966) and complemented by Bustos et al. (1990a, 1999) and by Concha and Bustos (1992).

Kynch's theory was so successful that there was a tendency to extend its validity. Several authors, among them Fitch (1983) and Font (1988), tried to modify Kynch's theory to account for the compressive effects. This approach encountered several problems that could not be solved within the theory and a different theory was needed.

At the beginning of the 1980s, mathematicians from the University of Concepción in Chile and the Darmstadt Technical University in Germany and later the University of Stuttgart in Germany, in collaboration with the Department of Metallurgical Engineering at the University of Concepción, worked on conservation laws. They started a comprehensive study of the mathematical aspects of sedimentation. Their results were published in the doctoral theses by Bustos (1984), Kunik (1990) and in several papers by Bustos, Bürger, Concha, Wendland and other collaborators and in the book by Bustos et al. (1999). The first step in their work was to establish a rigorous framework for the theory of *batch* sedimentation of ideal suspensions (Kynch's theory). Using the method of characteristics, Bustos and Concha (1988a, b) and Concha and Bustos (1991) constructed entropy weak solutions of Kynch's problem, in which zones of constant concentrations are separated by shocks, rarefaction waves or combinations of these. For flux density functions with two inflexion points, they presented five modes of sedimentation. These finding were complemented with two other *Modes of Sedimentation* in Bustos et al. (1999) and Bürger and Tory (2000).

Similar solutions can be constructed by the method of characteristics for *continuous* sedimentation of ideal suspensions (Bustos et al. 1990a; Bustos and

Concha 1996; Concha and Bustos 1992; Concha and Bürger 1998). The solution of this equation leads to three *Modes of Continuous Sedimentation*. For a detailed or a concise overview of the construction of weak solutions, see Bustos et al. (1999) or Concha and Bürger (1998) respectively. See also Chap. 5.

On the basis of the construction of weak solutions, Bustos et al. (1990b) formulated a simple control model for continuous sedimentation of ideal suspensions in an ideal continuous thickener. It shows that certain steady states can always be recovered after perturbation of the feed flux density by solving two initial and boundary value problems at known times.

Diehl (1997, 1999) and Bürger et al. (2001, 2002) studied the effect of vessels with varying cross section and boundary conditions. Diehl showed that basically the suspension concentration increases because of its conical shape. Bürger et al. (2002) indicated that because of the varying cross sections, characteristics and iso-concentration lines do not coincide; numerical methods must be used to obtain results. Chancelier et al. (1994), Diehl (1997) and Bürger et al. (2002) treated the feed, discharge and overflow mechanisms as discontinuities of the flux density function adding a source term at the feed level, thus making boundary conditions unnecessary.

8.1.5 Phenomenological Theory of Sedimentation-Consolidation

The experience of several authors, among them Yoshioka et al. (1957), Hassett (1958, 1964a, b, 1968), Shannon et al. (1963), and Scott (1968a, b), demonstrate that, while Kynch's theory accurately predicts sedimentation for suspension of equally sized rigid particles, this is not the case for suspensions of compressible materials.

Behn (1957) was the first writer to attempt to apply consolidation theory to the settling of compressible slurries, but it was Mompei Shirato et al. (1970) who first solved the combined settling-consolidation problem. Using material coordinates, they obtained settling curves and excess pore pressure profiles. It took another five years for Adorján (1975, 1976) to present an ad-hoc theory of sediment compression, giving the first satisfactory method of thickener design.

At about the same time, a group of researchers in Brazil gave the phenomenological sedimentation theory a proper framework. Important research was going on in Brazil in the 1970s on thickening and flows through porous media in general. At COPPE, the Graduate School of the Federal University of Rio de Janeiro, several researchers and graduate students were applying a newly developed mathematical tool, the *Theory of Mixtures* of continuum mechanics to particulate systems. The findings of Giulio Massarani, Affonso Silva Telles, Rubens Sampaio, I-Shih Liu, José Teixeira Freire and João D'Avila (D'Avila 1976, 1978; D'Avila and Sampaio 1977; D'Avila et al. 1978), to mention only a few, were presented at the Porous Media Symposia organized uninterruptedly every year since 1973 by Massarani and his group. This series was renamed the Brazilian Congress on Particulate Systems in 1996. Contributions presented at these meetings are well documented in yearly-published annals.

With strong ties to the Brazilian researchers, the author of this book worked in the same direction at the University of Concepción in Chile. Initial findings were presented by Bascur and Concha at the 1975 IMPC in Sao Paulo Brazil, by Bascur (1976) and Barrientos (1978) in their Engineering Theses, and Concha and Barrientos (1980) at the Engineering Foundation Conference on Particle Technology in New Hampshire, USA. Independently, Kos (1977) used the Theory of Mixtures to set up a boundary value problem for batch and continuous sedimentation. Thacker and Lavelle (1977) used the same theory for incompressible suspensions.

During the 1980s several papers, among them Buscall and White (1987), Auzerais et al. (1988), Landmann et al. (1988), Bascur (1989) and Davis and Russel (1989) showed that the phenomenological model based on the Theory of Mixtures was well accepted by the international scientific community.

The *phenomenological theory of sedimentation-consolidation* assumes that a particulate system composed of two superimposed continuous media with some restrictions obeys local mass and momentum balances and constitutive equations for stresses and forces. The result is a non-linear degenerate parabolic differential equation (Bürger and Concha 1998; Bürger et al. 1999; Bustos et al. 1999) describing the sedimentation and consolidation of flocculated suspensions.

Bürger et al. (2000b) and Bürger and Karlsen (2001) devised numerical methods for solving this equation. These numerical methods have built-in properties to appropriately reproduce discontinuities of the entropy solutions, especially the *suspension-sediment interface,* where the equation changes from parabolic to hyperbolic. This property makes it unnecessary to track the interface explicitly, that is, the scheme has the so-called *shock-capturing property* Bürger et al. (2000d). Garrido et al. (2000) showed the application of this method to several batch sedimentation processes published in the literature.

Concha and coworkers (Garrido et al. 2004, 2003) developed a thickener design and simulation procedure based on the numerical method of Bürger and Karlsen (2001). This procedure was the most comprehensive design method yet presented in the literature. Garrido (2005) and Concha et al. (2006a, b) presented methods and new instrumentation to determine thickening parameters and Segovia and Concha (2012) designed and constructed on-line instruments to determine these parameters. Using these instruments Segovia et al. (2011) and Betancourt et al. (2013) presented algorithms for the automatic control of thickeners.

A research group in Melbourne Australia developed steady state thickener models based on the same theory but with slightly different variables and parameters. Their results are equivalent to those of Concha and co-workers (Green 1997; Green et al. 1998; De Kretser et al. 2001; Usher et al. 2001; Usher 2002; Shane et al. 2005).

8.2 Equipment

The continuous thickener is a typical device that has not changed much since Dorr invented it in 1905. They have become larger and are built of different materials, such as wood, steel or cement and their raking system has been improved and modernized, but their elements continue to be the same. Figure 8.4 shows the original *Dorr thickener*. The cylindrical tank is the body of the equipment, with the feedwell, overflow launder, rakes and underflow discharge, all common elements in any modern thickener.

In small units of less than 30 m in diameter, the cylindrical tank is steel or wood, while tanks up to 150 m in diameter are made of concrete. The bottom of the thickener is made of the same material as the tank and has a cone in the center to improve sediment evacuation.

The feedwell is a small concentric cylinder designed to thoroughly mix incoming pulp with the flocculant and in some cases to dilute the feed and deliver it evenly into the thickener. Many thickeners have baffles to accomplish these tasks. See Figs. 8.5 and 8.6. We will discuss this subject further in Sect. 8.7.

Rakes transport the sediment from the bottom of the tank to the underflow discharge orifice. The rakes, which can have several supporting structures, rotate at a rate on the order of one revolution per hour. A secondary effect of the rakes is to produce channels in the sediment through which water can escape to the surface, thus increasing the pulp density of the underflow. The rakes can have a central motor as shown in Fig. 8.6 or a peripheral tracking system, Fig. 8.7.

All thickener models have mechanism to lift the rakes whenever the torque to move the sediment becomes excessive. See Figs. 8.8 and 8.9.

Supernatant water overflows at the top of the thickener through overflow launders as shown in Fig. 8.10. The overflow launder at the tank periphery receives water recovered from the pulp and evacuates it slowly to avoid dragging fine particles. A flow of about 0.1 (m^3/min/m of perimeter) is common.

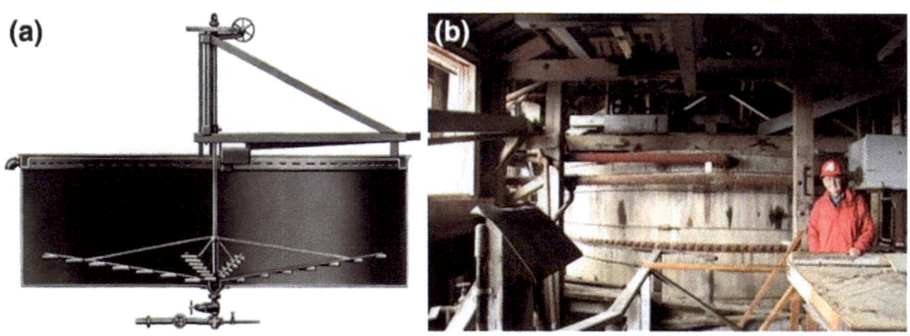

Fig. 8.4 Original Dorr thickener. **a** Schematic Dorr thickener invented in 1905 (Dorr 1936). **b** 6 m diameter Dorr thickener, Kennicott Alaska 1938 (conventional thickener)

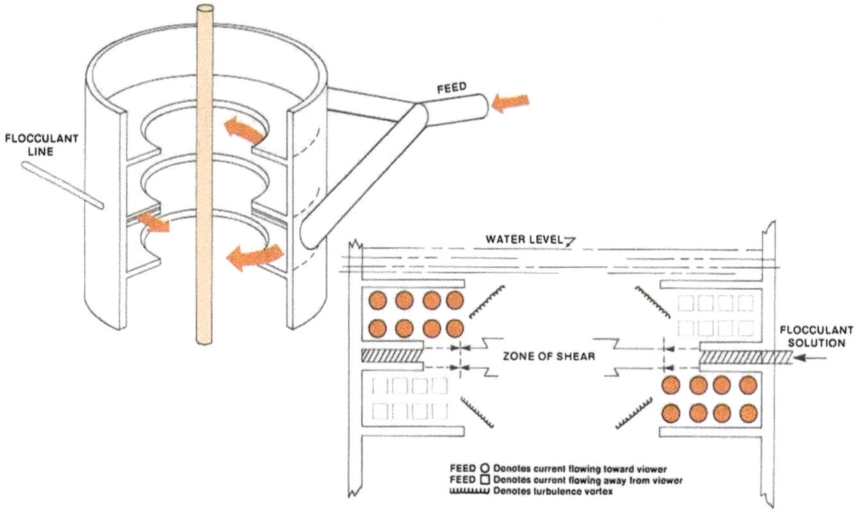

Fig. 8.5 Feedwell with two tangential entries and internal baffles (courtesy Dorr-Oliver)

Fig. 8.6 Small thickener with two tangential entries showing centrally driven rakes (courtesy Dorr-Oliver)

8.2.1 Conventional, High Rate, High Density and Paste Thickeners

Thickeners are usually classified into four types: conventional thickeners, high-rate thickeners, high-density thickeners and paste thickeners. See Fig. 8.11.

In a *Conventional Thickener* the feedwell is located in the upper part of the tank. When the feed enters it, it is diluted to the so-called *conjugate concentration*

Fig. 8.7 Thickener with peripheral tracking system (courtesy of Eimco process equipment)

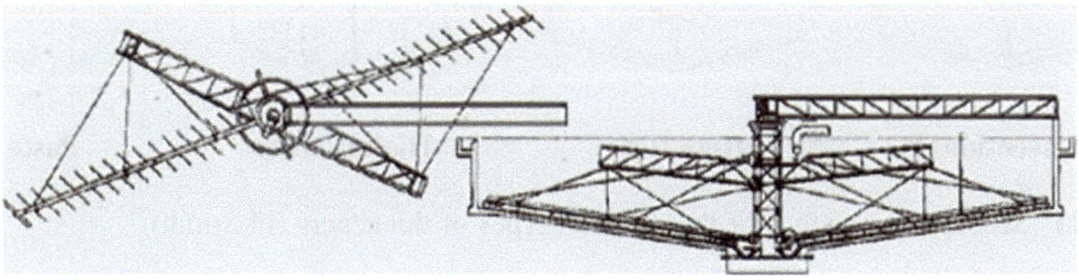

Fig. 8.8 Rakes with lifting system (courtesy of Dorr-Oliver)

Fig. 8.9 Rakes with lifting system (courtesy of Eimco process equipment)

by an upcoming flow of water. The diluted suspension settles at a constant velocity to form sediment at the bottom of the tank. Figure 8.12 shows a scheme of a conventional thickener.

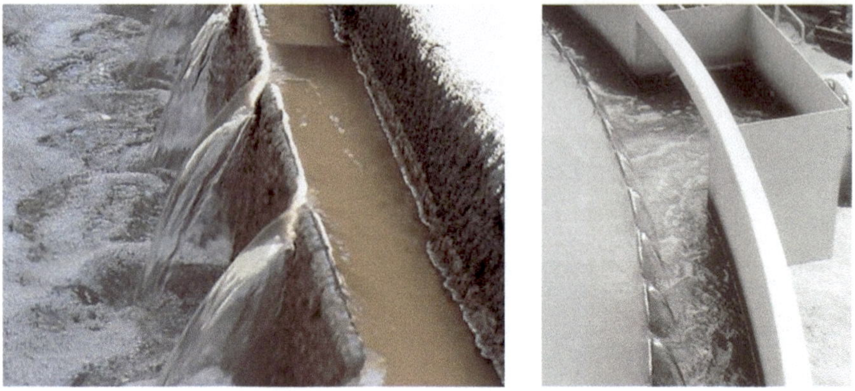

Fig. 8.10 Overflow launders (courtesy of Outokumpu Supaflo technologies)

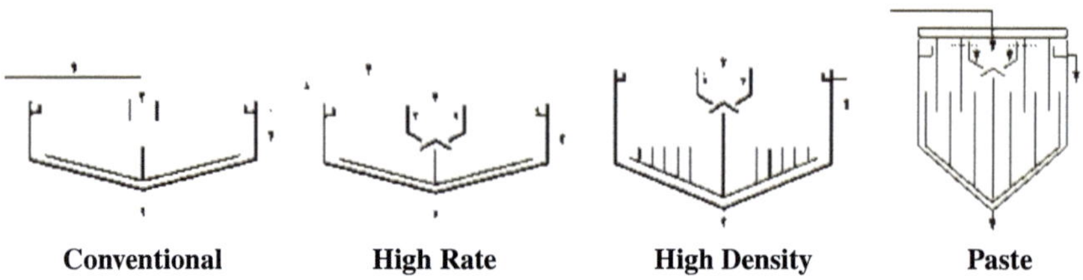

Fig. 8.11 Schematic drawings of the different types of thickeners (Fl Smith)

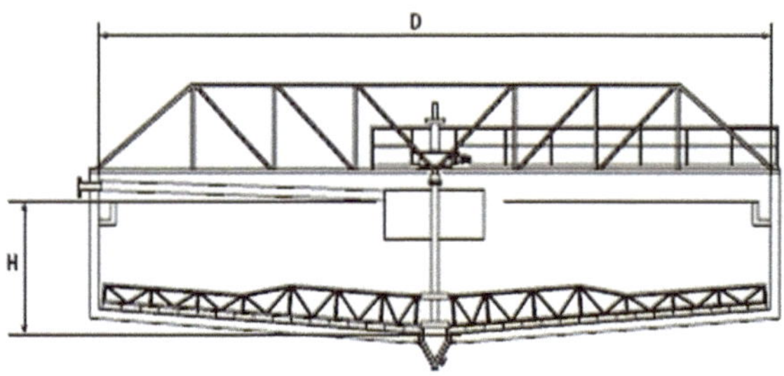

Fig. 8.12 Schematic drawings of a conventional thickener

In the early 1960s machines known as *high capacity* or *high rate thickeners* were introduced into the mining industry by various manufacturers (Eimco HI-CAP, Enviroclear High Capacity CT). See Fig. 8.13.

These thickeners have longer feedwells that deliver the feed directly into the sediment. When the feed is mixed with the high-density sediment, it increases in concentration forming a suspension with a concentration equal to or higher than the critical concentration. Therefore there is no settling zone in this type of thickener. Often the underflow of high capacity thickeners is recycled to the feed,

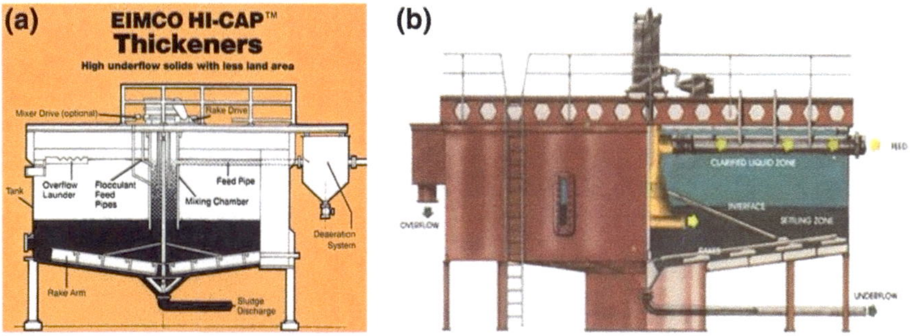

Fig. 8.13 Schematic view of high capacity thickeners. **a** EIMCO HI-CAP thickener (Eimco 2011). **b** EVIRO-CLEAR CT (Enviro Clear 2011)

producing a mixture of higher concentration that supposedly increases the equipment capacity.

Later in this chapter we will show that it is more appropriate to talk of conventional and high capacity *operations* than of conventional and high capacity *thickeners*, and furthermore, that if one chooses to operate a thickener in a high capacity mode, there is always a conventional operation that has equal or higher capacity. It is the flocculant dose, the feed dilution and the optimum shear rate that eventually defines the capacity of a thickener.

The term *high capacity* or *high rate thickeners* is used for small to medium sized thickener but also for large conventional thickeners processing very high tonnage due to the optimization of flocculation by careful choice of the flocculant dose and feed slurry concentration. See Fig. 8.14.

High-density and paste thickeners are similar to conventional or high rate thickeners, but with steeper cones and much higher cylindrical tanks. The additional height produces more pressure on the sediment at the bottom of the tank and therefore denser underflow. Picket fences are used in both types of thickeners to help consolidate the sediment. The only difference between them is that paste thickeners are much taller and slimmer than High Density thickeners. Figure 8.15 shows a schematic view and Figs. 8.16 and 8.17 show high-density thickeners.

Fig. 8.14 125 m high-rate thickener (Smith 2013)

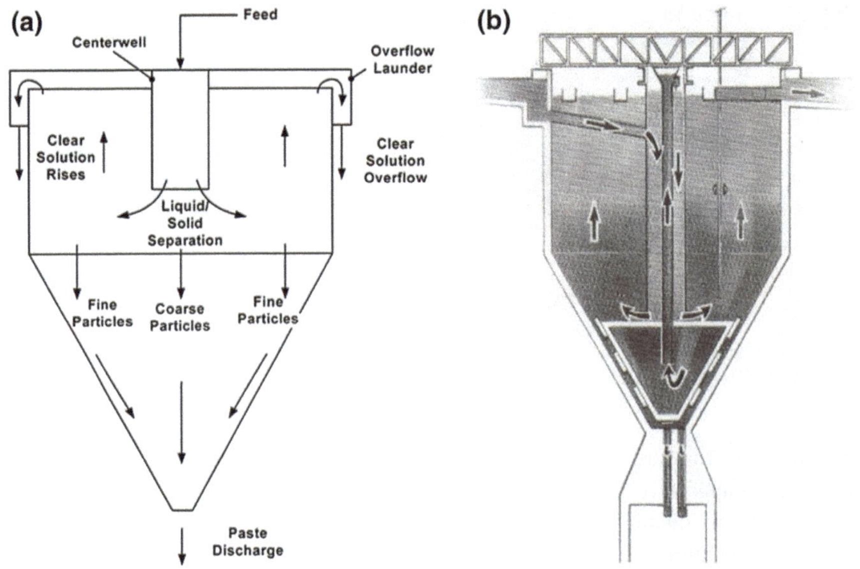

Fig. 8.15 Schematic views of paste thickeners. **a** Conventional mode of operation Innovat (2013). **b** High capacity mode of operation Delkor (2013)

Fig. 8.16 View of high density and paste thickeners (courtesy of Eimco process equipment)

Fig. 8.17 View of paste thickeners. (courtesy of Eimco process equipment)

8.3 Thickening Theory

Thickening consists of the superposition of two phenomena, sedimentation and consolidation, which behave differently.

During sedimentation, particles settle individually by collision among them and through the fluid by pressure and friction. See Fig. 8.18a. Sedimentation of individual particles and suspensions, which was analyzed in Chaps. 4 and 5, consist of gravity settling particles or flocs in a fluid, in our case, water. At a certain

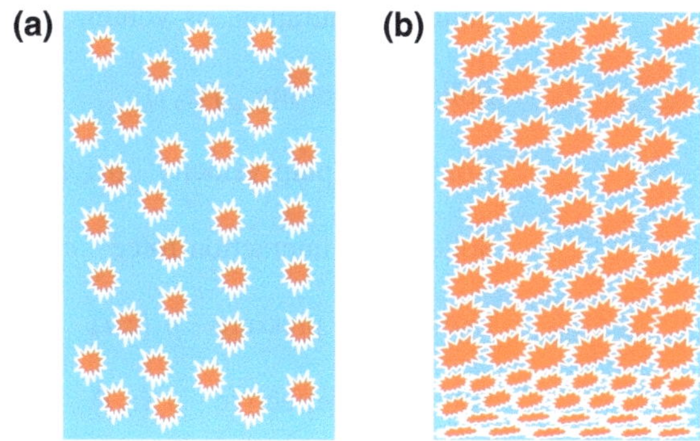

Fig. 8.18 Physical model of sedimentation and consolidation. **a** Sedimentation. **b** Consolidation

concentration, particles begin to touch each other transforming the suspension into a network of solid particles called *sediment*. Now, forces among particles are transmitted directly from particle to particle. If settling particles that reach the bottom of the vessel and lay one on top of the other are incompressible, such as glass beads, the whole process ends, but if they are compressible, as is the case with flocculated copper flotation tailings, the weight of the sediment compresses the flocs lying underneath expelling the water from the pores. This phenomenon of extracting water by compression is called *consolidation*. See Fig. 8.18b.

Consolidation is an important discipline in several fields besides thickening, for example in Geotechnique, which studies the behavior of the ground when buildings are constructed. In this case, the applied force on the consolidating material is the external weight of buildings. In thickening, the force compressing the sediment is its own weight, which makes a major difference in the theory of consolidation given the scale of the forces.

Sedimentation and consolidation are usually separated by an interface having a characteristic concentration. The concentration of this interface, where the sedimenting particles or flocs begin to touch each other, is called a *critical concentration*, also known as *compressive yield point* by authors from the field of colloidal science.

8.3.1 Dynamic Thickening Process

Field Equations

The phenomenological theory of sedimentation ignores the individuality and physical structure of particles and fluids, and considers the solid and the fluid as continuous media. The properties of the solid and fluid involved in thickening are those describing sedimentation and consolidation: (1) the concentration of the suspension; (2) the solid phase velocity, (3) the fluid phase velocity, (4) the solid–fluid interaction force, (5) the excess pore pressure and (6) the compressibility of the sediment. Properties (1) to (6) have the following associated *field variables*. See Chap. 3.

$$\text{Solid concentration, as volume fraction} \quad \varphi(z,t) \tag{8.1}$$

$$\text{Solid flux density} \quad f(z,t) \tag{8.2}$$

$$\text{Convective pulp velocity} \quad q(z,t) \tag{8.3}$$

$$\text{Solid--fluid interaction force} \quad m_d(z,t) \tag{8.4}$$

$$\text{Excess pore pressure} \quad p_e(z,t) \tag{8.5}$$

$$\text{Solid stress effective} \quad \sigma_e(z,t) \tag{8.6}$$

These variables constitute a *dynamic thickening process* in regions where the variables are continuous when they obey the following four *field equations*. See Chap. 3.

$$\frac{\partial \varphi}{\partial t} + \frac{\partial f}{\partial z} = 0 \tag{8.7}$$

$$\frac{\partial q}{\partial z} = 0, \quad \text{with} \quad q = v_s - (1-\varphi)v_r \tag{8.8}$$

$$\frac{\partial \sigma_e}{\partial z} = -\Delta\rho\varphi g + \frac{m_d}{1-\varphi} \tag{8.9}$$

$$\frac{\partial p_e}{\partial z} = -\frac{m_d}{1-\varphi} \tag{8.10}$$

where $f = \varphi v_s$ is the solid flux density function, $v_r = v_s - v_f$ is the relative solid–fluid velocity and m_d is the solid–fluid dynamic interaction force. To understand the conditions for these equations to be valid, see Chap. 3.

At discontinuities, the field equations are not valid and the following *jump conditions* should be used:

$$[f] = \sigma[\varphi]; \quad [q] = 0 \tag{8.11}$$

where σ is the displacement velocity of the discontinuity. For conditions that σ must obey, see Eqs. (5.12, 5.13).

Constitutive Equations

Since we have six field variables and only four field equations, two constitutive equations should be established to describe the relationship between the dynamic variables m_d and σ_e and the kinematic variables φ and v_r:

$$m_d = m_d(\varphi, \varphi_c v_r) \tag{8.12}$$

$$\sigma_e = \sigma_e(\varphi, \varphi_c, v_r) \tag{8.13}$$

Compressibility of the Sediment

Experience has demonstrated that the solid effective stress can be expressed solely as a function of the concentration. As we noted at the beginning of this chapter, the only way momentum is transferred directly from particle to particle during sedimentation is by particle collision. Therefore, for solid concentrations below the critical level where particles are suspended in the fluid, the effective stress is a constant. At concentration greater than the critical concentration, the network of particles formed transmits forces directly among particles in the sediment. The self-weight of the particles is transmitted through the network producing compression and increasing the solid concentration of the sediment. This phenomenon can be characterized by the following restriction for solid effective solid stress $\sigma_e(\varphi)$:

$$\sigma_e' = \frac{d\sigma_e}{d\varphi} = \begin{cases} 0 & \text{for } \varphi < \varphi_c \\ \geq 0 & \text{for } \varphi \geq \varphi_c \end{cases} \tag{8.14}$$

Two common expressions have been used for the solid effective stress:

$$\sigma_e(\varphi) = \begin{cases} \text{constant for} & \varphi < \varphi_c \\ \alpha \exp(\beta\varphi) \text{for} & \varphi \geq \varphi_c \end{cases} \tag{8.15}$$

$$\sigma_e(\varphi) = \begin{cases} \text{constant} & \text{for} \quad \varphi < \varphi_c \\ \sigma_0\left(\left(\frac{\varphi}{\varphi_c}\right)^n - 1\right) & \text{for} \quad \varphi \geq \varphi_c \end{cases} \tag{8.16}$$

Solid–Fluid Interaction Force During Sedimentation $(\varphi < \varphi_c)$

During settling, for $\varphi < \varphi_c$, the flocs move slowly through the fluid and the hydro-dynamical force can be represented as a linear function of relative solid–fluid velocity:

$$m_d = -\mu K(\varphi) v_r \tag{8.17}$$

where $K(\varphi)$ is the *translational solid–fluid resistance coefficient*. Replacing $K(\varphi)$ in Eq. (8.9), with $\sigma_e'(\varphi) = 0$, yields:

$$0 = -\Delta\rho\varphi g - \frac{\mu K(\varphi) v_r}{1 - \varphi}; \quad \varphi < \varphi_c$$

from which the relative solid–fluid velocity v_r is:

$$v_r = -\frac{\Delta\rho\varphi(1 - \varphi)g}{\mu K(\varphi)} \tag{8.18}$$

Replacing v_r from (8.18) in (8.8); from the result calculate v_s and, multiplying by φ, obtain:

$$f = q\varphi - \frac{\Delta\rho\varphi^2(1 - \varphi)^2 g}{\mu K(\varphi)} \tag{8.19}$$

Defining a parameter $f_{bk}(\varphi)$ in the form:

$$f_{bk}(\varphi) = -\frac{\Delta\rho\varphi^2(1 - \varphi)^2 g}{\mu K(\varphi)}, \tag{8.20}$$

Eq. (8.19) can be written in the form:

$$f = q\varphi + f_{bk}(\varphi), \quad \text{for } \varphi < \varphi_c \tag{8.21}$$

Finally, replacing (8.21) in (8.7) yields:

$$\frac{\partial \varphi}{\partial t} + \frac{\partial}{\partial z}(q\varphi + f_{bk}(\varphi)) = 0, \quad \text{for } \varphi < \varphi_c \tag{8.22}$$

Equation (8.22) describes the *sedimentation of particles, compressible or not, in a suspension of less than the critical concentration* $(\varphi < \varphi_c)$. It can be identified as Kynch's equation for continuous sedimentation of ideal suspensions. See Eqs. 5.8 and 5.10. The function $f_{bk}(\varphi)$, defined by (8.20), is the *Kynch solid flux density function* and can be considered the constitutive equation for the sedimentation process. The Kynch sedimentation process is completely determined once the constitutive equation and the initial conditions are established.

Adding Eqs. (8.9) and (8.10) gives:

$$\frac{\partial (p_e + \sigma_e)}{\partial z} = -\Delta\rho\varphi g$$

from which the excess pore pressure can be obtained once (8.22) is solved for $\varphi(z)$:

$$p_e(z) = -\left(\sigma_e + \Delta\rho g \int_z^L \varphi(\xi) d\xi\right) \tag{8.23}$$

Solid–Fluid Interaction Force During Consolidation $(\varphi \geq \varphi_c)$

During consolidation, the fluid moves slowly though the porous bed constituting the sediment and can be quantified by the permeability of the bed and the viscosity of the fluid. For the slow motion of a Newtonian fluid through a compressible porous bed, Darcy's equation is valid (see Chap. 6):

$$m_d = -\frac{\mu}{k(\varphi)}(1-\varphi)^2 v_r \tag{8.24}$$

where $k(\varphi)$ and μ are the permeability of the sediment and the fluid viscosity respectively. Replacing m_d from (8.24) in (8.9) and from this result calculate v_r:

$$v_r = -\frac{k(\varphi)}{\mu} \times \frac{\Delta\rho\varphi g}{(1-\varphi)}\left(1 + \frac{\sigma_e'(\varphi)}{\Delta\rho\varphi g}\frac{\partial \varphi}{\partial z}\right) \quad \text{for } \varphi \geq \varphi_c \tag{8.25}$$

Replacing (8.25) in (8.8) and multiplying by φ, gives us:

$$f = q\varphi - \frac{k(\varphi)}{\mu} \times \frac{\Delta\rho\varphi^2 g}{(1-\varphi)}\left(1 + \frac{\sigma_e'(\varphi)}{\Delta\rho\varphi g}\right)\frac{\partial \varphi}{\partial z} \quad \text{for } \varphi \geq \varphi_c \tag{8.26}$$

Parameter $f_{bk}(\varphi)$, for $\varphi \geq \varphi_c$, is defined in the form:

$$f_{bk}(\varphi) = -\frac{k(\varphi)}{\mu} \times \frac{\Delta\rho\varphi^2 g}{(1-\varphi)} \quad \text{for } \varphi \geq \varphi_c, \tag{8.27}$$

and note that Eq. (8.27) has the same role as Eq. (8.20) but now permeability $k(\varphi)$ is the parameter instead of the translational resistance coefficient $K(\varphi)$.

Equations (8.20) and (8.27) can be combined to define an extended Kynch solid flux density function for all values of φ:

$$f_{bk}(\varphi) = \begin{cases} -\dfrac{\Delta\rho\varphi^2(1-\varphi)^2 g}{\mu K(\varphi)} & \text{for } \varphi < \varphi_c \\ -\dfrac{k(\varphi)}{\mu}\Delta\rho\varphi^2 g & \text{for } \varphi \geq \varphi_c \end{cases}, \tag{8.28}$$

Substituting (8.28) on Eq. (8.8), we can write the solid flux density function for the whole range of concentration in the form:

$$f = q\varphi + f_{bk}(\varphi)\left(1 + \frac{\sigma_e'(\varphi)}{\Delta\rho\varphi g}\right)\frac{\partial\varphi}{\partial z} \quad \text{for } 0 < \varphi < 1 \tag{8.29}$$

Thickening Equation

Replacing (8.29) in (8.7) yields:

$$\frac{\partial\varphi}{\partial t} + \frac{\partial}{\partial z}\left(q\varphi + f_{bk}(\varphi)\left(1 + \frac{\sigma_e'(\varphi)}{\Delta\rho\varphi g}\frac{\partial\varphi}{\partial z}\right)\right) = 0, \quad \text{for } 0 < \varphi < 1 \tag{8.30}$$

which can be written in the form

$$\frac{\partial\varphi}{\partial t} + \frac{\partial}{\partial z}(q\varphi + f_{bk}(\varphi)) = \frac{\partial}{\partial z}\left(-\frac{f_{bk}(\varphi)\sigma_e'(\varphi)}{\Delta\rho\varphi g}\frac{\partial\varphi}{\partial z}\right), \quad \text{for } 0 < \varphi < 1 \tag{8.31}$$

Equation (8.31) is a *degenerate parabolic partial differential equation*. The name comes from the fact that for values of $\varphi < \varphi_c$; $\sigma_e(\varphi) = 0$ the equation becomes *hyperbolic*:

$$\frac{\partial\varphi}{\partial t} + \frac{\partial}{\partial z}(q\varphi + f_{bk}(\varphi)) = 0, \quad \text{for } \varphi < \varphi_c \tag{8.32}$$

This result shows that for *compressible flocculated suspensions Kynch's equation is still valid in those regions where the concentration is less than critical.*

Still another form of expressing Eq. (8.31) is obtained by defining the *diffusion coefficient* $\mathfrak{D}$ in the form:

$$\mathfrak{D} = \begin{cases} 0 & \text{for } \varphi < \varphi_c \\ -\dfrac{f_{bk}(\varphi)\sigma_e'(\varphi)}{\Delta\rho\varphi g} & \text{for } \varphi \geq \varphi_c \end{cases} \tag{8.33}$$

then replacing (8.33) in (8.31) the following alternative of the thickening equation is obtained as a *convective-diffusion equation*:

$$\frac{\partial \varphi}{\partial t}+\frac{\partial}{\partial z}(q\varphi+f_{bk}(\varphi))=\frac{\partial}{\partial z}\left(\mathfrak{D}(\varphi)\frac{\partial \varphi}{\partial \mathfrak{z}}\right),\quad \text{for } 0\leq\varphi<1 \tag{8.34}$$

8.3.2 Batch Thickening

To obtain a description of batch thickening, we will repeat and complete the stages in the sedimentation process described in Chap. 5.

(a) Before sedimentation starts, the suspension is flocculated and homogenized by agitation so that its concentration is constant.
(b) At the beginning of sedimentation, all flocs settle at the same speed forming a well-defined water-suspension interface in the upper part of the column. This interface descends at the same speed as the flocs. This stage is called *hindered settling*. A diffuse interface indicates incomplete flocculation, especially of the fine particles.
(c) When the flocs reach the bottom of the column, they rapidly occupy the entire available area forming a *sediment*. From then on, the flocs accumulate one on top of another, compressing those lying underneath. We say the sediment is under *compression* or *consolidation*.
(d) The interface between the sediment and the settling suspension has no flocs lying on top of it and therefore suffers no compression. The concentration at which this occurs is called *critical concentration*.
(e) Following a given constant concentration φ over a period of time, for example, setting an X Ray instrument such as that described by Been and Sills (1981) at a given concentration, it will move upwards during sedimentation from the bottom of the column at a constant characteristic speed as time passes. This upward motion is termed the *wave of constant concentration* φ.
(f) At a certain instant, the water-suspension and the suspension-sediment interfaces meet. The coordinates at this time are called *critical height* and *critical time* and they define the *critical point* where hindered sedimentation ends and flow in the porous media and consolidation prevails. In time, consolidation ends and a characteristic concentration profile is established in the column. Water in the upper part is followed by a concentration gradient from the critical concentration at the top to the sediment to a maximum concentration at the bottom of the column. Figure 8.19 is a typical sedimentation curve showing the interfaces and constant concentration lines.

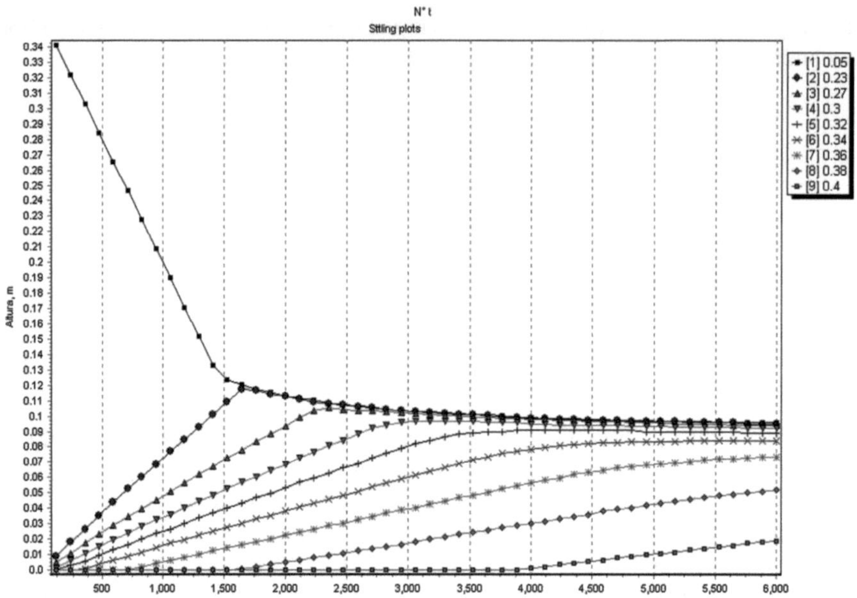

Fig. 8.19 Sedimentation *curve* for a copper flotation tailing, showing the water-suspension interface ($\varphi = 0.05$) and the suspension-sediment interfaces ($\varphi = 0.23$) and several *lines* of constant concentration. Data from Becker (1982)

(a) **Initial and boundary value problem**

In batch sedimentation there is no convective flow, therefore $q = 0$ and (8.31) reduce to:

$$\frac{\partial \varphi}{\partial t} + \frac{\partial}{\partial z}\left(f_{bk}(\varphi)\left(1 + \frac{1}{\Delta\rho\varphi g}\frac{\partial \sigma_e(\varphi)}{\partial z}\right)\right) = 0, \quad 0 \leq z \leq L; \text{ for } 0 < \varphi < 1 \quad (8.35)$$

From stage (a) of batch sedimentation, the initial condition is known and is set at $\varphi(z,t) = \varphi_0$ for $0 \leq z \leq L$, and the boundary condition at $z = L$ is $\varphi(L,t) = 0$.

At the bottom of the column, where $z = 0$, the velocity of the solid is zero and therefore the solid flux $f_{bk}(0,t) = 0$. From (8.29), for $t \geq 0$, the boundary condition is $\partial\varphi/\partial z|_{z=0} = \Delta\rho\varphi g/\sigma_e'(\varphi)$. Then the initial-boundary condition can be expresses as:

$$\varphi(z,0) = \varphi_0, \quad \text{for } 0 \leq z \leq L \quad (8.36)$$

$$\varphi(L,t) = 0, \quad \text{for } t > 0 \quad (8.37)$$

$$\left.\frac{\partial \varphi}{\partial z}\right|_{z=0} = \frac{\Delta\rho\varphi g}{\sigma_e'(\varphi)}, \quad \text{for t} \geq 0 \quad (8.38)$$

where L and φ_0 are the initial height and initial concentration of the suspension.

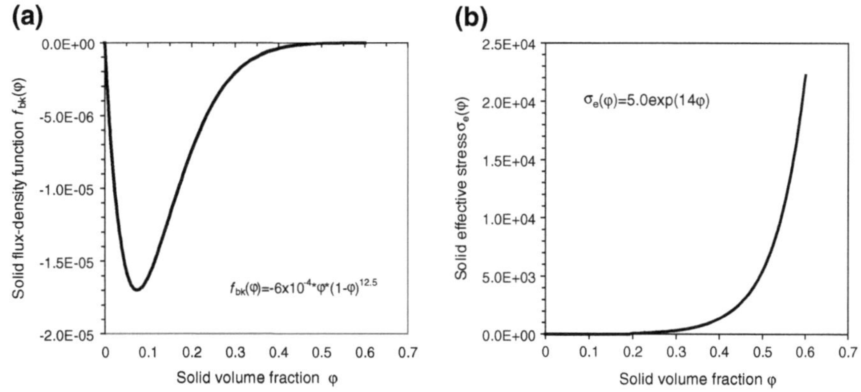

Fig. 8.20 Thickening parameters for a flocculated copper flotation tailing. Data from Becker (1982). **a** Solid flux density function. **b** Solid effective stress

Figure 8.20 shows the typical form of parameters $f_{bk}(\varphi)$ and $\sigma_e(\varphi)$.

$$f_{bk}(\varphi) = u_\infty \varphi (1-\varphi)^c \quad \text{and} \quad \sigma_e(\varphi) = \begin{cases} 0 & \text{for} \quad \varphi \leq \varphi_c \\ \alpha \exp(\beta\varphi) & \text{for} \quad \varphi > \varphi_c \end{cases} \tag{8.39}$$

where u_∞, c, α and β are empirical parameters.

Mathematical analysis by Bürger et al. (2000b) imply that the initial-boundary value problem, given by Eqs. (8.35) to (8.39), has a unique solution depending on φ_0 and is therefore well posed, even in cases where $\sigma'_e(\varphi)$ is discontinuous at $\varphi = \varphi_c$.

(b) **Numerical solution**

The initial-boundary value problem is solved numerically using a finite-difference operator splitting method described by Bürger and Concha (1998). The equation is split into a second order diffusion equation, a linear convective equation and a non-linear first order hyperbolic equation, which are solved numerically for each time step by an implicit finite difference method, a second order upwind method and a second order total-variation diminishing method, respectively. For details see Bürger and Concha (1997, 1998), Bustos et al. (1999) and Bürger et al. (2000c). The algorithm for the solution of the equation is given in Fig. 8.21.

The solution to this problem, with parameters shown in Fig. 8.20, is given by the settling curve in Fig. 8.19 and the concentration profile in Fig. 8.22.

(c) **Simulation and comparison with published experimental results**

Several cases published in the literature were simulated with Eq. (8.35) and compared with the experimental results (Bürger et al. 1999, 2000a; Garrido et al. 1999). All simulations of batch sedimentation with the phenomenological model

Fig. 8.21 Algorithm for the simulation of batch sedimentation according to Garrido et al. (2001)

$$L, t, \varphi_0, \varphi_c, \rho_s, \rho_f, u_\infty, c, \alpha, \beta \text{ or } \sigma_0, n$$

$$\Downarrow$$

Determine Δt romthe CFL condition

Solve the mdel functions

$$\frac{\partial \varphi}{\partial t} + \frac{\partial f_{bk}(\varphi)}{\partial z} = -\frac{\partial}{\partial z}\left(\frac{f_{bk}(\varphi) \times \sigma_e'(\varphi)}{\Delta \rho \varphi g} \frac{d\varphi}{dz} \right)$$

Using the model functions

$$f_{bk}(\varphi) = u_\infty \varphi (1-\varphi)^c$$

$$\sigma_e(\varphi) = \alpha \exp(\beta \varphi)$$

by the generalized upwind method with extrapolation.

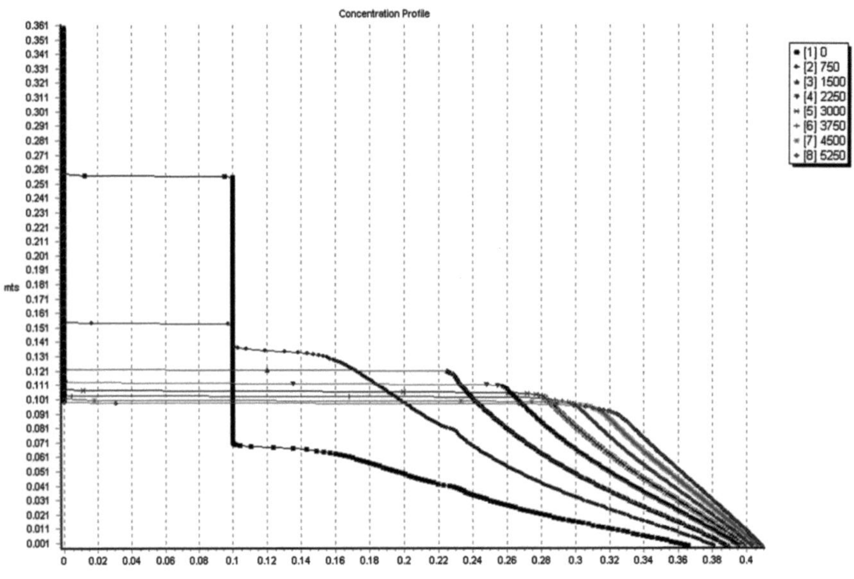

Fig. 8.22 Concentration profile for batch sedimentation with parameters given in Fig. 8.20

closely approximate the experimental results for the settling curve, concentration profile and excess pressure profile. These results affirm the value of the model presented in this work. See Figs. 8.22, 8.23, 8.24, 8.25, 8.26, 8.27.

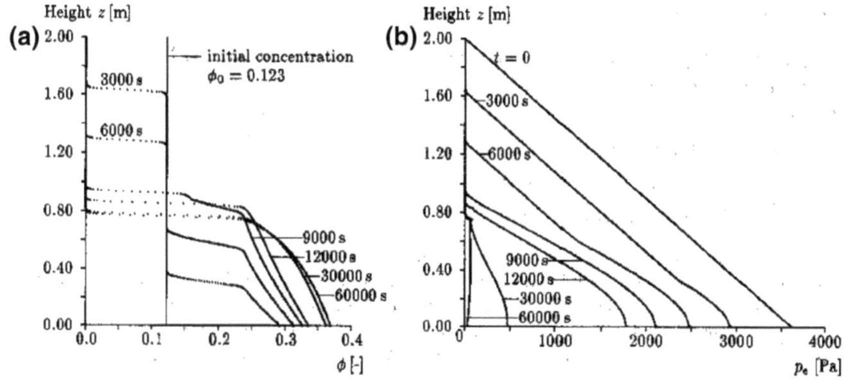

Fig. 8.23 Simulation of the sedimentation of a flotation copper tailing (Bürger and Concha 1998). **a** Concentration profiles. **b** Excess pore pressure profiles

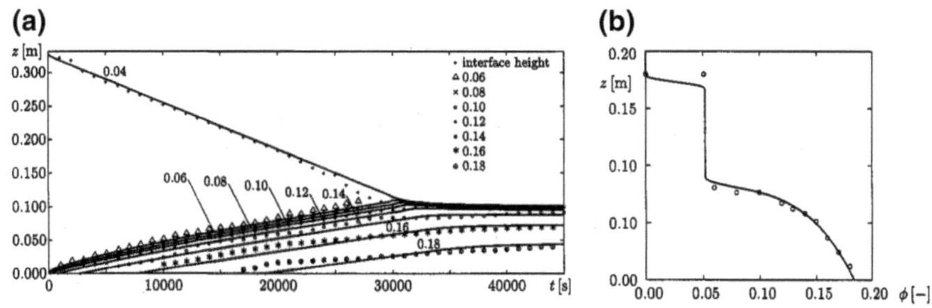

Fig. 8.24 Simulation and experimental sedimentation data of Holdich and Butt (1997) (Garrido et al. 2000). **a** Settling curve. **b** Concentration profile for 22.000 s

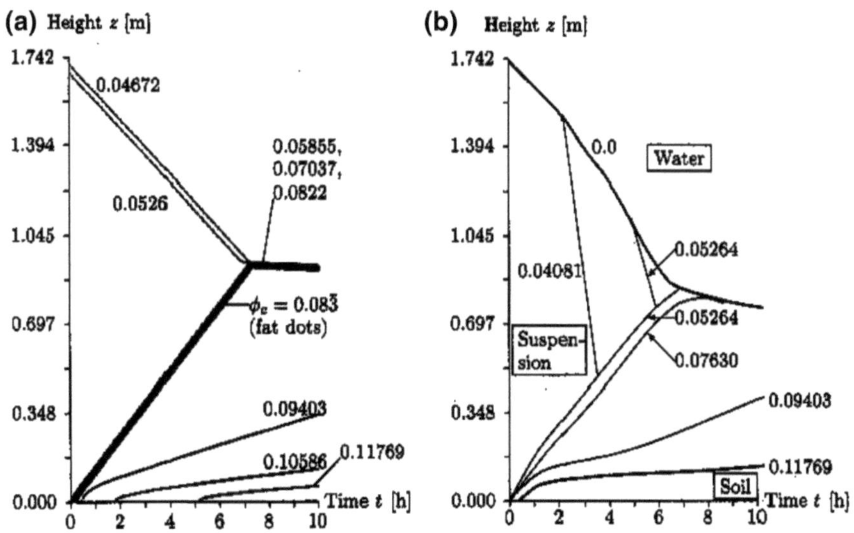

Fig. 8.25 Comparison of simulated and experimental results of Been and Sills (1981) for the settling and consolidation of a soil. **a** Simulation. **b** Experimental

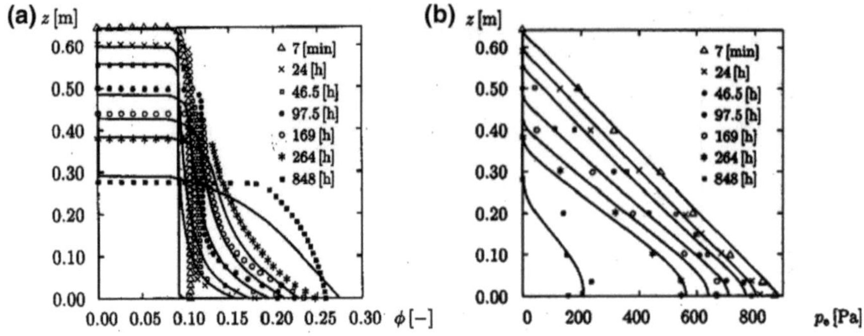

Fig. 8.26 Simulation and experimental data from Been y Sills (1981) experiment N° 11, (Bürger et al. 2000a). **a** Concentration profile. **b** Excess pore pressure profile

8.3.3 Model of Conventional Thickening

To analyze continuous conventional thickening it is useful to study separately steady state and transient behavior. For steady state, the field equations are significantly simplified with $\partial\varphi/\partial t = 0$ and there is no need to solve the entire degenerate parabolic equation numerically.

(a) **Field equations at steady state**

Equations (8.35) to (8.39) represent the transient evolution of the field variables in a continuous thickener. Eliminating the time dependence of these equations, the steady state in a continuous thickener is obtained for regions where the variables are continuous:

$$\frac{\partial f}{\partial z} = 0 \tag{8.40}$$

$$\frac{\partial q}{\partial z} = 0, \quad \text{with } q = v_s - (1-\varphi)v_r \tag{8.41}$$

$$\frac{\partial \sigma_e}{\partial z} = -\Delta\rho\varphi g + \frac{m_d}{1-\varphi} \tag{8.42}$$

$$\frac{\partial p_e}{\partial z} + \frac{\partial \sigma_e}{\partial z} = -\Delta\rho\varphi g \tag{8.43}$$

Equations (8.40) and (8.41) indicate that the solid flux density f and the volume average velocity q are constant in the thickener and are determined by boundary conditions. At discontinuities, fluxes are continuous and $\sigma[\varphi^+, \varphi^-] = 0$ and $[q] = 0$.

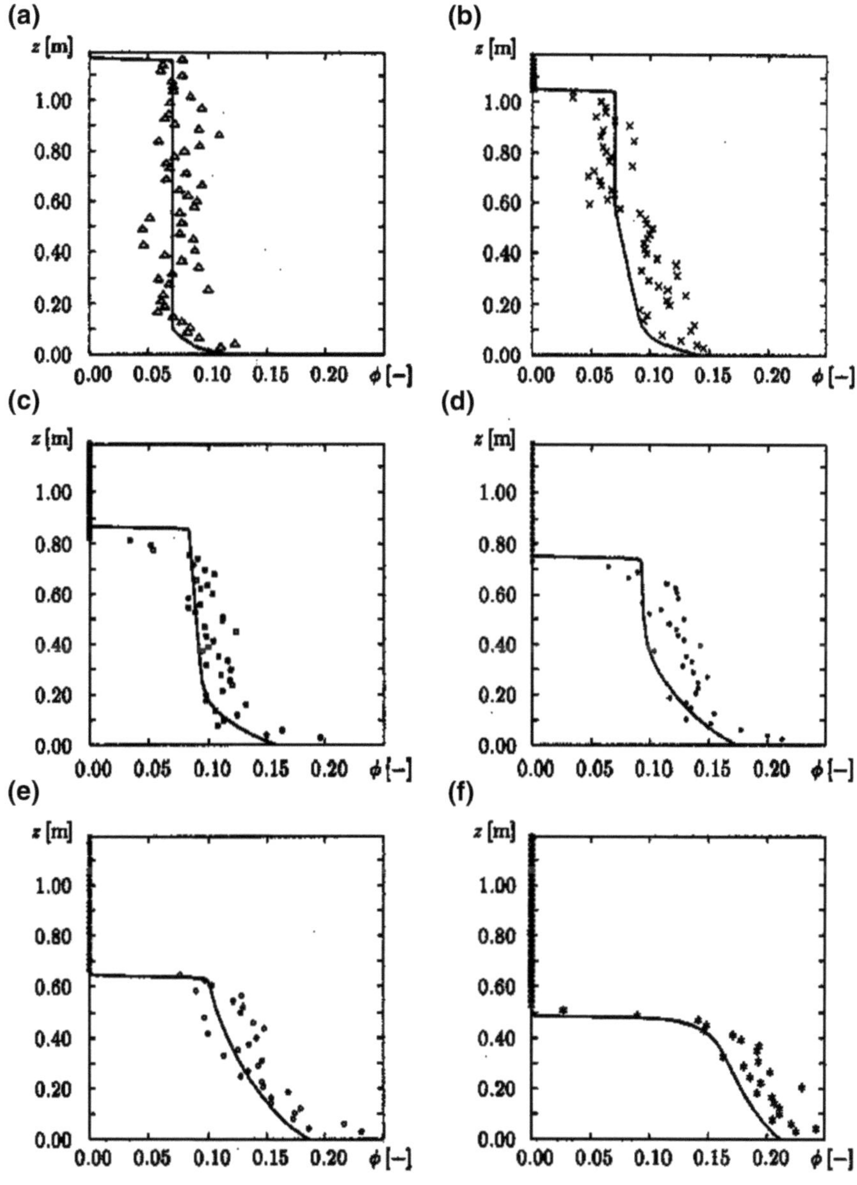

Fig. 8.27 Comparison of simulated results with experimental results of kaolin suspensions. Dreher (1997): **a** t = 0.052 days, **b** t = 0.312 days, **c** t = 0.87 days, **d** t = 2.13 days, **e** 3.91 days and f = 13.27 days (Bürger et al. 2000)

(b) **Solution of the boundary value problem**

Feed

The feed to the thickener is assumed totally flocculated in the feedwell and is mixed with the upcoming water as it enters the thickener body at $z = L$. It dilutes rapidly and spreads evenly throughout the thickener's cross-section. The only source of solids in the thickener is the feed and, since no solid passes to the overflow, the solid flux density is continuous at the feed level. This phenomenon is modeled by assuming a *surface source* of strength f_F at the feed level L. Ideal Continuous Thickener (ICT), (see Sect. 5.3.1 for the definition). If we choose $Q_F > 0$, the volume flow of the feed pulp of solid concentration φ_F, the solid feed flux density f_F is defined by:

$$f_F = \frac{-Q_F \varphi_F}{S} < 0$$

where S is the thickener's cross-sectional area. Then, by (8.40):

$$f(z) = f_F$$

Due to the dilution of the feed, the concentration at the feed level φ_L, known as *conjugate concentration* can be obtained by solving the implicit equation:

$$f(L) = f_F = q\varphi_L + f_{bk}(\varphi_L)$$

The thickener will show a zone I with water (see Fig. 8.28), a zone II with the constant conjugate concentration φ_L, a zone III with varying concentration from φ_c to φ_L, depending on the *Mode of Continuous Sedimentation* (see Chap. 5, Sect. 5.2.2) and a zone IV of sediment.

Underflow

At the underflow, $z = 0$, the pulp of concentration φ_D is evacuated by gravity or by pumping through an orifice at the bottom of the thickener at a volume flow rate of Q_D without mixing with any other source of solid or water. We model this as a surface sink f_D defined by:

$$f_D = \frac{-Q_D \varphi_D}{S}$$

The volume average velocity at the underflow is:

$$q = -\frac{Q_D}{S} = \frac{f_D}{\varphi_D}$$

The solid flux density at the underflow is then:

$$f_D = f_F = q\varphi_D \tag{8.44}$$

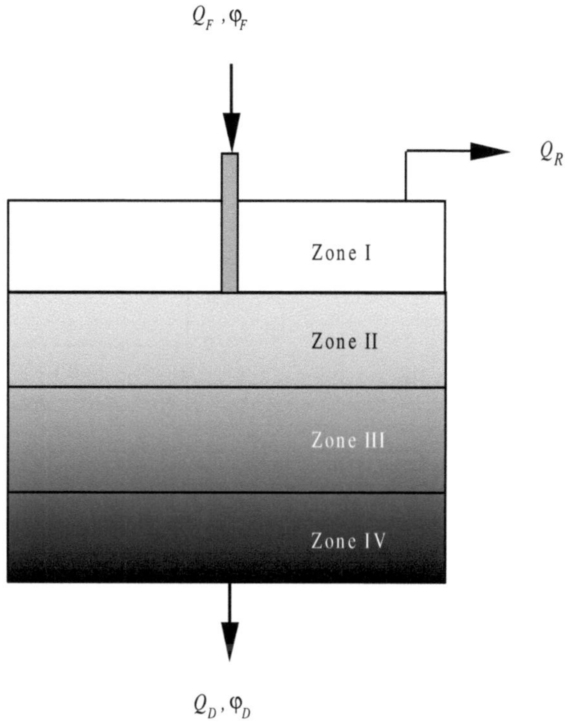

Fig. 8.28 Ideal conventional thickener (*ICT*). Zone I water, Zone II suspension of constant concentration, Zone III suspension of variable concentration and Zone IV sediment

Eq. (8.41) shows that q is independent of z, therefore:

$$q(z) = q = \frac{f_F}{\varphi_D} \tag{8.45}$$

A zone IV of increasing concentration will form below zone III, from the critical concentration φ_c to the underflow concentration φ_D. Figure 8.28 shows the four zones in a thickener at a steady state. The concentration profile $z = z(\varphi)$ in zone IV at steady state can be obtained from Eq. (8.42) $\partial z/\partial\varphi$:

$$\frac{\partial z(\varphi)}{\partial\varphi} = \frac{f_{bk}(\varphi)\sigma'_e(\varphi)}{\Delta\rho\varphi g}\frac{1}{(f_F - f_{bk}(\varphi) - q\varphi)} \tag{8.46}$$

Integrating this expression with $q = f_F/\varphi_D$ and boundary condition $z = 0$ and $\varphi = \varphi_D$ results in:

$$z(\varphi) = \int_{\varphi_D}^{\varphi} \frac{f_{bk}(\xi)\sigma'_e(\xi)}{\Delta\rho\xi g(f_F(1 - \xi/\varphi_D) - f_{bk}(\xi))} d\xi \tag{8.47}$$

Figure 8.29 shows a typical concentration profile of the sediment in a continuous thickener.

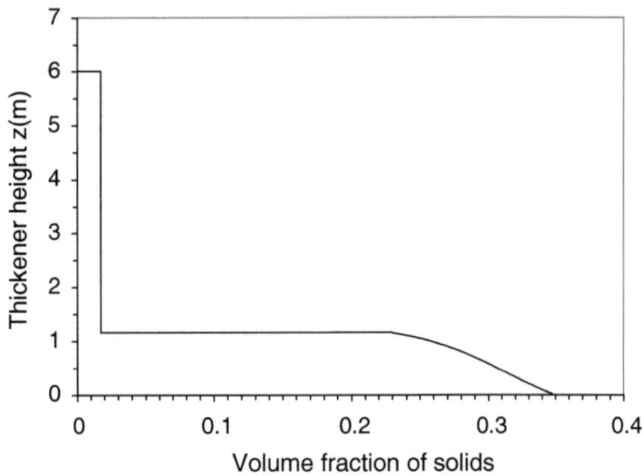

Fig. 8.29 Typical concentration profile in a continuous thickener at steady state

Overflow

In a conventional thickener at steady state no solid particle passes to the overflow and therefore solids are restricted to $0 \leq z \leq L$. If the overflow of water is designated by $Q_R > 0$, the macroscopic balance of pulp, solid and water in the thickener is:

$$\text{Solid}: \quad Q_F\varphi_F = Q_D\varphi_D$$

$$\text{Pulp}: \quad Q_R = Q_F - Q_D \tag{8.48}$$

$$\text{Water}: \quad Q_R = Q_F\left(1 - \frac{\varphi_F}{\varphi_D}\right) \equiv Q_D\left(\frac{\varphi_D}{\varphi_F} - 1\right) \tag{8.49}$$

(c) **Existence of a steady state**

The concentration in a thickener is either constant or increases downwards, that is, the concentration gradient is zero or negative, $\partial\varphi/\partial z \leq 0$. From Eq. (8.46):

$$\frac{\partial\varphi}{\partial z} = -\underbrace{\frac{\Delta\rho\varphi g}{\sigma_e'(\varphi)f_{bk}(\varphi)}}_{>0}\underbrace{(q\varphi + f_{bk}(\varphi) - f_F)}_{\leq 0} \leq 0 \tag{8.50}$$

Since $f_{bk} < 0$, the first term of (8.50) is positive and the term in round braked must be less than or equal to zero, that is:

$$f_k = q\varphi + f_{bk}(\varphi) \leq f_F \tag{8.51}$$

The Inequality (8.51) indicates that *at steady state in a flux-density function versus concentration plot, in the range $\varphi_L \leq \varphi \leq \varphi_D$, the term $q\varphi + f_{bk}(\varphi)$ should always be below the straight horizontal line representing the feed flux density f_F.*

Figure 8.30a shows a valid steady state where f_F lies above the continuous flux density curve and an invalid steady state, where the feed flux density is under the continuous flux density curve. Figure 8.30b shows the corresponding concentration profiles. For an invalid steady state the thickener overflows.

Problem 8.1 Consider a flotation tailing 35[%] solid in concentration with the following thickening parameters: $f_{bk}(\varphi) = -6.05 \times 10^{-4}\varphi(1-\varphi)^{12.09}$ m/s and $\sigma_e(\varphi) = 5.35\exp(17.9\varphi)$. The solid and fluid densities are $\rho_s = 2.65$ t/m^3 and $\rho_f = 1.0$ t/m^3 respectively. Plot the solid flux density function versus concentration and the concentration profiles for the following three solid feed fluxes $F = 178$ tph, $F = 200$ tph and $F = 260$ tph in a thickener of $D = 53$ m in diameter and an underflow of $w_D = 57.3$ % solid by weight.

Data			
F (tph)	178.0	200.0	260.0
D (m)	43.0	43.0	43.0
S (m^2)	1,452.2	1,452.2	1,452.2
ρ_s (ton/m^3)	2.65	2.65	2.65
ρ_f (ton/m^3)	1.0	1.0	1.0
w_F (%weight)	35.0	35.0	35.0
w_D (%weight)	57.3	57.3	57.3
Results			
φ_F	0.169	0.169	0.169
φ_D	0.336	0.336	0.336
ρ Pulp (ton/m^3)	1.28	1.28	1.28
Q_F (m^3/h)	139.2	156.4	203.3
Q_D (m^3/h)	199.82	224.51	291.87
q (m/s)	−3.82E−05	−4.29E−05	−5.58E−05
f_F (m/s)	−1.28E−05	−1.44E−05	−1.88E−05

Figure 8.31 shows two valid and one invalid steady state. Figure 8.32 gives the corresponding concentration profiles. The profile corresponding to the invalid steady state tends to infinite and the thickener could not handle that feed rate without overflowing.

8.3.4 Model of Conventional Thickening in Vessels with Varying Cross-Section

Conventional thickeners have two sections with different areas, a cylinder at the top and conical at the bottom to aid to the underflow discharge. The ideal conventional thickener used for modeling a real thickener ignores these facts and assumes a constant cross section for the equipment. In the following section, we will introduce a model for an ideal varying cross-section thickener.

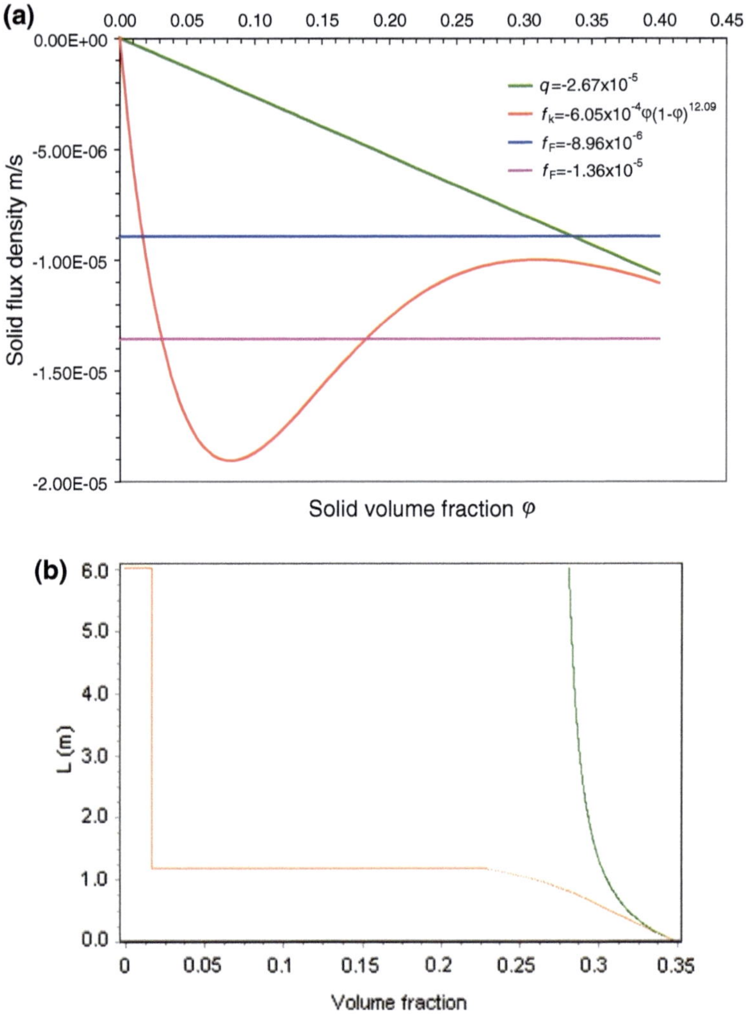

Fig. 8.30 **a** Valid solid flux density ($f = -8.96 \times 10^{-6}$ m/s) and invalid solid flux density ($f = -1.36 \times 10^{-5}$ m/s) versus concentration at steady state. **b** Concentration profiles for examples in Fig. 8.31a

Consider the settling of a flocculated suspension with constant concentration $\varphi(z,t)$ on each horizontal cross-section in a vessel with varying cross-sections sections $S(z)$, where $0 \leq z \leq L$ is the vertical coordinate. The solid and fluid particles are subjected to the effect of stresses, gravity, buoyancy and drag force.

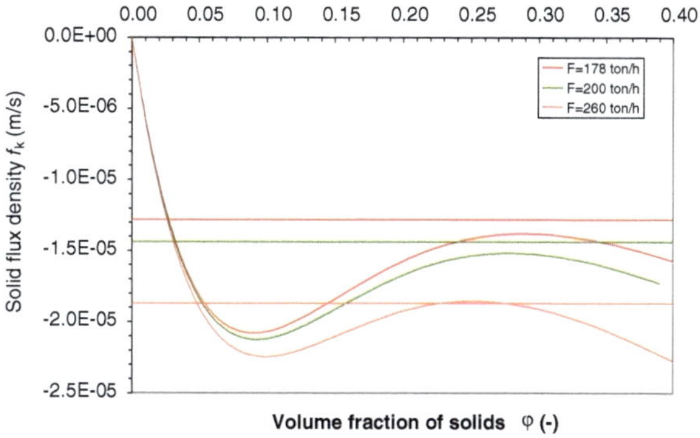

Fig. 8.31 Solid flux density function versus concentration for Problem 8.1. Two valid steady states for $F_1 = 178$ (tph) and $F_2 = 200$ (tph) and one invalid steady state for $F_3 = 260$ (tph), with the same feed and underflow concentration

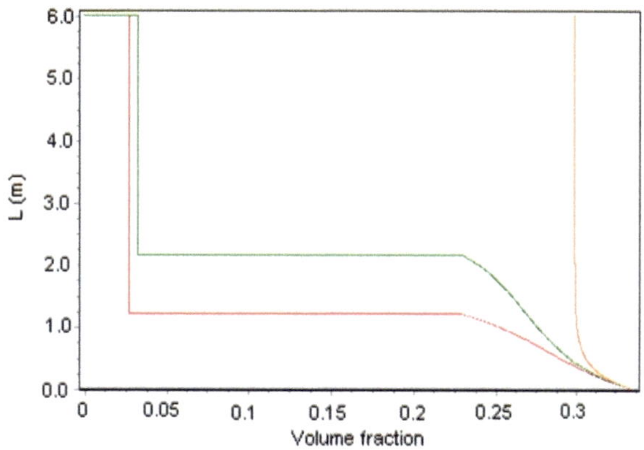

Fig. 8.32 Concentration profile for problem 8.1

Continuity Equation

If the suspension is considered a continuous superimposed two-component medium, the conservation of mass for the solid and fluid components are:

$$\frac{\partial \varphi}{\partial t} + \frac{1}{S(z)}\frac{\partial}{\partial z}(S(z)\varphi v_s) = 0, \quad 0 \le z \le L,\ t > 0 \tag{8.52}$$

$$-\frac{\partial \varphi}{\partial t} + \frac{1}{S(z)}\frac{\partial}{\partial z}\left(S(z)(1-\varphi)v_f\right) = 0, \quad 0 \le z \le L,\ t > 0 \tag{8.53}$$

where t is time and v_s and v_f are the solid and fluid velocity components respectively and $S(z)$ is the varying cross-sectional area.

Adding these two equations, the mass balance of the suspension is obtained:

$$\frac{\partial}{\partial z}(S(z)q(z,t)) = 0 \tag{8.54}$$

where $q = v_s - (1-\varphi)v_r$ is the volume average velocity and v_r is the relative solid–fluid velocity. Note that:

$$-S(z)q(t) = Q(z,t) > 0 \tag{8.55}$$

$Q(z,t)$ is the volume flowrate of suspension through the thickener, which according to (8.55) is independent of z:

$$Q(z,t) = Q(0,t) = Q_D(t), \quad 0 \le z \le L,\ t > 0 \tag{8.56}$$

where $Q_D > 0$ is the prescribed suspension volume underflow rate. From $q = v_s - (1-\varphi)v_r$,

$$\varphi v_s = q\varphi + \varphi(1-\varphi)v_r = -\frac{Q_D(t)}{S(z)}\varphi + \varphi(1-\varphi)v_r \tag{8.57}$$

Substituting equation into (8.52) yields:

$$\frac{\partial\varphi}{\partial t} + \frac{1}{S(z)}\frac{\partial}{\partial z}\left(-Q_D(t)\varphi + S(z)\varphi(1-\varphi)v_r\right) = 0, \quad 0 \le z \le L,\ t > 0 \tag{8.58}$$

We saw that the relative solid–fluid velocity v_r can be written in terms of the Kynch flux density function $f_{bk}(\varphi)$ and the solid effective stress $\sigma_e(\varphi)$ in the form:

$$v_r = \frac{f_{bk}(\varphi)}{\varphi(1-\varphi)}\left(1 + \frac{\sigma_e'(\varphi)}{\Delta\rho\varphi g}\frac{\partial\varphi}{\partial z}\right) \quad \text{for } 0 \le \text{z} \le \text{L} \tag{8.59}$$

Therefore replacing in Eq. (8.58) gives:

$$\frac{\partial\varphi}{\partial t} + \frac{1}{S(z)}\frac{\partial}{\partial z}\left(-Q_D(t)\varphi + S(z)f_{bk}(\varphi)\left(1 + \frac{\sigma_e'(\varphi)}{\Delta\rho\varphi g}\frac{\partial\varphi}{\partial z}\right)\right) = 0$$

which can be written for $0 \le z \le L$; $t > 0$ in the form:

$$\frac{\partial\varphi}{\partial t} + \frac{1}{S(z)}\frac{\partial}{\partial z}\left(-Q_D(t)\varphi + S(z)f_{bk}(\varphi)\right) = \frac{1}{S(z)}\frac{\partial}{\partial z}\left(S(z)\frac{-f_{bk}(\varphi)\sigma_e'(\varphi)}{\Delta\rho\varphi g}\frac{\partial\varphi}{\partial z}\right) \tag{8.60}$$

Equation (8.60) is a degenerate parabolic partial differential equation representing the sedimentation of a flocculated suspension in a thickener with varying cross-sections. This equation reduces to a hyperbolic equation representing the sedimentation of an ideal suspension when the concentration is less than or equal to the critical, $\varphi < \varphi_c$:

$$\frac{\partial\varphi}{\partial t} + \frac{1}{S(z)}\frac{\partial}{\partial z}\left(-Q_D(t)\varphi + S(z)f_{bk}(\varphi)\right) = 0, \quad \text{for } 0 \le z \le L,\ t > 0 \tag{8.61}$$

Initial and Boundary Conditions

Consider Eq. (8.61). At $t = 0$, the initial concentration distribution is known:

$$\varphi(z,t) = \varphi_0(z), \quad \mathit{for}\ 0 \le z \le L \tag{8.62}$$

At $z = z_F$, the thickener is fed with a suspension of concentration $\varphi_F(t)$ at a volume flow rate of $Q_F(t) \ge 0$. Therefore, the solid flux at the feed is $Q_F(t)\varphi_F(t) \ge 0$:

$$Q_F(t)\varphi_F(t) = Q_D(t)\varphi_D - S(L)f_{bk}(\varphi_L), \quad \text{for } t \ge 0 \tag{8.63}$$

where $\varphi_L = \varphi(L,t)$. At $z = 0$, the solid flux density f_D reduces to its convective part, therefore:

$$f(0,t) = q\varphi_D$$

and the boundary condition is:

$$\left.\frac{\partial\varphi}{\partial z}\right|_{z=0,t} = -\left.\frac{\Delta\rho\varphi g}{\sigma_e'(\varphi)}\right|_{z=0,t}, \quad \text{for } t \ge 0 \tag{8.64}$$

The difficulty in solving Eq. (8.60) is the degeneracy into a non-linear hyperbolic equation for $\varphi < \varphi_c$. Since this equation is discontinuous for $\varphi < \varphi_c$, it has weak solutions that need an additional entropy condition to single out the physically relevant solution. Such a solution for (8.60) was proven by Bürger and Karlsen (2001).

Steady State

At steady state, Eq. (8.60) becomes:

$$S(z)f(z) = S_F f_F = -Q_F\varphi_F, \quad \text{for } 0 \le z \le L \tag{8.65}$$

Since at $z = 0$ the *solid flux* is $Q_D\varphi_D$,

$$Q_F\varphi_F = Q_D\varphi_D \tag{8.66}$$

On the other hand, for a constant underflow rate of Q_D, the steady state solution of Eq. (8.60) is:

$$-Q_D\varphi(z) + S(z)f_{bk}(\varphi) = -S(z)\frac{f_{bk}(\varphi)\sigma_e'(\varphi)}{\Delta\rho\varphi g}\frac{d\varphi}{dz} + C \tag{8.67}$$

where C is an integration constant. Using the boundary condition (8.64) for a desired underflow concentration φ_D yields $C = -Q_D\varphi_D$. Replacing this value in (8.67) we have:

$$-Q_D(\varphi(z) - \varphi_D) + S(z)f_{bk}(\varphi) = -S(z)\frac{f_{bk}(\varphi)\sigma_e'(\varphi)}{\Delta\rho\varphi g}\frac{d\varphi}{dz}, \quad \text{for } 0 \le z \le L \tag{8.68}$$

Existence of a Steady State

We have a steady state if the concentration increases downwards in the thickener, that is, if:

$$\frac{d\varphi}{dz} \leq 0 \quad for\ 0 \leq z \leq z_c \tag{8.69}$$

where z_c is the top of the sediment layer at $\varphi = \varphi_c$. Since the right side of (8.68) is negative, we have:

$$-Q_D(\varphi(z) - \varphi_D) + S(z)f_{bk}(\varphi) \leq 0, \quad \text{for}\, 0 \leq z \leq z_c \tag{8.70}$$

Concentration Profile

(a) In the hindered settling region, where $\varphi \leq \varphi_c$, Eq. (8.70) becomes:

$$-Q_D(\varphi(z) - \varphi_D) + S(z)f_{bk}(\varphi) = 0$$

and, since $Q_D\varphi_D = Q_F\varphi_F$, the concentration in Kynch's region can be obtained by solving $\varphi(z)$ for values $z_c \leq z \leq L$, from the implicit algebraic equation:

$$-Q_D\varphi(z) + S(z)f_{bk}(\varphi(z)) = -Q_F\varphi_F \tag{8.71}$$

(b) In the sediment, where $0 \leq z \leq z_c$, we can write from (8.60):

$$\frac{d\varphi}{dz} = \frac{\Delta\rho\varphi g}{f_{bk}(\varphi)\sigma_e'(\varphi)}\left(-\frac{Q_D}{S(z)}(\varphi(z) - \varphi_D) + f_{bk}(\varphi(z))\right), \quad 0 \leq z \leq z_c \tag{8.72}$$

This equation can be solved with the boundary condition $\varphi(0) = \varphi_D$.

8.3.5 Clarifier-Thickener Model

In the 1970s and 1980s, equipment called *High Capacity Thickeners* were introduced to the mining industry by various manufacturers such as Eimco and Enviroclear, among others. See Fig. 8.13a and b.

These devices were promoted as having smaller unit area requirements than conventional installations. The distinct feature of this device was that the feed was introduced via a deep feedwell below the top of the sediment. The claim was that by eliminating the settling zone it was possible to reduce space requirements.

Although the conventional thickener model, as described above, is the best tool available today to design, simulate and control industrial thickeners, it has theoretical and practical drawbacks. From a theoretical point of view, there is deficient interpretation of the global conservation principle because it does not consider the overflow stream in the solid balance equation. From a practical point of view, the model does not permit the suspension to rise over the feeding level, which is sometimes the case in high capacity thickeners. This is not very relevant in conventional thickeners, but is especially important in thickener fed under the sediment level. These problems led to the development of the clarifier-thickener model (Diehl 1995, 1996, 1997, 2000 2001; Bürger et al. 2001, 2003).

In spite of the fact that the Conventional Thickener model, as shown in the previous section, is the best tool available today to design, simulate and control industrial thickeners, it has theoretical and practical draw backs. From a theoretical point of view, there is a deficient interpretation of the global conservation principle by not consider the overflow stream in the solid balance equation. From a practical point of view, the model does not permit the suspension to rise over the feeding level, which is sometimes the case in High Capacity Thickeners. This fact has little relevance in conventional thickeners, but is especially important in thickener fed under the sediment level. These problems lead to the development of the Clarifier-Thickener Model (Diehl 1995, 1996, 1997, 2000, 2001; Bürger et al. 2001, 2003).

Consider a thickener and divide the vessel into two zones: (1) $z \geq z_F$, where the solids can flows upwards and $z \leq z_F$, where the solid flows downwards. $z = z_F$ represents the feeding level, $z = 0$ represents the underflow level, $z = z_c$, is the surface of the sediment level and $z = z_O$ is the overflow level. See Fig. 8.33.

The feed volume flowrate of pulp and solid are introduced as a singular surface sources of strength $Q_F(t)$ and $Q_F(t)\varphi_F(t)$ at $z = z_F$. The underflow volume flux of pulp and solid, $Q_D(t)$ and $Q_D(t)\varphi_D(t)$ respectively, leave as a singular surface sinks at $z = 0$. $Q_O(t)$ and $Q_O(t)\varphi_O(t)$ are the overflow volume flux and solid overflow volume flux at $z = z_O$, with $Q_i > 0$; with $i = F, D, O$.

Macroscopic mass balances yields:

$$Q_F(t) = Q_D(t) + Q_O(t) \tag{8.73}$$

$$Q_F(t)\varphi_F(t) = Q_D(t)\varphi_D(t) + Q_O(t)\varphi_O(t) \tag{8.74}$$

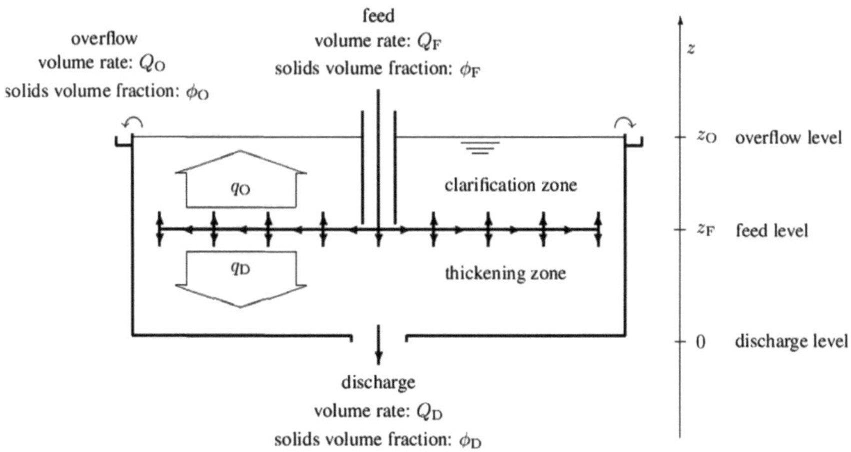

Fig. 8.33 Schematic view of a clarifier-thickener

Defining:

$$q_F = -\frac{Q_F}{S}, \quad q_D = -\frac{Q_D}{S}, \quad q_O = \frac{Q_O}{S}$$

which can be written as

$$q_F(t) = q_D(t) - q_O(t) \tag{8.75}$$

$$q_F(t)\varphi_F(t) = q_D(t)\varphi_D(t) - q_O(t)\varphi_O(t) \tag{8.76}$$

The spacial velocities for both zones are

$$q(t) = \begin{cases} q_O(t) & \text{for} \quad z_F < z < z_O \\ q_D(t) & \text{for} \quad 0 < z < z_F \end{cases} \tag{8.77}$$

and the solid flux density $\tilde{f}_k(z,t,\varphi)$ for both zones including the singular surface $q_F(t)\varphi_F(t)$ is

$$\tilde{f}_k(z,t,\varphi) = \begin{cases} q_O(t)(\varphi - \varphi_F) & \text{for} \quad z \geq z_O \\ q_O(t)(\varphi - \varphi_F) + f_{bk}(\varphi) & \text{for} \quad z_O > z > z_F \\ q_D(t)(\varphi - \varphi_F) + f_{bk}(\varphi) & \text{for} \quad z_F \geq z > 0 \\ q_D(t)(\varphi - \varphi_F) & \text{for} \quad z \leq 0 \end{cases} \tag{8.78}$$

Figure 8.34 shows the flux density functions for $z_O > z > z_F$ and $z_F \geq z > 0$.

With these variables, the phenomenological sedimentation-consolidation equation reads:

$$\frac{\partial \varphi}{\partial t} + \frac{\partial}{\partial z}\tilde{f}_k(z,t,\varphi) = -\frac{\partial}{\partial z}\left(\frac{f_{bk}(\varphi)\sigma'_e(\varphi)}{\Delta\rho\varphi g}\frac{\partial \varphi}{\partial z}\right) \tag{8.79}$$

The numerical solution to this problem was presented by Bürger et al. (2003). This numerical scheme ensures that the solution generate converges to physically relevant one.

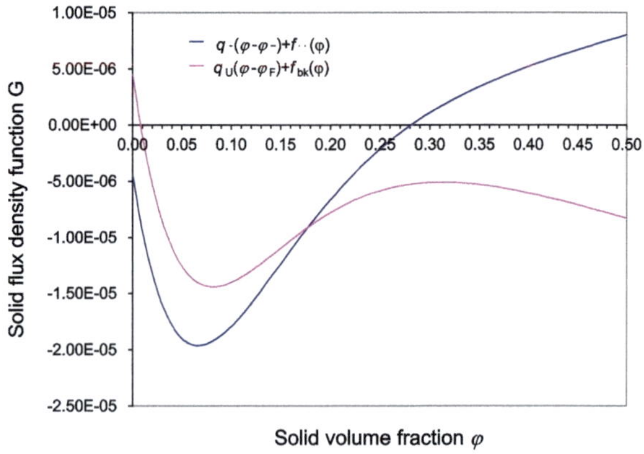

Fig. 8.34 Solid flux density functions in the clarification and thickening zones

Steady States

At steady state yields

$$\tilde{f}_k(z,\varphi) = -\left(\frac{f_{bk}(\varphi)\sigma_e'(\varphi)}{\Delta\rho\varphi g}\frac{\partial\varphi}{\partial z}\right) \tag{8.80}$$

The conditions to obtain a steady state are:

- That $d\varphi/dz<0$ for $0\leq z\leq z_c$, where $z=z_c$ is the location of the critical concentration φ_c.
- That at discontinuities the jump condition $\sigma(+.-) = \left(f_k^+ - f_k^-\right)/(\varphi^+ - \varphi^-)$ is valid and that the concentration should satisfy the jump stability (entropy of the jump).
- The term in bracket at the right side of (8.80) is continuous function of z.

Modes of Operation

Betancourt and Concha (2011) argue that it is better to talk about a high capacity mode of operation instead of high capacity thickener. It is possible to establish two modes of operation for the clarifier-thickener depending on the location of the critical concentration φ_c: if φ_c occurs in the thickening zone $0<z<z_F$, we have the *Conventional Mode* of operation, and if φ_c occurs in the clarification zone $z_F<z<z_O$ we say that the unit is operated in *high capacity* mode.

Conventional Mode of Operation

This mode of operation is characterized by a continuous concentration profile from $z=0$ to $z=z_c<z_F$.

$$\begin{aligned}\frac{d\varphi}{dz} &= \frac{(q_D(\xi-\varphi_D)+f_{bk}(\xi))\Delta\rho\xi g}{f_{bk}(\xi)\sigma_e'(\xi)} \quad \text{for} \quad 0\leq z\leq z_c\\ \varphi(0) &= \varphi_D\end{aligned} \tag{8.81}$$

In the region $z_c<z<z_F$ the concentration takes the constant value φ_L, called *conjugated concentration.* The profile is calculated by integrating with boundary condition $\varphi(0)=\varphi_D$, provided $q_D(\xi-\varphi_D)+f_{bk}(\xi)<0$. The conjugate concentration φ_L is calculated from

$$q_D\varphi_D = q_D\varphi_L + f_{bk}(\varphi_L) \tag{8.82}$$

considering the entropy stability conditions (see Bürger and Narváez 2007):

$$\begin{aligned} q_O\varphi + f_{bk}(\varphi) &\leq 0 && \text{for} \quad \varphi\in(0,\varphi_L)\\ q_D(\varphi_L-\varphi)+f_{bk}(\varphi_L)-f_{bk}(\varphi) &\geq 0 && \text{for} \quad \varphi\in(\varphi_L,\varphi_c)\end{aligned} \tag{8.83}$$

High Capacity Mode of Operation

The High capacity mode of operation is characterized by a continuous concentration profile from $z=0$ to $z_c > z_F$, followed by a concentration $\varphi = 0$ for $z_c < z \le z_O$. The concentration profile for this mode of operation is calculated from:

$$\frac{d\varphi}{dz} = \begin{cases} \dfrac{(q_D(\varphi-\varphi_D)+f_{bk}(\varphi))\Delta\rho\varphi g}{f_{bk}(\varphi)\sigma_e'(\varphi)} & \text{for} \quad z_D \le z < z_F \\ \dfrac{(q_O\varphi+f_{bk}(\varphi))\Delta\rho\varphi g}{f_{bk}(\varphi)\sigma_e'(\varphi)} & \text{for} \quad z_F < z \le z_c \end{cases} \tag{8.84}$$
$$\varphi(0) = \varphi_D$$

provided the following entropy conditions holds

$$\begin{array}{ll} q_D(\varphi-\varphi_D)+f_{bk}(\varphi) \le 0; & \text{for } \varphi(z_F) \le \varphi \le \varphi_D \\ q_O\varphi+f_{bk}(\varphi) \le 0; & \text{for } 0 < \varphi \le \varphi(z_F) \end{array} \tag{8.85}$$

Problem 8.2 Consider a flotation tailing with the following thickening parameters: $f_{bk}(\varphi) = -6.05\times10^{-4}\varphi(1-\varphi)^{12.09}$ and $\sigma_e = 5.35\exp(17.9\varphi)$. The material, with solid a density 2.65 t m^3 and a fluid density of 1.0 t/m^3, will be treated in a Clarifier-Thickener with $D = 53$ m in diameter and 5 m in height. The feedwell delivers the feed al $z_F = 1.2$ m and the sediment should not rise above 2 m. If the feed and underflow concentrations are $w_F = 35.0$ % and $w_D = 57.3$ % solid by weight respectively. (1) What is the capacity of the thickener, the volume average velocity and de volume overflow velocity? (2) Draw the solid flux density functions and (3) draw the concentration profile.

Solution

$F = 179.337$ tph solid; $q_D = 2.5859\text{E}{-5}$; $q_O = 2.5108\text{E}{-5}$

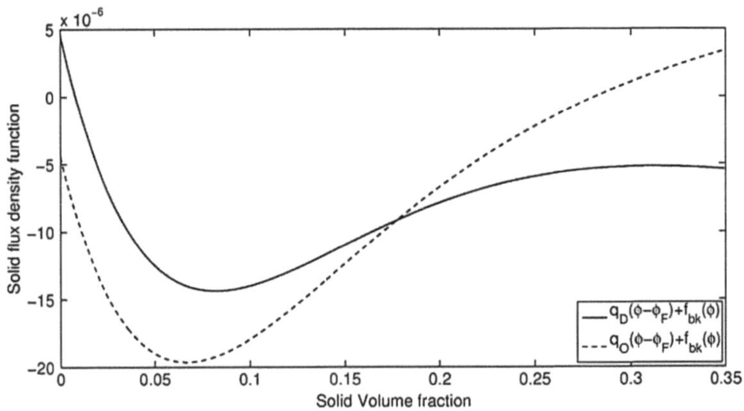

Flux density function fro problem 8.2.

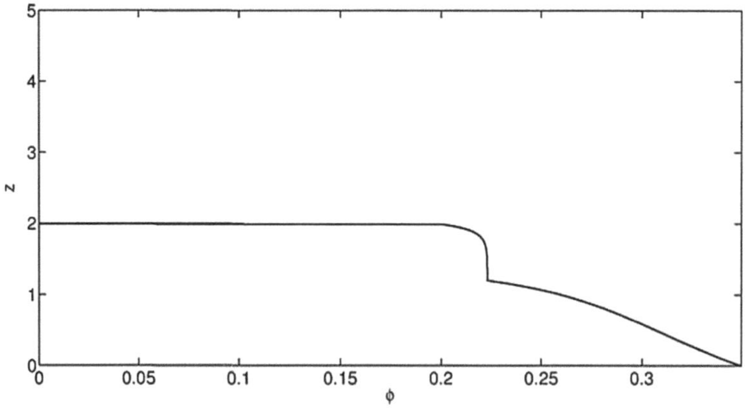

Concentration profile for problem 8.2.

8.4 Thickening Parameters and Their Determination

There is agreement today that conventional thickeners can be designed based on laboratory experiments. Nevertheless, one of the difficulties in thickening is laboratory determination of parameters. Although the thickening process is described by both sedimentation and consolidation, to ignore consolidation leads to errors. Most experimental work to determine parameter today is solely related to sedimentation.

8.4.1 Relevant Parameters

Parameters describing the thickening process are the solid flux-density function $f_{bk}(\varphi)$ and the solid effective stress $\sigma_e(\varphi)$. The solid flux density function is related to the resistance coefficient $K(\varphi)$ for concentrations below the critical level, and to the sediment permeability $k(\varphi)$ for concentrations equal to or greater than the critical. The solid effective stress $\sigma_e(\varphi)$ is related to the compressibility of the sediment.

Initial Settling Velocity

Two simple models are often used to describe the initial settling velocity as a function of concentration, one with two parameters u_∞ and c, presented by Richardson and Zaki in 1954, and the other with three parameters u_∞ , φ_m , n, presented by Michels and Bolger in (1962):

$$u(\varphi) = \begin{cases} u_\infty (1-\varphi)^c & \text{Richardson and Zaki or} \\ u_\infty \left(1-\frac{\varphi}{\varphi_m}\right)^n & \text{Michels and Bolger} \end{cases} \tag{8.86}$$

Solid Flux Density Function

The solid flux density function is given for values less and greater than the critical concentration by:

$$f_{bk}(\varphi) = \begin{cases} \varphi \times u(\varphi) & \text{for} \quad \varphi < \varphi_c \\ k_0 \left(\frac{\varphi_c}{\varphi}\right)^m & \text{for} \quad \varphi \geq \varphi_c \end{cases} \tag{8.87}$$

where the *coefficient of resistance* $K(\varphi)$ is given by

$$K(\varphi) = -\left(\frac{\Delta\rho\varphi^2(1-\varphi)^2 g}{\mu \times f_{bk}(\varphi)}\right) \text{for } \varphi < \varphi_c \tag{8.88}$$

and the permeability of the sediment $k(\varphi)$ by:

$$k(\varphi) = \frac{\mu}{\Delta\rho\varphi^2 g} f_{bk}(\varphi) \tag{8.89}$$

Initial Settling Velocity

Settling experiments are usually performed with different solid concentrations in one or two-liter capacity graduate cylinders, with an initial suspension height of around 35 cm. For each concentration, the height of the water suspension interface is recorded as a function of time.

Problem 8.3 Calculate the initial settling velocities for a suspension of calcium carbonate with solid and fluid densities of $\rho_s = 2{,}710\,(\text{kg/m}^3)$ and $\rho_f = 1{,}000\,(\text{kg/m}^3)$ respectively.

Table 8.1 gives the experimental height of the water-suspension interface. Settling plots are drawn with these data, see Fig. 8.35. The initial settling velocities are obtained from the linear sections of the settling plots for short times. A straight line is drawn for each concentration, see Fig. 8.36. The slopes of these lines yield the settling velocities shown at the bottom of Table 8.1.

From Fig. 8.36 the initial settling velocities as a function of φ are:

$$u(0.017) = -2.91 \times 10^{-4}\,\text{m/s}, \quad v_s(0.044) = -7.97 \times 10^{-5}\,\text{m/s}$$

$$u(0.103) = -3.07 \times 10^{-5}\,\text{m/s}, \quad v_s(0.265) = -1.16 \times 10^{-6}\,\text{m/s}$$

Table 8.1 Settling velocity data for a calcium carbonate suspension

$\varphi = 0.265$		$\varphi = 0.103$		$\varphi = 0.044$		$\varphi = 0.017$	
Time (s)	Height (m)	Time (s)	Height (m)	Time (s)	Height (m)	Time (s)	Height (m)
0	0.338	0	0.338	0	0.338	0	0.338
1,692	0.336	1080	0.292	1,080	0.210	468	0.128
2,304	0.335	1,368	0.284	1,332	0.186	720	0.076
3,204	0.331	1,692	0.274	1,620	0.162	1,008	0.051
6,804	0.329	1,980	0.265	1,908	0.144	1,908	0.044
10,404	0.326	2,556	0.247	2,520	0.112	2,520	0.041
15,804	0.319	3,132	0.230	3,096	0.096	5,508	0.037
23,004	0.312	6,120	0.172	6,120	0.079	9,108	0.030
37,404	0.301	9,720	0.162	9,720	0.073	14,508	0.029
51,804	0.291	15,120	0.150	15,120	0.064	21,708	0.029
66,204	0.277	22,320	0.139	22,320	0.061	36,108	0.028
80,604	0.274	36,720	0.127	36,720	0.057	50,508	0.028
95,004	0.264	51,120	0.122	51,120	0.056	64,908	0.027
109,404	0.257	65,520	0.118	65,520	0.056	79,308	0.027
123,804	0.252	79,920	0.117	79,920	0.054	93,708	0.027
138,204	0.247	94,320	0.115	94,320	0.054	108,108	0.026
152,604	0.243	108,720	0.115	108,720	0.054	122,508	0.026
167,004	0.238	123,120	0.115	123,120	0.054	136,908	0.025
181,404	0.235	137,520	0.113	137,520	0.054	151,308	0.025
195,804	0.232	151,920	0.113	151,920	0.054	165,708	0.025
210,204	0.230	166,320	0.113	166,320	0.054	180,108	0.025
v_s (m/s) =	−1.160E−06	−3.070E−05		−7.970E−05		−2.910E−04	

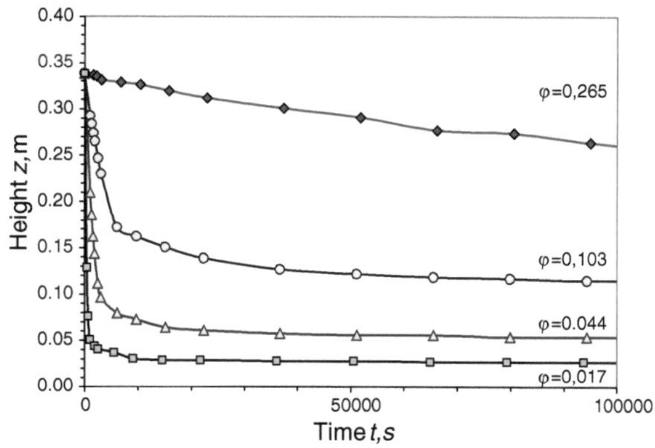

Fig. 8.35 Settling curve for calcium carbonate suspensions at four different concentrations. Data in Table 8.1

The initial settling velocities are correlated with concentration using Richardson and Zaki's equation $u(\varphi) = u_\infty(1 - \varphi)^c$. The result is shown in Fig. 8.37, obtaining the following parameters: $u_\infty = -1.72 \times 10^{-4}$ and $c = 15.6$.

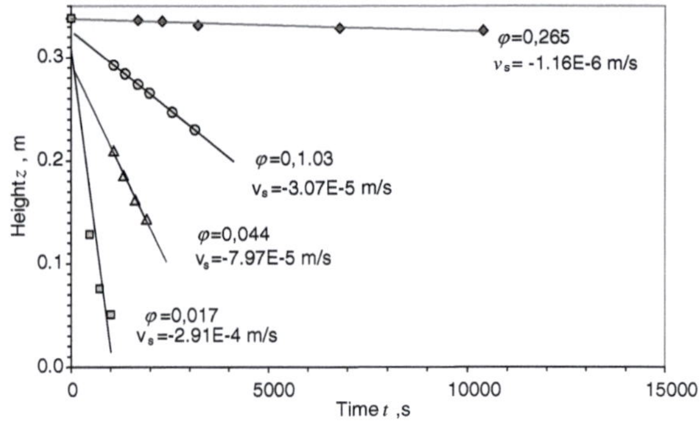

Fig. 8.36 Initial settling velocities for cases of Table 8.1 and Fig. 8.35

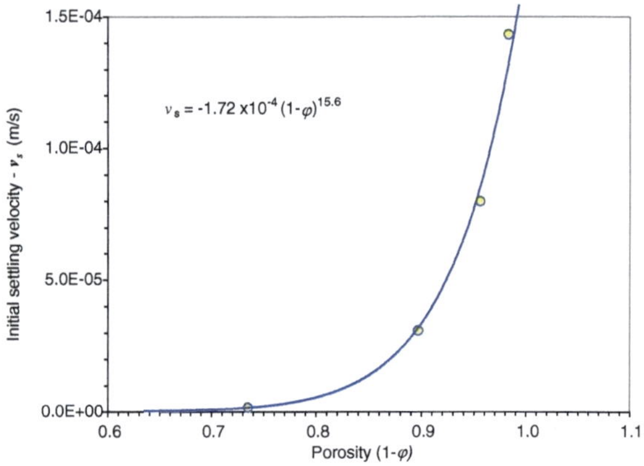

Fig. 8.37 Initial settling velocity function according to Richardson and Zaki for the cases in Table 8.1 and Fig. 8.35

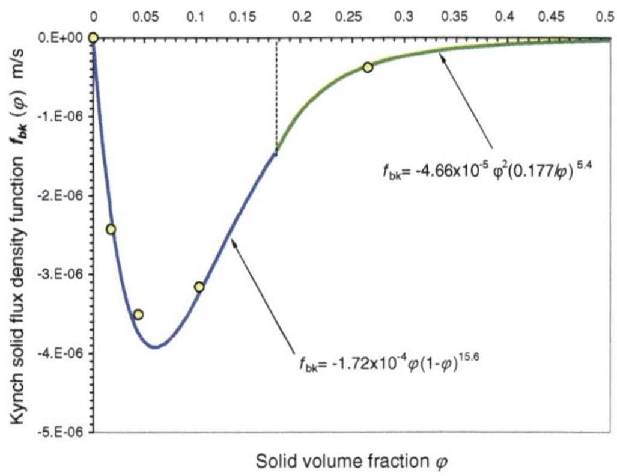

Fig. 8.38 Kynch solid flux density function for the cases in Table 8.1 and Richardson and Zaki's equation

Flux Density Function

The solid flux density function is given in Eq. (8.90) and Fig. 8.38, where the critical concentration was $\varphi_c = 0.177$ and Eq. (8.87) was used to obtain $k_0(\varphi_c)$ and m:

$$f_{bk}\varphi = \begin{cases} -1.72 \times 10^{-4}\varphi(1-\varphi)^{15.6} & \text{for} \quad \varphi < \varphi_c \\ -4.66 \times 10^{-5}\varphi^2(0.177/\varphi)^{5.4} & \text{for} \quad \varphi \geq \varphi_c \end{cases} \tag{8.90}$$

The effect of the parameters c and u_∞ on the flux density function are shown in Figs. 8.39 and 8.40.

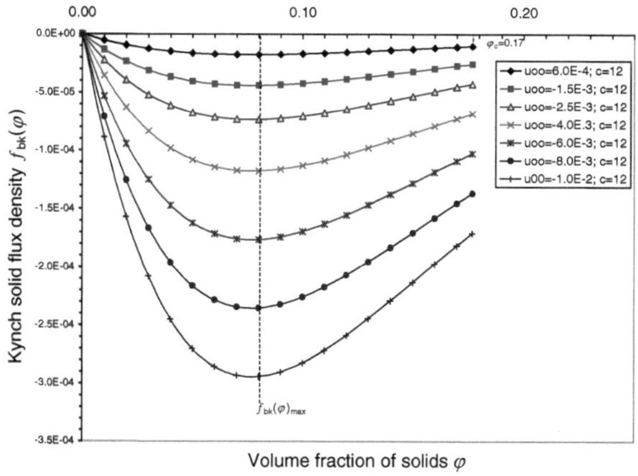

Fig. 8.39 Effect of the u_{oo} parameter on the solid flux density function. The *dotted line* joins the points of maximum solid flux densities

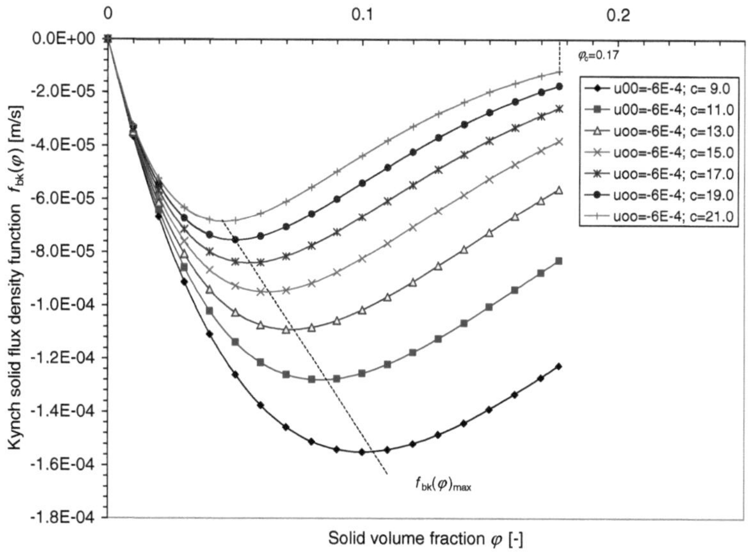

Fig. 8.40 Effect of the c parameter on the solid flux density function. The *dotted line* joins the points of maximum solid flux densities

Resistance Coefficient and Permeability

The solid–liquid resistance coefficient $K(\varphi)$ and the permeability $k(\varphi)$ can be obtained from Eqs. (8.88) and (8.89) and from the solid flux density function $f_{bk}(\varphi)$ on (8.90) (Fig. 8.41, 8.42):

$$K(\varphi) = -9.75 \times 10^{-5}\varphi/(1-\varphi)^{15.6} \tag{8.91}$$

$$k(\varphi) = -4.66 \times 10^{-5}\varphi^2(0.177/\varphi)^{5.4} \tag{8.92}$$

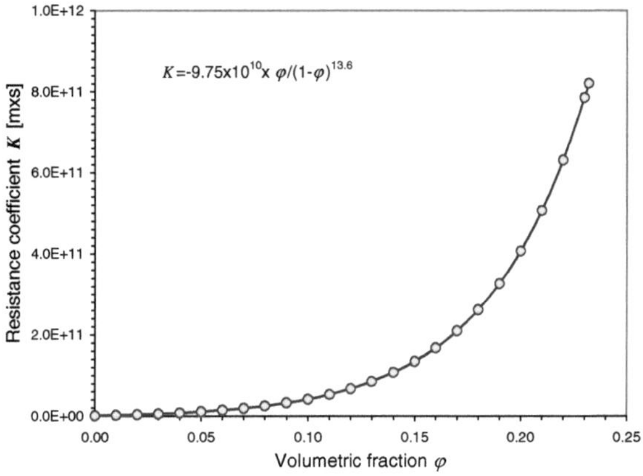

Fig. 8.41 Resistance coefficient for the cases in Table 8.1

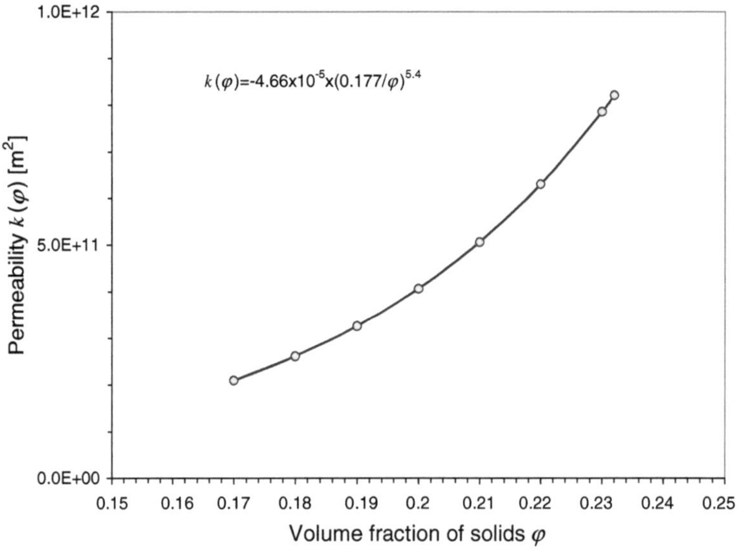

Fig. 8.42 Permeability

Critical Concentration

Roberts (1949) proposed the following equation for consolidation of compressible suspensions:

$$\frac{d\mathfrak{D}}{dt} = -\mathfrak{K}(\mathfrak{D} - \mathfrak{D}_\infty), \text{ for } \quad \mathfrak{D}_c \geq \mathfrak{D} \geq \mathfrak{D}_\infty \tag{8.93}$$

where $\mathfrak{D}$ is the dilution, that is, the ratio of mass of water to mass of solid in the suspension, $\mathfrak{K}$ is a rate constant and t is time. The subscript c and ∞ refer to the critical and the equilibrium dilution. He considered $\mathfrak{D} - \mathfrak{D}_\infty$ proportional to $z - z_\infty$, where z_∞ is the final sediment height (Tory and Shannon 1965) then, $\mathfrak{D}$ can be substituted by z in equation and Robert's equation may can be written in the integrated form:

$$\frac{z - z_\infty}{L - z_\infty} = \exp(-\mathfrak{K}t), \quad \text{for} \quad \mathrm{L} \geq \mathrm{z} \geq \mathrm{z}_\infty \tag{8.94}$$

Obviously, Eq. (8.94) is not valid for the whole sedimentation range, but only for consolidation. Robert showed that, in plotting Eq. (8.94) for the whole sedimentation range, three straight lines can be drawn with the solution of this equation, one for hindered settling, one for the intermediate range and the last for consolidation. See Fig. 8.43. The critical point in this figure is the intersection of the second and the third lines. With the value of the critical time t_c, the critical height z_c can be obtained from the settling plot.

Problem 8.4 Calculate the critical concentration for data for Problem 8.3. Figure 8.43 shows the plot of $(z - z_\infty)/(L - z_\infty)$ versus t for a concentration $\varphi = 0.044$.

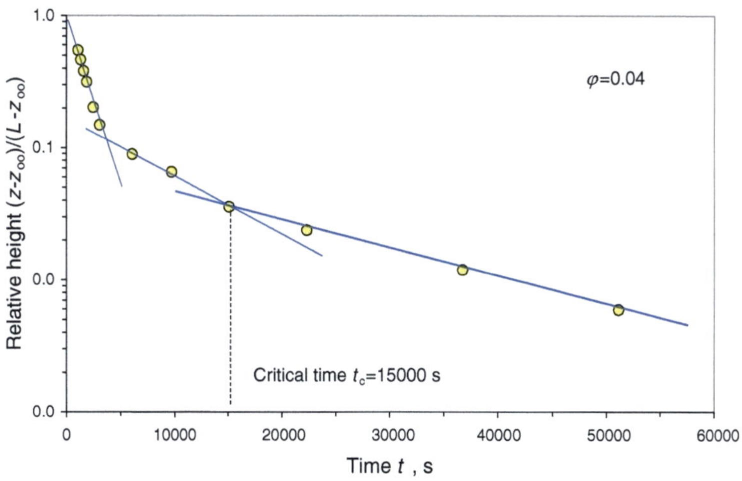

Fig. 8.43 Settling data plotted according to Robert's equation for a suspension with an initial concentration of $\varphi_0 = 0.044$

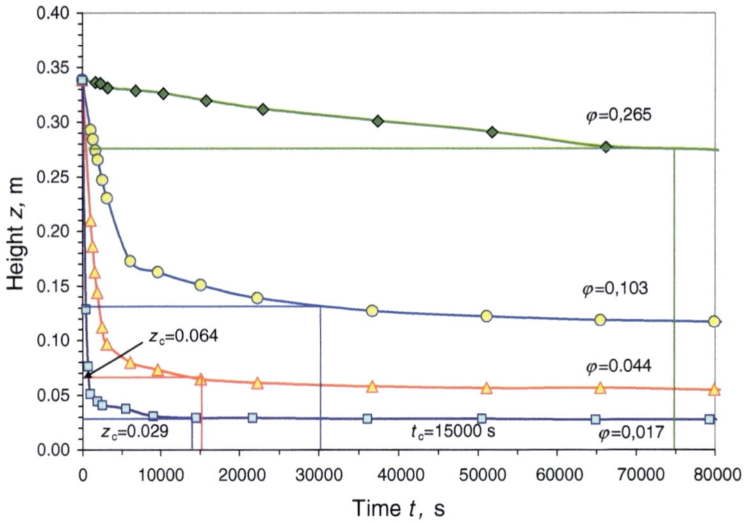

Fig. 8.44 Calculation of the critical height and critical time for the concentration $\boldsymbol{\varphi} = 0.044$ are $z_c = 0.064$ m, $t_c = 15{,}000$ s

From Fig. 8.43, the critical time for the concentration $\varphi = 0.044$ is $t_c = 15{,}000$ s. The critical height $z_c = 0.064$ m is obtained from the settling plot for $t_c = 15{,}000$ s, see Fig. 8.44.

In Kynch's theory, φ_c is the final concentration, therefore φ_c can be obtained from the volume balance at $t = 0$ and $t = t_c$:

$$L\varphi_o = z_C\varphi_c \tag{8.95}$$

For the case of Fig. 8.44:

$$\varphi_c = 0.338 \times 0.044/0.064 = 0.232 \tag{8.96}$$

The critical concentration obtained in this way depends on the initial suspension concentration φ_0, because Eq. (8.95) is valid only for Kynch's regime at the end of sedimentation. It converges to a constant value as the initial concentration tends to zero (Mode of sedimentation I, see Chap. 5). Therefore, for every test with a different initial concentration, a critical concentration is calculated and plotted against the initial concentration, as shown in Table 8.2 and Fig. 8.45. Extrapolation to $\varphi_0 = 0$ gives the critical concentration $\varphi_c = 0.177$.

A simpler and more accurate method for calculating the critical concentration is based on the conservation of the total solid mass before and after sedimentation, similar to (8.95), but now with compression. Before settling, the pulp is homogenized with a solid mass $\rho_s LS\varphi_0$, where S is the cross sectional area of the column, L is the initial height of the suspension and ρ_s and φ_0 are respectively its density

Table 8.2 Initial versus critical concentrations

φ	L	t_c	z_c	φ_c
0.017	0.338	14,500	0.029	0.198
0.044	0.338	15,000	0.064	0.232
0.103	0.338	30,000	0.130	0.268
0.265	0.338	75,000	0.275	0.326

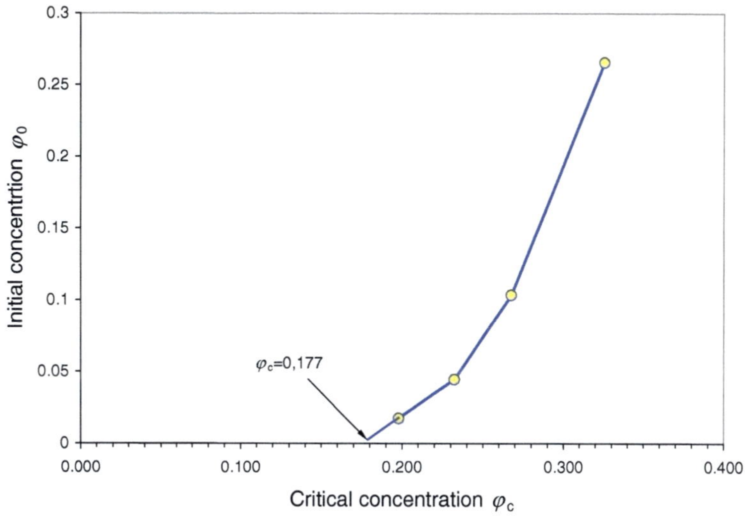

Fig. 8.45 Critical versus initial concentration. The final value of the critical concentration is $\varphi_c = 0.177$

and concentration. After sedimentation all the solid material is contained in the sediment, so the following equation is valid:

$$L\varphi_0 = \int_{\varphi_D}^{\varphi_c} \varphi d\varphi \tag{8.97}$$

Solid Effective Stress

Consolidation occurs due to the self-weight supported by the sediment layers and its effect is seen in the concentration profile. Therefore, the constitutive equation for *effective solid stress* can be calculated by using the measured values of the concentration profile at the end of a batch sedimentation-consolidation process.

Consider equation:

$$\frac{\partial \sigma_e}{\partial z} = -\Delta\rho\varphi g - \frac{\mu}{k(\varphi)}(1-\varphi)v_r, \quad \text{for } \varphi \geq \varphi_c \tag{8.98}$$

At the end of the consolidation process, all velocities are zero. Therefore $v_r = 0$ in the previous equation. Writing $\partial\sigma_e/\partial z = \partial\sigma_e/\partial\varphi \times d\varphi/dz$, Eq. (8.98) can be written in the form:

$$\frac{\partial \sigma_e}{\partial \varphi} = -\left.\frac{\Delta\rho\varphi g}{d\varphi/dz}\right|_{t\to\infty}, \quad \text{for} \quad 0 \leq z \leq z_c \tag{8.99}$$

Measuring the concentration profile at the end of the consolidation process, the functional form of the solid effective stress can be obtained by integrating this equation in the range $\varphi_D \leq \varphi \leq \varphi_c$:

$$\sigma_e(\varphi) = -\int_{\varphi_D}^{\varphi} \frac{\Delta\rho\xi g}{(d\xi/dz)} d\xi \tag{8.100}$$

As we have already seen, there are two models in common use for the solid effective stress $\sigma_e(\varphi)$, an exponential function and a potential function:

$$\sigma_e(\varphi) = \begin{cases} \alpha \exp(\beta\varphi) & \text{or} \\ \sigma_0\left(\left(\frac{\varphi}{\varphi_c}\right)^m - 1\right) & \text{for } \varphi \geq \varphi_c \end{cases} \tag{8.101}$$

Determination of the Critical Concentration

Since at the top of the sediment, $z = z_c$, and the concentration is $\varphi = \varphi_c$, to find the critical concentration, the concentration profile of the sediment should be integrated until Eq. (8.100) is satisfied. Using the exponential equation of (8.101) for the solid effective stress, from (8.99) we have:

$$d\varphi/dz = -\frac{\Delta\rho\varphi g}{\alpha\beta \exp(\beta\varphi)}; \quad \text{for } 0 \leq z \leq z_c \tag{8.102}$$

Problem 8.5 Defining $A_1 = -\frac{\alpha\beta}{\Delta\rho g}$ and $A_2 = \alpha$, the values of α, β and φ_c can be simultaneously determined by means of the following numerical integration of the mass balance (Fig. 8.46):

$$\varphi_0 L = \int_{\varphi_D}^{\varphi_c} z(\varphi) d\varphi \tag{8.103}$$

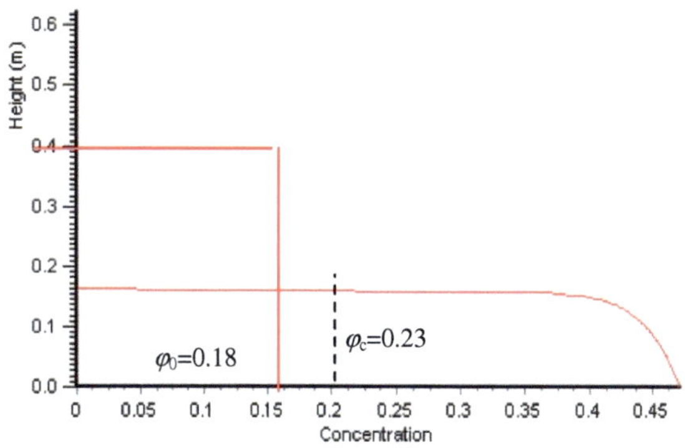

Fig. 8.46 Calculation of the consolidation parameters and the critical concentration

$$\varphi_0 L = \varphi_c\left(A_1\ln(\varphi_c) + A_2\varphi_c + \frac{(A_2\varphi_c)^2}{2\times 2!} + \cdots + \frac{(A_2\varphi_c)^n}{n\times n!}\right) + A_3$$
$$+\left(A_1\left(\varphi\ln(\varphi) - \varphi + \frac{A_2\varphi}{2} + \frac{(A_2\varphi)^2\varphi}{2\times 3!} + \cdots + \frac{(A_2\varphi)^n\varphi}{n\times(n+1)!}\right) + A_3\varphi\right)\Bigg|_{\varphi_c}^{\varphi_D} \tag{8.104}$$

where φ_D and φ_c are respectively the underflow and critical concentration. For example, for $L = 0.4\,(\mathrm{m})$ and $\varphi_0 = 0.18$, the application of (8.104) yields the values $\alpha = 2.35$, $\beta = 17.9$ and $\varphi_c = 0.23$.

Permeability

The determination of the permeability for the region $\varphi \geq \varphi_c$ can be done once the critical concentration has been determined.

For values greater than the critical concentration, the following empirical equation can be used:

$$f_{bk}(\varphi) = k_0\left(\frac{\varphi_c}{\varphi}\right)^m \tag{8.105}$$

For example, in Fig. 8.47, for $\varphi = \varphi_c = 0.23$, $f_{bk}(\varphi_c) = k_0 = 6.67\times 10^{-12}$ and the best value for the parameter $m = 0.8$ is obtained by non-linear regression.

8.5 Instrumentation for the Automatic Determination of Thickening Parameters

It is surprising that until recently the Marcy balance and one-liter graduate cylinders were the off-line instruments to routinely measure pulp concentrations and thickening parameters in the laboratory and in plants. These artifacts have been

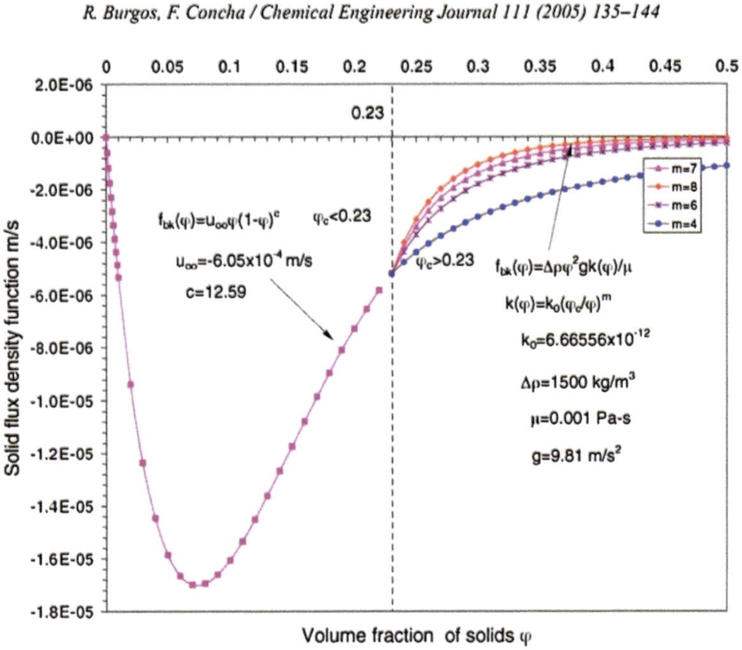

Fig. 8.47 Solid flux-density function. From Burgos and Concha (2005)

used for most of the 20th century. They are obsolete but are still used by many laboratories in mineral processing. One of the aims of my academic and professional work has been to update theory and instrumentation in this area.

8.5.1 Measuring Suspension Concentrations

Controlling pulp density in a concentrator is a daily task. Knowledge of this parameter is essential in processes as diverse as grinding, flotation, solid–liquid separation, and concentrate and trailing transport. The Marcy balance is used in laboratories and plants despite its limitations and imprecise readings. Two new instruments have been developed by Concha et al. (2011) called the *DensiTest* and *Industrial Picnometer*, which use the same principles as the Marcy balance, but with a digital strain gauge and a robust picnometer for larger samples.

The advantage of the DensiTest, beside its precision, is that the vessel can have any volume since it can be introduced to the instrument as input data. The results can be obtained in all the units used in mineral processing, such as pulp density, % solid by weight, volume fraction and dilution (Fig. 8.48).

The *Industrial Picnometer* can be used to determine solid, liquid and suspension densities. The instrument consists of a balance and a touch screen notebook

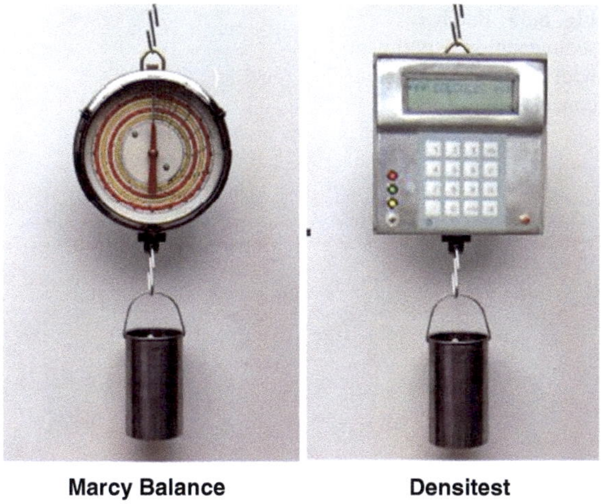

Fig. 8.48 Suspension concentration measurement

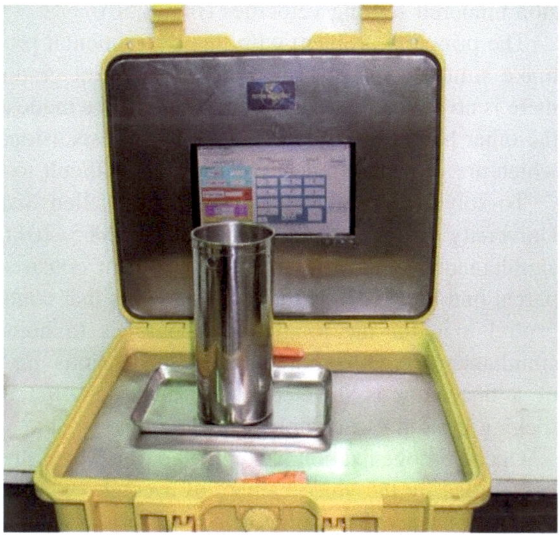

Fig. 8.49 Industrial picnometer (Cettem instruments)

concealed in a robust valise. A simple software program directs the operator during measurement (Fig. 8.49).

8.5.2 *Solids Flux Density Function*

Traditionally thickener *unit areas* were obtained by settling tests with suspensions of several concentrations in one or two-liter graduate cylinders, measuring the water-suspension interface at regular intervals. In a plot of the water-suspensions

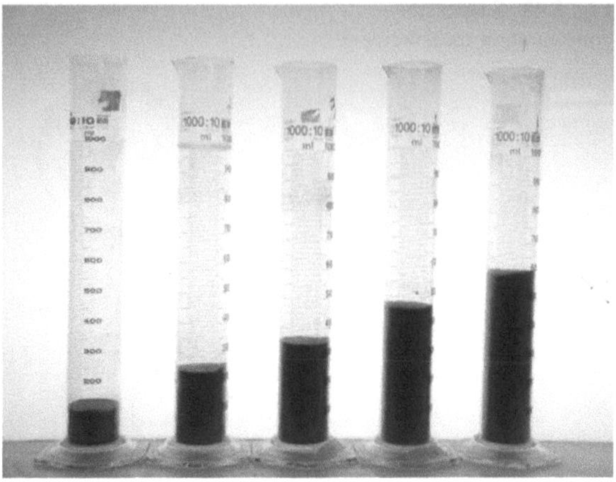

Fig. 8.50 Batch sedimentation experiment with different suspension concentrations

interfaces versus time, the initial straight line was chosen to indicate the suspension hindered settling velocities (Fig. 8.50).

The principal problem with this experimental procedure is that it is difficult to make simultaneous measurements in several graduate cylinders and, therefore, there is no certainty that all measurements are made under the same conditions. On the other hand, measurements in diluted suspensions at the beginning of the test, which are of primordial importance, are difficult to read by the naked eye.

To solve this problem an instrument called *SediRack* was developed at the University of Concepción (Concha et al. 2005, 2011). SediRack can perform five simultaneous settling tests. The instrument consists of a frame, with five transparent one-liter tubes with rubber stoppers that contain the suspension. There is a central axis for the rotation of the tubes to homogenize all five suspensions simultaneously. Once homogenous suspensions are attained, an appropriate dose of a flocculant is injected and further agitation as necessary. Alternatively a rocking motion can complement rotation.

Once homogenization is completed, the water-suspension interface in the tubes is recorded by the video camera of a notebook. The data are processed automatically to produce the settling velocities for the suspension concentrations. Using an optimization procedure, the software calculates the solid flux density function and Coe and Clevenger thickener unit area. See Figs. 8.51 and 8.52.

Figures 8.53 and 8.54 show the settling curves and the calculated solid flux density function and thickener unit area obtained by Coe and Clevenger's method. (See the next section).

SediRack On-line (PSE)

Although it is important for a mineral processing plant to perform periodic laboratory batch tests to determine thickening parameters and in this way adjust operations, it is even more important to be able to make on-line determinations of these parameters. For this purpose, the University of Concepción developed an on-

Fig. 8.51 SediRack instrument to measure the solid flux density function

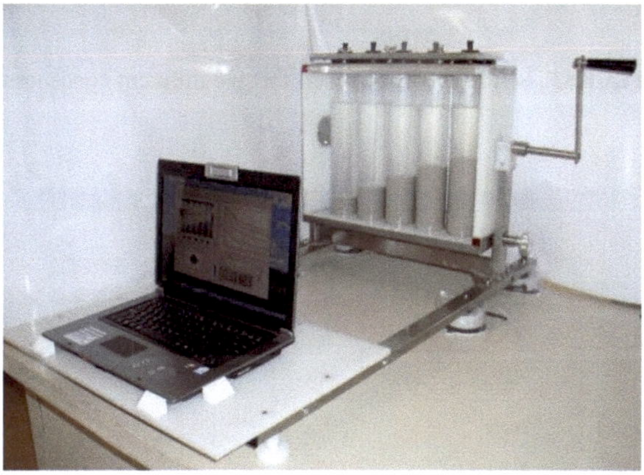

Fig. 8.52 *SediRack* complete with notebook with webcam

line version of the SediRack, called (PSE). See Fig. 8.55 (Segovia and Concha 2012 patent pending).

PSE is an instrument to determine the settling velocity of suspensions in an industrial thickener. The instrument is based on the laboratory instrument SediRack that was developed and patented by the University of Concepción and used successfully in laboratories and concentrators in Chile. After its validation in Chilean copper concentrators, it was able to:

1. Determine the settling velocities of flocs coming out of the feedwell, which is an important parameter to design control strategies.
2. Determine the dilution in the feedwell.
3. Determine the dilution of the feed entering the thickener.
4. Determine the optimal flocculation dose for a pulp entering the thickener.

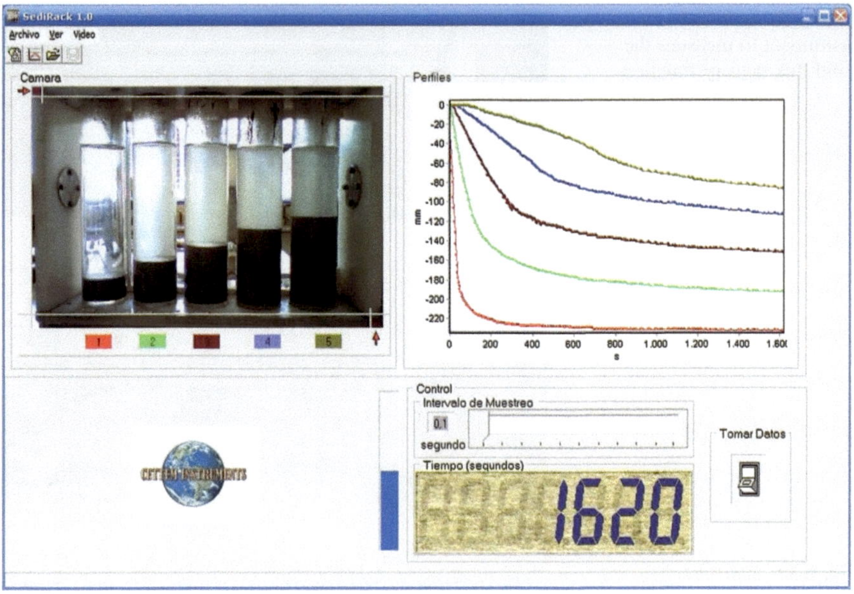

Fig. 8.53 Settling plots obtained for five different concentrations

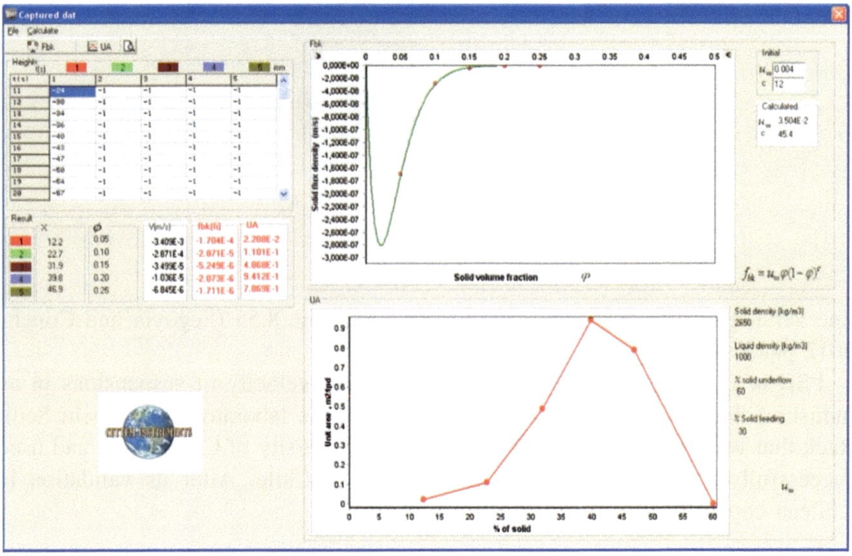

Fig. 8.54 Solid flux density and thickening unit area obtained

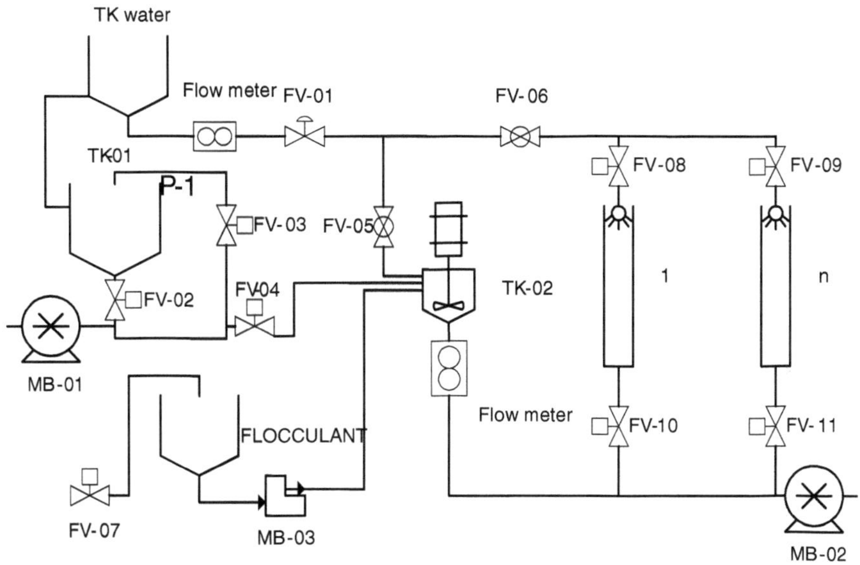

Fig. 8.55 Scheme of the PSE instrument (SediRack on-line)

The instrument has six main components:

- Tank for storing and homogenizing the pulp.
- Tank for storing and preparing the flocculant.
- Set of settling tubes to measure the settling velocity.
- An ultra-flocculation reactor (See Chap. 7).
- Flow meters, pumps, and valves.
- Video camera to record the water-suspension interface.

Figure 8.56 shows the complete PSE system.

A complete test takes about 10 min. The settling curves and the numerical results of the initial settling velocities for different suspension concentrations are given in Fig. 8.57.

8.5.3 Solids Effective Solid Stress

An instrument called *SediTest* was developed by Concha et al. (2008) The instrument consists of a settling column with 16 conductivity sensors. It measures the conductivity profile when consolidation is complete. Figure 8.58 shows the

Fig. 8.56 PSE instrument. **a** General view. **b** Sedimentation tubes. **c** Flocculant tank and dose pumps. **d** Pulp mixing tank

first prototype of the instrument. Figure 8.59 shows the evolution of conductivity during the settling experiment.

The conversion between conductivity and concentration is done with an empirical equation based on Maxwell's work (Garrido 2005):

$$\varphi = \frac{(k_s + 2k_f)(1 - K)}{(k_f - k_s)(K + 2)} \tag{8.106}$$

where k_f and k_s are the concentration and the conductivity of the suspension, k_f is the conductivity of the fluid, k_s is the conductivity of the solid and $K = k_s/k_f$. The

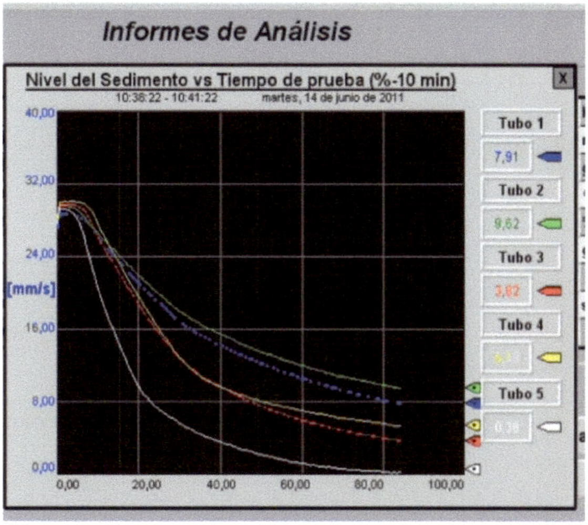

Fig. 8.57 Initial settling velocities from PSE

Fig. 8.58 Prototype of seditest instrument to measure sediment compressibility

computer software chooses long-term conductivities and converts them into concentrations and plots them, as shown in Fig. 8.60.

Equation (8.106) reference goes here converts into concentration profile into the solid effective stress $\sigma_e(\varphi)$ (Fig. 8.61).

SediTest On-line (SMC)

For the compressibility of the sediment, represented by the solid effective stress $\sigma_e(\varphi)$, an on-line instrument was also developed at the University of Concepción by (Segovia and Concha 2012; patent pending). See Figs. 8.62 and 8.63.

The SMC On-line determines the compressibility of sediment by measuring the concentration profile in a thickener. It consists of the following components:

- Airflow controller to maintain constant flow in the instrument.
- Level detection of the detector.

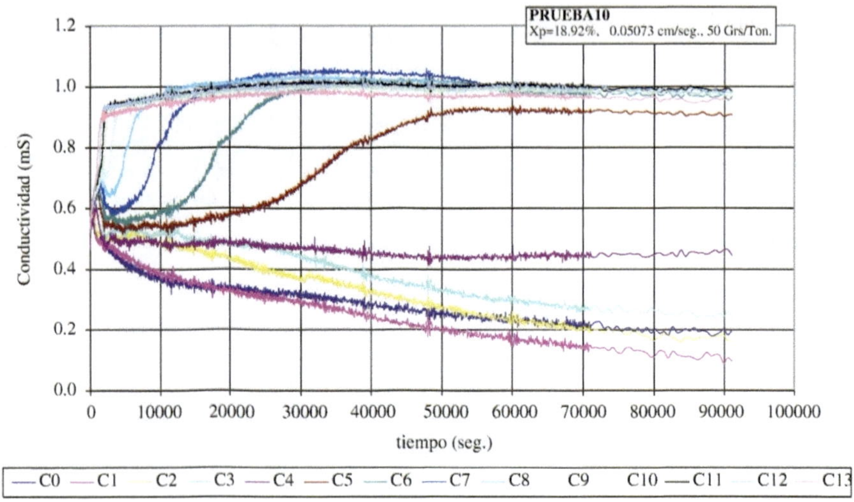

Fig. 8.59 Conductivity values at different heights in the settling column as a function of time

Fig. 8.60 Concentration profile at the end of sedimentation

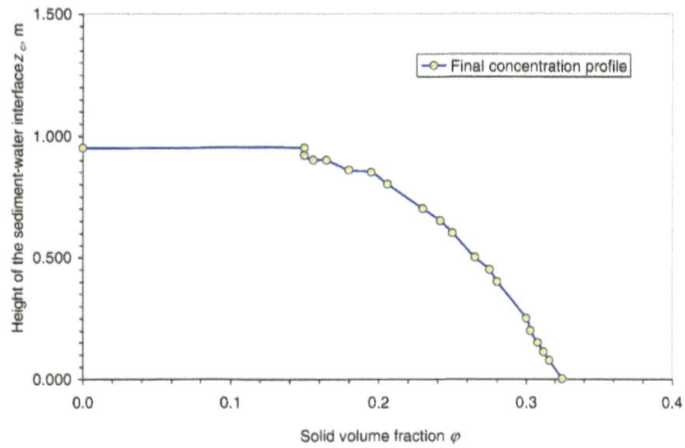

Fig. 8.61 Solid effective stress $\sigma_e(\varphi)$ versus φ

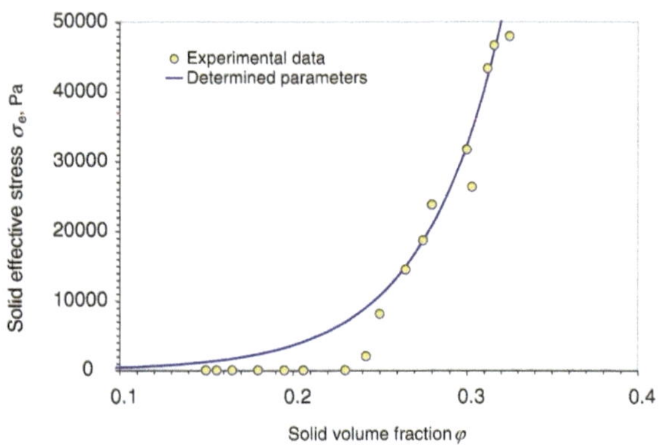

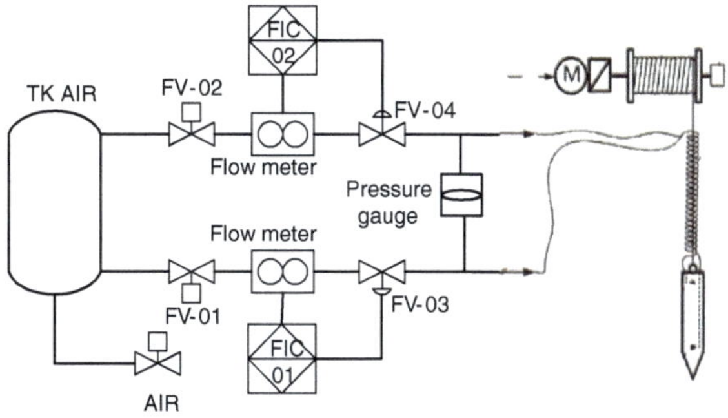

Fig. 8.62 Schematic view of the SMC (SediTest on-line)

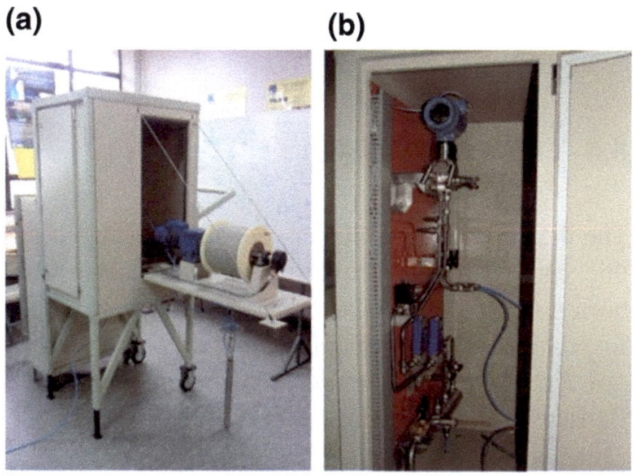

Fig. 8.63 Prototype of SMC instrument. **a** General view of prototype. **b** Sensor and instrumentation

- Descent detention.
- Differential pressure measurement.
- Washing the cable and detector.
- Data processing.

The instrument measures the pulp concentration as solid volume fraction in the thickener on the basis of differential pressure measurements following the equation:

$$\varphi_{pulp} = \frac{\Delta P}{(\rho_{solid} - \rho_{Liquid}) \cdot g \cdot \Delta h} - \frac{\rho_{Liquid}}{(\rho_{solid} - \rho_{Liquid})} \tag{8.107}$$

Figure 8.64 shows concentration profiles measured in an industrial thickener of a Chilean copper concentrator. The horizontal axis represents the solid concentration and the vertical axis represents the height of the thickener.

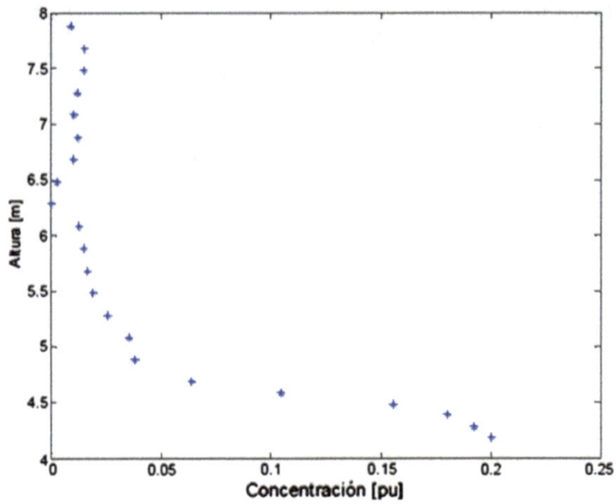

Fig. 8.64 Concentration profile in an industrial copper tailing thickener obtained by the SMC

8.6 Thickener Design

The design of a new thickener or the determination of the capacity of an existing thickener requires calculations at steady state. Since 1912, several methods have been proposed for thickener design. They can be classified based on their physical foundation as: *macroscopic balances, batch Kynch kinematical process, continuous Kynch kinematic process and dynamic sedimentation process* (Concha and Barrientos 1993). In this section, we discuss their advantages and limitations.

8.6.1 Methods Based on Macroscopic Balances

Mishler (1912) and Coe and Clevenger (1916) proposed the first methods of thickener design based on macroscopic balances.

(a) **Mishler's method**

Consider a thickener at steady state, with a solid mass feed flow of F (MT^{-1}), a concentration $\mathfrak{D}_F$ $[-]$, expressed as a *dilution*, defined as the ratio of liquid to solid mass, an underflow mass rate and concentration D (MT^{-1}) and $\mathfrak{D}_D$ $[-]$ respectively and a mass flow rate of water in the overflow of O (MT^{-1}). A solid and water mass balances gives:

$$\text{Solid}: \quad F = D \tag{8.108}$$

$$\text{Water}: \quad F\mathfrak{D}_F = D\mathfrak{D}_D + O \tag{8.109}$$

If ρ_f is the water density, the volume overflow rate of water is $Q_O = O/\rho_f$, then:

$$Q_O = \frac{F(\mathfrak{D}_F - \mathfrak{D}_D)}{\rho_f} \tag{8.110}$$

According to Mishler (1912) the overflow rate of water Q_O within the thickener is equal to the volume flow rate of water formed in a settling column during sedimentation for a suspension of the same concentration as that of the feed (Fig. 8.65).

Since the rate of water appearance, $R > 0$ in the batch test is equal to the rate of descent of the water suspension interface, the water volume flow rate is $Q_O = S \times R$, and from Eq. (8.110):

$$S = \frac{F(\mathfrak{D}_F - \mathfrak{D}_D)}{\rho_f R} \tag{8.111}$$

Mishler used the units of (short tons/day) for F, (feet/min) for R and (pounds/cubic feet) for ρ_f, so that the area in square feet is:

$$S = 0.0222S = \frac{F(\mathfrak{D}_F - \mathfrak{D}_D)}{\rho_f R(\mathfrak{D}_F)} \text{ in } (\text{ft}^2) \tag{8.112}$$

The method consists of measuring the settling velocity $R(\mathfrak{D}_F)$ of the water-suspension interface in a settling column, with the dilutions of the feed to the thickener, and using Eq. (8.112). This balance has the implicit assumption that the concentration in zone II in a thickener is that of the feed to the thickener.

The unit area, defined as $UA = S/F$, is given by:

$$UA = 0.0222S = \frac{(\mathfrak{D}_F - \mathfrak{D}_D)}{\rho_f R(\mathfrak{D}_F)} \text{ in } (\text{ft}^2/\text{stpd}) \tag{8.113}$$

Problem 8.6 Use Mishler's method to calculate the area of a thickener to treat 1,200 (tpd) of calcium carbonate with a density of 3.6 t/m^3 and feed concentration of $w_F = 35\,\%$ solids by weight. The settling velocity of a sample of feed is $R = 8.24 \times 10^{-6}$ m/s and the density of the water is 1.0 t/m^3. An underflow concentration of 52.4 % solid by weight is needed.

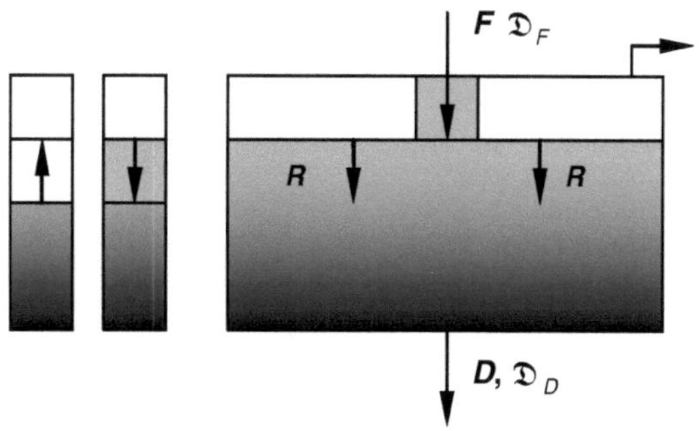

Fig. 8.65 Macroscopic mass balance in a continuous thickener according to Mishler (1912)

The relationship between dilution and % solid by weight is:

$$\mathfrak{D}_F = \frac{100 - w_F}{w_F} = \frac{100 - 35.0}{35.0} = 1.857$$

$$\mathfrak{D}_D = \frac{100 - w_D}{w_D} = \frac{100 - 52.4}{52.4} = 0.908$$

$$UA_M = \frac{\mathfrak{D}_F - \mathfrak{D}_D}{\rho_f R} = \frac{1.857 - 0.9084}{1.0 \times 8.24 \times 10^{-6} \times 24 \times 3600} = 1.33\ (\mathrm{m}^2/\mathrm{tpd})$$

$$S = UA_M \times F = 1.3326 \times 1.200 = 1599\,(\mathrm{m}^2)$$

$$D = (4 \times S/\pi)^{1/2} = 45.1\,(\mathrm{m})$$

The following table shows the results.

w_F % solid weight	$\mathfrak{D}_F$	$\mathfrak{D}_D$	R m/s	*UA* m^2/tpd	F tpd	S m^2	D m
35	1.8570	0.9080	8.24E−06	1.333	1,200	1,599	45

(b) **Coe and Clevenger's method**

Coe and Clevenger (1916) used the same mass balance as Mishler, but indicating that in a thickener a range of pulp concentrations, each with a specific settling velocity, will appear from the feed to the discharge the concentration. See Fig. 8.66.

The feed with pulp dilution $\mathfrak{D}_F$ passes through several dilutions $\mathfrak{D}_k$ before leaving the thickener with dilution $\mathfrak{D}_D$. The pulp with the lowest settling velocity predominates in the settling zone. A mass balance at the level of dilution $\mathfrak{D}_k$ gives:

$$\text{Solid :} \quad F = D \tag{8.114}$$

$$\text{Water :} \quad F\mathfrak{D}_k = D\mathfrak{D}_D + O \tag{8.115}$$

Calculating the mass flow rate of water O and the volume flow rate Q_O results in:

$$Q_o = \frac{F(\mathfrak{D}_{\mathfrak{f}} - \mathfrak{D}_{\mathfrak{D}})}{\rho_f}$$

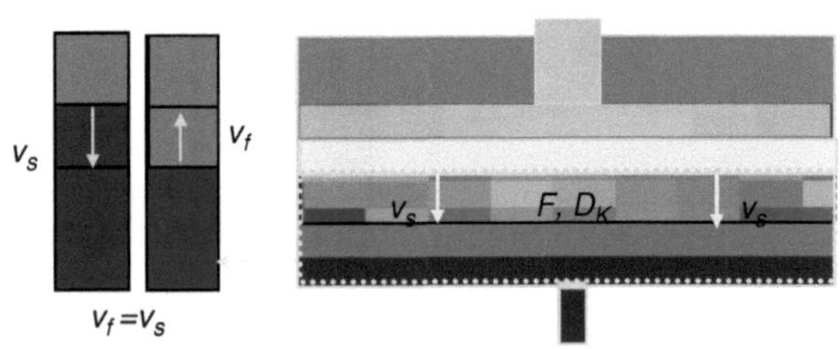

Fig. 8.66 Macroscopic mass balance in a continuous thickener according to Coe and Clevenger (1916)

With the same argument given by Mishler, the unit area *UA* is given by:

$$UA_{CC} = \frac{(\mathfrak{D}_k - \mathfrak{D}_D)}{\rho_f R(\mathfrak{D}_k)} \tag{8.116}$$

Since settling velocities at different dilutions give different *UAs*, several settling experiments should be made with dilutions ranging from the feed to the discharge dilution, calculating the respective unit area. The maximum unit area should be chosen. Coe and Clevenger used the following units: liquid density $\rho_f = 62.4\,(\text{lb/ft}^3)$, settling velocity *R* in (ft/h) and thickener capacity *F* in (lb/h/ft^2), with the result that the unit area is given in (ft^2/lb/h) per pounds:

$$UA_{CC} = \max_{\mathfrak{D}_F \geq \mathfrak{D}_k \geq \mathfrak{D}_D} \left(0.01604 \frac{\mathfrak{D}_k - \mathfrak{D}_D}{R(\mathfrak{D}_k)} \right) (\text{ft}^2/\text{lb/h}) \tag{8.117}$$

Taggart (1927) and Dalstrohm and Fitch (1985) used $\rho_f = 62.4\,(\text{lb/ft}^3)$ and the settling velocity R_k in (ft/h), giving the Unit Area in (ft^2/stpd):

$$UA_{TDF} = \max_{\mathfrak{D}_F \geq \mathfrak{D}_k \geq \mathfrak{D}_D} \left(1.33 \frac{\mathfrak{D}_k - \mathfrak{D}_D}{R(\mathfrak{D}_k)} \right) (\text{ft}^2/\text{stpd}) \tag{8.118}$$

Expressing the densities ρ in [kg/m^3], the concentrations in volume fraction of solids $\varphi = \rho_f/(\rho_f + \rho_s \mathfrak{D})$ and the solid flux density $f_{bk}(\varphi) = -\varphi R(\varphi) < 0$ in (m/s), equation can be written in the form:

$$UA_{CC} = \max_{\varphi_F \leq \varphi_k \leq \varphi_D} \left(1.1574 \times 10^{-2} \times \frac{1}{\rho_s f_{bk}(\varphi_k)} \left(\frac{\varphi_k}{\varphi_D} - 1 \right) \right) \quad (\text{m}^2/\text{tpd}) \tag{8.119}$$

Problem 8.7 Use Coe and Clevenger's method to calculate the area of a thickener to treat 1,200 tpd of calcium carbonate with a density if 2.53 t/m^3 and feed concentration of w = 35 % solids by weight. The density of the water is 1 t/m^3. The settling velocity of pulps ranging from the feed to the underflow concentration is given in the next table. An underflow concentration of 52.4 % solid by weight is needed.

Experiments at different concentrations gave the following settling velocities:

Concentration as % solids by weight	Concentration as dilution	Settling velocity *R* in m/s
35.0	1.857	8.24×10^{-6}
40.0	1.500	4.27×10^{-6}
45.0	1.222	2.06×10^{-6}
50.0	1.000	8.9×10^{-7}
52.4	0.908	5.76×10^{-7}

Using equation in metric units, we have for $\mathfrak{D}_k = 1.857$:

$$UA(1.857) = \frac{1.857 - 0.908}{8.24 \times 10^{-6} \times 24 \times 3600} = 1.333 \text{ m}^2/\text{tpd}$$

Calculating all concentrations of the table above, the following results were obtained:

W_k % sol	φ_k	D_k	D_D	R m/s	AU m²/tpd	F tpd	S m²	D m
35.0	0.177	1.8570	0.9080	8.24E−06	1.3330	1,200.00	1,599.6	45
40.0	0.211	1.5000	0.9080	4.27E−06	1.6046	1,200.00	1,925.6	50
45.0	*0.247*	*1.2220*	0.9080	*2.06E−06*	*1.7642*	*1,200.00*	*2,117.0*	*52*
50.0	0.286	1.0000	0.9080	8.90E−07	1.1964	1,200.00	1,435.7	43
52.4	0.306	0.9080	0.9080	5.70E−07	0.0000	1,200.00	0.0	0

The area of the thickener is $S = 2117\,(\text{m}^2)$ and $D = 52\,(\text{m})$.

Coe and Clevenger argued that the height of the thickener is important only if the underflow concentration is greater than the critical concentration. In that case, space should be provided for the sediment to have a sufficient residence time to reach the underflow concentration.

Call t^* the time to reach the required underflow concentration in a batch test. Divide the time interval $[0, t^*]$ in n intervals $\Delta t_i = t_{i-1} - t_i$. The height z_i of each interval is calculated as $z_i = V_i/S$ with $i = 1,2,\ldots n$, where V_i is the volume of the sediment with average pulp density $\bar{\rho}_i$ and S is the column cross-section area. The volume V_i is $V_i = F\Delta t_i/\varphi_i\bar{\rho}_i$, where F is the mass flow rate to the thickener. Then:

$$\begin{aligned} z_i &= \frac{F\Delta t_i}{\varphi_i\bar{\rho}_i S}, \quad i = 1,2,\ldots,n \\ &= \frac{1}{UA}\frac{\Delta t_i}{\bar{\varphi}_i(\Delta\rho\bar{\varphi}_i + \rho_f)} \end{aligned}$$

and the total height is:

$$\begin{aligned} z_c &= \sum_i z_i \\ &= \frac{1}{UA_{CC}}\sum_i \frac{\Delta t_i}{\bar{\varphi}_i(\Delta\rho\bar{\varphi}_i + \rho_f)}, \quad i = 1,2,\ldots,n \end{aligned} \tag{8.120}$$

In these expressions, $\bar{\varphi}_i = \varphi_{i-1} - \varphi_i$. For a height z_c, Coe and Clevenger recommended the addition of 0.50–1.00 m for clear water.

Problem 8.8 For the data of Problem 8.7, calculate the height of the thickener.

The unit area in problem 8.7 was $UA_0 = 1.76\ \text{m}^2/\text{tpd}$ for an underflow concentration of $w_D = 52.4\,[\%]$ solids by weight. In terms of solid volume fraction, the underflow concentration is $\varphi_D = 0.306$ (Figs. 8.67, 8.68).

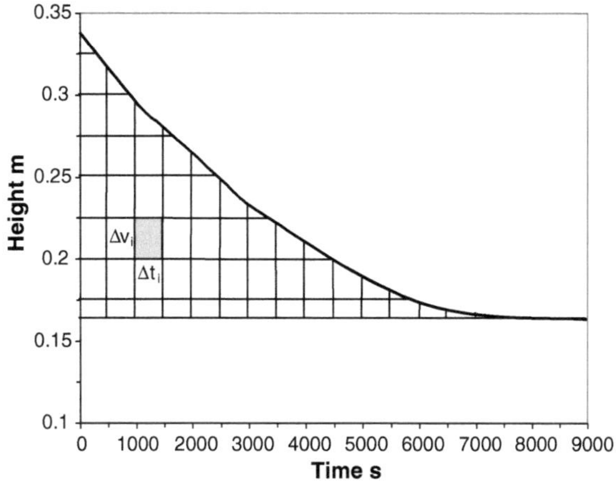

Fig. 8.67 Calculation of the height of a thickener according to Coe and Clevenger (1916) (Concha and Barrientos 1993)

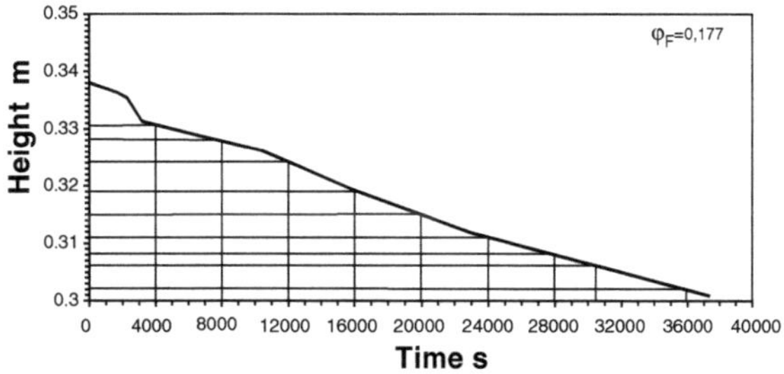

Fig. 8.68 Settling plot to calculate the thickener's height according to Coe and Clevenger (1916)

Applying Eq. (8.120) the sediment height should be 0.813 m. Adding 1.0 m for the clear water zone, the total height of the thickener must be more than 1.813 m. See the following table.

Z_i cm	φ_i	Δt_i s	ΔZ_i m
0.338	0.177	0	0
0.331	0.181	10	0.000
0.328	0.182	8,000	0.221
0.324	0.185	12,000	0.109
0.319	0.188	16,000	0.107
0.315	0.190	20,000	0.105
0.311	0.192	24,000	0.104
0.308	0.194	28,000	0.103
0.306	0.196	30,500	0.064
			H_c (m) = 0.813

8.6.2 Methods Based on Kynch Sedimentation Processes

Kynch's theory led several researchers, including Talmage and Fitch (1955), Wilhelm and Naide (1979), Oltmann (), Hassett (1958, 1961, 1964a, b, 1968) and Yoshioka et al. (1957) to use this theory as the basis for thickener design. The method used universally until then was that of Coe and Clevenger (1916) but it was felt that it required too many laboratory tests. Researchers looked for a methodology that required less effort in the laboratory and Kynch's theory was the answer.

To describe the different methods developed by this group of researchers, divide them into those based on the *Kynch Batch Sedimentation Processes* and on the *Kynch Continuous Sedimentation Process*. To review Kynch's theory see Chap. 5.

(a) **Methods based on Batch Kynch Sedimentation Processes**

Consider an ideal suspension subjected to batch sedimentation. Kynch's theory shows that the suspension concentration $\varphi(z,t)$ can be obtained from a hyperbolic conservation law for regions where the variables are continuous, by the theory of characteristics:

$$\frac{\partial \varphi}{\partial t}+\frac{\partial f}{\partial z}=0 \tag{8.121}$$

For discontinuities, jump conditions replace Eq. (8.121):

$$\sigma(\varphi^{+},\varphi^{-})=\frac{[f]}{[\varphi]},\ \text{with}\ [\circ]=(\circ)^{+}-(\circ)^{-} \tag{8.122}$$

The necessary conditions to obtain a solution to Eq. (8.122) are: to have a constitutive equation for the flux density function $f=f(\varphi)$ and initial and boundary conditions.

In a settling curve $z=z(t)$, as in Fig. 8.69, tangents such as Z/T to the settling curve are the rate of displacement of the discontinuity of a given concentration in the water/suspension, and represents the sedimentation velocity of suspensions of concentration $Z/T=v_s(\varphi_k)$:

$$\sigma_I(\varphi)=\sigma(0,\varphi)=\frac{f_{bk}(\varphi)-f_{bk}(0)}{\varphi-0}=\frac{f_{bk}(\varphi)}{\varphi}=v_s\varphi\equiv\frac{Z}{T} \tag{8.123}$$

The straight lines starting at $(0,0)$ and ending at (z_k,t_k), for several values of k, are characteristics with constant concentration φ_k and slope $f'_{bk}(\varphi)$.

The volume of solids per unit cross sectional area in the settling column is $\varphi_0 L$. While traveling from $(0,0)$ to (z_k,t_k) the wave of concentration φ_k crosses the total volume of solids, then:

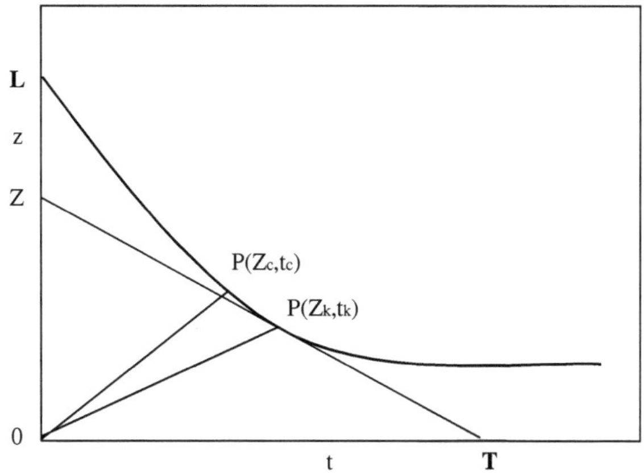

Fig. 8.69 Settling curve

$$\varphi_0 L = \int_0^{t_k} \varphi_k \left(v_s(\varphi_k) + f'_{bk}(\varphi) \right) dt$$

$$= \varphi_k \left(\frac{Z}{T} + \frac{z_k}{t_k} \right) t_k$$

From Fig. 8.69, we see that $Z/T = (Z - z_k)/t_k$, therefore

$$\varphi_0 L = \varphi_k Z \tag{8.124}$$

Therefore, the value of the concentration φ_k can be found as:

$$\varphi_k = \varphi_0 \frac{L}{Z} \tag{8.125}$$

From Kynch's theory, many pars $(\varphi_k, v_s(\varphi_k))$ can be obtained from a settling curve by drawing tangents as shown in Fig. 8.70.

Since $\varphi_k = \varphi_0 L/Z$ and $v_s(\varphi_k) = -Z/T$, then $f_{bk}(\varphi) = \varphi_k v_s(\varphi_k) = -\varphi_0 L/T$. The unit area can be calculated from Coe and Clevenger's equation:

$$UA_{CC} = \max_{\varphi_F \leq \varphi_k \leq \varphi_D} \left(1.1574 \times 10^{-2} \times \frac{1}{\rho_s f_{bk}(\varphi_k)} \left(\frac{\varphi_k}{\varphi_D} - 1 \right) \right) \text{m}^2/\text{tpd}$$

From (8.125) $\varphi_k/\varphi_D = Z_D/Z$ and $f_{bk}(\varphi_k)$ from yields the unit area determined from Kynch'e equation:

$$UA_K = \max_{\varphi_F \leq \varphi_k \leq \varphi_D} \left(1.1574 \times 10^{-2} \times \frac{T}{\rho_s \varphi_0 L} \left(1 - \frac{Z_D}{Z} \right) \right) \text{m}^2/\text{tpd} \tag{8.126}$$

Problem 8.9 Use Kynch's theory for batch sedimentation to calculate the area of a thickener to treat 1,200 tpd of calcium carbonate with a density if 2.5 t/m^3 and feed concentration of w = 35.0 % solids by weight. The density of the water is

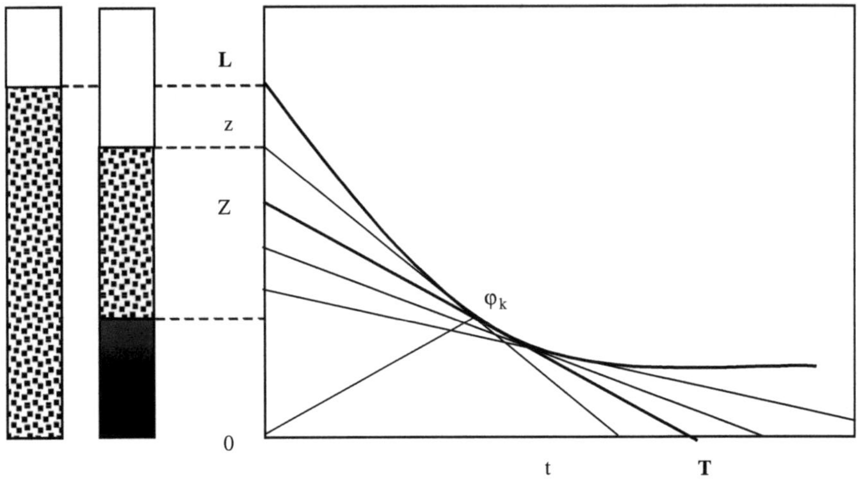

Fig. 8.70 Determination of the concentration and settling velocity of a suspension according to Kynch's method of batch sedimentation

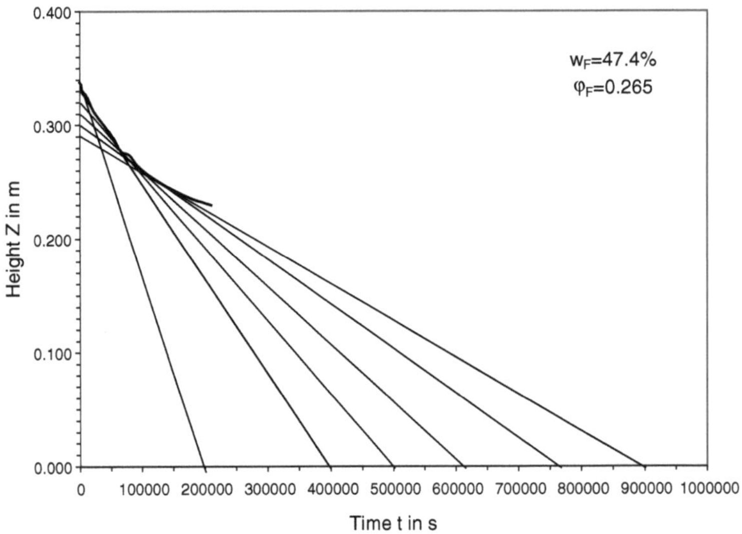

Fig. 8.71 Tangents in the settling curve from Kynch's theory

1 t/m^3. A settling curve is given for a suspension with 10.3 % solid by weight. An underflow concentration of 52.4 % solid by weight is needed.

The table below and Fig. 8.71 show the values of T and Z obtained from the tangents and the tangent drawn to the curve.

Z_k cm	T_k s	L cm	φ_0	φ_k	$f_{bk}(\varphi)$ m/s	AU_O m₂/tpd	F tpd	S m^2	D m
0.338	20,0000	0.338	0.265	0.265	4.4785E−07	1.3762797	1,200	1,652	46
0.330	*395,000*	*0.338*	*0.265*	*0.271*	*2.2676E−07*	*2.2891053*	*1,200*	*2,747*	*59*
0.320	500,000	0.338	0.265	0.280	1.7914E−07	2.1805432	1,200	2,617	58
0.310	610,000	0.338	0.265	0.289	1.4684E−07	1.7290118	1,200	2,075	51
0.300	760,000	0.338	0.265	0.299	1.1786E−07	0.9165819	1,200	1,100	37
0.293	850,000	0.338	0.265	0.306	1.0538E−07	0	1,200	0	0

According to this method, the area of the thickener is $S = 2{,}747\,\mathrm{m}^2$ and the diameter is $D = 59\,\mathrm{m}$.

Talmage and Fitch method

Talmage and Fitch (1955) wrote from Fig. 8.72:

$$\frac{Z}{T} = \frac{Z - Z_D}{t_u}$$

where Z_D is the tangent to the settling curve for the underflow concentration φ_D, and t_u is the intersection of the tangent *Z-T* with the horizontal drawn from Z_D. The coordinate Z_D is obtained from $Z_D = \varphi_0 L / \varphi_D$.

Replacing into (8.126), and observing that the maximum of t_u is t_U, the following simple equation is obtained:

$$AU = \underbrace{\max}_{Z}\left(\frac{t_u}{\rho_s \varphi_0 L}\right) \equiv \frac{t_U}{\rho_s \varphi_0 L} \tag{8.127}$$

$$AU_{TF} = 1.1574 \times 10^{-2} \frac{t_U}{\rho_s \varphi_0 L}, \; (\mathrm{m}^2\mathrm{tpd}) \tag{8.128}$$

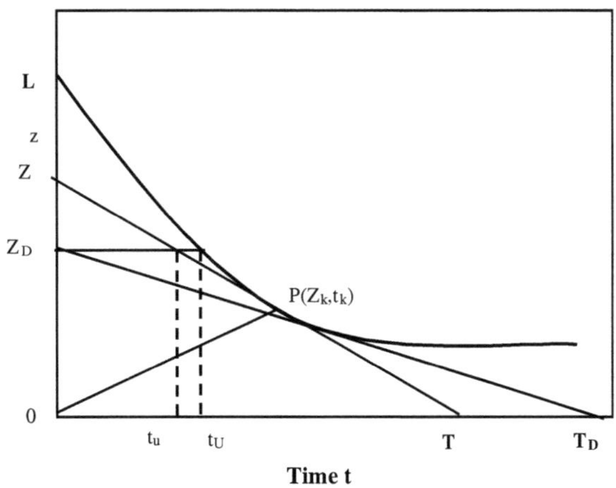

Fig. 8.72 Talmage and Fitch construction (1955)

Time (s)	Height (m)
0	0.338
1692	0.336
2304	0.335
3204	0.331
6804	0.329
10404	0.326
15804	0.319
23004	0.312
37404	0.301
51804	0.291
66204	0.277
80604	0.274
95004	0.264
109404	0.257
123804	0.252
138204	0.247
152604	0.243
167004	0.238
181404	0.235
195804	0.232
210204	0.230

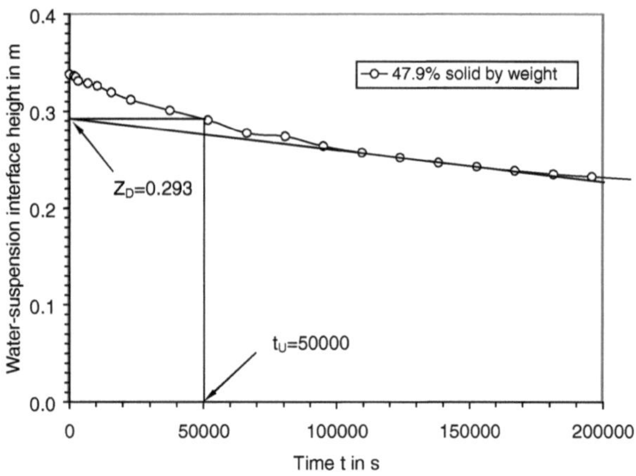

Fig. 8.73 Talmage and Fitch construction (1955)

Talmage and Fitch method consist of:

1. Performing a settling test with an "intermediate "concentration. (we will discuss this further on).
2. Drawing the settling curve.
3. Determining the height Z_D from $Z_D = \varphi_0 L/\varphi_D$.
4. Drawing a horizontal line from Z_D to the settling curve. The intersection defines t_U.
5. If metric units are used, using Eq. (8.128) to calculate the unit area.

Problem 8.10 Use Talmage and Fitch's method to calculate the area of a thickener to treat 1,200 tpd of calcium carbonate with a density if $\rho_s = 2.5\,\text{t/m}^3$ and feed concentration of $w_F = 35\,\%$ solids by weight. The density of the water is $\rho_f = 1\ \text{t/m}^3$. A settling curve is given for a suspension with $w_0 = 47.9\,\%$ solid by weight. The underflow concentration is $w_D = 52.4\,\%$ solid by weight. The critical concentration is $w_c = 42.7\,\%$ by weight.

According to Kynch's theory $Z_D = \varphi_0 \times L/\varphi_D = 0.265 \times 0.338/0.303 = 0.293$. Figure 8.73 should be constructed.

Applying Eq. (8.128), the unit area UA_{TF} can be calculated. The results can be summarized in the following table:

t_U	50,000
AU_0	2.251
S	2,702
D	59

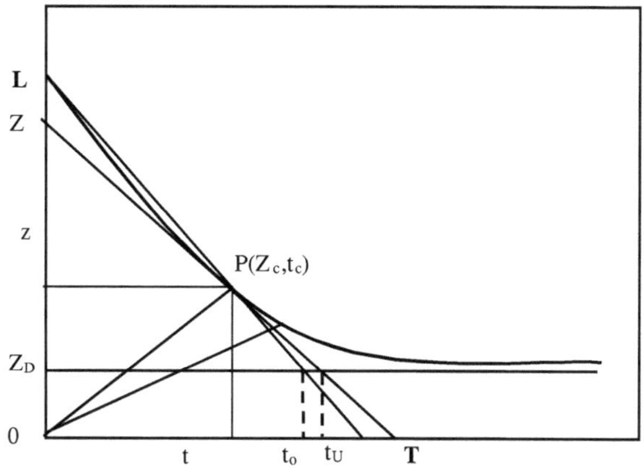

Fig. 8.74 Talmage and Fitch II (1955) and Oltman's construction (Fitch and Stevenson 1976)

Since Talmage and Fitch's method is used for all types of pulps, compressible and incompressible, in some cases the horizontal line drawn from Z_D does not cut the settling curve. In these cases, Talmage and Fitch recommend drawing the tangent *Z-T* for the critical concentration φ_c and obtaining t_U as the intersection of this line with the horizontal line. See Fig. 8.74.

Oltman method

From experience, Fitch and Stevenson (1976) found that thickener area obtained from the critical concentration was too great and adopted an empirical modification proposed by Oltman, an engineer at Dorr Oliver. Without justification, Oltman proposed substituting the tangent to the settling curve by the straight line drawn from point $(L, 0)$ passing through the point (z_c, t_c). See Fig. 8.74.

Talmage and Fitch and Oltman's method are based on the knowledge of the critical concentration and on the assumption (erroneously) that a compressible pulp at the underflow concentration follows Kynch's theory.

8.6.3 Methods Based on Continuous Kynch Sedimentation Processes

In Chap. 5 Sect. 5.2.3 Kynch's theory for a continuous process was studied. Figure 5.14a and c show that Unit Area of an ideal thickener treating an ideal suspension is given by:

$$UA_0 = \frac{1}{\rho_s f_{bk}(\varphi_M^{**})}\left(\frac{\varphi_M^{**}}{\varphi_M} - 1\right)$$

where φ_M and φ_M^{**} are the concentration of the maximum point in the flux-density curve and the conjugate concentration, respectively. The underflow concentration is $\varphi_D = \varphi_M$. These methods imply knowledge of the flux-density function. See Fig. 5.14a and c.

Yoshioka and Hassett Method

Yoshioka et al. (1957) proposed a method of thickener design based on the knowledge of the batch flux-density function and its relationship with the continuous flux density and the underflow concentration (see Chap. 5, Sect. 5.2.4). If $f_{bk}(\varphi)$ is the batch fluxdensity and $f_k(\varphi)$ is the continuous flux density, we can write at steady state:

$$f_F = f_k = q\varphi + f_{bk}(\varphi) \tag{8.129}$$

(8.129) is written in the form:

$$f_{bk}(\varphi) = f_F - q\varphi \tag{8.130}$$

In a graph of $f_{bk}(\varphi)$ versus φ, Eq. (8.130) represents a straight-line tangent to $f_{bk}(\varphi)$ at $\varphi = \varphi_M$ with slope $q = f_F/\varphi_D$. Figure 8.75 shows these functions with the ordinate axis as $(-f_k(\varphi))$ and $(-f_{bk}(\varphi))$.

The design method can be summarized as follows:

1. Draw the given batch flux-density function.
2. Select the underflow concentration.
3. For the given batch flux-density function, draw a straight line from the point, $(0, \varphi_D)$ tangent to the batch flux density curve at point $\varphi = \varphi_M$.
4. The line cuts the ordinate at point $(f_F, 0)$.

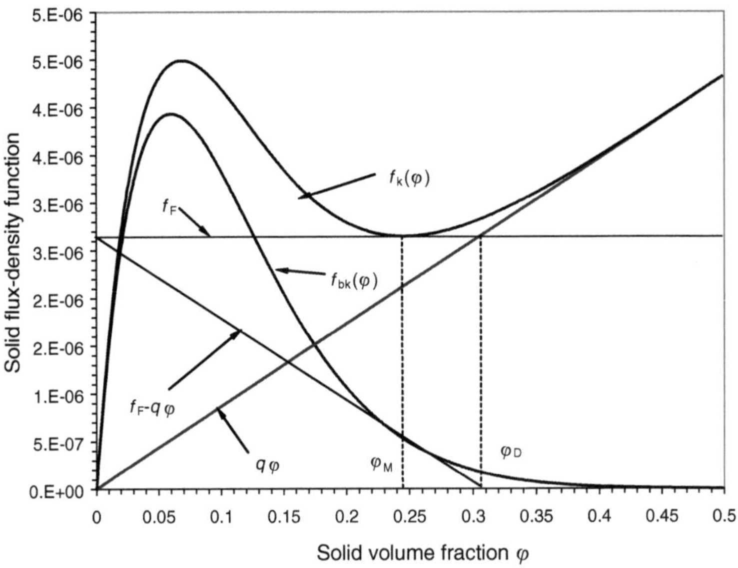

Fig. 8.75 Yoshioka and Hassett design method

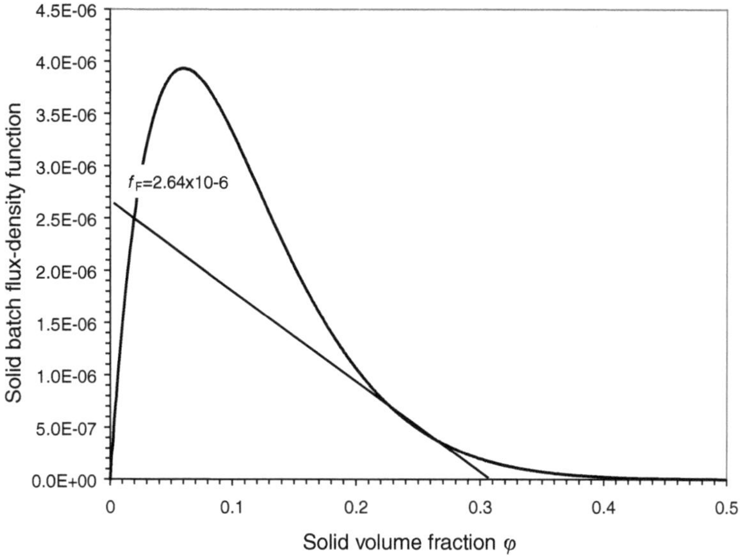

Fig. 8.76 Yoshioka and Hasset construction

Problem 8.11 Using the method of Yoshioka and Hassett, design a thickener to process 1,200 tpd of calcium carbonate with a feed concentration of 35 % solids by weight. The solid density is 2.5 t/m^3 and the desired underflow concentration is 52.4 % solids by weight. The solid batch flux-density function is given in Fig. 8.76.

The feed and underflow concentrations are:

$$\varphi_F = \frac{1 \times 35.0}{2.5 \times (100 - 35.0) + 1 \times 35.0} = 0.177$$
$$\varphi_D = \frac{1 \times 52.1}{2.5 \times (100 - 52.4) + 1 \times 52.4} = 0.306$$

In Fig. 8.76 a straight line starting from and tangent to the curve give a value of $-f_F = 2.64 \times 10^6\,(\text{m/s})$. Then the thickener capacity is given by:

$$S = \frac{F}{-f_F \times \rho_s \times 3600 \times 24} = \frac{1200}{2.64 \times 10^{-6} \times 2.5 \times 3600 \times 24} = 2104\,(\text{m}^2)$$
$$UA_0 = \frac{S}{F} = \frac{2104}{1200} = 1.753\,(\text{m}^2/\text{tpd}); \quad D = 52\ (\text{m})$$

Hassett (1958) noted a problem in the interpretation of the underflow concentration. Figure 8.76 shows that according to Kynch's theory only two concentrations exist in the thickener at steady state, the underflow concentration φ_M and the conjugate concentration φ_M^{**}. Hassett said:

> … In this way the theory predicts the absence of the feed concentration and the underflow concentration in the thickener, which means that an *abrupt growth of concentration* must exist in the underflow….

The theory developed in Chap. 5 shows that Hassett's reasoning is wrong because he tried to apply Kynch's theory to compressible suspensions. The theory is correct for ideal but not for compressible suspensions.

Wilhelm and Naide Method

Wilhelm and Naide (1979) also started from the knowledge of the flux-density function for continuous thickening, like Yoshioka and Hasset, and wrote similar equation at steady state:

$$f_F = q\varphi + f_{bk}(\varphi) = q\varphi_D$$

Differentiating with respect φ results:

$$0 = q + f'_{bk}(\varphi_M)$$

from which:

$$q = -f'_{bk}(\varphi_M) \tag{8.131}$$

Substituting in the previous expression yields:

$$-f'_{bk}(\varphi_M)\varphi + f_{bk}(\varphi) = -f'_{bk}(\varphi_M)\varphi_D$$

This equation should apply for $\varphi = \varphi_M$, then:

$$-f'_{bk}(\varphi_M)\varphi_M + f_{bk}(\varphi_M) = -f'_{bk}(\varphi_M)\varphi_D \tag{8.132}$$

Now, if we assume that the solid settling velocity can be expressed in the form:

$$v_s(\varphi) = -a\varphi^{-b}, \tag{8.133}$$

the solid flux-density function $f_{bk}(\varphi)$ would be $f_{bk}(\varphi) = -a\varphi^{1-b}$ and its derivative at $\varphi = \varphi_M$:

$$f'_{bk}(\varphi) = -a(1-b)\varphi_M^{-b} \tag{8.134}$$

Substituting (8.134) with (8.132), and calculating φ_M results in:

$$\varphi_M = \frac{b-1}{b}\varphi_D \tag{8.135}$$

Relacing φ_M from (8.135) in (8.134) and the result in (8.131) yields:

$$q = -a(1-b)\left(\frac{b-1}{b}\right)^{-b}\varphi_D^{-b} \tag{8.136}$$

Since at steady state $f_F = q\varphi_D$, using (8.136) we reach the result:

$$f_F = -a(1-b)\left(\frac{b-1}{b}\right)^{1-b}\varphi_D^{1-b} \tag{8.137}$$

and the area and unit area AU are:

$$\begin{aligned} S &= \frac{F}{-f_F \times \rho_s \times 3600 \times 24} = \frac{F}{\rho_s \times 3600 \times 24}\,\frac{\left(\frac{b-1}{b}\right)^{b-1}\varphi_D^{b-1}}{ab(1-b)}\ (\mathrm{m}^2) \\ UA &= \frac{S}{F} = \frac{1}{-f_F \times \rho_s \times 3600 \times 24} = \frac{1}{\rho_s \times 3600 \times 24}\,\frac{\left(\frac{b-1}{b}\right)^{b-1}\varphi_D^{b-1}}{a(1-b)}\ (\mathrm{m}^2/\mathrm{tpd}) \end{aligned} \tag{8.138}$$

The expression $v_s(\varphi) = -a\varphi^{-b}$ in (8.133) represents the settling velocity solely in a narrow range of concentrations, therefore the complete function $AU(\varphi)$ must be replaced by a series of expressions valid in this narrow range. In this way, expression (8.138) is valid for each region, as shown in Fig. 8.77. Wilhelm and Naide conclude that the maximum value found for $AU(\varphi)$ must be chosen as the correct value for the Unit Area.

Problem 8.12 Using Wilhelm and Nadie's method of thickener design to process 1,200 tpd of calcium carbonate with a feed concentration of 35 % solids by weight. The solid density is 2.5 t/m^3 and the desired underflow concentration is 52.4 % solids by weight. From a table of data, the settling velocity was calculated and plotted in Fig. 8.77. The feed and underflow concentrations are respectively:

$$\begin{aligned} \varphi_F &= \frac{1 \times 35.0}{2.5 \times (100 - 35.0) + 1 \times 35.0} = 0.177 \\ \varphi_D &= \frac{1 \times 52.1}{2.5 \times (100 - 52.4) + 1 \times 52.4} = 0.306 \end{aligned}$$

For each region, these equations can be applied to obtain the following values for the unit area, the highest of which should be chosen. See Fig. 8.78.

Wilhelm and Nadie's design method gives a value of $UA = 1.815\,\mathrm{m}^2/\mathrm{tpd}$ and $S = UA \times F = 1.815 \times 1200 = 2178\ \mathrm{m}^2$; $D = 53\ \mathrm{m}^2$.

8.6.4 Methods Based on the Phenomenological Theory

In Sect. 8.3.6 we demonstrate that, for a steady state to exist in a continuous thickener, it is necessary that the following inequality be obeyed:

$$f_F \geq q\varphi + f_{bk}(\varphi), \quad \text{for } \varphi_L \leq \varphi \leq \varphi_D \tag{8.139}$$

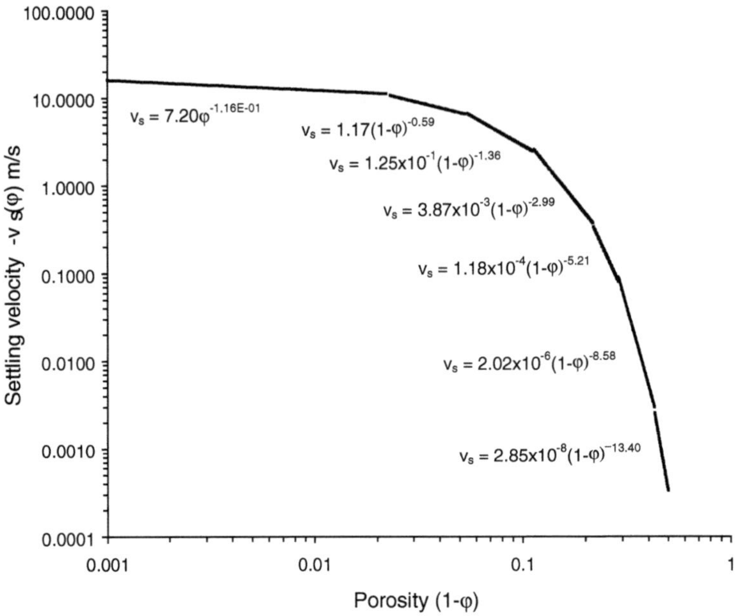

Fig. 8.77 Settling velocity by regions fitting expression

Fig. 8.78 Thickener design according to Wilhelm and Nadie (1979)

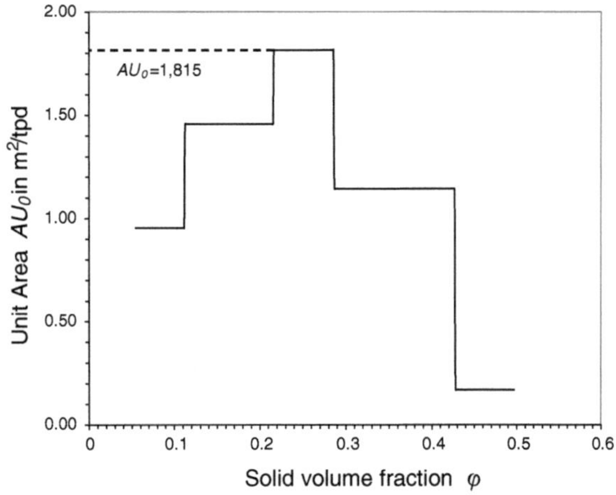

where f_F is the feed solid flux-density function, and $f_{bk}(\varphi)$ is Kynch's batch flux density function. Since $q = f_F/\varphi_D$, substituting (8.139), the following is valid for the steady state: $f_F \geq \frac{f_F}{\varphi_D}\varphi + f_{bk}(\varphi)$, for $\varphi_L \leq \varphi \leq \varphi_D$.

Dividing by φ and rearranging yields:

$$f_F \geq \frac{f_{bk}(\varphi)}{1-\frac{\varphi}{\varphi_D}}, \quad \text{for } \varphi_L \leq \varphi \leq \varphi_D \tag{8.140}$$

Flux Density

By definition $F = -\rho_s f_F S$, where F is the thickener capacity measured as solid mass flux, and S is the thickener's cross sectional area. Then, substituting into (8.140) gives:

$$\frac{F}{S} \leq \frac{\rho_s f_{bk}(\varphi)}{1-\frac{\varphi}{\varphi_D}}, \text{ in ML}^{-2}\text{T}^{-1}, \text{ for } \varphi_L \leq \varphi \leq \varphi_D \tag{8.141}$$

Unit Area

Since the unit area is the reciprocal of solid flux density:

$$\frac{S}{F} \geq \frac{1}{\rho_s f_{bk}(\varphi)}\left(1-\frac{\varphi}{\varphi_D}\right), \text{ in L}^{-2}\text{M}^{-1}\text{T}, \quad \text{for} \varphi_L \leq \varphi \leq \varphi_D \tag{8.142}$$

Defining the *unit area* function $UA(\varphi)$:

$$UA(\varphi) := \left(-\frac{1}{\rho_s f_{bk}(\varphi)}\right)\left(1-\frac{\varphi}{\varphi_D}\right), \quad \text{for } \varphi_L \leq \varphi \leq \varphi_D \tag{8.143}$$

And the *unit area* is:

$$UA = \max_{\varphi_L \leq \varphi \leq \varphi_D} (UA(\varphi)) \tag{8.144}$$

Height of the Sediment

The height of the sediment is obtained by integrating the equation:

$$z_c = \int_{\varphi_D}^{\varphi_c} \frac{\sigma_e'(\xi) f_{bk}(\xi)}{\Delta\rho\xi g(f_F - q\xi - f_{bk}(\xi))} d\xi \tag{8.145}$$

where φ_D and φ_c are the underflow and the critical concentrations respectively.

To the height of the sediment "z_c" in the thickener, a height "h" for the sedimentation zone and a height "c" for the clear water must be added. The last two zones have arbitrary depth.

$$H = c + h + z_c \tag{8.146}$$

Solid Inventory in the Thickener

Solid inventory is the amount of solid stored as sediment and suspension in the thickener at any time. This material consists of a pulp of the conjugate concentration φ_L and a sediment of variable concentration from φ_c to φ_D. Then, the solid inventory can be expressed as:

$$I = \int_0^L \rho_s \varphi(z) S dz = \rho_s \varphi_L S(L - z_c) + \int_0^{z_c} \rho_s \varphi(z) S dz \tag{8.147}$$

where L is the height of the outlet of the feedwell.

(a) **Adorjand's Method**

Adorjan (1975, 1976) was the first author to utilize Eq. (8.144) for thickener design. Although his deduction differs from the one used here. Adorjan argues that a thickener operated under limiting condition, that is, designed using Eq. (8.143) with the equal sign, requires a considerable pulp depth and therefore it must be operated at only a fraction of the limiting rate. This fraction he called the *loading factor*, and he designed it by $\lambda = F/F_0$, where F is the actual feed rate and F_0 is the limiting feed rate. In terms of unit area $UA = UA_0/\lambda$, with $0 \le \lambda \le 1$. Adorjan adopted as criteria to select λ a safety factor in the design, so that a certain deviation from the design would be possible.

Problem 8.13 Using Adorjan's method, design a thickener to process 1,200 (tpd) of calcium carbonate with a feed concentration of 35 % solids by weight. The solid density is 2.5 t/m^3 and the desired underflow concentration is 52.4 % solids by weight. Constitutive equations for the settling velocity and the solid effective stress are available with critical concentration at 42 % solids by weight:

$$f_{bk}(\varphi) = -1.72 \times 10^{-4} \varphi \times (1 - \varphi)^{15.6} \text{ m/s}$$
$$\sigma_e(\varphi) = 2.0 \exp(22\varphi) \text{ Pa}$$

The feed and underflow concentrations are respectively $\varphi_F = 0.177$ and $\varphi_D = 0.306$.

Replacing the corresponding numerical values in Eq. (8.144), the following is obtained:

$$UA_F(\varphi) = -\frac{1}{2.5 \times 3{,}600 \times 24 \times -1.72 \times 10^{-4} \times \varphi(1 - \varphi)^{15.6}} \left(\frac{\varphi}{0.306} - 1 \right)$$

Plotting the unit area function $UA(\varphi)$, gives Fig. 8.79, with a maximum at $\varphi = 0.256$ and a unit area of $1.756\,\text{m}^2/\text{tpd}$. Using the safety factor $\lambda = 0.95$,

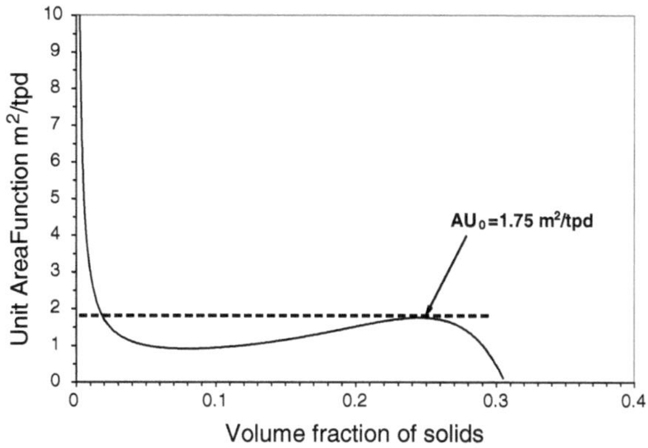

Fig. 8.79 Unit area function versus the concentration with the chosen unit area AU_0

Adorjan's Unit Area becomes: $UA = 1.75/0.95 = 1.84\,\text{m}^2/\text{tpd}$. Finally, the area and diameter of the thickener are: $S = 1.84 \times 1,200 = 2,208\,\text{m}^2$ and $D = 53\,\text{m}$.

(b) **Cettem's Method**

In a recent work, Concha and collaborators (Garrido et al. 2003) developed an algorithm to design and simulate continuous thickeners. Equations (8.143) and (8.144) are also the bases of this method:

$$UA(\varphi,\varphi_D):=\frac{1}{\rho_s f_{bk}(\varphi)}\left(\frac{\varphi}{\varphi_D}-1\right),\quad \text{for}\quad \varphi_L\leq\varphi\leq\varphi_D$$
$$UA=\max_{\varphi_L\leq\varphi\leq\varphi_D}(UA(\varphi))$$

It is interesting to note that Eq. (8.144) has the same form as Coe and Clevenger's equations. But there is a small but important difference. Coe and Clevenger assumed that the range of concentration where the value of $UA(\varphi,\varphi_D)$ should be maximized was the feed and the underflow concentrations $\varphi_F\leq\varphi\leq\varphi_D$, while Cettem's method used the known fact that the feed to a thickener is always diluted to the conjugate concentration φ_L when entering the thickener, therefore the range of concentration should be $\varphi_L\leq\varphi\leq\varphi_D$. Figure 8.80 shows the difference.

In designing a thickener, the desired feed flow rate F and underflow concentration φ_D, together with the thickening parameters $f_{bk}(\varphi)$, $\sigma_e(\varphi)$, φ_c and the solid and liquid densities ρ_s and ρ_f, must be known. With these values, the function $UA(\varphi,\varphi_D)$ can be calculated for any value of $\varphi_L\leq\varphi\leq\varphi_D$.

The first step in the design procedure makes use of the solution to the phenomenological model at steady state plotting the function $UA(\varphi,\varphi_D)$ in the interval $\varphi_L\leq\varphi\leq\varphi_D$, see Fig. 8.80.

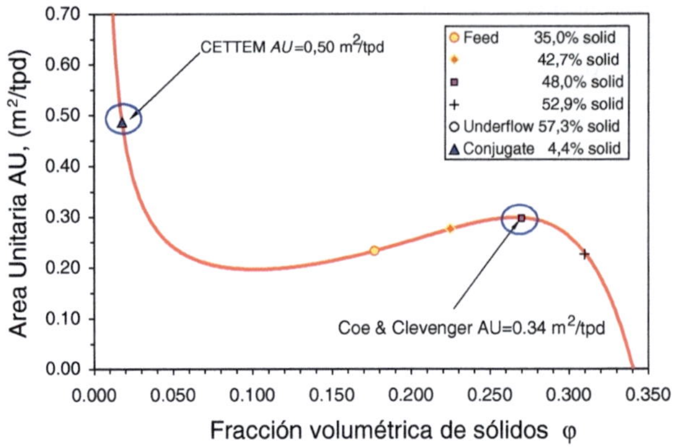

Fig. 8.80 Comparison of CETTEM's and Coe and Clevenger's methods for the determination of the unit area for a copper flotation tailing

The problem to calculate the maximum indicated in the figure is that the conjugate concentration φ_L is unknown until the value of $f_F = -F/\rho_s S$ is calculated, and to do so, we need to know the thickener cross sectional area *S*, which is the result we are seeking. Therefore, the problem is undetermined and to solve it we must assume a value of φ_L, for example $\varphi_L \approx 0.01$. Once φ_L is chosen, the maximum of $UA(\varphi, \varphi_D)$ is obtained, and the area *S* and diameter *D* can be calculated. Now, f_F and q can be obtained and, therefore, the concentration profile in the thickener can be calculated from Eq. (8.145) and the value of the sediment height z_c is obtained.

The next step in the design is to set a target for the value of the sediment height z_c in the thickener. This value is a restriction imposed by the motor of the rakes, which allows a given maximum torque. For example, 20 % of the available height *L* of the thickener is a reasonable value. Then, an iteration changing the value of *D* obtains the desired z_c value.

Finally, once the thickener diameter is obtained, it is approximated to the next integer (or to the next commercially available thickener diameter) and the calculation is repeated to obtain: (1) the final thickener diameter, (2) the thickener cross sectional area *S*, (3) the Unit Area *UA*, (4) the mass and water balance in the thickener, (5) the solid inventory *I*, (6) the concentration profile in the thickener, (7) the sediment height z_c and (8) the correct value the conjugate concentration φ_L (Fig. 8.81).

Problem 8.14 Use CETTEMS's method to process 1,200 tpd of calcium carbonate with a feed concentration of 35[%] solids by weight. The solid density is 2.5 (ton/m^3) and the desired underflow concentration is 52.4[%] solids by weight. Constitutive equations for the settling velocity and the solid effective stress are available with critical concentration at 42[%] solids by weight. Consider the design of the thickener with $L = 6$ (m) height and a sediment height of 20 [%] of *L*, knowing the following properties of the tailings:

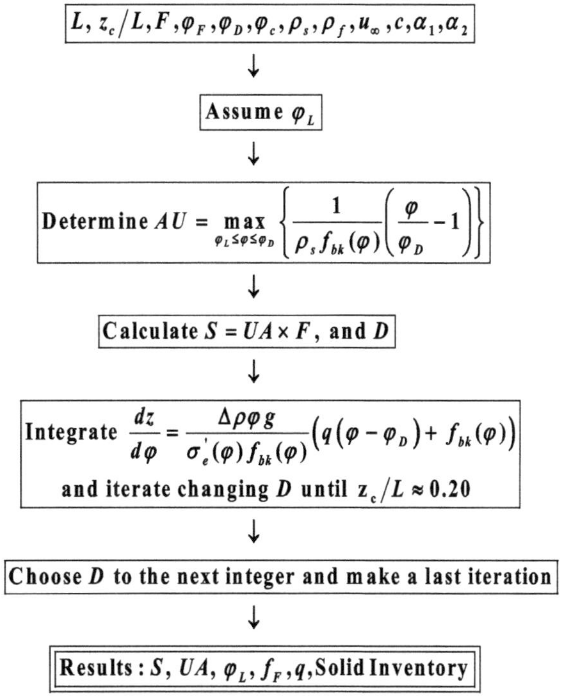

Fig. 8.81 Algorithm for the design of continuous thickeners by CETTEM's method (Concha et al. 2003)

Solid density $\rho_s = 2{,}500\,(\mathrm{kg/m^3})$

Liquid density $\rho_f = 1{,}000\,(\mathrm{kg/m^3})$

Feed concentration $w_F = 35.0\,[\%]$ solid by weight

Critical concentration $w_c = 40.2\,[\%]$ solid by weight

Conjugate concentration $\varphi_L = 0.01$

Underflow concentration $w_D = 52.4\,\%$ solid by weight

Settling parameter $f_{bk}(\varphi) = -1.72 \times 10^{-4} \times \varphi \times (1-\varphi)^{15.6}\,(\mathrm{m/s})$

Compression parameter $\sigma_e(\varphi) = 2.0\exp(22\varphi)\,(\mathrm{Pa})$

To perform the calculation we use the software SimEsp developed by Concha and collaborators (Garrido et al. 2003, 2004; Burgos and Concha 2005) (Fig. 8.82).

(a) **Design step**

Choose the *Design Module* and enter: the solid feed rate $F = 50$ (tph), $(1200/24)$, the underflow concentration $w_u = 52.4[\%]$ solid by *weight*, the critical concentration $w_c = 42.8\,[\%] \to \varphi_c = 0.23$, assume a conjugate concentration $\varphi_L \approx 0.01$ and enter the parameters of the flux density function $u_\infty = -1.72 \times 10^{-4}\,(\mathrm{m/s})$ and $c = 15.6$. Press the *calculate button* to obtain the

Fig. 8.82 Industrial thickener simulator SimEsp

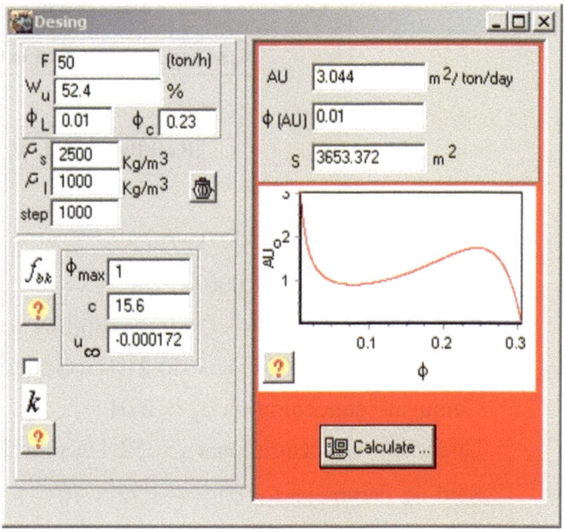

Fig. 8.83 Design module

Unit Area function $UA(\varphi)$ and the unit area $AU(\varphi_L = 0.01) = 3.044\,\mathrm{m}^2/\mathrm{tpd}$ giving a thickener cross sectional area of $S = 3653.4\,\mathrm{m}^2$ for the assumed conjugate concentration. See Fig. 8.83.

- **Simulation step 1**

To continue, press the *Simulation button* to obtain Fig. 8.84. The information from the Design Module is transferred automatically to the Simulation Module. Add the feed concentration $w_F = 35\,\%$, the sediment height in terms of the available thickeners height $z_c = 20\,\%$, the consolidation parameters $\alpha_1 = 2.0\,\mathrm{Pa}$ and $\alpha_2 = 22$. Choose between Case 1, for a flat bottom thickener or

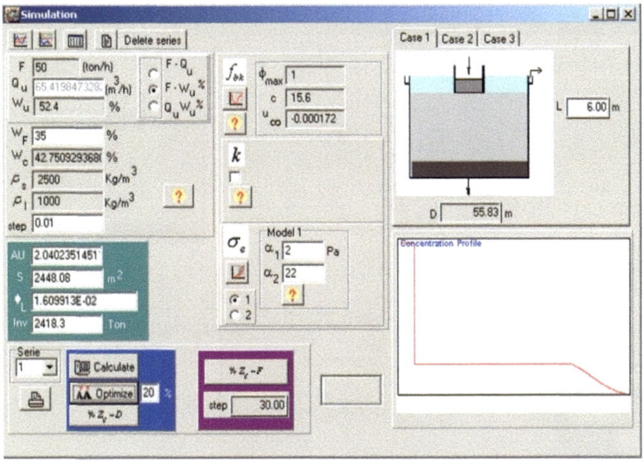

Fig. 8.84 Simulation and optimization module

Case 2 or 3 for conical bottom thickeners. For this problem use Case 1. Now press the *Optimize button* to start the iteration and obtain the thickener diameter: $D = 55.84$ m.

Set the thickener diameter to the next greater integer in meters $D = 56\,\text{m}$, and press the *Calculate* button to obtain the final result in Fig. 8.85.

$$f_F = -2.256 \times 10^{-6}\,\text{m/s},\ q = -7.378\,\text{m/s},\ z_c = 1.18\,\text{m},\ \varphi_L = 0.016,$$
$$I = 2385.6\,\text{tons},\ AU = 2.05\,\text{m}^2/\text{tpd},\ S = 2{,}463\,\text{m}^2,\ D = 56\,\text{m},\ Q_U = 65.42\,\text{m}^3/\text{h}$$

Figure 8.86 gives a material balance around the thickener and Fig. 8.87 shows the concentration profile in the thickener.

Figure 8.88 shows the continuous solid flux density and the feed flux density. The interception of the feed flux density function with the convective flux $q\varphi$ gives the underflow concentration φ_D and with the continuous flux density function gives the conjugate concentration φ_L.

It is possible to design the thickener for 50 tph feed with different values for the sediment height. Press *button* ($\%z_c$-*D*) and the results are given in Fig. 8.89, which shows that the higher the sediment is allowed, the smaller the thickener diameter required.

- **Simulation step 2**

Once the thickener is designed, it is interesting to study its flexibility to other feed rates. For example, for the designed thickener of $D = 56$ m, choose *button* ($\%z_c$-F) and see the concentration profiles for each the thickener capacity in Fig. 8.90. Figure 8.91 shows the height of the sediment for each thickener capacity.

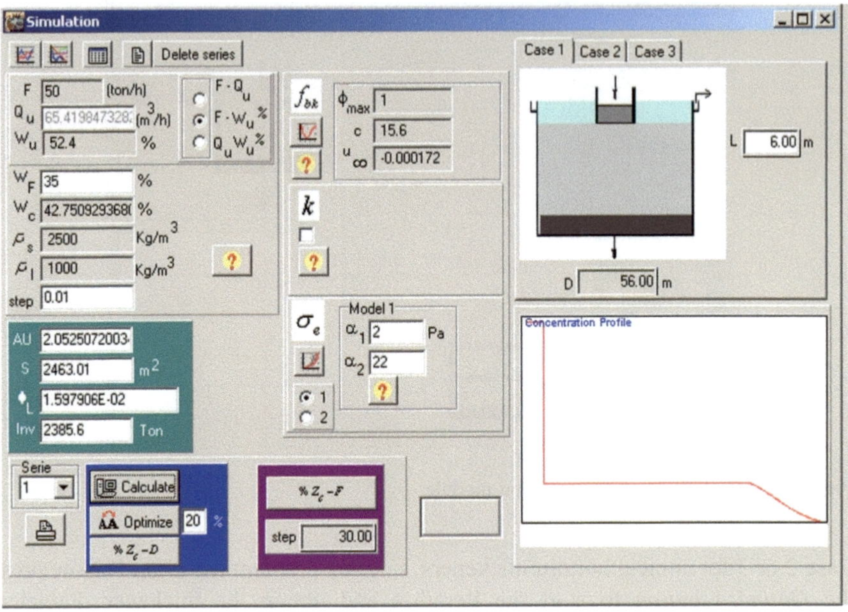

Fig. 8.85 Simulation and optimization module

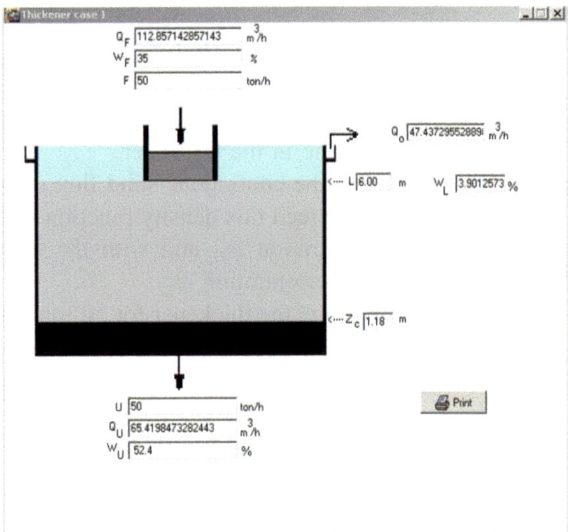

Fig. 8.86 Mass balance in the designed thickener

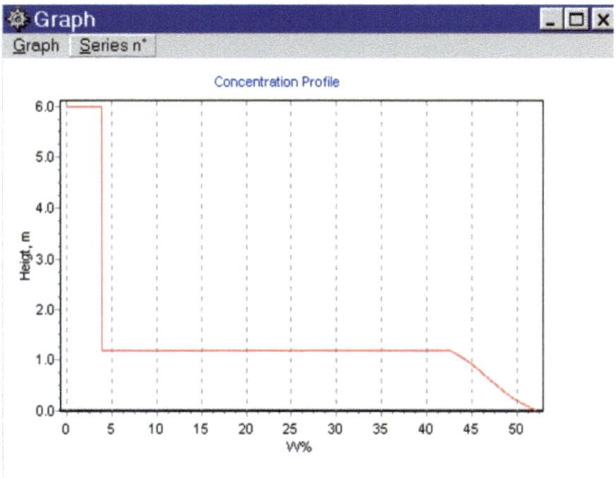

Fig. 8.87 Concentration profile in the designed thickener for sediment with 20 % of the total height

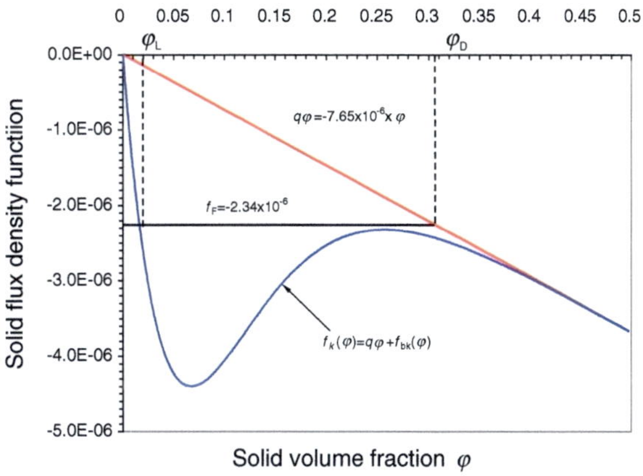

Fig. 8.88 Solid flux density functions

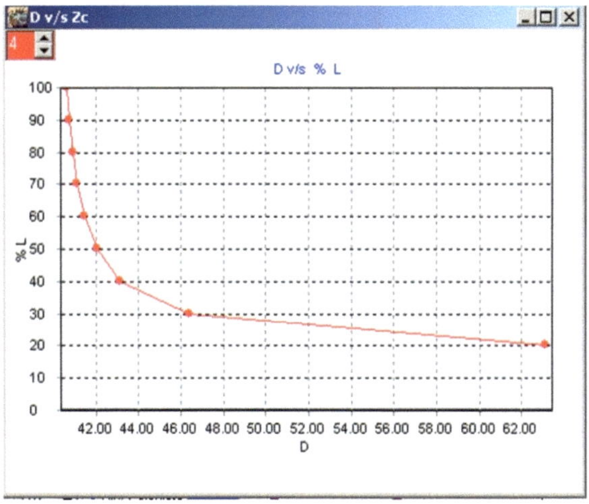

Fig. 8.89 Thickener diameter for several allowed sediment heights

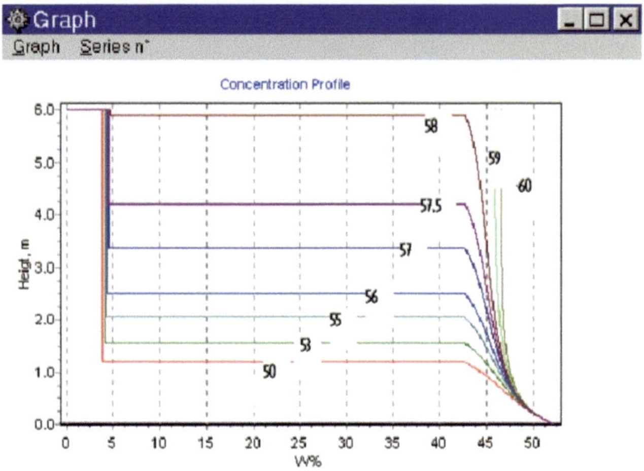

Fig. 8.90 Simulation of the concentration profile for the designed thickener for several thickener feed rates and constant underflow concentration

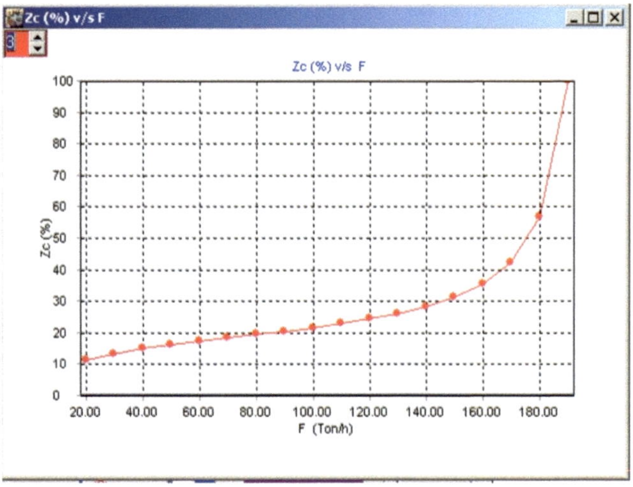

Fig. 8.91 Simulation of the sediment height for several thickener capacities at constant underflow concentration

8.6.5 Comparison of Thickener Design Methods

The following table compares the results of the diverse thickener design methods.

Thickener design	Unit area	Area in m^2	Diameter in m
Mishler	1.333	1,599	45
Coe y Clevenger	1.764	2,117	52
Kynch	2.289	2,747	59
Talmage and Fitch	2.251	2,702	59
Yoshioka and Hasset	1.783	2,104	52
Wilhelm and Naide	1.815	2,178	53
Adorjan	1.840	2,208	53
CETTEM	2.05	2,463	56

The smallest thickener diameter is obtained with Mishler's design and the largest with methods based on Kynch sedimentation processes. The phenomenological methods have the additional advantage that fitting the best equation to the experimental data eliminates experimental errors. Finally, CETTEM's method takes into account the desired height of the sediment in the thickener and provides a plot of thickener capacity versus sediment height.

It is interesting to point out that, although the most recent phenomenological models validated Mishler and Coe and Clevenger's equations for the unit area, these older methods gave the smallest thickener areas. The error in Coe and Clevenger's method is to assume that the concentration in the settling zone of the thickener is that of the feed concentration and, therefore, performing laboratory experiments above this concentration. The phenomenological model gave a value of $\varphi_L = 0.016$ as the limiting concentration for a unit area of $UA = 2.05\,(\mathrm{m}^2/\mathrm{tpd})$. This concentration is far lower than the feed concentration of $\varphi = 0.177$. This is why the laboratory experiments should include smaller concentrations in the range of $\varphi \approx 0.01$.

Figure 8.92 shows the experimental points used by Coe and Clevenger's and the additional point by CETTEM's methods. Figure 8.93 also explains the weakness of Adorjan's method. According to the calculations he obtained $AU = 1.756\,(\mathrm{m}^2/\mathrm{tpd})$, similar to Coe and Clevenger, and with a safety factor of 0.95 he reached $AU = 1.84\,(\mathrm{m}^2/\mathrm{tpd})$. A comparison to Cettem's methods shows the safety factor should have been 0.75.

Problem 8.15 Consider a flocculated suspension of a copper tailing defined by the flux density function $f_{bk}(\varphi) = -6.05 \times 10^{-4}\varphi(1-\varphi)^{12.59}\,(\mathrm{m/s})$ critical concentration of $\varphi_c = 0.23$ and a solid effective stress of $\sigma_e(\varphi) = 5.35\exp(17.9\varphi)\,(\mathrm{Pa})$. Assume feed and underflow concentrations of $w_F = 35.0\,[\%]$ and $w_D = 57.3\,[\%]$ by weight. The solid feed is 178 (tph) and the solid and liquid densities are 2,600 (kg/m^3) and 1,000 (kg/m^3) respectively.

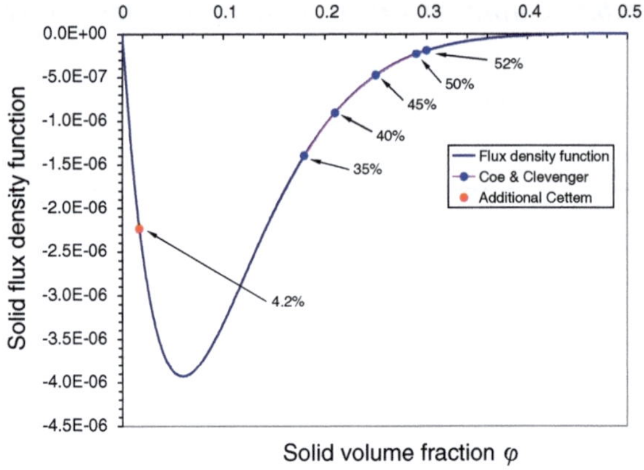

Fig. 8.92 Solid flux density function showing Coe and Clevenger's experimental data and the additional concentration of CETTEM's method

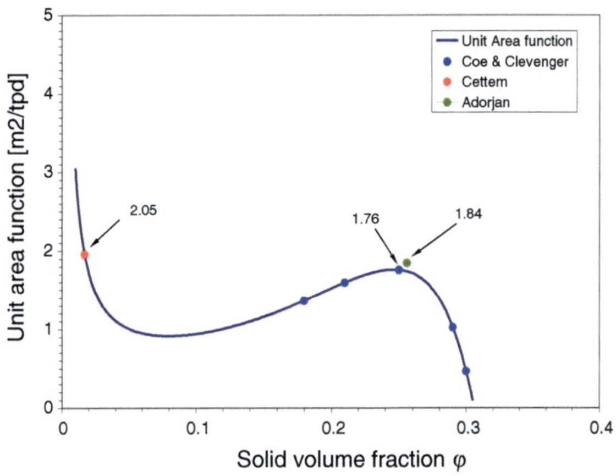

Fig. 8.93 Unit area function showing Coe and Clevenger's results and the unit area by Cettem's and Adorjan's method for a copper flotation tailing

(1) Design a cylindrical thickener with $L = 6$ (m) to handle the feed rate and (2) design a cylindrical-conical thickener to handle the feed rate with 5 (m) and 1 (m) for the cylindrical and conical height respectively. Allow a sediment height of 40 % L.

Using SimEsp we obtain:

Type of thickener	Area m^2	Diameter m	Unit area m^2/tpd	φ_L	Inventory tons
Cylindrical	1,385	42	0.324	3.05×10^{-2}	2,739
Cylindrical-conical	1,521	44	0.417	9.7×10^{-3}	2,993

Figure 8.94 shows the comparison between the concentration profile in the cylindrical and the cylindrical-conical thickeners.

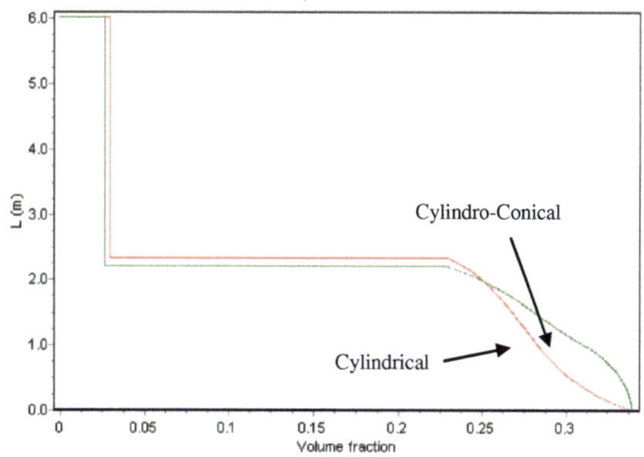

Fig. 8.94 Concentration profile in a flat-bottomed and a conical-bottomed thickener for the same capacity

8.7 Operational Strategies and Metallurgical Control

8.7.1 Steady State

In Sect. 8.3 we showed that the condition for obtaining a steady state in a thickener is that the extended flux density function should lie below the solid feed flux density, in the region of concentration between the conjugate and the underflow concentrations:

$$f_k(\varphi) \leq f_F, \quad \text{for} \quad \varphi_L \leq \varphi \leq \varphi_D$$

At steady state $f_F = q\varphi_D$, where $f_F = -Q_F\varphi_F/S = -F/\rho_s S$, is the solid feed flux density function, $q = -Q_D/S$ is the volume average velocity, Q_F and Q_D are the feed and underflow volume flow rates, F is the solid mass flow rate and S is the thickener cross sectional area.

The sediment height is obtained from (8.145), which can be written in the following form:

$$z_c = -\int_{\varphi_D}^{\varphi_c} \frac{\sigma_e'(\varphi)d\varphi}{\Delta\rho\varphi g\left(\frac{f_F(1-\varphi/\varphi_D)}{f_{bk}(\varphi)} - 1\right)}$$

Varying Feed Rates

If a thickener, provided with an underflow pump with variable speed, is operating at steady state, the solid underflow rate balances the solid feed rate. If the solid feed rate suddenly increases and we do not change the underflow volume flow rate, the excess solid material accumulates in the thickener, increasing the pulp inventory. The increase in sediment height, and weight produces an increase in the

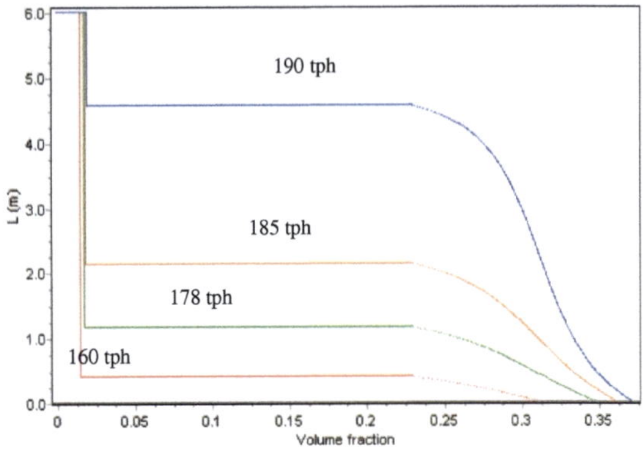

Fig. 8.95 Concentration profile in a copper tailing thickener for increasing feed rate, with a volume underflow rate of $Q_D = 203.9$ m³/h

underflow concentration so that eventually the solid underflow rate will again balance the solid feed rate. Figure 8.95 shows this effect for copper tailings.

Problem 8.16 Consider a 53 m diameter by 6 m high thickener treating a copper tailing with solid and fluid densities of $\rho_s = 2{,}500$ and $\rho_f = 1{,}000\ (\mathrm{kg/m^3})$ and several solid feed rates: 160; 178; 185 and 190 tph. The following thickening parameters are given:

$$f_{bk}(\varphi) = -6.05 \times 10^{-4} \times \varphi \times (1-\varphi)^{12.59}$$

$$\sigma_e(\varphi) = \begin{cases} 0 & \text{for } \varphi \le \varphi_c = 0.23 \\ 5.35\exp(17.9\varphi)\ \mathrm{N/m^2} & \text{for } \varphi \le \varphi_c = 0.23 \end{cases}$$

If the pulp feed is 203.9 m³/h.

Using the SimEsp simulator, the results shown in Figs. 8.95 and 8.96 and the following table are obtained.

Results of the simulation of Problem 8.16

F tph	f_F m/s	Q_D m³/h	q m/s	W_D % sol weight	Sediment height m
160	-8.06×10^{-6}	203.4	-2.56×10^{-5}	53.4	0.40
178	-8.97×10^{-6}	203.4	-2.56×10^{-5}	57.4	1.20
185	-9.32×10^{-6}	203.4	-2.56×10^{-5}	58.8	2.20
190	-9.57×10^{-6}	203.4	-2.56×10^{-5}	59.9	4.96

8.7.2 Underflow Concentration Control

During the operation of a thickener, the underflow concentration is expected to remain constant under changes in feed rates. An increase in the solid feed flux F and/or the feed concentration φ_F, is offset by a corresponding increase in the

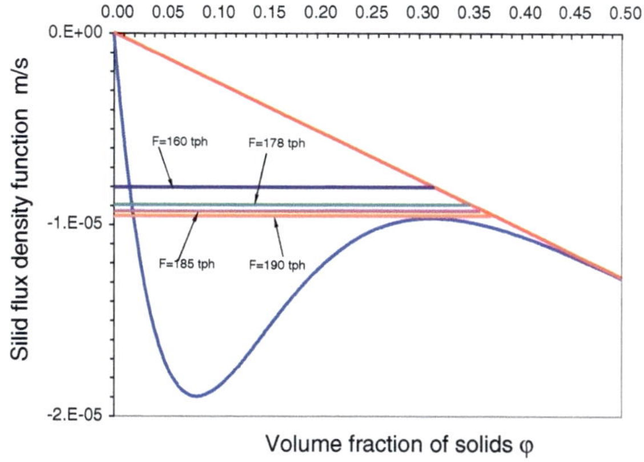

Fig. 8.96 Solid flux density function versus concentration showing the four feed flux density functions expressed as tons per hour

underflow volume rate Q_D to maintain a steady state. In thickeners provided with variable speed underflow pumps, the underflow concentration control is accomplished by changing the pump velocity. If the thickener has gravity discharge, changing underflow discharge rings makes changes in the volume underflow rate.

Usually, the feed rate is not measured in thickeners. In these cases, indirect internal variables must be used. The underflow concentration is the direct effect of the self-weight of the sediment; therefore, the solid pressure at the bottom of the tank or the sediment height can be used as internal variables.

The total pressure P in a suspension of volume fraction φ, is the sum of the solid effective stress $\sigma_e(\varphi)$ and the fluid pore pressure p. Therefore:

$$\sigma_e(\varphi) = P - p \tag{8.148}$$

and depending on the constitutive equation of the solid effective stress, the concentration can be obtained by inverting (8.148). For example, if the solid effective stress is expressed as $\sigma_e(\varphi) = \alpha_1 \exp(\alpha_2\varphi)$, the concentration is:

$$\varphi = \frac{1}{\alpha_2}\ln\left(\frac{P - p}{\alpha_1}\right) \tag{8.149}$$

or if the solid effective stress is expressed as $\sigma_e(\varphi) = \sigma_0((\varphi/\varphi_c)^n - 1)$, the concentration is:

$$\varphi = \varphi_c\left(\frac{P - p + \sigma_0}{\sigma_0}\right)^{1/n} \tag{8.150}$$

These equations show that the concentration of the sediment can be inferred by measuring the pore pressure and the total pressure. In this case, the measurement is at the bottom of the tank.

As we have already seen, the sediment height z_c can be calculated from equation:

$$z_c = -\int_{\varphi_D}^{\varphi_c} \frac{\sigma_e'(\varphi)d\varphi}{\Delta\rho\varphi g\left(\frac{f_F(1-\varphi/\varphi_U)}{f_{bk}(\varphi)} - 1\right)}$$

therefore, z_c can be used to infer the necessary volume underflow rate to maintain a constant underflow concentration under a variable feed rate. For the case of the calcium carbonate, the following relationship exists between the sediment height and the volume underflow rate (Fig. 8.97).

8.7.3 Feed Dilution

The feed enters a thickener, mixes with the upcoming water and dilutes to the *conjugate concentration* φ_L at $z = L$. The conjugate concentration can be obtained by solving the implicit Kynch equation at $z = L$:

$$f_F = q\varphi_L + f_{bk}(\varphi_L) \tag{8.151}$$

The conjugate concentration does not depend directly on the feed concentration φ_F, but only on the solid feed rate f_F, the convective pulp velocity in the thickener q and the pulp nature through the Kynch flux density function $f_{bk}(\varphi)$.

Problem 8.17 Consider a copper tailing with the following settling properties: 2.500 kg/m^3 in density and a solid flux density $f_{bk}(\varphi) = -6.05 \times 10^{-4} \times \varphi \times (1-\varphi)^{12.59}$. For a 53 m diameter thickener with a feed rate of 185 tph, a feed concentration of 21.7 % and underflow concentration of 58.8 % solid by weight, calculate the conjugate concentration.

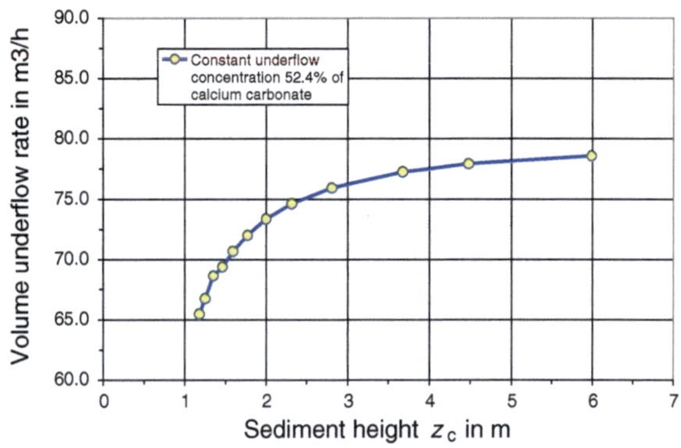

Fig. 8.97 Underflow volume flow rate versus sediment height to obtain a constant underflow concentration of $w_D = 52.4[\%]$ by weight with variable feed rate

From Eq. (8.151):

$$S = \frac{\pi D^2}{4} = \frac{3.14157 \times 53^2}{4} = 2206\,\mathrm{m}^2$$

$$f_F = -\frac{F}{3600 \times \rho_s \times S} = \frac{185}{3600 \times 2.5 \times 2206} = -9.32 \times 10^{-6}\,\mathrm{m/s}$$

$$\varphi_D = \frac{\rho_f \times w_U}{\rho_s \times (100 - w_U) + \rho_f \times w_U} = \frac{1000 \times 58.8}{2500 \times (100 - 58.8) + 1000 \times 58.8} = 0.363$$

$$q = \frac{f_F}{\varphi_D} = \frac{9.32 \times 10^{-6}}{0.363} = -2.57 \times 10^{-5}\,(\mathrm{m/s})$$

$$f_F = q\varphi_L + f_{bk}(\varphi_L)$$

$$-9.32 \times 10^{-6} = -2.57 \times 10^{-5} \times \varphi_L - 6.05 \times 10^{-4} \times \varphi_L \times (1 - \varphi_L)^{12.59}$$

Solving this implicit equation we get: $\varphi_L = 0.0185$, $w_L = 4.5[\%]$ solid by weight

This problem can also be solved graphically as in Fig. 8.98. The value of φ_L is obtained at the intersection of f_F with the continuous solid flux density function $f_k(\varphi)$.

Since the feed to a thickener dilutes naturally, the question arises whether there is any advantage in diluting the pulp before feeding it to the thickener. There are two possible advantages. Firstly, a concentrated pulp entering a more diluted zone behaves like a *density current* that reaches deeper into the tank, requiring more time for its homogeneous distribution. This effect can be minimized if the concentration of the entering feed is similar to that of the receiving zone. Secondly, the flocculant is added to the feed stream and as we will see later, flocculation is more effective for diluted suspensions.

Water is usually very costly around concentrators; therefore mechanisms are introduced into the feed wells to permit *auto-dilution* with the rising overflow water. The most commonly used are; (1) dilution by overflow through ports in the upper part of the feedwell, and (2) dilution by jet-suction.

Fig. 8.98 Feed dilution to the conjugate concentration

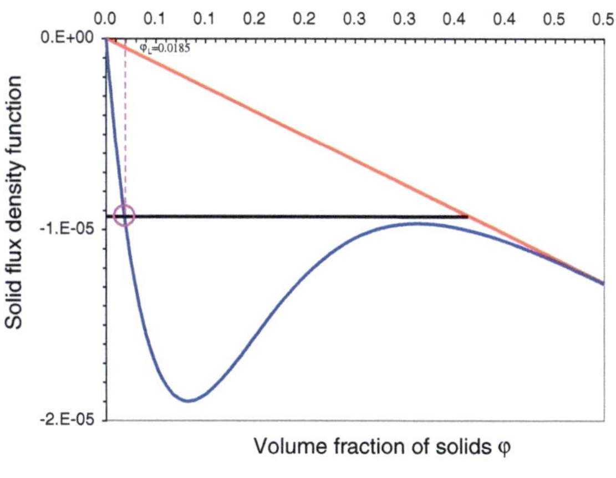

1. **Dilution ports and overflow**

Consider the feedwell shown in Fig. 8.99. We can make the following mass balance for a feed with density $\rho_F = 1.1\,(\text{ton/m}^3)$ diluted to $\rho_S = 1.05\,(\text{ton/m}^3)$ when entering the thickener, then:

$$\begin{aligned}\rho_F z_F &= \rho_S z_S + \rho_W z_W\\ 1.1 \times z_F &= 1.05 \times z_S + 1.0 \times z_W\\ z_F &= 0.95 \times z_S + 0.91 \times z_W\\ z_F &< z_S + z_W\end{aligned}$$

where ρ_F, ρ_S and ρ_W are the pulp density and z_F, z_S and z_W are the depth of the feed, the suspension and the water respectively.

Because the feed pulp and the conjugate suspension are denser than the water $\rho_W = 1.0$, the water level outside the feedwell is always higher then that of the feed, as shown in Fig. 8.99. This principle can be used for dilution by overflow or through ports in the feedwell. Figure 8.100 shows a feedwell with water ports.

2. **Dilution by jet suction**

When a fluid is injected as a high velocity jet into a tank containing a secondary stationary fluid, the secondary fluid is suctioned into the boundary layer of the first fluid mixing with it. The amount of secondary fluid depends on the jet diameter and velocity. This principle is used to dilute the feed in a thickener, as is shown in Fig. 8.101.

Since the pulp is fed by gravity, the height of the head tank controls the speed of the jet and therefore, the amount of water suctioned. The inner and outer pipes diameters are design variables. Figure 8.102 shows this type of feeding mechanism implemented by a thickener manufacturer.

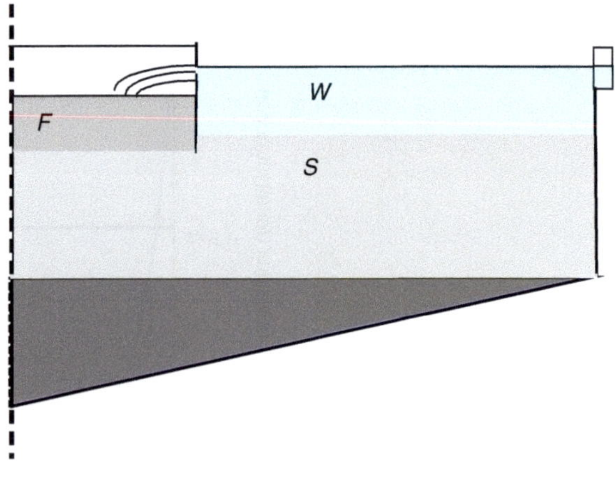

Fig. 8.99 Dilution ports in a feedwell

Fig. 8.100 Feedwell with water ports to dilute the feed (Supaflo technologies)

Fig. 8.101 Dilution of the thickener's feed by jet eductor

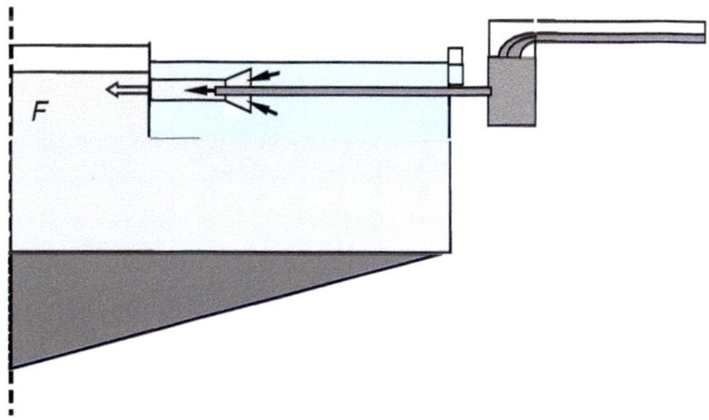

Figures 8.103 and 8.104 show the velocity magnitude and velocity vectors in a typical E-Duc System and Fig. 8.105 shows the solid concentration distribution in a typical feedwell (Köck and Concha 2003).

Fig. 8.102 E-Duc from EIMCO process equipment

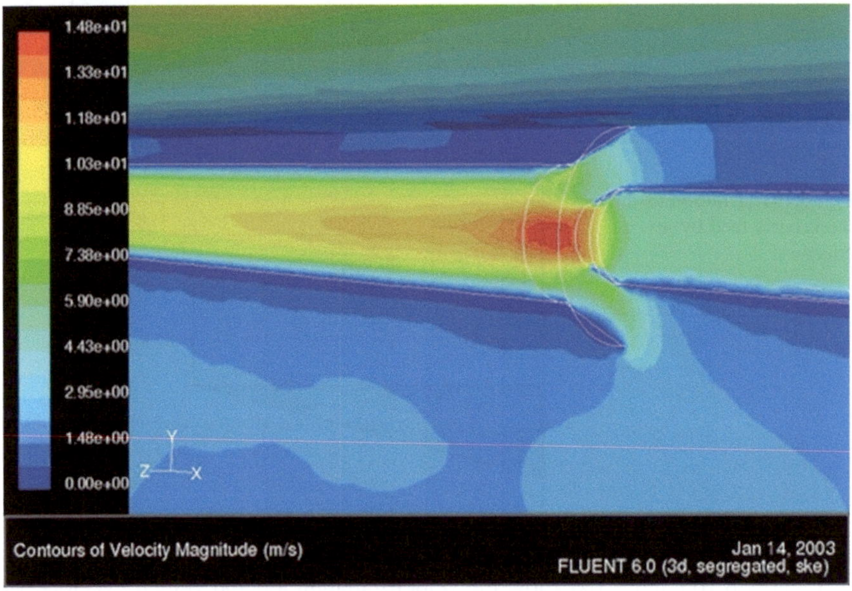

Fig. 8.103 Velocity distribution for the E-Duc dilution system, Köck and Concha 2003

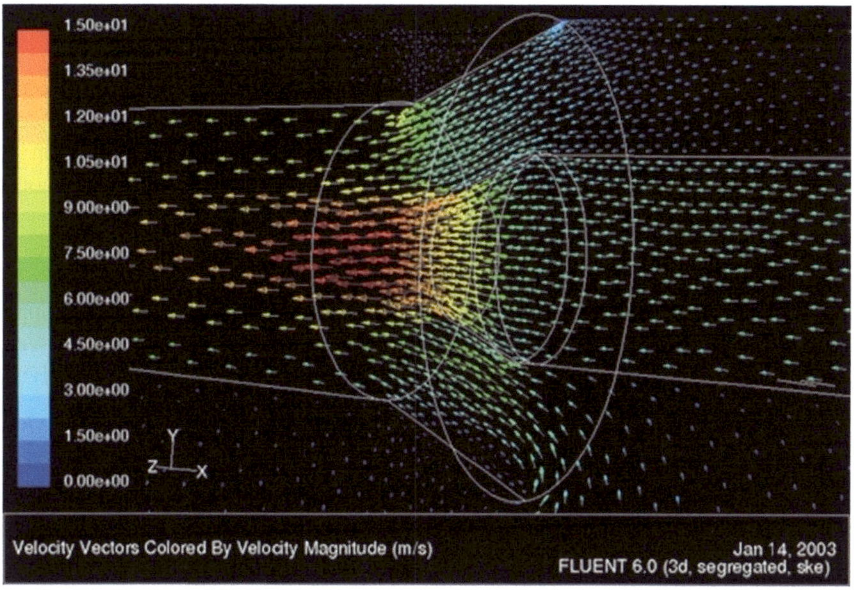

Fig. 8.104 Velocity vector for the E-Duc dilution system, Köck and Concha 2003

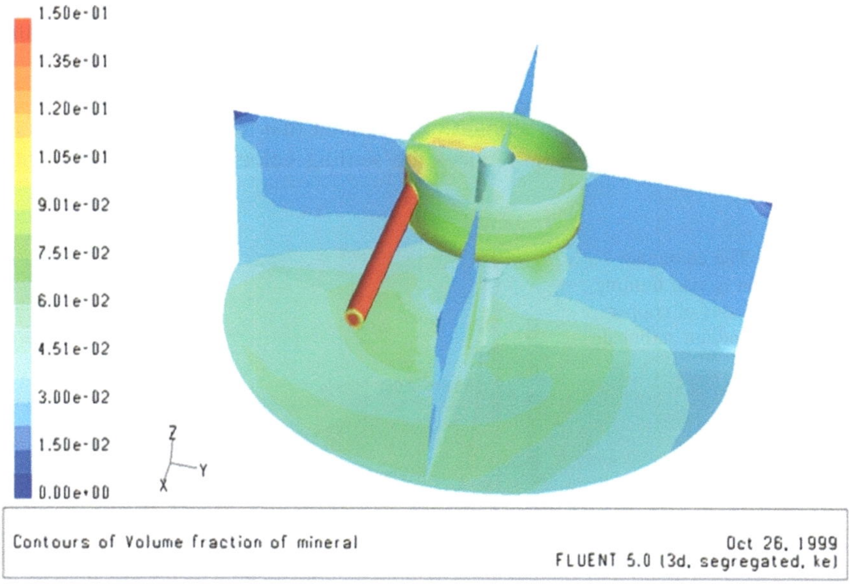

Fig. 8.105 Concentration distribution at the inlet and outlet of a feedwell in a thickener, Köck and Concha 2003

8.7.4 Limiting Concentration

According to Coe and Clevenger, the limiting concentration in a thickener establishes the settling rate and lies between the feed and the underflow concentration. We have seen that the feed is diluted to the conjugate concentration on entering the thickener and therefore the range for the limiting concentration should be extended to include the conjugate concentration.

The limitations are determined by the flux density function and by the values of the conjugate and the underflow concentrations. In most cases, the conjugate concentration is the limiting concentration. See Fig. 8.106 for the cases of copper tailings and calcium carbonate, where the conjugate concentrations are $\varphi_L = 0.0185$ and $\varphi_L = 0.0160$. They are also the limiting concentrations, giving unit areas of $AU = 0.516\,(\mathrm{m}^2/\mathrm{tpd})$ and $AU = 2.05\,(\mathrm{m}^2/\mathrm{tpd})$ respectively.

8.7.5 Effect of the Flocculant Dose on Thickener Capacity and Fines Control

Flocculants are added to a thickener to increase the settling velocity of the particles by forming flocs. A floc is an aggregate of many particles of different sizes having a greater overall size, but a lower density. Flocculated pulps leave clear water and sharp water-suspension interfaces when settling. For more information on this subject see Chap. 7.

The flocculant dose is an important factor from a technical and economic point of view. There is an optimum amount of flocculant that has to be experimentally determined. An excessive dose reduces the settling velocities, and consequently

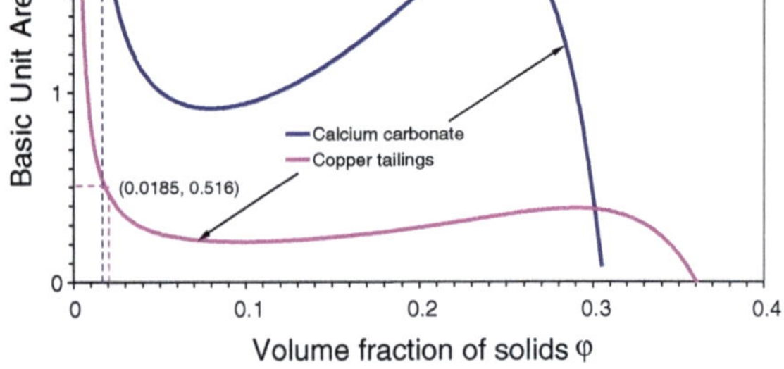

Fig. 8.106 The conjugate concentration is also limiting for the two cases of copper tailings and sodium carbonate

the thickener capacity, and increases the operating cost. There are several ways to optimize the flocculant addition as a thickener, the most commonly used is to perform settling experiments with different pulp dilutions and flocculant dose. There is also the possibility of selecting the dose according to floccules properties such as flocculant molecules size and density.

Problem 8.18 Consider settling tests performed on copper tailings for several flocculant doses. The result shows that the solid flux density function and the solid effective stress can be written in the form: $f_{bk}(\varphi, gpt) = u_\infty(gpt) \times \varphi \times (1-\varphi)^{12.59}$, m/s and $\sigma_e(\varphi) = 5.35\exp(17.9\varphi)$, Pa.

Calculate the capacity of a 57 m diameter thickener for a solid density of $\rho_s = 2.500\,\text{kg/m}^3$, flocculant dose of 2, 4, 6, 8 and 10 gpt for an underflow concentration of 52 % solid. The feed concentration is 49.3 % solid and the critical concentration is 40.6 [% solid] by weight.

The calculated solid flux density functions and initial settling velocities, for different flocculant dose, are described in Figs. 8.107 and 8.108.

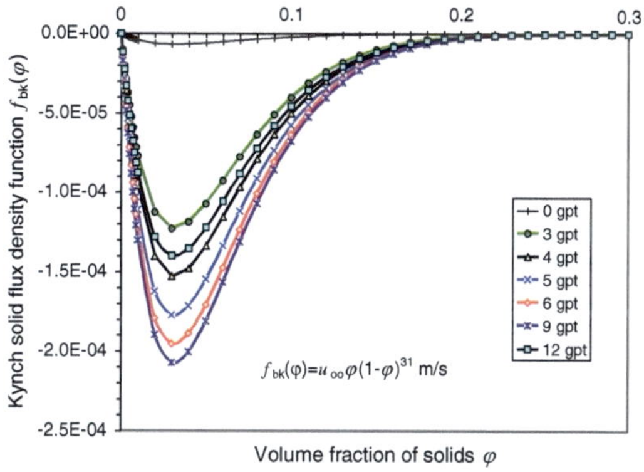

Fig. 8.107 Effect of the flocculant dose on the solid flux density function

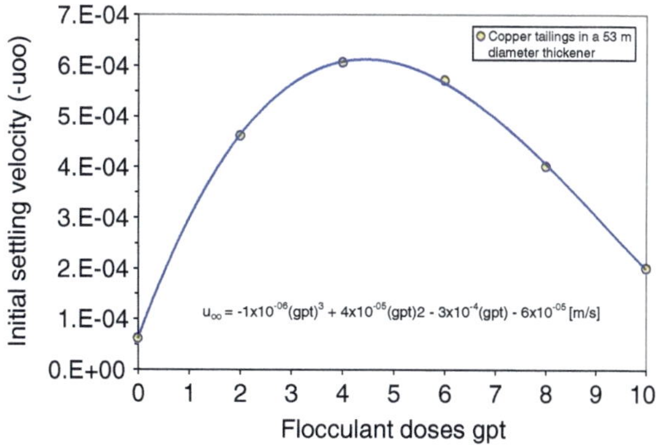

Fig. 8.108 Effect of the flocculant dose on the initial settling velocity

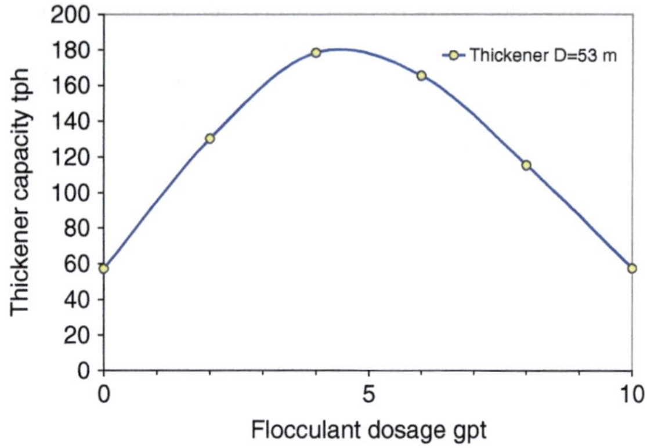

Fig. 8.109 Effect of the flocculant dose on the thickener capacity

With the above information, the capacity of the thickener can be calculated for each flocculant dose, to give the same sediment height, that is, 20 % of the total available thickener height. The result is given in Fig. 8.109.

The fact that a change in the flocculant dose drastically affects the capacity of a thickener shows that the addition of flocculant as a variable to control fines in the overflow water must be taken carefully. In general flocculant is added in proportion of the thickener's solid feed rate. But, when fines are controlled by flocculant addition, an excess of flocculant can be detrimental for the thickener's capacity. This effect was dealt with in Chap. 7.

8.7.6 Effect of the Shear Rate on Flocculation and on Thickener Capacity

The flocculant is injected directly into a thickener in the feedwell or into the pipe feeding the thickener. In both cases the exact position of injection is in the zones of greatest shear. Farrow et al. (2001) presented shear rate distribution in a feedwell calculated with CFD and predicted the favorable positions for flocculant injection. Reports of sampling of industrial thickener by Farrow et al. (1999) confirmed that CFD predictions and modeling resulted in significant improvement in industrial thickener operation in Australia. Kahane et al. (1997) reported on work in a plant in Australia were the relocation of the addition points of flocculant to the lower part of the feedwell, where the natural dilution occurs, doubled the thickening capacity of the plant.

In Chap. 7 Köck and Concha (1999) presented a CFD model of a typical feedwell of a copper tailing flotation thickener, concluding that the average shear rate in the feedwell was very low, on the order of 1 s^{-1}. Köck and Concha (1999)

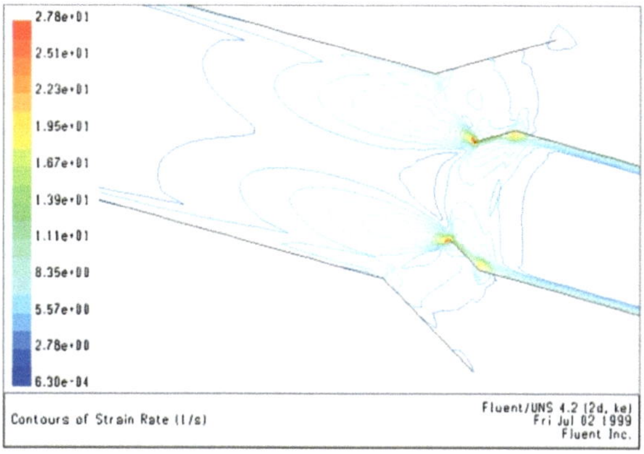

Fig. 8.110 Distribution of shear rate in an E-Duc feeding system to a thickener (Köck and Concha 1999)

also used CFD to establish the properties of an E-Duc system of a copper tailing flotation thickener, Fig. 8.110. It was established that the average shear rate in the tube was $\bar{\dot{\gamma}} \sim 8\,\mathrm{s}^{-1}$ and that in a small region it reached $\dot{\gamma} \sim 27\,\mathrm{s}^{-1}$.

8.7.7 Optimum Flocculation of Thickener Feeds

Farrow et al. (2001) indicated that for a specific solid-flocculant system under high shear conditions, the average size of the flocs increases rapidly from 30 to 130 μm in 15 s before the rupture of flocs occur. Under low shear conditions, flocs slowly increase in size from 30 to 100 (μm) in 50 (s) without any floc rupture. This difference in growth velocity and floc is extremely important in industrial thickeners. The key for a good feedwell design is to provide appropriate hydrodynamic conditions and sufficient residence time for floc growth.

Rulyov (2004) and Concha et al. (2012) showed that optimum flocculation is obtained by a proper combination of feed dilution, high shear rate $\approx 500\,\mathrm{s}^{-1}$ for short periods $\approx 5\,\mathrm{s}$ and a flocculant dose for the collision of particles and flocculant macro-molecules, and low shear floc growth ($\approx 60\,\mathrm{s}$), leading to maximum floc size and density. Unfortunately this combination cannot be obtained within a feedwell. See Chap. 7.

The auto-dilution of a suspension in a feedwell is established once the type of dilution system of the feed has been chosen. The shear rate during flocculation depends on the thickener feeding system or feedwell design, which is determined for a given thickener. Therefore, under a change of mineralogy or particle size distribution, the only control variable available is the flocculant dose so that with operational variables it is not possible to maintain an optimum flocculation in a thickener. The solution to this problem is to flocculate the suspension in a specialized reactor before feeding it to the thickener. This specialized industrial reactor has not yet been developed for copper concentrators.

References

Adorjan, L. A. (1975). *A theory of sediment compression* (pp. 1–22). Calgary: XI IMPC.

Adorjan, L. A. (1976). Determination of thickener dimensions from sediment compression and permeability test results. *Transactions of the Institution of Mining and Metallurgy Section C, 85*, 157–163.

Agricola, G. (1994). 1494–1994, Bergwelten, Edition Glückauf GmbH, Essen.

Ashley, H. E. (1909, June 12). *Theory of the settlement of slimes, mining*. Scientific Press.

Auzerais, F. M., Jackson, R., & Russel, W. B. (1988). The resolution of shocks and the effects of compressible sediments in transient settling. *Journal of Fluid Mechanics, 195*, 437–462.

Barrientos, A. (1978). Parameters determination in the phenomenological model of sedimentation. Engineering Thesis, University of Concepción (in Spanish).

Bascur, O. (1976). Phenomenological model of suspensions in sedimentation. Engineering Thesis, University of Concepción (in Spanish).

Bascur, O. (1989). A unified solid-liquid separation framework. *First Regional Meeting, American Filtration Society, Houston, Texas*, October 30–November 1, 1–11.

Becker, R. (1982). Continuous thickening: Design and simulation of thickeners. Engineering Thesis, University of Concepción (in Spanish).

Beens, K. & Sills, G. C. (1981). Self-weight consolidation of soils: an experimental and theoretical study, Geotechnique 31, 519–535.

Behn, V. C. (1957, October). Settling behavior of waste suspensions. *Proceeding of the ASCE, Journal of the Sanitary Engineering Division*, 1423-5–1423-20.

Betancourt, F. & Concha, F. (2011). Unveiling the myth of high capacity thickeners, 8the International Seminar od Mineral Processing Procemin 20011, Santiago, Chile. (In Spanish).

Betancourt, F., Concha, F., & Sbarbaro, D. (2013). Simple mass balance controllers for continuous sedimentation. *Computers and Chemical Engineering*, 34–43.

Bürger, R., & Concha, F. (1997). Simulation of the transient behavior of flocculated suspensions in a continuous thickener. *Proceedings of the XX International Mineral Processing Congress, Achen*, September 4, 21–26 (pp. 91–101).

Bürger, R., & Concha, F. (1998). Mathematical model and numerical simulation of the settling of flocculated suspensions. *International Journal of Multiphase Flow, 24*(6), 1005–1023.

Bürger, R., & Karlsen, K. H. (2001). On some upwind difference schemes for the phenomenological sedimentation-consolidation model. *Journal of Engineering Mathematics, 41*, 145–166.

Bürger, R., & Tory, E. M. (2000). On upper rarefaction waves in batch settling. *Powder Technology, 108*, 74–87.

Bürger, R., Bustos, M. C., & Concha, F. (1999). Settling velocities of particulate systems: 9. Phenomenological theory of sedimentation processes: Numerical solution of the transient behavior of flocculated suspensions in an ideal batch or continuous thickener. *International Journal of Mineral Processing, 55*(4), 267–282.

Bürger, R., Concha, F., & Tiller, F. M. (2000a). Applications of the phenomenological theory to several published experimental cases of sedimentation process. *Chemical Engineering Journal, 80*(1–3), 105–117.

Bürger, R., Evje, S., Karlsen, K. H., & Lie, K.-A. (2000b). Numerical methods for the simulation of the settling of flocculated suspensions. *Chemical Engineering Journal, 80*, 91–104.

Bürger, R., Evje, S., & Karlsen, K. H. (2000c). On strongly degenerate convection-diffusion problems modeling sedimentation-consolidation processes. *Journal of Mathematical Analysis and Applications, 247*, 517–556.

Bürger, R., Wendland, W., & Concha, F. (2000d). Modelling equations for gravitational sedimentation-consolidation processes. *ZAMM, 80*, 1–18.

Bürger, R., Karlsen, K. H., Klingerberg, C., & Risebro, N. H. (2003). A front tracking approach to a model of continuous sedimentation in ideal clarifier. Thickener units. *Nonlinear Analysis: Real World Applications, 4*, 457–481.

Burgos, R., & Concha, F. (2005). Further development of software for the design and simulation of industrial thickeners. *Chemical Engineering Journal, 111*, 135–144.

Buscall, R., & White, L. R. M. (1987). The consolidation of concentrated suspensions. *Journal of the Chemical Society, Faraday Transactions, 1*(83), 873–891.

Bustos, M. C. (1984). On the existence and determination of discontinuous solutions to hyperbolic conservation laws in the theory of sedimentation. Doctoral Thesis, Mathematical Department, University of Darmstadt.

Bustos, M. C., & Concha, F. (1988a). On the construction of global weak solutions in the Kynch theory of sedimentation. *Mathematical Methods in the Applied Sciences, 10*, 245–264.

Bustos, M. C., & Concha, F. (1988b). Simulation of batch sedimentation with compression. *AIChE Journal, 34*(5), 859–861.

Bustos, M. C., & Concha, F. (1996). Kynch theory of sedimentation. In E. M. Tory (Ed.), *Sedimentation of small particles in a viscous fluid. Advances in fluid mechanics* (Chap. 2, Vol. 7, pp. 7–49). Southampton: Computational Mechanics Publications.

Bustos, M. C., Concha, F., & Wendland, W. (1990a). Global weak solutions to the problem of continuous sedimentation of an Ideal Suspension. *Mathematical Methods in the Applied Sciences, 13*, 1–22.

Bustos, M. C., Paiva, F., & Wendland, W. L. (1990b). Control of continuous sedimentation of ideal suspensions as an initial and boundary value problem. *Mathematical Methods in the Applied Sciences, 12*, 533–548.

Bustos, M. C., Concha, F., Bürger, R., & Tory, E. M. (1999). *Sedimentation and thickening: Phenomenological foundation and mathematical theory*. Dordrecht, The Netherlands: Kluwer Academic Publishers. 285 p.

Clark, A. (1915). A note on the settling of slimes. *Engineering and Mining Journal, 99*(9), 412.

Coe, H. S., & Clevenger, G. H. (1916). Methods for determining the capacity of slime settling tanks. *Transactions, AIME, 55*, 356–385.

Comings, E. W. (1940). Thickening calcium carbonate slurries. *Industrial and Engineering Chemistry, 32*(5), 663–668.

Comings, E. W., Pruiss, C. E., & De Bord, C. (1954). Continuous settling and thickening. *Industrial & Engineering Chemistry Design Processing Development, 46*, 1164–1172.

Concha, F., & Barrientos, A. (1980). Phenomenological theory of thickening. *Engineering Foundation Conference on Particle Science and Technology, New Hampshire, USA*, 7–11 July.

Concha, F., & Barrientos, A. (1993). A critical review of thickener design methods. *Kona, Powder and Particles, 11*, 79–104.

Concha, F., & Bürger, R. (1998). Wave propagation phenomena in the theory of sedimentation. In E. F. Toro & J. F. Clarke (Eds.), *Numerical methods for wave propagation phenomena* (pp. 173–196). Dordrecht, The Netherlands: Kluwer Academic Publishers.

Concha, F., & Bustos, M. C. A. (1991). Settling velocities of particulate systems. Part 6. Kynch sedimentation processes: Batch settling. *International Journal of Mineral Processing, 32*, 193–212.

Concha, F., & Bustos, M. C. A. (1992). Settling velocities of particulate systems. Part 7. Kynch sedimentation processes: Continuous thickening. *International Journal of Mineral Processing, 34*, 33–51.

Concha, F., Burgos, R., & Garrido, P. (2005). Sistema para el diseño y simulación de espesadores industriales, XII Symposium de MolyCop en Procesamiento de Minerales, Termas de Chillán, noviembre 14–18.

Concha, F., Segovia, J. P., Sbarbaro, D., & B̈rger, R. (2008). XXIV Int. Mineral Procces. Congress, Beiging 2008.

Concha, F., Sbarbaro, D., Segovia, J. P., & Bürger, R. (2011). *Integrated system of parameter determination, thickener design, simulation and control*. Beijing, China: IMPC. 2008.

Concha, F., Rulyov, N. N., & Laskowski, J. S. (2012). Settling velocities of particulate systems 18: Solid flux density determination by ultra flocculation. *International Journal of Mineral Processes, 104–105*, 53–57.

D'Avila, J. (1976). An analysis of Kynch theory of sedimentation. *Brazilian Journal of Technology, 7*, 447–453. (in Portuguese).

D'Avila, J. (1978). A mathematical model of sedimentation. Ph.D. Thesis, COPPE, Federal University of Río de Janeiro (in Portuguese).

D'Avila, J., & Sampaio, R. (1977). Equations for the solid pressure. *Brazilian Journal of Technology, 8*, 177–191. (in Portuguese).

D'Avila, J., Concha, F., & Telles, A. S. (1978). A phenomenological model for two dimensional continuous sedimentation. *VI Porous Media Seminar, Río Claro, Brazil*, October 11–13, III (pp. 1–19) (in Portuguese).

Davis, K. E., & Russel, W. B. (1989) An asymptotic description of transient settling and ultra-filtration of colloidal suspensions. *Physics of Fluids A, 1*, 82–100.

De Haas, R. D. (1963). Calculation of the complete batch settling behavior for rigid spheres in a Newtonian liquid. M.Sc Thesis, Purdue University.

De Kretser, R. G., Usher, S. P., Scales, P. J., Bolger, D. V., & Landmann, K. A. (2001). Rapid filtration measurements of dewatering design and optimization parameters. *AIChE Journal, 47*(8), 1758–1769.

Diehl, S. (1995). On scalar conservation laws with point source and discontinuous function modeling continuous sedimentation, *SIAM Journal on Applied Mathematics*, 1425–1451

Diehl, S. (1996). A conservation law with point source and discontinuous flux function modeling continuous sedimentation, *SIAM Journal on Applied Mathematics*, *56* (2), 388–419.

Diehl, S. (1997). Dynamic and steady state behaviour of continuous sedimentation. *SIAM Journal on Applied Mathematics, 57*(4), 991–1018.

Diehl, S. (1999). On boundary conditions and solutions for ideal clarifier-thickener units. *Chemical Engineering Journal, 80*(1–3), 119–133.

Diehl, S. (2000). On boundary conditions and solutions for ideal clarifier-thickener units, *Journal of Chemical Engineering*, *80*, 119–133.

Diehl, S. (2001). Operating charts for continuous sedimentation I: Control steady states, *Journal of Engineering Mathematics*, *41*, 117–144.

Dorr, J. V. N. (1915). The use of hydrometallurgical apparatus in Chemical Engineering. *Journal of Industrial and Engineering Chemistry, 7*, 119–130.

Dorr, J. V. N. (1936). *Cyanidation and concentration of gold and silver ores* (p. 117). New York: McGraw-Hill Book Co. Inc.

Egolf, C. B., & McCabe, W. L. (1937). Rate of sedimentation of flocculated particles. *Transactions AIChE, 33*, 620–640.

Farrow, J. B., Fawell, P. D., Johnston, R. R. M, Nguyen, T. B., Rudman, M., Simic, K., & Swift, J. D. (2001). Técnicas de floculación y metodologías para la optimización de espesadores, in Concha, F., Manual de Filtración y Separación, Cettem, Concepción, Chile, 347–362.

Farrow, J. B., Johnstone, R. R. M, Nguyen, T. B., Rudman, M, Simic, K., Swift, J. D. (2001). Téchniques of flocculation and methodologies to optimize thickners, In Concha, Filtration and Separation, Cettem, Concepción, Chile.

Fitch, B. & Stevenson, D. G. (1977). Gravity Separation Equipment, In Solid Sepazration Equipment Scale up, Ed. D.B. Purchased, Upland Press Inc., Croydon, Englnd.

Fitch, E. B. & Stevenson, D. G. (1997). Gravity Separation Equipment Scale-up, Ed. D.B. Purchas, Uplands Press Ltd., Croydin, England.

Fitch, B. (1983). Kynch theory and compression zones. *AIChE Journal, 29*, 940–947.

Font, R. (1988). Compression zone effect in batch sedimentation. *AIChE Journal, 34*, 229–238.

Forbes, D. L. H. (1912). The settling of mill slimes. *Engineering and Mining Journal, 93*, 411–415.

Free, E. E. (1912). Rate of slime settling. *Engineering and Mining Journal*, *101*, 243, 429, 509, 681.

Garrido, P. (2005). Determinación de los parámetros de espesamiento, Tesis de doctorado, Departamento de Ingeniería Metalúrgica, Universidad de Concepción, Concepción, Chile.

Garrido, P, Bürger, R., & Concha, F. (2000). Settling velocities of particulate systems: 11. Comparison of the phenomenological sedimentation-consolidation model with published experimental results. *International Journal of Mineral Processing*, *60*(3–4), 213–227.
Garrido, P., Burgos, R., Concha, F., & Bürger, R. (2003). Software for the design and simulation of gravity thickeners. *Minerals Engineering, 16*(2), 85–92.
Garrido, P., Burgos, R., Concha, F., & Bürger, R. (2004). Settling velocities of particulate systems: 13. A simulator for batch and continuous sedimentation of flocculated suspensions. *International Journal of Mineral Processing, 73*, 131–144.
Garrido, P. & Concha, F. Dosificzción de Flaculante y su relación a la operación y control de espesadores continuos, III Coloquio Nacional de Filtración y Separación, Santiago, Chile. (In Spanish).
Green, M. D. (1997). Characterization of suspensions in settling and compression (p. 246). PhD Thesis, Particulate Fluids Processing Centre, Department of Chemical Engineering, The University of Melbourne, Melbourne, Australia.
Green, M. D., Landman, K. A., de Kretser, R. G., & Bolger, D. V. (1998). Pressure filtration technique for complete characterization of consolidating suspensions. *Industrial and Engineering Chemistry Research, 37*(10) 4152–4156.
Hassett, N. J. (1958). Design and operation of continuous thickeners. *The Industrial Chemist*, *34,* 116–120; 169–172; 489–494.
Hassett, N. J. (1961, January). Theories of the operation of continuous thickeners. *The Industrial Chemist*, 25–28.
Hassett, N. J. (1964a). Mechanism of thickening and thickener design. *Transactions Institution of Mining and Metallurgy, 74*, 627–656.
Hassett, N. J. (1964, January). Concentrations in a continuous thickener. *Industrial Chemist*, 29–33.
Hassett, N. J. (1968). Thickening in theory and practice. *Mining Science Engineering, 1*, 24–40.
Holdich, R. G., & Butt, G. (1997). Experimental and numerical analysis of a sedimentation forming compressible compacts. *Separation Science and Technology, 32*, 2149–2171.
Kahane, R. B., Schwarz, P., & Johnston, R. R. M. (1997). *International Conference on Computational Fluid Dynamics in Minerals and Metals Processing and Power Generation* (pp. 109–117), 3–5 July, Melbourne, Australia.
Kammermeyer, K. (1941). Settling and thickening of aqueous suspensions. *Industrial and Engineering Chemistry, 33*, 1481–1491.
Kos, P. (1977). Fundamentals of gravity thickening. *Chemical Engineering Progress, 73*(99–105), 1975.
Kynch, G. J. (1952). A theory of sedimentation. *Transactions of the Faraday Society, 48*, 166–176.
Landmann, K. A., White, L. R., & Buscall, R. (1988). The continuous-flow gravity thickener: Steady state behavior. *AIChE Journal, 34*(2), 239–252.
Mishler, R. T. (1912, October 5). Settling slimes at the Tigre Mill. *Engineering and Mining Journal*, *94*(14), 643–646.
Nichols, H. G. (1908a). A method of settling slimes, as applied to their separation from solution in cyanide treatment. *Transactions Institution of Mining and Metallurgy, London, 17*, 293–329.
Nichols, H. G. (1908, July 11). *Theory of settlement of slimes* (pp. 54–56). Mining and Scientific Press.
Petty, C. A. (1975). Continuous sedimentation of a suspension with a nonconvex flux law. *Chemical Engineering Science, 30*, 1451–1458.
Ralston, O. C. (1916). The control of ore slimes. *Engineering and Mining Journal*, *101*, 763, 890.
Roberts, E. J. (1949, March). Thickening: Art or Science? *Transactions AIME*, 61–64; *Mining Engineering*, *1*, 61.
Rulyov, N. N. (2004). *Ultra-flocculation, theory, experiments and applications* (pp. 197–2014). Hamilton, Canada: CIMM Metallurgical Society.

Scott, K. J. (1968a). Theory of thickening: Factors affecting settling rate of solids in flocculated pulps. *Transactions Institution of Mining and Metallurgy, 77*, C85–C97.

Scott, K. J. (1968b). Experimental study of continuous thickening of flocculated silica slurry. *IEC Fundamentals, 7*, 582–595.

Segovia, J. P., & Concha, F. (2012). AMIRA Final Report, June 2012.

Shannon, P. T., Stroupe, E., & Tory E. M. (1963). Batch and continuous thickening. *Industrial & Engineering Chemistry Fundamentals, 2*(3), 203–211.

Shannon, P. T., & Tory, E. M. (1965). Settling of slurries. *Industrial and Engineering Chemistry, 57*(2), 18–25.

Shannon, P. T., Tory, E. M. (1966, December). The analysis of continuous thickening. *Transactions SME, 235*, 375–382.

Shirato, M., Kato, H., Kobayashi, K., & Sakazahi, H. (1970). Analysis of settling of thick slurries due to consolidation. *Journal of Chemical Engineering Japan, 3*, 98–104.

Stewart, R. F., & Roberts, E. J. (1933). The sedimentation of fine particles in liquids. *Transactions. Institute of Chemical Engineers, 11*, 124–141.

Stroupe, E. (1962). Batch sedimentation of rigid spheres in a Newtonian liquid. M.Sc. Thesis, Purdue University, West Lafayette, USA.

Talmage, W. P., & Fitch, E. B. (1955). Determining thickener unit areas. *Industrial and Engineering Chemistry, Engineering Design and Development, 47*(1), 38–41.

Thacker, W. C., & Lavelle, J. W. (1977). Two-phase flow analysis of hindered settling. *The Physics of Fluids, 20*(9), 1577–1579.

Tory, E. M. (1961). Batch and continuous thickening of slurries. Ph.D. Thesis, Purdue University, West Lafayette, USA.

Tory, E. M., & Shannon, P. J. (1965). Reappraisal of the concept of settling in compression. *I&EC Fund, 4*, 194–204.

Usher, S. P. (2002). Suspension dewatering: Characterization and optimization (p. 347). PhD Thesis, Particulate Fluids Processing Centre, Department of Chemical Engineering, The University of Melbourne, Melbourne, Australia.

Usher, S. P., de Kretser, R. G., & Scales, P. J. (2001). Validation of a new filtration technique for dewaterability characterization. *AIChE Journal, 47*(7), 1561–1570.

Wilhelm, J. H., & Naide, Y. (1979). Sizing and operating thickeners. *SME, AIME annual meeting*, Preprint N°79-30, New Orleans, (pp. 18–22).

Wilson, A. J. (1994). *The living rock*. Cambridge: Woodhead Pub. Ltd.

Work, L. T., & Kohler, A. S. (1940). Sedimentation of suspensions. *Industrial and Engineering Chemistry, 32*(10), 1329–1334.

Yoshioka, N., Hotta, Y., Tanaka, S., Naito, S., & Tsugami, S. (1957). Continuous thickening of homogeneous flocculated slurries. *Chemical Engineering Japan, 21*, 66–75.

Chapter 9
Filtration

Abstract Filtration is the process whereby a solid separates from a fluid by making the suspension pass through a porous bed, known as a filter medium. The bed retains the particles while the fluid passes through the filter medium and becomes a filtrate. To establish a flow of filtrate, it is necessary to apply a pressure difference, called a pressure drop, across the filter medium. There are several ways to do this depending on the driving force, for example: (1) gravity, (2) vacuum, (3) applied pressure, (4) vacuum and pressure combined, (5) centrifugal force, and (6) a saturation gradient. Usually the different driving forces require different filtration equipment called filters. Two main dewatering stages are studied, cake formation and dehumidification, which are studied as mono-phase flow and two-phase flow of a fluid through rigid porous medium, respectively. Field variables and constitutive equations are deduced from the chapter on flow in porous media. Methods of filtration, cake porosity, permeability, capillary curves and relative permeabilities are presented. Finally models of continuous filters are developed.

9.1 Definition, Equipment and Operation

Filtration is the process whereby a solid separates from a fluid by making the suspension pass through a porous bed, known as a *filter medium*. The bed retains the particles while the fluid passes through the filter medium and becomes a *filtrate*.

To establish a flow of filtrate, it is necessary to apply a pressure difference, called a *pressure drop*, across the filter medium. There are several ways to do this depending on the driving force, for example: (1) gravity, (2) vacuum, (3) applied pressure, (4) vacuum and pressure combined, (5) centrifugal force, and (6) a saturation gradient. Usually the different driving forces require different filtration equipment called *filters*.

We can distinguish three classes of filtration: (a) filtration with cake-formation, (b) filtration without cake-formation and (c) deep filtration.

F. Concha A, *Solid–Liquid Separation in the Mining Industry*,
Fluid Mechanics and Its Applications 105, DOI: 10.1007/978-3-319-02484-4_9,

9.1.1 Filtration with Cake Formation

During filtration with cake-formation, the filter medium retains the solid of the suspension on the surface of the filter medium as a layer called a *filter cake*. This layer forms naturally when the pores of the filter medium are smaller than the particles. When this is not the case, it is necessary to cover the filter medium with a thin sheet of a fibrous material, called a *filter aid* that blocks the particles from passing to the filter medium. This type of filtration, in which the flow is perpendicular to the filter medium surface, is the most commonly used in the mineral industry (see Fig. 9.1).

Filtration without cake formation

When the suspension flow is parallel to the filter medium surface, the medium retains the particles and allows the fluid pass through. However, the flow produces a high shear at the solid surfaces that prevents the formation of a solid layer over the filter medium, returning the particles to the suspension. In this way, the filtrate crosses the filter medium while the particles increase the suspension concentration with time. This type of filtration, which is called *cross flow filtration*, is useful when suspensions are to be concentrated and there is no need for a dry solid product. Although filtration without cake formation is also used in solid–liquid filtration, it is mainly used in solid–gas separation (see Fig. 9.2).

Deep bed filtration

To filter fine particles in diluted suspensions, filter media are used. The filters have pores that are larger than the particles they retain. Since they have greater depths, particles penetrate the interior of the filter medium and are captured by the fibers or particles forming the medium. This type of filter loses its properties with time, and it is necessary to clean it to eliminate particles from its interior or replace the filter with a new one. Two examples of deep filtration are sand filters and car air filters (see Fig. 9.3).

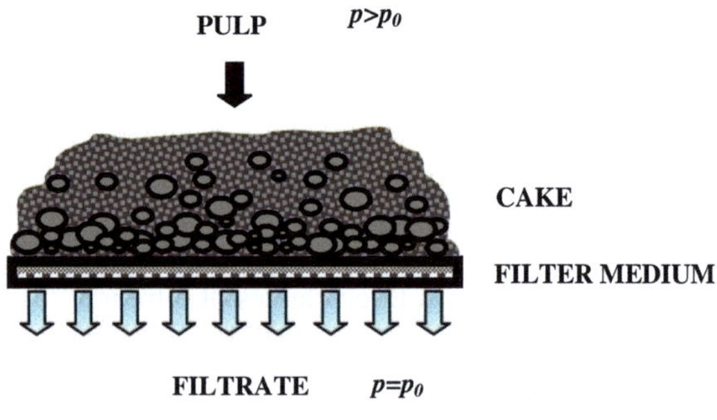

Fig. 9.1 Filtration with cake formation

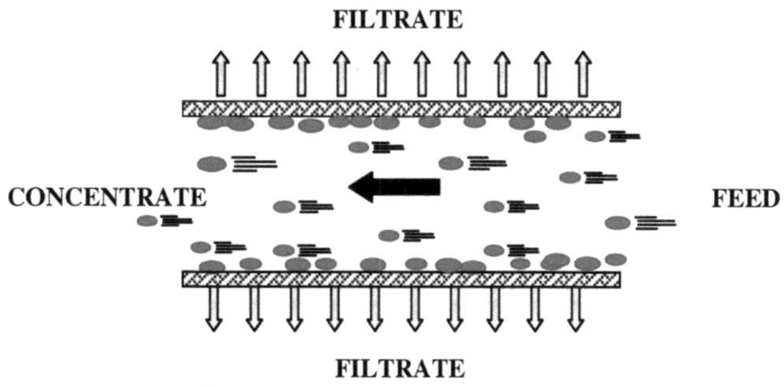

Fig. 9.2 Filtration without cake formation

Fig. 9.3 Deep bed filtration

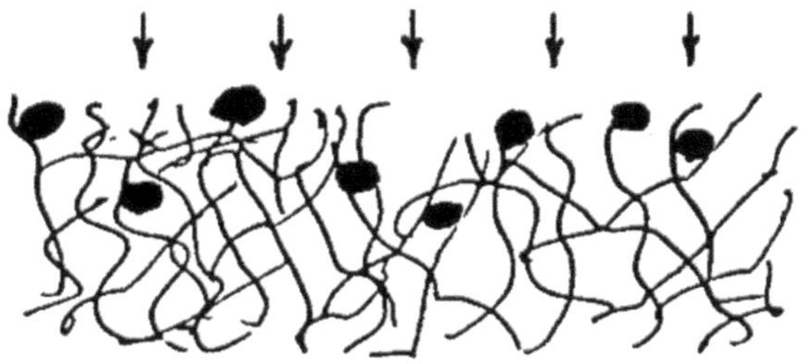

A filtration process depends on many factors:

- fluid properties, such as density and viscosity,
- nature of the solid, such as its size, shape and size distribution,
- properties of the suspension, such as concentration and compressibility,
- filter capacity,
- commercial value of the material and whether the solid or the fluid is the valuable material,
- whether it is necessary to wash the cake,
- whether it is important to keep the product from contamination.

9.1.2 Operating Variables

The principal variables in a filtration process are presented in the (Fig. 9.4).

Inlet variables	Feed mass flow $F(t)$ and feed % solid $w_F(t)$.
Outlet variables	Cake discharge mass flow $m(t)$ and cake humidity $h(t)$.
Design variables	Filtration area S and pressure drop Δp.
Control variables	Cake formation time t_1, cake washing time t_2 and cake demoisturizing time t_3.
Parameters	Cake porosity ε, permeability $k(\varepsilon)$ and compressibility $\sigma_e(\varphi)$. Filtrate density ρ_f and viscosity μ_f, solid density ρ_s and sphericity ψ.
Perturbations	Particle size and particle size distribution; Agitation.

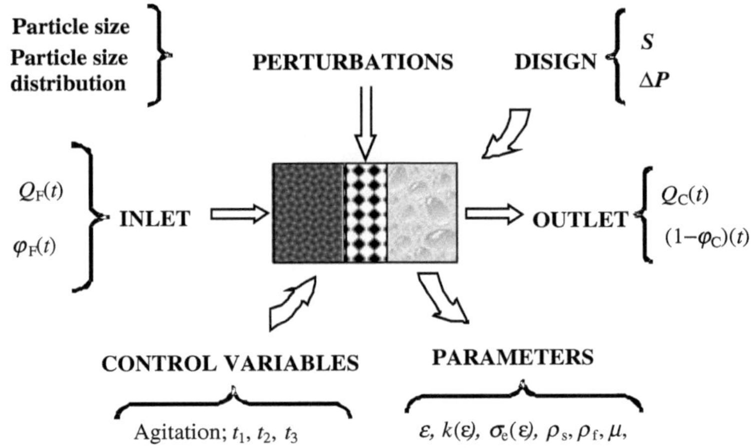

Fig. 9.4 Variables of the filtration process

Filtration cycles

All filtration equipment, whether batch or continuous, operate in cycles of cake formation, washing, drying and discharge.

- **Cake formation**: The first step in a filtration process is the formation of the cake. A pump feeds the pulp into the filter chamber for pressure filtration, or the suspension of solid particles is suctioned through the filter medium during vacuum filtration. The magnitude of the deposited material depends on the pressure gradient, on the suspension concentration and on the filtration time. In this stage of the cycle there is a continuous flow of filtrate across the cake and filter medium.
- **Cake washing**: When it is necessary to eliminate impurities from the filter cake, washing is part of the process. Washing implies calculating the minimum amount of water necessary to displace the liquor from the cakes pores and the time necessary to do this.
- **Cake drying**: Drying is a key part of the filtration process. Usually the overall requirement is a cake with a small amount of moisture, for example 8 % by weight for copper concentrates. Drying is accomplished by blowing dry air over the filter cake until enough water is displaced from the pores to obtain a given humidity. To control this part of the process it is necessary to know the amount of water retained in a saturated filter cake and the tolerated residual humidity in the product. Generally, this is a technical and an economic choice.
- **Cake discharge**: The separation of the cake from the filter medium and its discharge are important steps for efficient filtration. In vacuum filtration, blades scrape the filter cloth and discharge the cake by gravity. In hyperbaric, or pressure filtration, removing the dried cake is complicated because of the need to maintain pressure in the filtration chamber. Valves pressurize and de-pressurize the discharge area, depending on the filtration cycle.

9.2 Filtration Equipment

Vacuum filters had the advantage over pressure filters of their simple design and operation, which made them very popular in the mining industry for most of the twentieth century. *Pressure filters* began displacing vacuum filters in the 1980s due to the limitation in pressure drop with vacuum filters, which depends on local atmospheric pressure. This restriction is fundamental in mines in high altitudes, in some cases of more than 4,000 m.a.s.l. The limitations of vacuum filters and the great advances in control mechanism for pressure filtration have made the latter the favorite of the mining industry today.

An interesting alternative is the combination of vacuum and pressure equipment in a single unit. If a traditional vacuum filter is introduced into a pressure chambers so that the pressure drop is increased, a *hyperbaric* filter is obtained.

9.2.1 Vacuum Filters

There are four types of vacuum filters: drum, disc, pan and band. Whereas the first three produce cakes with 12–18 % humidity, a band filter reaches humidity levels of 8–10 %. In the following section we will briefly describe this equipment.

Drum filter

The drum filter consists of a rotating drum with the lower part submerged in a pool containing the suspension to be filtered. The drum surface is covered with a filter medium called a *filter cloth.* The suspension is suctioned from the drum interior, which is maintained under vacuum. While the drum rotates, the filtrate is suctioned into the drum interior and the solid is retained, forming a cake on the submerged surface of the filter cloth. This surface eventually emerges from the pool where air is suctioned through the cake displacing the water from the pores. During the rotation, it is possible to wash and dry the cake. Finally, a scrapping mechanism separates the cake from the filter cloth and discharges it on a chute before the surface of the drum submerges again into the suspension pool. These operations complete a cake forming-drying-washing-drying and discharge filtration cycle for the drum filter (see Fig. 9.5).

Disc filter

The disc filter consists of a horizontal shaft mounted on two main bearings. The shaft supports and connects a certain number of discs with the vacuum. The bottom of each disc is submerged in a pool with a suspension from which the filtrate is suctioned by the vacuum while the solid forms a cake on the disc surface. Each disc presents several *sectors* that are individually connected to the vacuum

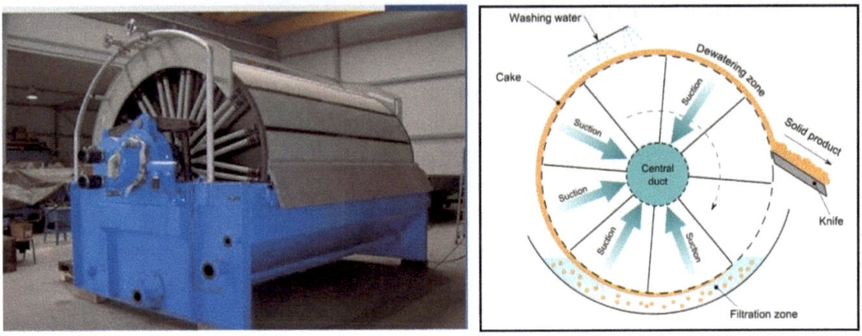

Fig. 9.5 Drum filter

chamber. The sectors are covered with a filter cloth. The discs rotate producing the different filtration cycles: cake formation-drying-washing-drying-discharge. The advantage of this equipment over the drum filter is its greater filtration surface per unit of floor area, which is because both surfaces of the disc are operative. Another advantage is its modular structure in sectors that permits the changing only the damaged filter cloths (see Figs. 9.6 and 9.7).

A special type of disc filter uses micro-pore ceramic sectors instead of steel covered with filter cloth. These vacuum filters are called *ceramic filters*.

There are two types of ceramic material. The first has 1.5-micron pores, with a capillary entry pressure of 1.6 bars, while the second has 2.0-micron pores with capillary entry pressure of 1.2 bars (for the meaning of entry pressure see Problem (9.9) in Sect. 9.5.1). When a ceramic disc is submerged in a pool containing a suspension, the *capillary action* initiates suction without any external force (see Fig. 9.8).

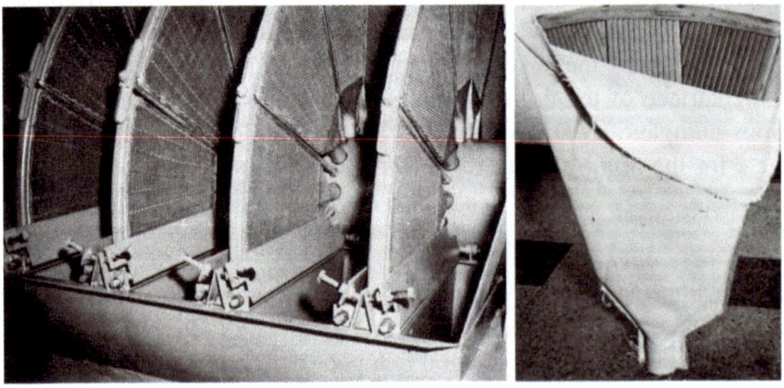

View of the discs **Detail of a filter sector**

Fig. 9.6 Disc filters

Fig. 9.7 Ceramic filters

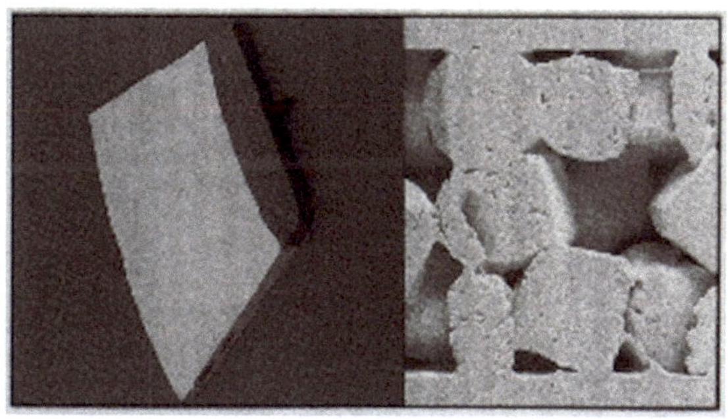

Fig. 9.8 Ceramic plate of a disc filter

The solid material accumulates at the disc surface and the dewatering continues as long as liquid is present. This process is called *capillary filtration*, which combines the advantages of the conventional disc filter with the capillary effect. Ceramic filters are used to filter copper concentrates and industrial minerals.

Pan filter

The pan filter consists of a series of horizontal trays rotating around a central vertical axis. Each tray is a trapezoidal sector slightly inclined from the central axis. All the trays are connected to a common valve. Filter cakes form on the trays and are washed with a liquid jet. At a given point in the filter, a screw conveyor drags the cake to the center of the tray and discharges it. Alternatively, each tray is tilted to discharge the cake. The main disadvantage of the pan filter is that there is only one filter surface and therefore the capacity per unit of floor is limited (see Fig. 9.9).

Fig. 9.9 Horizontal pan filter

Horizontal belt filter

The horizontal belt filter uses a continuously moving rubber belt to support and transport the filter cloth (see Fig. 9.10). A vacuum is applied below the filter cloth via a stationary vacuum box running the length of the filter. The advantage of this filter is the flexibility to choose the length of the stages in a filtration-washing-drying cycle. Belt filters are used in some concentrators, especially in gold mines, to recover water from mineral tailings, with recovery up to 80 %.

9.2.2 Pressure Filters

Nowadays vacuum filters like the drum, disc, pan and belt filters are less accepted in the mining industry because of the high moisture content of their products, which require thermal drying to reach an adequate humidity level of around 8 %. Pressure filters are presently the most reliable approach to filter concentrates in the mineral industry. They directly deliver cakes with 8 % humidity or less. As vacuum filters, they operate in cycles, but have to stop operating to feed the suspension and to discharge the cake, making the operation discontinuous.

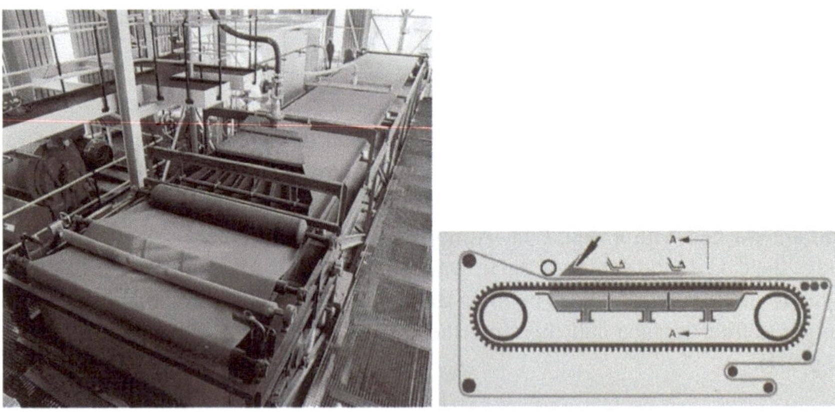

Fig. 9.10 Horizontal belt-filter and a schematic view

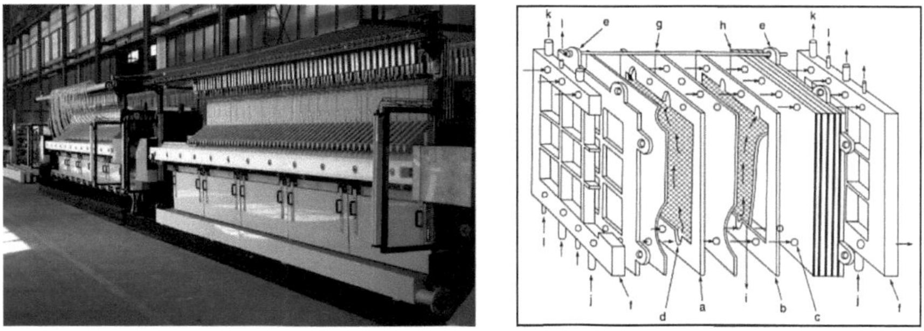

Fig. 9.11 Vertical filter press

Three types of pressure filters are commonly used: vertical filter presses, horizontal filter presses and candle filters.

(a) **Vertical filter press**

In vertical filter press the separation takes place in chambers between plates. The plates have openings for feeding and draining the filtrate. The plates are held together by hydraulic pressure. They are mounted between two lateral bars that are fixed at one end and connected to a hydraulic system at the other (see Fig. 9.11).

Filtration stage

A high-pressure hydraulic pump locks the pressure filter plate-pack. Feed slurry enters the filter chambers through the top-feed ports. Filtration begins immediately on both sides of the chamber. The filtrate drains through the ports of each chamber. The double-sided filtration gives speedy buildup of the filter cake, making the filtration part of the cycle short.

Compression stage

When the cake has formed, a rubber membrane is inflated by compressed air on one side of each cake and eliminates water by compression.

Blowing stage

Compressed air flows to the surface of the filter cake displacing the water in the cake to the filtrate discharge. The membrane remains inflated for a certain period to maintain good cake stability. The duration of the air blowing depends on the material to be dewatered but is typically 1–4 min.

Cloth washing stage

With the plate-pack still in the open position, the cake chute door (drip tray) is closed and spray nozzles rinse the cloth. Cloth vibrators actuate during the cloth washing. This sequence takes about 30 s, after which the filter is closed and the cycle begins again.

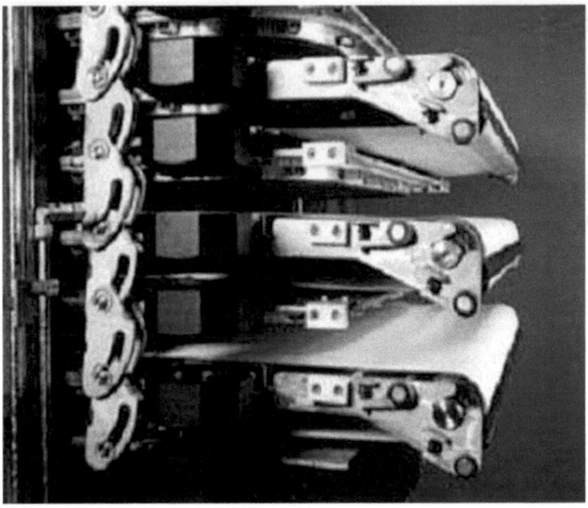

Fig. 9.12 Horizontal plates filter press

Cake discharge stage

When the cakes are ready for discharge, the cake chute door (drip tray) is retracted and the filter opens by actuating the high capacity hydraulic pump. The filter opens at a rate exceeding one chamber per second. The cloths hang freely from the suspension bar and the cakes are released at the same rate. During the fully open position, the cloth is vibrated to ensure release of any cake residue.

(b) **Horizontal filter press**

The horizontal filter press has become very popular recently. It has the combination of features that the process industry is looking for, low cake humidity and high capacity in a small area (see Fig. 9.12).

The equipment consists of a series of horizontal chambers one on top of the other on a continuous belt. In this way, the filter area per floor area is multiplied. The chambers are fixed in space and do not move during operation. The belt consists of a high resistance moving filter cloth mounted over driving rods, which move the filter belt during the discharge operation. In the upper part of the chambers are flexible rubber diaphragms that compress the cake during the expression cycle. The filtration cycle has the following stages:

1. **Cake formation**
   The slurry is pumped into all the filter chambers simultaneously. The solids begin to form as the filtrate displaces more slurry enters the chamber. As the solids build up, the pumping pressure increases. The filtrate is forced through the cloth until the required solid thickness is achieved.

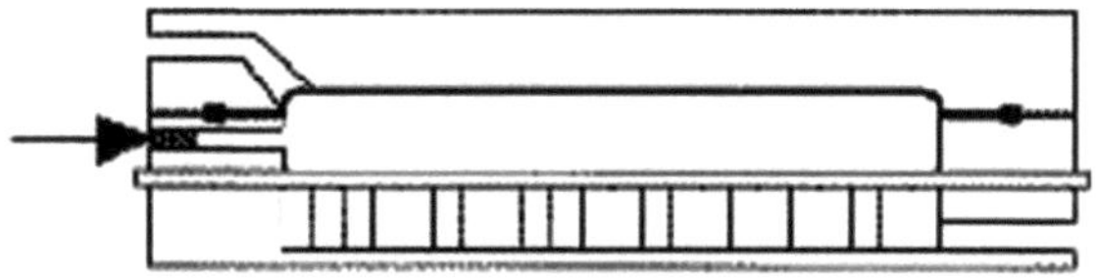

Slurry flows into chambers

2. **Diaphragm Pressing I**
   High-pressure air or water automatically inflates the diaphragm located at the top of each chamber, reducing the chamber volume and squeezing the solids to remove additional filtrate. The tightly woven filter cloth produces clear filtrate. High pressure of over 5 bars maximizes efficiency. Pressing the diaphragm produces homogenous saturated solids of uniform thickness, which assists the air blowing step.

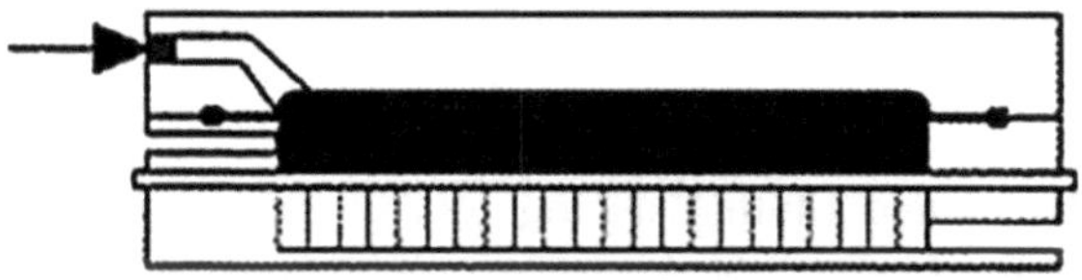

Pressurized water fills diaphragms

3. **Solid Washing**
   Pressure filters can wash dewatered solids in situ to maximize solute removal or to recover mother liquor with minimal dilution. The wash liquid distributes evenly on the homogenous solid since the filter plates are horizontal. The wash liquid flows through the solids, displacing the mother liquid with minimal mixing.

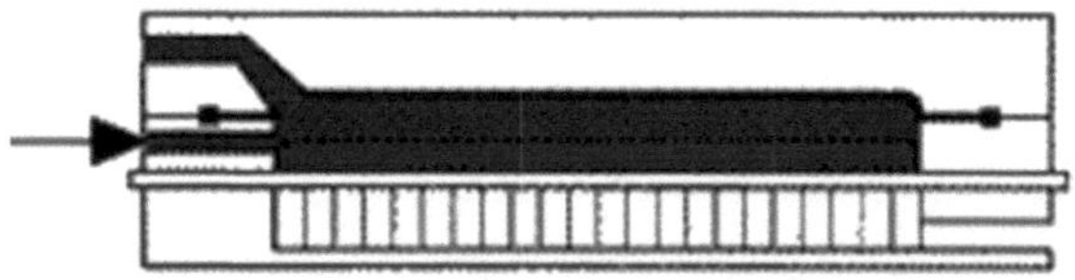

Wash water is pumped in

4. **Diaphragm Pressing II**
   The diaphragms are re-inflated, forcing the wash liquid uniformly through the solids. This produces a washing efficiency of over 95 %, with consistent saturated solids and minimum wash liquid consumption.

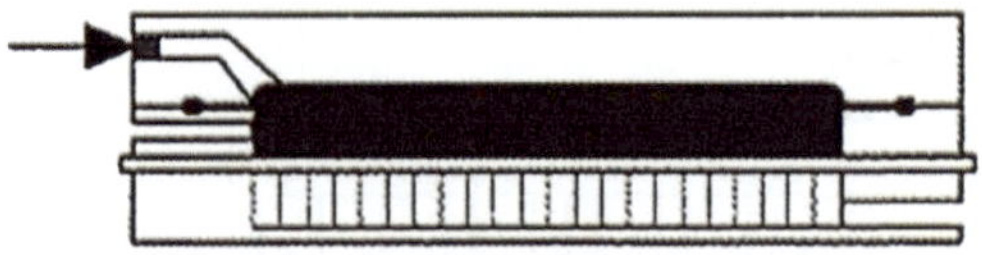

5. **Air Blowing**
   Compressed air blows through the solids for final dewatering. The moisture content reaches values around 8 % and is controlled by adjusting the pressure and duration of the blowing air.

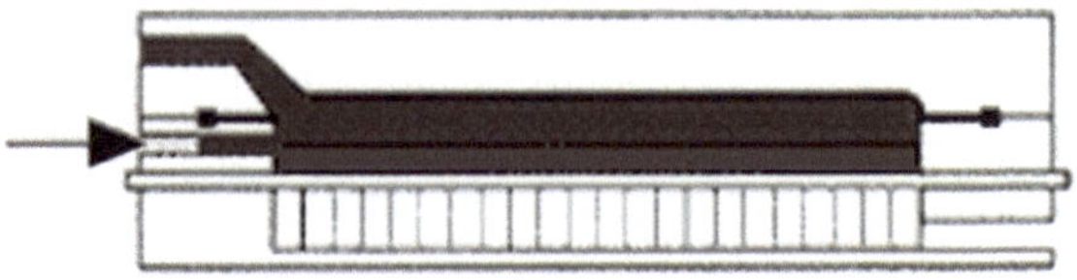

6. **Solids Discharge and Cloth Washing**
   After the plate-pack opens, the dewatered solid is conveyed out of the chambers on the moving filter cloth (see Fig. 9.13).

(c) **Candle filter**

Filtration takes place in a pressure vessel. The filter consists of a series of porous tubes on the outside of which the cake forms. The filtrate flows inside the tubes and is collected at the bottom of the vessel. After completion of the cake formation cycle, the cake adhering to the filtration element is washed with a suitable medium. To ensure the driest possible product, the washed cake is dried for as long as necessary, with any of the following media: ambient air, hot air, nitrogen or steam. Figure 9.14 shows a candle filter.

Fig. 9.13 Discharge of the filter cake

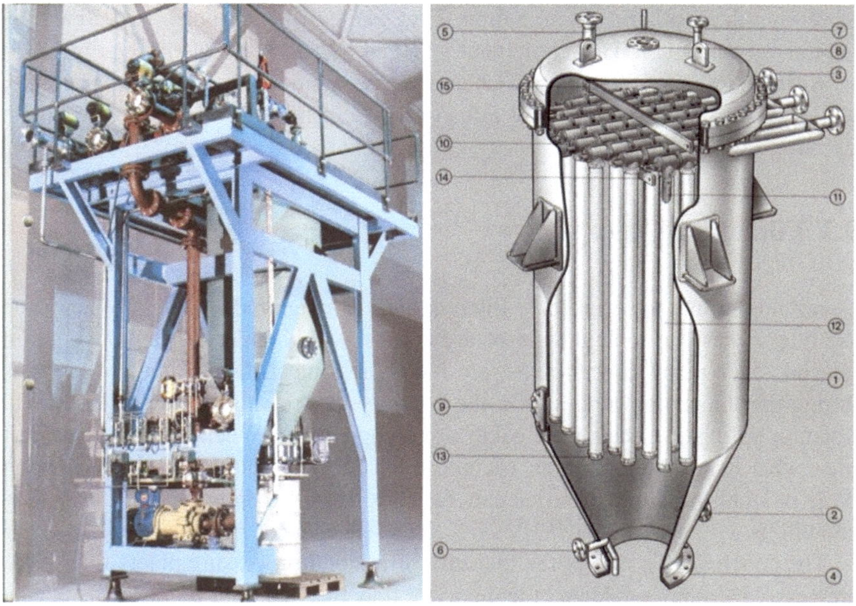

Fig. 9.14 Candle pressure filter

Fig. 9.15 Hyperbaric filter showing the pressure chamber with a disc filter

9.2.3 Hyperbaric Filter

Vacuum filters have 0.8 atmospheres as limiting pressure drop under the most favorable conditions, that is, at sea level. Using them in high mountains, where mines are usually found, the pressure drops drastically.

In the late 1980s Professor Dr. Werner Stahl and co-workers introduced vacuum filters, such as disk and drum filters, in a pressure vessel, in this way, increasing the available pressure drop. They can easily add four or more bars to a natural vacuum filter. This equipment is called a *hyperbaric filter* (see Fig. 9.15).

As in all pressure filters, the discharge is a problem in this new filter. Humidity, which in vacuum filters does not fall to 15 %, reaches 8 % in hyperbaric filters.

9.3 Filtration Theory

Depending on the material to be filtered, and the magnitude of the pressure drop, the filter cakes can remain rigid or are compressed. Copper concentrates, and in general other metal concentrates, are incompressible if flocculants are not used during the thickening process. On the contrary, flocculants are always used with flotation tailings, so their filter cakes are compressible. When filtering these tailings, what is not that common, vacuum filter are used and therefore the pressure drop is small and the material can be considered incompressible under these conditions.

The conclusion is that in the majority of mineral processing plants filter cakes are nearly incompressible and the theory of flow through rigid porous beds is valid as the basis to develop a filtration system. If in some instances this is not the case, compression must be introduced in the theory.

Filtration with incompressible cakes

Consider a filtration process under the following restrictions:

1. The properties of the suspension, the cake and the filtrate are constant.
2. The cake formed is incompressible.
3. The filtration surface is plane.
4. The percolation velocity of the filtrate across the filter cake and the filter medium is slow.

Restriction (1) permits eliminating the densities from the material balances. From restriction (2), the filter cake is assumed to be rigid and the solid component of the particulate system is immobile. In cases of curved surfaces, for example in drum filters, restriction (3) requires that the diameter of the drum is much greater than the cake thickness and restriction (4) permits using Darcy's law as a constitutive equation for the percolation velocity.

Filtration as a batch or a continuous process works in cycles that begin with the entry of pulp into the filter and ends with the discharge of the filter cake. A continuous filter is capable of making several thousand cycles before changing filter media or mechanical maintenance is necessary. In each cycle we can distinguish two fundamental stages: (a) the *cake formation stage*, which includes pumping the suspension into the filter chambers and compression of the suspension on the filter medium by means of pressured air, a rubber diaphragm or a piston, and (b) *cake dewatering* by blowing or suctioning air through the filter cake. In special cases, other stages are added, for example, if the filter cake is compressible a (c) *expression stage* is used to compress the cake to reduce its porosity and eliminate some of the water saturating the cake. If the solid material is to be free of contaminants, a (d) *washing stage* is used, followed by another dewatering stage.

9.3.1 Cake Formation

Consider the following field variables [see Eqs. (6.39) and (6.40)]: cake porosity $\varepsilon(z,t)$, percolation velocity $q(z,t)$ and excess pore pressure $p_e(z,t)$. These variables constitute a *simple filtration process* if the following field equations are satisfied:

$$\varepsilon(z,t) = \varepsilon_0 \quad \text{and} \quad q(z,t) = q_0 \tag{9.1}$$

$$\frac{\partial p_e}{\partial z} = \frac{\mu}{k(\varepsilon_0)} q_0 \tag{9.2}$$

where ε_0 is the constant porosity of the filter cake and q_0 and $k(\varepsilon_0)$ are the percolation velocity and the cake permeability, both constant.

Consider Fig. 9.16 representing a filtration process with a plane rigid filter cake. Call $p_0 = p_e(-\ell_m), p_m = p_{em}(\ell_m)$ and, $p_\ell(t) = p_e(\ell(t))$, then the pressure drop $\Delta p_1 = p_\ell - p_m > 0$ and $\Delta p_2 = p_m - p_0 > 0$.

The boundary conditions are:

$$p_e(-\ell_m) = p_0 \tag{9.3}$$

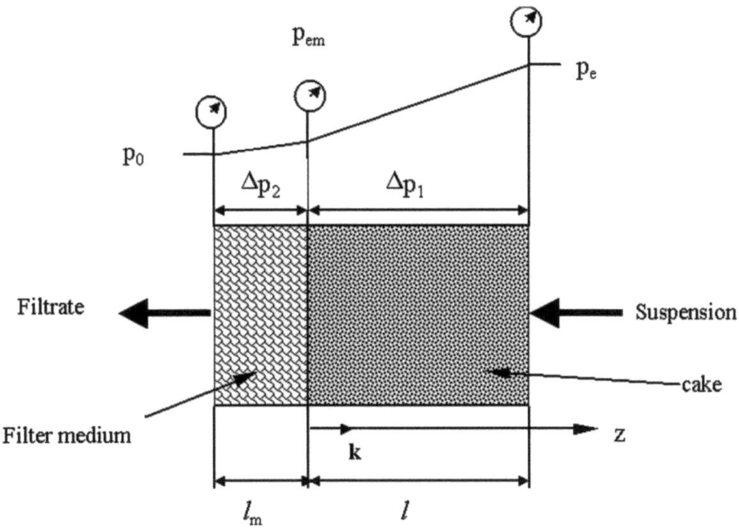

Fig. 9.16 Filtration with plane incompressible cake

$$p_e(\ell(t)) = p_\ell \tag{9.4}$$

The term *simple* refers to the fact that these equations represent the simplest case of a filtration process.

In vacuum filtration, the filter element is partially submerged in the suspension, which is fed with a pump. The suspension is kept homogeneous by agitation. In this case, the *pressure drop is constant* and depends on the available vacuum and local atmospheric pressure.

Pressure filters are fed with centrifugal pumps. At the beginning of the feeding process, the resistance of the cake and filter medium is low and the volume flow rate is high and relatively constant. As the cake resistance increases, the pressure drop also increases, while the volume flow rate decreases depending on the pump characteristic curve. Figure 9.17 shows the characteristic curves of a centrifugal pump and the curve of the system (see Chap. 11).

Once the filter chambers are full of suspension, the feed is stopped and a constant pressure is applied by a rubber membrane. This step of the cake formation stage is called *expression*.

These considerations lead to the study of the filtration process in the special case of constant pressure drop.

(a) **Cake formation with constant pressure drop**

Consider the case of a constant pressure drop across the filter cake and the filter medium:

$$\Delta p = \Delta p_1 + \Delta p_2 = p_\ell - p_0 > 0 \tag{9.5}$$

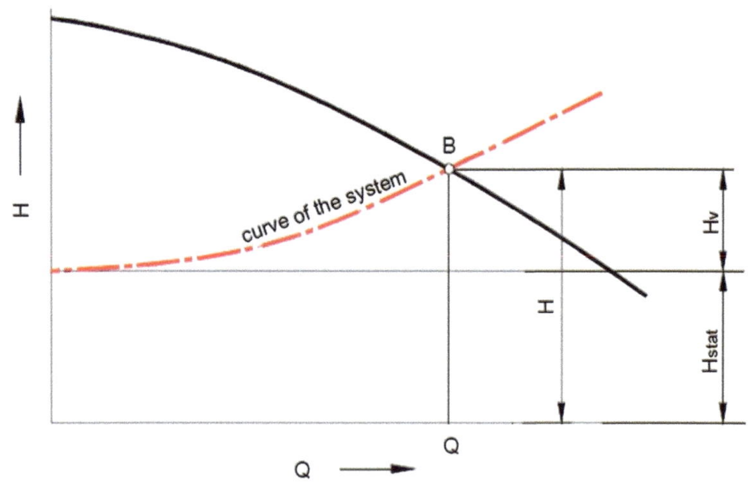

Fig. 9.17 Characteristic curves of a centrifugal pump

Each pressure drop is a positive constant, therefore integration of Eq. (9.2) for a given instant t leads to:

$$\int_{p_m}^{p_\ell} dp_e = \int_0^{\ell(t)} \frac{\mu}{k(\varepsilon_0)} q_0 dz \quad \Rightarrow \quad p_\ell - p_m = \Delta p_1 = \frac{\mu \ell(t)}{k(\varepsilon_0)} q_0 \tag{9.6}$$

$$\int_{p_0}^{p_m} dp_e = \int_{-\ell_m}^{0} \frac{\mu}{k_m} q_0 dz \quad \Rightarrow \quad p_m - p_0 = \Delta p_2 = \frac{\mu \ell_m}{k_m} q_0 \tag{9.7}$$

where k_m and ℓ_m and $k(\varepsilon_0)$ and $\ell(t)$ are the permeability and thickness of the filter medium and of the filter cake, and q_0 is the percolation velocity, that is, the volume flow rate of filtrate per unit filter area and ε_0 is the porosity of the filter cake. Adding Eqs. (9.6) and (9.7) yields:

$$\Delta p = \mu \left(\frac{\ell_m}{k_m} + \frac{\ell(t)}{k(\varepsilon_0)} \right) q_0 \tag{9.8}$$

We define the filter medium *resistance* $R_m = \ell_m / k_m$, so that (9.8) becomes:

$$\Delta p = \mu \left(R_m + \frac{\ell(t)}{k(\varepsilon_0)} \right) q_0 \tag{9.9}$$

The volume flow rate of filtrate Q_f can be expressed in terms of the percolation velocity q_0, and in terms of the volume of filtrate $V_f(t)$.

$$Q_f = q_0 S := \frac{dV_f(t)}{dt} \tag{9.10}$$

where S is the filtration area. Q_f has a simple relationship with the cake thickness ℓ; see Eq. (9.14). Substituting (9.10) in the previous equation gives:

$$\Delta p = \frac{\mu}{S} \left(R_m + \frac{\ell(t)}{k(\varepsilon_0)} \right) \frac{dV_f(t)}{dt} \tag{9.11}$$

Cake mass and thickness, filtrate volume and cake formation time

We want to obtain a practical cake formation equation involving the mass of solid $m_s(t)$, the volume of filtrate obtained $V_f(t)$ and the thickness of the filter cake $\ell(t)$, all as functions of time t. The volume fraction of solids φ_0 is written as:

$$\varphi_0 = \frac{\text{solid volume}}{\text{suspension volume}} = \frac{\text{solid volume}}{\text{solid volume} + \text{liquid volume}}$$

The liquid volume is equal to the filtrate volume produced, plus the liquid retained in the saturated filter cake, then:

$$\varphi_0 = \frac{m(\varepsilon_0, t)/\rho_s}{m(\varepsilon_0, t)/\rho_s + \left(V_f(t) + S\ell(t)\varepsilon_0\right)} \tag{9.12}$$

The solid mass is the volume of the solid in the filter cake times its density:

$$m(\varepsilon_0, t) = \rho_s(1 - \varepsilon_0)S\ell(t) \tag{9.13}$$

Substituting Eq. (9.13) with Eq. (9.12) and calculating $\ell(t)$ or $m(\varepsilon_0, t)$ we get:

$$\ell(\varphi_0, t) = \frac{\varphi_0}{1 - \varphi_0 - \varepsilon_0}\frac{V_f(t)}{S} \quad \rightarrow \quad m(\varphi_0, \varepsilon_0, t) = \frac{\varphi_0(1 - \varepsilon_0)}{1 - \varphi_0 - \varepsilon_0}\rho_s V_f(t) \tag{9.14}$$

Expression (9.14) gives the relationship between the thickness of the filter cake and the mass of material filtered, with the volume of filtrate at any given time t. Substitution with Eq. (9.11) yields:

$$\Delta p S = \mu\left(R_m + \frac{\varphi_0(1 - \varepsilon_0)}{S(1 - \varphi_0 - \varepsilon_0)k(\varepsilon_0)}V_f(t)\right)\frac{dV_f(t)}{dt} \tag{9.15}$$

The only variable in this expression is the filtrate volume $V_f(t)$, therefore integrating the time interval 0 to t gives:

$$\begin{aligned} \int_0^t \frac{\Delta p S}{\mu} d\eta &= \int_0^{V_f} R_m d\xi + \int_0^{V_f} \frac{\varphi_0}{S(1 - \varphi_0 - \varepsilon_0)k(\varepsilon_0)}\xi d\xi \\ \frac{\Delta p S}{\mu} t &= R_m V_f + \frac{\varphi_0}{2S(1 - \varphi_0 - \varepsilon_0)k(\varepsilon_0)}V_f^2 \end{aligned} \tag{9.16}$$

Filtrate volume:

Solving the quadratic Eq. (9.16) yields the volume of filtrate over time:

$$V_f(t) = Sk(\varepsilon_0)\frac{1 - \varphi_0 - \varepsilon_0}{\varphi_0}\left\{\left[\left(\frac{\ell_m}{k_m}\right)^2 + \frac{2\Delta p_e}{\mu k(\varepsilon_0)}\frac{\varphi_0}{(1 - \varphi_0 - \varepsilon_0)}t\right]^{1/2} - \frac{\ell_m}{k_m}\right\} \tag{9.17}$$

Cake thickness

Substitution of $V_f(t)$ from (9.14) with (9.16) gives the algebraic equation in terms of the thickness $\ell(t)$ of the cake:

$$\ell^2 + 2k(\varepsilon_0)\frac{\ell_m}{k_m}\ell - 2\frac{k(\varepsilon_0)}{\mu}\Delta p\frac{\varphi_0}{(1 - \varphi_0 - \varepsilon_0)}t = 0 \tag{9.18}$$

$$\ell(t) = k(\varepsilon_0)\left\{\left[\left(\frac{\ell_m}{k_m}\right)^2 + \frac{2\Delta p}{\mu k(\varepsilon_0)}\frac{\varphi_0}{(1-\varphi_0-\varepsilon_0)}t\right]^{1/2} - \frac{\ell_m}{k_m}\right\} \tag{9.19}$$

Cake mass:

Substituting Eq. (9.13) with (9.19) yields the cake mass formed at time t:

$$m(\varepsilon_0, t) = \rho_s S(1-\varepsilon_0)k(\varepsilon_0)\left\{\left[\left(\frac{\ell_m}{k_m}\right)^2 + \frac{2\Delta p}{\mu k(\varepsilon_0)}\frac{\varphi_0}{(1-\varphi_0-\varepsilon_0)}t\right]^{1/2} - \frac{\ell_m}{k_m}\right\} \tag{9.20}$$

Cake formation time:

Inverting Eq. (9.20) the time to form a cake is:

$$t = \left\{\left(\frac{m_s}{\rho_s S(1-\varepsilon_0)k(\varepsilon_0)} + \frac{\ell_m}{k_m}\right)^2 - \left(\frac{\ell_m}{k_m}\right)^2\right\}\left\{\frac{2\Delta p}{\mu k(\varepsilon_0)}\frac{\varphi_0}{1-\varphi_0-\varepsilon_0}\right\}^{-1} \tag{9.21}$$

All these expressions become simpler if the filter medium *specific resistance* $R_m = \ell_m/k_m$ is neglected.

Filtrate volume:

$$V_f(t) = S\left(\frac{2}{\mu}\right)^{1/2}\left(\frac{1-\varphi_0-\varepsilon_0}{\varphi_0}\right)^{1/2}(k(\varepsilon_0))^{1/2}\Delta p^{1/2}t^{1/2} \tag{9.22}$$

Cake mass:

$$m(\varepsilon_0, t) = \rho_s(1-\varepsilon_0)S\left(\frac{2}{\mu}\right)^{1/2}\left(\frac{\varphi_0}{1-\varphi_0-\varepsilon_0}\right)^{1/2}(k(\varepsilon_0))^{1/2}\Delta p^{1/2}t^{1/2} \tag{9.23}$$

Cake thickness:

$$\ell(t) = \left(\frac{2}{\mu}\right)^{1/2}\left(\frac{\varphi_0}{1-\varphi_0-\varepsilon_0}\right)^{1/2}(k(\varepsilon_0))^{1/2}\Delta p^{1/2}t^{1/2} \tag{9.24}$$

Cake formation time:

$$t = \left(\frac{\mu}{2\rho_s^2 S^2}\right)\left(\frac{1-\varphi_0-\varepsilon_0}{\varphi_0(1-\varepsilon_0)^2}\right)\left(\frac{1}{\Delta p k(\varepsilon_0)}\right)m_s^2 \tag{9.25}$$

It is important to take note that the expressions for filtrate volume, cake mass and cake thickness are proportional to the square root of the product of the pressure drop and time.

Equations developed in the previous section are valid for cake formation *in a steady state*, but also for a dynamic process where the amount of solid deposited per time unit is small compared to the solid already forming the cake. This occurs throughout the main part of the cake formation stage, except at the beginning of the process. For example, during the feeding of a pressure filter, the flow is variable, but the process becomes steady when the compression starts. It is obvious that all the equations developed are valid for the washing stage of the filter cake.

Figure 9.18 shows the experimental curve of cake formation in the laboratory. The circles represent experimental points, while the curve was drawn with Eq.(9.17). For about 25 % of the filtration time, the filtration volume $V_f(t)$ produced is proportional to the time t. From that time on, $V_f(t)$ is proportional to $t^{1/2}$.

In effect, Darcy's law is valid at any instant t, therefore:

$$\Delta p = \mu\left(\frac{\ell_m}{k_m} + \frac{\ell}{k(\varepsilon_0)}\right)q \tag{9.26}$$

For the first instance, the cake thickness ℓ can be ignored, giving:

$$\Delta p = \mu\frac{\ell_m}{k_m}q; \quad \text{since} \quad q = V_f(t)/St:$$

$$\Delta p = \mu\frac{\ell_m}{k_m}\frac{V_f(t)}{St}, \quad \Rightarrow \quad V_f(t) = \frac{S\Delta p}{\mu}\frac{k_m}{\ell_m}t \tag{9.27}$$

and the filtrate volume is a linear function for short times.

These considerations are important when filtration parameters are to be obtained in the laboratory.

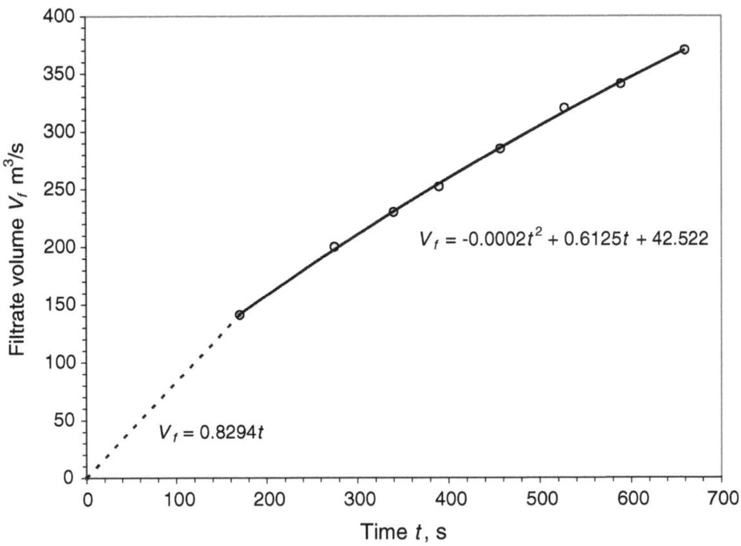

Fig. 9.18 Modeling of the cake formation step during pressure filtration

(b) **Cake formation with constant volume**

Filtration with constant volume occurs when the pulp is fed with a positive displacement pump. In this case, the volume flow rate is constant. Equation (9.8) is still valid, but now q is a constant and Δp is variable:

$$\Delta p(t)dt = \mu\left(\frac{\ell_m}{k_m} + \frac{d\ell(t)}{k(\varepsilon_0)}\right)q$$

$$\Delta p(t)Sdt = \mu\left(\frac{\ell_m}{k_m} + \frac{d\ell(t)}{k(\varepsilon_0)}\right)Q$$

where Q is the volume flow rate of filtrate. The thickness of the cake and the mass of solid in the cake is then:

$$\ell(t) = \frac{Sk(\varepsilon_0)}{\mu Q}\int_0^t \Delta p(\xi)d\xi - \frac{\ell_m k(\varepsilon_0)}{k_m} \tag{9.28}$$

$$m_s(t) = \frac{S\rho_s(1-\varepsilon_0)k(\varepsilon_0)}{\mu Q}\int_0^t \Delta p(\xi)d\xi - \frac{\ell_m\rho_s(1-\varepsilon_0)k(\varepsilon_0)}{k_m} \tag{9.29}$$

Values for $\Delta p(t)$ should be obtained from the pump curve.

Problem 9.1 (Wakeman and Tarleton 1999a) A filtration experiment is performed with a 45 cm^2 area pressure filter at 70 kPa. The ratio of moist to dry cake is 1.34. The densities of the solid, filtrate and feed pulp are respectively: $\rho_s = 2{,}640$ kg/m^3, $\rho_f = 1{,}000$ kg/m^3 and $\rho_0 = 1{,}320$ kg/m^3. The viscosity of the filtrate is $\mu = 0.01$ poises. The filtrate production in time is given in Table 9.1. Determine the time necessary to obtain a filter cake with a thickness of 4.5 cm.

Volume fraction of the suspension is $\varphi_0 = \dfrac{\rho_0 - \rho_f}{\rho_s - \rho_f} = \dfrac{1{,}320 - 1{,}000}{2{,}640 - 1{,}000} = 0.195$

Cake porosity:

$$\varepsilon_0 = \frac{V_f}{V_f + V_s} = \frac{m_f/\rho_f}{m_f/\rho_f + m_s/\rho_s}$$

Table 9.1 Filtrate production over time

Time t, s	Volume V_f, cm^3	Time t, s	Volume V_f, cm^3
0	0	457	285
170	141	527	320
275	200	589	341
340	230	660	370
390	252		

$$\varepsilon_0 = \frac{1}{1 + \frac{m_s \rho_f}{m_f \rho_s}}$$

but:

$$\frac{m_f + m_s}{m_s} = 1.34; \quad \frac{m_s}{m_f} = \frac{1}{0.34}$$

then:

$$\varepsilon_0 = \frac{1}{1 + \frac{\rho_f}{0.34\rho_s}} = \frac{1}{1 + \frac{1.00}{0.34 \times 2.64}} = 0.473$$

There is a direct relationship between the mass in the cake, the suspension concentration, cake porosity and filtrate volume. From (9.14), we have:

$$\frac{\ell(t)}{V_F(t)} = \frac{\varphi_0}{1 - \varphi_0 - \varepsilon_0} \frac{1}{S} = \frac{0.195}{1 - 0.195 - 0.473} \frac{1}{45} = 0.0131$$

$$V_F(t) = \frac{\ell(t)}{0.0131} = \frac{4.5}{0.0131} = 344.75$$

From the table we can obtain the correlation between time versus the volume of filtrate (see Fig. 9.19).

$$t(V_F) = -3 \times 10^{-6} V_F^3 + 3.8 \times 10^{-3} V_F^2 + 7.397 \times 10^{-1} V_F - 3.142 \times 10^{-1}$$

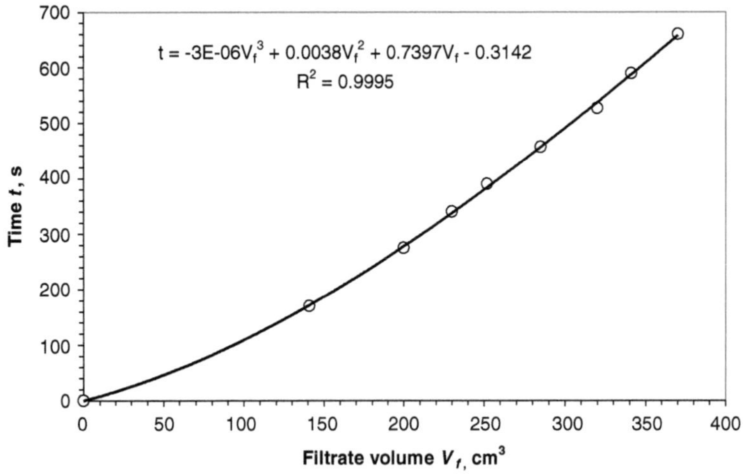

Fig. 9.19 Filtrate volume in time, from the previous table

Substituting $V_F(t)$ in the equation for $t(V_F)$ results in:

$$\begin{aligned} t(V_F) = & -3 \times 10^{-6}(344.27)^3 + 3.8 \times 10^{-3}(344.27)^2 + 7.397 \\ & \times 10^{-1}(344.27)V_F - 3.142 \times 10^{-1} \\ = & \, 500\,\text{s} \end{aligned}$$

9.3.2 Cake Dehumidification

During the filtration process, the cake formation stage removes the main part of the water from the suspension. At the end of this stage, the suspension covering the filter cake disappears and the filter cake looks dry but its pores are *saturated* with water. Saturation s is the fraction of pore volume in a porous medium filled by water. The saturation varies between zero and one, $0 \leq s \leq 1$. If $s = 1$, the cake pores are full of water and if $s = 0$, the cake is dry. The only way to eliminate water from a saturated filter cake is by *expression* and/or by *dehumidification*.

For compressible cakes, it is possible to eliminate some water by expression, that is, by applying pressure and squeezing the filter cake like a sponge. In this case, the mechanism of expression is the reduction of the pore size by compression.

In dehumidification, the water is displaced from the filter cake by air. During vacuum filtration, air is suctioned through the filter cake, while in pressure filtration the air is blown into the filter cake. During dehumidification, air and water flow simultaneously through the filter cake. In general, the solid particles attract water, wetting the solid skeleton. The air, on the contrary, displaces the water from the pores, but leaves a thin film adjacent to the solid surfaces held by capillary forces. These phenomena make the flow of water different in a water saturated and unsaturated porous medium.

To calculate the water and airflow through an unsaturated filter cake, the theory of two-phase flow through a porous medium is used (see Chap. 3). Darcy's equation can be used in this case but here the permeabilities of the liquid and air are not functions of the cake porosity only, but are also a function of saturation, $k_i(\varepsilon_0, s), i = \ell, a$, called *effective permeabilities*, where ℓ and a refers to liquid and air respectively:

$$q_{\ell,a}(s) = \frac{k_{\ell,a}[(\varepsilon_0, s)]}{\mu}\frac{\partial p_e}{\partial z} \tag{9.30}$$

Relative permeabilities of the liquid $k_{r\ell}(\varepsilon_0, s)$ and air $k_{ra}(\varepsilon_0, s)$ are defined as the quotient between the effective permeabilities $k_{\ell,a}[(\varepsilon_0, s)]$ and the permeabilities $k_{\ell,a}[(\varepsilon_0)]$ for the flow $q_{l,a}$ under the same pressure gradient $\partial p_e/\partial z$:

$$k_{ri}(\varepsilon_0, s) = \frac{k_i(\varepsilon_0, s)}{k(\varepsilon_0)}; \quad \text{where} \quad i = \ell, a \tag{9.31}$$

With these definitions of relative permeability, Darcy's equation for the water and airflow in a filter cake are:

$$q_\ell(s) = \frac{k_{r\ell}(\varepsilon_0, s) \times k(\varepsilon_0)}{\mu_\ell} \frac{\partial p_e}{\partial z} \tag{9.32}$$

$$q_a(s) = \frac{k_{ra}(\varepsilon_0, s) \times k(\varepsilon_0)}{\mu_a} \frac{\partial p_e}{\partial z} \tag{9.33}$$

The relative permeabilities must be determined experimentally. Figure 9.20 shows the relative permeabilities of a wetting fluid (water) and a non-wetting fluid (air) in a filter cake.

When air is blown through the filter cake, it displaces the water from the pores and water and air leaving simultaneously. When the cake saturation reaches the residual saturation s_∞, only air will flow out. The rest of the water is retained in the cake by capillary forces.

Functional forms for the relative permeabilities must be postulated to introduce them in Eqs. (9.32) and (9.33)

$$k_{ri}(s) = f_i(s, s_\infty), \quad \text{with} \quad i = \ell, a \tag{9.34}$$

where s and s_∞ are the instantaneous and the residual saturations, and i refers to the type of fluid.

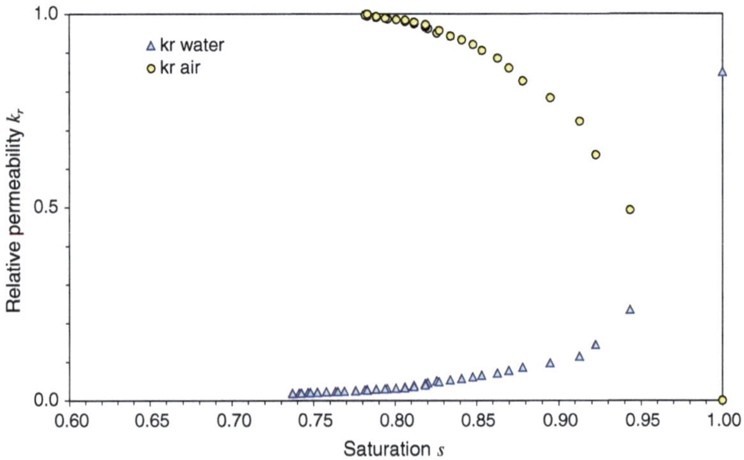

Fig. 9.20 Experimental relative permeability curves for the flow of water and air through a copper concentrate filter cake

Blowing time

Taking into consideration that $q_f(t) = \partial V_f(t)/\partial t$, for the blowing stage Eq. (9.11) is written in the form:

$$\frac{\Delta p}{\mu_\ell} = \left(R_m + \frac{\ell}{k(\varepsilon_0)k_{r\ell}(s)}\right)\frac{dV_f(t)}{dt} \tag{9.35}$$

The thickness of the cake is constant and equal to $\ell = m_s/S(1-\varepsilon_0)$ and the water in the cake at time t is $S\ell\varepsilon_0 s(t)$, where $s(t)$ diminishes with time. The filtrate volume is $V_f(t) = S\ell\varepsilon_0(1-s(t))$, so that: $dV_f/dt = -S\ell\varepsilon_0 ds/dt$. Introducing these relationships in Eq. (9.35) yields an equation in terms of the saturation:

$$\frac{\Delta p}{\mu_\ell} = -\left(R_m + \frac{\ell}{k(\varepsilon_0)k_{r\ell}(s)}\right)S\ell\varepsilon_0\frac{ds}{dt}$$

Integrating with respect to s:

$$\int_0^t \frac{\Delta p}{\mu_\ell}d\xi = -\int_1^s S\ell\varepsilon_0\left(R_m + \frac{\ell}{k(\varepsilon_0)k_{r\ell}(\eta)}\right)d\eta$$

Integrating, the blowing time to reach a saturation s is obtained:

$$t = \frac{\ell\varepsilon_0\mu_\ell S}{\Delta p}\left(R_m(1-s) + \frac{\ell}{k(\varepsilon_0)}\int_s^1 \frac{d\eta}{f_\ell(\eta)}\right) \tag{9.36}$$

The integral in (9.36) can be calculated once the function $f(s)$ is known.

Airflow

Since in pressure filtration the pressure gradient is much greater than the saturation gradient, the latter can be neglected (see Sect. 6.5.4) and the air flow necessary to reach a given saturation can be obtained by directly integrating Eq. (9.33):

$$\frac{\Delta p}{\mu_a} = \left(R_m + \frac{\ell}{k(\varepsilon_0)k_{ra}(s)}\right)q_a(t)$$

The air flow rate is $Q_a = Sq_a$, therefore:

$$Q_a(t) = \left(\frac{S\Delta p}{\mu_a}\right)\left(R_m + \frac{\ell}{k(\varepsilon_0)k_{ra}(s)}\right)^{-1} \tag{9.37}$$

In these equations, R_m and S are equipment parameters, $\rho_s, k(\varepsilon_0)$ and ε_0 are characteristic parameters of the porous medium, ρ_ℓ and μ_ℓ are properties of the fluid and $\Delta p, \ell$ and s(or t) are operational parameters and variables.

Humidity is the ratio of water mass to total mass expressed as percentage. Equation (9.38) gives the relationship between saturation and humidity:

$$h = 100\frac{\rho_\ell \varepsilon_0 s}{\rho_\ell \varepsilon_0 s + \rho_s(1-\varepsilon_0)} \quad \text{and} \quad s = \frac{\rho_s}{\rho_\ell}\frac{1-\varepsilon_0}{\varepsilon_0}\frac{h}{100-h} \tag{9.38}$$

9.3.3 Cake Washing

The same equations and methods used for cake formation and blowing can be used for cake washing, with a restriction on the concentration of the element that is to be eliminated.

9.4 Filtration Parameters and Their Measurements

Figure 9.4 shows the several variables and parameters that influence the filtration process. *Inlet variables* are feed pulp flow and concentration. O*utlet variables* are filter capacity, mass of solid filtered per time unit and the thickness and humidity of the filter cake. *Design variables* are filtration area and pressure drop in the equipment. *Control variables* are applied pressure, times for cake formation, washing, expression and blowing, temperature, pH, additives and pulp agitation. *Perturbations* are the type of material, particle size and size distribution, impurities in the feed material. *Parameters* are porosity, permeability and compressibility of the cake, relative permeabilities for air and water and residual saturation.

In the previous section, we developed relationships between these variables and parameters. To complete the necessary information to design and simulate a filtration process, it is necessary to determine the dependence of the parameters on the properties of the solid and the liquid.

9.4.1 Filtration Parameter Measurements

Several companies that provide equipment have laboratory instruments to measure filtration parameters that are based on similar principles but differ in the amount of sample they can take and the quality of their instrumentation. Here we will describe *FILTRATEST*, an instrument designed by Bokela GmbH.

The core of *FILTRATEST* is a stainless steel pressure vessel that supports pressures up to 10 bars (150 ψ) and therefore can simulate a vacuum and most pressure filtration processes. The instrument has a filtration area of 19.63 cm^2 and a water jacket that permits operation at controlled temperatures. A set of rotameters of different sizes measures the airflow rate. A digital manometer controls air pressure. A beaker sitting on a digital balance receives and weighs the filtrate. The portability of the instrument is fundamental to take measurements at plant sites. The Fig. (9.21) shows the FILTRATEST.

Software especially designed for this instrument registers the exact time and filtrate produced during cake formation, expression and blowing cycles. From these data, functional forms are postulated for the filtration rate and time, from which the permeabilities of the cake and the specific resistance of the filter medium are calculated. The relative permeabilities for the water and air are also calculated.

(a) **Cake porosity**

The cake porosity is a function of the size distribution of the particles forming the porous bed. A bed formed of particles of only one size will have the same porosity irrespective of the size of the particles. The case of sphere packing illustrates this. Table 9.2 (Wakeman and Tarleton 1999b) shows the porosity for different types of sphere packing.

We can see that a three-fold value of the porosity of spheres is possible (0.26–0.78) for different types of packing. The nominal value used for porosity in filter cakes is $\varepsilon_0 = 0.4$, a bit lower than the average in Table 9.2.

In the case of filter cakes, it is more useful to determine the porosity experimentally. There are three ways to do that: by drying and weighing the cake, by measuring the depth of the saturated cake, and by a water balance at the end of the bed formation time. When the cake is saturated, the water has disappeared from the suspension and the difference between the total water in the suspension and the water in the filtrate gives the volume of water saturating the cake, which is equal to the volume of pores in the cake.

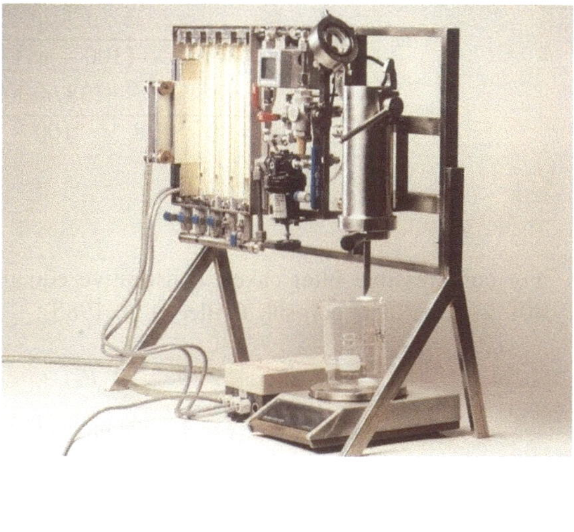

Fig. 9.21 FILTRATEST instrument

Table 9.2 Porosity versus type of packing for spheres

Types of packing	Coordination number	Porosity ε_0
	3	0.7766
	4	0.6599
	5	0.5969
Cubic	6	0.4764
	7	0.4388
Orto-rombic	8	0.3955
	9	0.3866
Tetragonal	10	0.3019
	11	0.2817
Romohedral	12	0.2595

Problem 9.2 A sample of copper concentrates was taken from a Larox pressure filter with a cylinder of 5 cm in diameter. The thickness of the sample was 35 mm. The weight of the sample after drying was 147 g and the density of the solid 3.87 g/cm^3.

The cake porosity then was:

$$\varepsilon_0 = 1 - \frac{147/3.87}{(\pi \times 5^2/4) \times 3.5} = 0.45$$

Problem 9.3 In a laboratory test 138.7 g of copper concentrate, with a density of 4,300 kg/m^3, was filtered from a suspension of 72.2 % solid by weight. The cake formation time was $t_1 = 42.7$ s and, during that time, 20.6 cm^3 of filtrate was recovered. Calculate the saturated cake porosity.

$$1 - \varepsilon_0 = \frac{V_s}{V_s + V_{fq}} = \frac{V_s}{V_s + V_f - V_F}$$

$$\varepsilon_0 = 1 - \frac{m_s/\rho_s}{m_s/\rho_s + m_s \times (100 - w_s)/w_s - V_F}$$

$$\varepsilon_0 = 1 - \frac{138/4.30}{138/4.30 + 138 \times (100 - 72.2)/72.2 - 20.6}$$

$$\varepsilon_0 = 0.50$$

For compressible filter cakes, constitutive equations, such as Eqs. (9.39) and (9.40) describe their porosity (Tiller et al. 1985):

$$\varepsilon = 1 - p_s^{\beta} \tag{9.39}$$

$$\varepsilon = 1 - \alpha \ln(p_s/\beta) \tag{9.40}$$

Problem 9.4 Laboratory tests were performed with 80 g of a copper flotation tailing in the FILTRATEST having 19.63 cm^2 of filter area, giving a cake with a porosity $\varepsilon = 0.493$ at a pressure drop of 0.75 bars. Tests at different pressure drops yielded the result shown in Table 9.3. Determine the constitutive equation for the compressibility of the cake as a function of porosity.

The correlation between the pressure p_s and thickness height is, see Fig. 9.22

$$\ell = 2.8708\, p_s^{-0.0519}\ \text{cm}.$$

The cake thickness for a pressure of 0.75 bars is

$$\ell = 2.8708(0.75)^{-0.0519} = 2.914\ \text{cm}$$

Calculating the cake volume V_c and the solid volume V_s from the filter area and the cake thickness, for a pressure of $p = 0.75$, we obtain:

$$\begin{aligned} V_c &= S \times \ell \\ &= 19.63 \times 2.914 = 57.20\ \text{cm}^3 \end{aligned}$$

$$\begin{aligned} V_s &= (1-\varepsilon) \times V_{\text{cake}} \\ &= (1-0.493) \times 57.202 = 29.0\ \text{cm}^3 \end{aligned}$$

Finally, the values of the cake porosity for the various pressures p_s are calculated from:

$$1 - \varepsilon(p_s) = \frac{V_s}{V_{\text{cake}}(p_s)} = \frac{29.0}{19.63 \times 2.9 \times p_s^{-0.0519}}$$

Figure 9.23 shows a plot of the solid volume fraction versus the applied pressure.

Table 9.3 Filter cake compression

Pressure drop. bars	Cake height (cm)
0.8	2.90
1.0	2.87
1.5	2.81
2.0	2.77
2.7	2.73
3.0	2.71
3.5	2.70
4.0	2.68
4.7	2.66
5.0	2.63
5.5	2.62
6.0	2.61

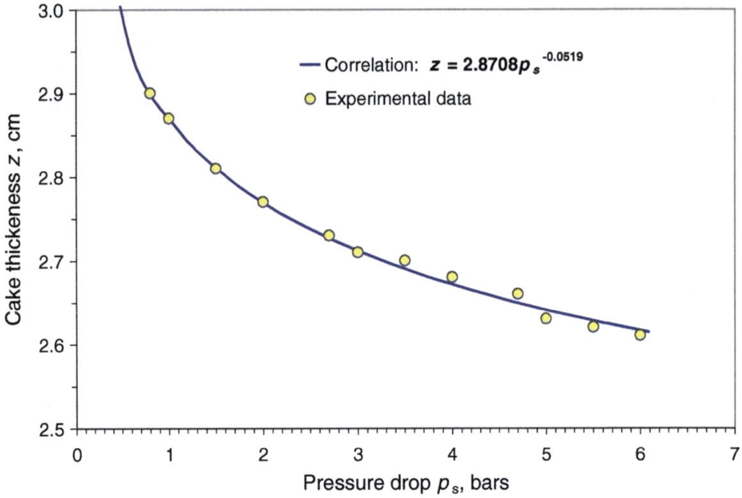

Fig. 9.22 Cake thickness heights versus pressure drop for a copper flotation tailing

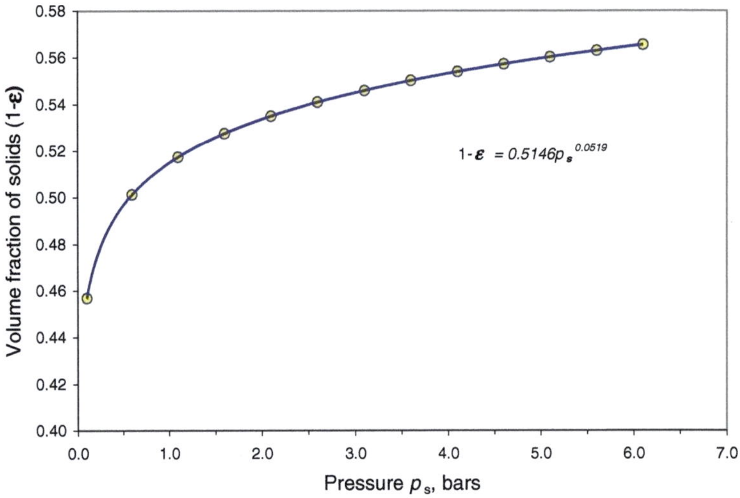

Fig. 9.23 Solid volume fraction $(1 - \varepsilon)$ versus pressure drop for a copper flotation tailing

From Fig. 9.23 the following constitutive equations can be obtained by non-linear fitting:

$$\varepsilon = 1 - 0.514\, p_s^{0.0519} \tag{9.41}$$

The constitutive equation for the solid pressure is:

$$p_s = \left(\frac{1 - \varepsilon}{0.514}\right)^{19.27} \tag{9.42}$$

Problem 9.5 A flotation tailing is filtered in a vacuum band filter at 0.75 bars, obtaining a cake with $\varepsilon_0 = 0.493$ in porosity. An experimental test to determine the compressibility of the cake gave the following constitutive equation:

$$\varepsilon = 1 - 0.5146\, p_s^{0.0519}$$

What would the porosity be if the same material were filtered in a hyperbaric filter with 3 bars of overpressure and in a Larox PF filter at 6 bars.

For the different filters, we have:

$$\text{Larox PF} \quad \varepsilon(6\text{ bars}) = 1 - 0.5146 \times (6)^{0.0519} = 0.435$$

$$\text{Hiperbaric flilter} \quad \varepsilon(3.75\text{ bars}) = 1 - 0.5146(3.75)^{0.0519} = 0.449$$

(b) **Filter medium resistance and cake permeability**

The filter medium is an important component in the filtration process. It is a medium with pores of different sizes and geometry, the structure of which can cause variations in the way in which the particles are deposited and in the distributions of the filtrate flow. A filter cloth must not only retain the solid particles and produce a clean filtrate, but must also resist the stresses imposed by the equipment. Therefore in addition to the *specific resistance* to filtration, we must consider the *mechanical resistance* as another parameter.

Experimental determination

To determine the specific resistance of a filter medium R_m and the permeability of the filter cake $k(\varepsilon)$ Eq. (9.17) is written in the form:

$$\frac{t}{V_f(t)} = \frac{R_m \mu_f}{S\Delta p} + \frac{\mu_f}{2S^2\Delta p} \frac{\varphi_0}{(1-\varphi_0-\varepsilon_0)k(\varepsilon_0)} V_f(t) \tag{9.43}$$

where $R_m = \ell_m / k_m$, and ℓ_m and k_m are the thickness and permeability of the filter medium, information that is usually not known.

Write Eq. (9.43) in the form:

$$\frac{t}{V_f(t)} = \underbrace{\frac{R_m \mu_f}{S\Delta p}}_{a} + \underbrace{\frac{\mu_f}{2S^2\Delta p} \frac{\varphi_0}{(1-\varphi_0-\varepsilon_0)k(\varepsilon_0)}}_{b} V_f(t) \tag{9.44}$$

$$\frac{t}{V_f} = a + bV_f$$

From the plot of t/V_f versus V_f, the value of R_m and $k(\varepsilon_0)$ can be obtained from the intercept "a" on the ordinate and from the slope "b" of the straight line (see Fig. 9.27):

$$R_m = a \times \frac{\Delta p S}{\mu_f} \quad \text{and} \quad k(\varepsilon_0) = \frac{1}{b} \times \frac{\mu_f}{2S^2 \Delta p} \frac{\varphi_0}{1 - \varphi_0 - \varepsilon_0} \tag{9.45}$$

Problem 9.6 (Massarani 1978) Calculate the filtration area necessary to treat 10,000 $1/\text{h}$ of a 5 % by weight calcium carbonate suspension. The solid density is 2,500 kg/m^3. The filter operates at 20 °C and 40 ψ of pressure. Laboratory experiments with a filter with 500 cm^2 area and 40 ψ pressure give a cake 3.2 cm in thickness. The weight percentage of solids of the cake is 60.2 % and the viscosity is $\mu = 1$ cp. Figure 9.24 shows a plot of t/V_f versus V_f.

Parameters:

From Fig. 9.24 the ordinate intercept $a = 0.0071$ and the slope is $b = 1 \times 10^{-6}$ and from Eq. (9.44), the intercept and slope are given by:

$$a = 0.0071 = \frac{\mu}{S\Delta p} \frac{\ell_m}{k_m} \quad \text{and} \quad b = 1 \times 10^{-6} = \frac{1}{2S} \frac{\varphi_0}{(1 - \varphi_0 - \varepsilon_0)k(\varphi_0)}$$

$$\varphi_0 = \frac{5}{2.5 \times (100 - 5) + 5} = 0.0206 \quad \text{(Suspension)}$$

$$\varphi_{\text{cake}} = 1 - \varepsilon_0 = \frac{60.2}{2.5 \times (100 - 60.2) + 60.2} = 0.374 \quad \text{(Cake)}$$

$$\varepsilon_0 = 1 - 0.209 = 0.791(\text{Cake})$$

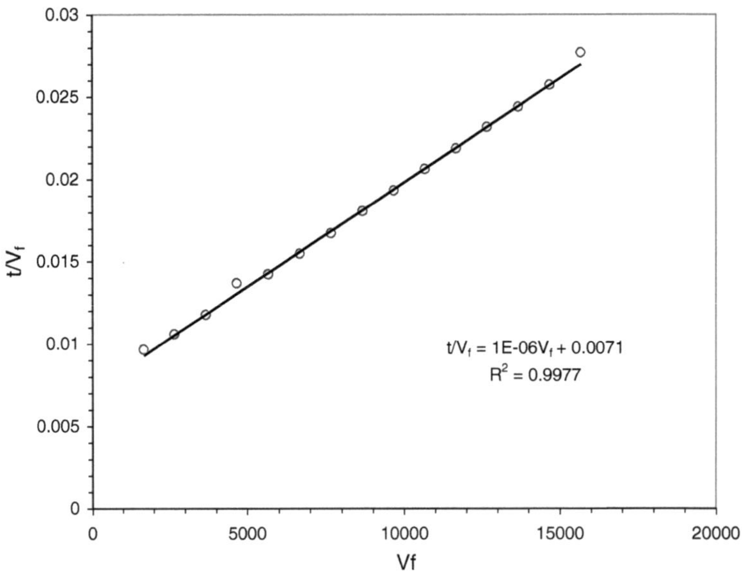

Fig. 9.24 Plot of t/V_f versus V_f to calculate filtration parameters

$$\Delta p = 40 \times 6.897 \times 10^4 = 2.76 \times 10^6 \text{ dyns/cm}^2$$

$$\mu = 0.01 \text{ g/cm-s}$$

From (9.45):

$$R_m = a \times \frac{\Delta p S}{\mu_f} = 0.0071 \times \frac{40 \times 6.897 \times 10^4}{0.01} = 9.79 \times 10^8 \text{ cm}^{-1}$$

$$\begin{aligned} k(\varphi_0) &= \frac{\mu}{2S^2 \Delta p} \times \frac{\varphi_0}{1 - \varphi_0 - \varepsilon_0} \times \frac{1}{b} \\ &= \left(\frac{0.01}{2 \times 500^2 \times 2.76 \times 10^6} \right) \left(\frac{0.0206}{1 - 0.0206 - 0.209} \right) \left(\frac{1}{10^{-6}} \right) \\ &= 1.18 \times 10^{-16} \text{ cm}^2 \end{aligned}$$

$$Q_{Filtrate} = 10{,}000 \times \left(\frac{10^3}{3.6 \times 10^3} \right) = 2{,}777.8 \text{ cm}^3/\text{s}$$

From Eq. (9.11) the filtration area is:

$$\begin{aligned} S &= \mu \times \left(R_m + \frac{\ell(t)}{k(\varphi_0)} \right) \frac{Q_{Filtrate}}{\Delta p} \\ &= 0.01 \times \left(9.67 \times 10^8 + 3.2/\left(1.94 \times 10^{-11}\right)\right) \times \left(\frac{2{,}777.8}{2.76 \times 10^6 \times 10^5} \right) \\ &= 1.7584 \times 10^5 \text{ cm}^2 \equiv 77.87 \text{ m}^2 \end{aligned}$$

Problem 9. 7 Calculate the permeability and resistance of the filter medium of a material filtered in the laboratory at 70 kPa through a filter area of 45 cm^2. The ratio of the wet and dry cake weight was 1.34. The solid, filtrate and feed pulp densities is $\rho_s = 2{,}640 \text{ kg/m}^3$, $\rho_f = 1{,}000 \text{ kg/m}^3$ and pulp density $\rho = 1{,}320 \text{ kg/m}^3$. The filtrate viscosity is $\mu_f = 0.01$ poises. Table 9.4 shows the filtrate production over time:

Calculate the permeability and filter medium resistance.
Volume fraction in the feed:

$$\varphi_0 = \frac{1{,}320 - 1{,}000}{2{,}640 - 1{,}000} = 0.195$$

Table 9.4 Filtration production in time

Time t in s	Volume in cm^3	Time t in s	Volume in cm^3
0	0	457	285
170	141	527	320
275	200	589	341
340	230	660	370
390	252		

The ratio of wet to dry cake is the inverse of the solid fraction of the cake, therefore:

$$X = \frac{m_s}{m_s + m_f} = \frac{1}{1.34} = 0.746$$

Volume fraction of solids in the cake:

$$\varphi_c = \frac{\rho_f X}{\rho_s(1 - X) + \rho_f X} = \frac{1{,}000 \times 0.746}{2{,}640 \times (1 - 0.746) + 1{,}000 \times 0.746} = 0.527$$

Cake porosity $\varepsilon = 1 - \varphi_{ck} = 1 - 0.527 = 0.473$
Plotting t/V_f versus V_f, we obtain the Fig. 9.25.
From the figure a = 0.88946 and b = 0.00246, so that:

$$R_m = a \times \frac{\Delta p S}{\mu_f} = 0.88946 \times \frac{7.0 \times 10^5 \times 45}{0.01} = 2.80 \times 10^9 \mathrm{cm}^{-1}$$

$$k(\varepsilon) = \frac{1}{b} \times \frac{\mu_f}{2S^2 \Delta p} \times \frac{\varphi_0}{1 - \varphi_0 - \varepsilon} = \frac{1}{0.00246} \times \frac{0.01}{2 \times 45^2 \times 7 \times 10^5} \times \frac{0.195}{1 - 0.195 - 0.473} = 8.44 \times 10^{11}\ \mathrm{cm}^2$$

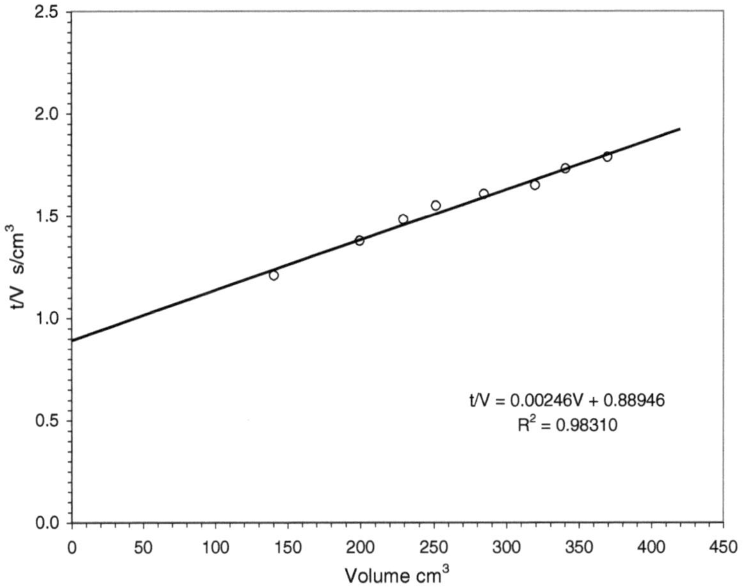

Fig. 9.25 Determination of the cake permeability and filter medium resistance

Specific resistance of the cake is:

$$\alpha = \frac{\ell}{k(\varepsilon)} = \frac{0.784}{8.44 \times 10^{-10}} = 5.33 \times 10^{-1}$$

Problem 9.8 Calculate the specific resistance of the filter medium and the cake permeability for a copper concentrate filtered in a Larox PF filter. A laboratory test was performed in the FILTRATEST having a filtration area of 19.63 cm^2. A sample of 156.78 g of solid was filtered at 2 bars for 39.76 s, time at which the cake was formed. The volume of filtrate formed at any time was registered and is shown Fig. 9.26. (Fig. 9.27)

The following data are known: (Tables 9.5, 9.6, 9.7)
Plotting t/V$_f$ versus V$_f$ in Fig. 9.28, we get the correlation:

$$t/V_f = t/(-0.0029 \times t^2 + 0.480 \times t - 2.841)$$

From Fig. 9.28 the values of a and b are $a = 3.0855$ s/cm^3 and $b = 0.029$ s/cm^6, then:

$$R_m = \frac{a \times \Delta p \times S}{\mu_f} = \frac{3.0855 \times 2 \times 10^5 \times 19.63}{0.012} = 1.0095 \times 10^9 \text{ cm}^{-1}$$

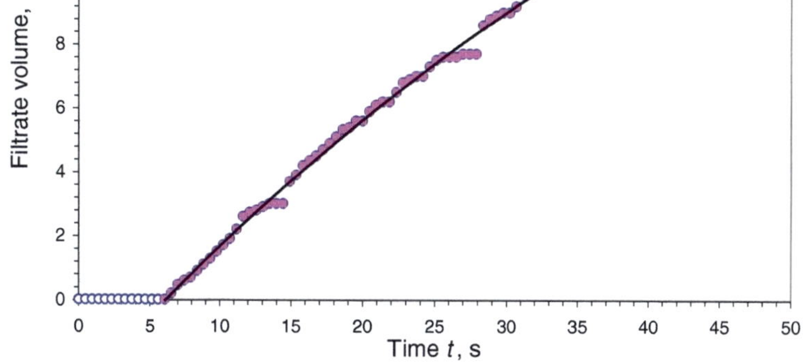

Fig. 9.26 Filtrate volume versus time for the laboratory experiment in a FILTRATEST. The *line* gives the polynomial correlation

Table 9.5 Data for the filtration of a copper concentrate in a Larox PF filter

Parameters	Larox PF filter
Solid density, g/cm^3	3.87
Filtrate density, g/cm^3	1.0
Feed pulp concentration, % by weight	78.0
Filtrate viscosity, kg/m-s	0.0012
Particle size	Size analysis
Feed pulp temperature, °C	20
Feed pulp pH	9.1
Pressure during cake formation, bar	2
Air pressure during blowing, bar	4
Cake thickness, mm	3.6

Table 9.6 Particle size analysis

Mesh size	Average size x, μm	Weight retained, g	$f_3(x)$
35/48	365	2.32	0.024
48/65	252	0.77	0.008
65/100	178	1.87	0.019
100/150	126	7.75	0.079
150/200	89	16.22	0.165
200/270	63	27.15	0.277
270/400	45	23.73	0.242
−400	31	18.27	0.186
SUMM		98.15	1.000

Table 9.7 Laboratory data

Cake thickness, mm	36
Weight of wet cake, g	168.28
Weight of dry cake, g	156.78
Cake formation time t_1, s	39.76
FILTRATEST area, cm^2	19.63

Pulp concentration 78 % solid

Feed volume fraction $\varphi_0 = \frac{78}{3.87 \times (100-78)+78} = 0.478$

Cake porosity

$$k(\varepsilon_0) = \frac{1}{b} \times \frac{\mu_f}{2S^2\Delta p} \frac{\varphi_0}{1-\varphi_0-\varepsilon}$$

$$\varepsilon = 1 - \frac{156.78/3.78}{19.63 \times 3.6} = 0.43 \qquad = \frac{1}{0.029} \frac{0.012}{2 \times (19.63)^2 \times 2 \times 10^6}$$

$$\frac{0.48}{1-0.478-0.43} = 5.74 \times 10^{-9}\ \text{cm}^2$$

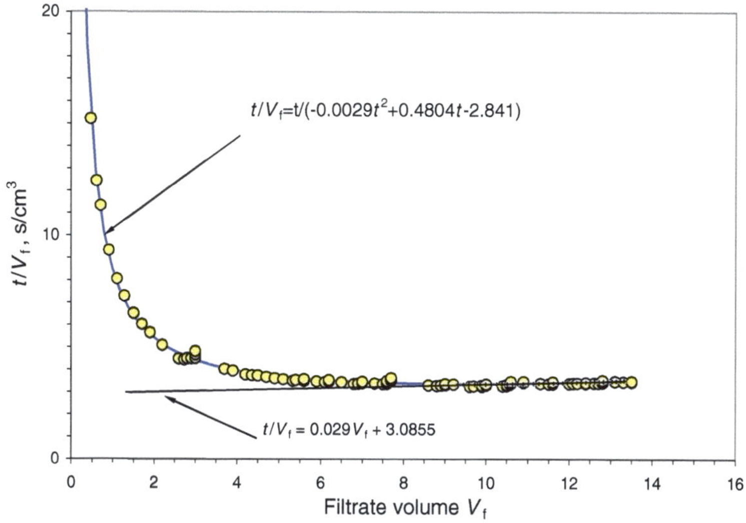

Fig. 9.27 Classical t/V_f versus V_f curve for the determination of the permeability and the filter medium resistance

Effect of the particle size

The Kozeny–Carman equation for the permeability of porous media was given in Sect. 6.3, Eqs. (6.29)–(6.33):

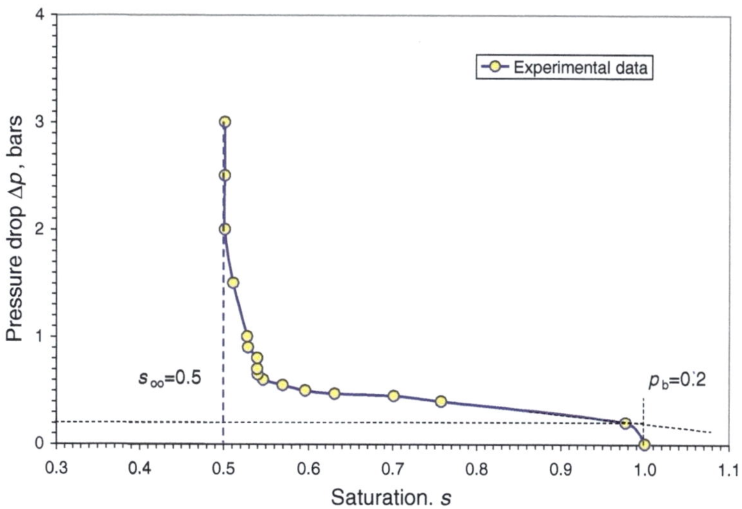

Fig. 9.28 *Capillary curve*: pressure drop versus saturation

$$k(\varepsilon)=\left(\frac{d_p^2}{36\beta}\right)\frac{\varepsilon^3}{(1-\varepsilon)^2};\quad k(\varepsilon)=\left(\frac{d_e^2\psi^2}{36\beta}\right)\frac{\varepsilon^3}{(1-\varepsilon)^2};$$
$$k(\varepsilon)=\left(\frac{\bar{x}_{12}^2}{\beta(\alpha_s/\alpha_v)^2}\right)\frac{\varepsilon^3}{(1-\varepsilon)^2} \tag{9.46}$$

These equations show that the permeability of a filter cake depends on the characteristics of the particles through two particle parameters, the average particle size and the particle shape, both squared, and on the porosity through the function $\varepsilon^3/(1-\varepsilon)^2$. This last function, as we have already seen, depends on size distribution, the packing factor and particularly the applied pressure.

Problem 9.9 For the data of problem 9.8, use the Kozeny–Carman equation to predict the permeability of a filter cake formed at 6 bars of applied pressure.

At 4 bars of cake formation pressure, the porosity was $\varepsilon=0.43$. Since the copper concentrate is practically incompressible, we can assume that at 4 and 6 bars the porosity will be the same. Assume a sphericity of $\psi=0.5$ for the concentrate. The particle size distribution is given in Table 9.8.

The surface volume average is $\bar{x}_{12}=\frac{1}{1.85\times10^{-2}}=54.18$ μm and the equivalent volume diameter $d_e=0.69\times\bar{x}_{12}=0.69\times54.18=37.39$ μm (Concha et al. 1973). After Coulson and Richardson (1968), $\beta=5$, therefore:

$$k(\varepsilon)=\frac{(d_e\psi)^2}{36\beta}\frac{\varepsilon^3}{(1-\varepsilon)^2}=\frac{(37.39\times10^{-4}\times0.5)^2}{36\times5}\times\frac{0.43^3}{(1-0.43)^2}$$
$$=4.75\times10^{-9}\ \text{cm}^2$$

(c) Residual saturation and capillary curve

The entry pressure and the residual saturation are two important parameters to determine the operating conditions of an industrial filter. *Entry pressure* is the

Table 9.8 Particle size distribution

Mesh size	Average size x, μm	Weight retained, g	$f_3(x)$	$xf_3(x)$	$f_3(x)/x$
35/48	365	2.32	0.024	8.89	6.67E−5
48/65	252	0.77	0.008	1.98	3.11E−5
65/100	178	1.87	0.019	3.39	1.07E−4
100/150	126	7.75	0.079	9.95	6.27E−4
150/200	89	16.22	0.165	14.71	1.86E−3
200/270	63	27.15	0.277	17.43	4.39E−3
270/400	45	23.73	0.242	10.88	5.37E−3
−400	31	18.27	0.186	5.77	6.00E−3
Sum		**98.15**	**1.000**	**72.99**	**1.85E−2**

minimum pressure at which a saturated filter cake begins dewatering. It is the lower boundary for the blowing pressure. *Residual saturation* is the value at which the saturation does not decrease with increased pressure.

Problem 9.9 Calculate the entry pressure and the residual saturation for a copper concentrate, with a density $\rho_s = 3.87\ \mathrm{g/\ cm^3}$, having the data of Table 9.9, plotted in Fig. 9.28 showing the values $p_b = 0.20, \quad s_\infty = 0.50$.

The figure shows that it is unnecessary to blow air at pressures over 2.5 bars, and that the minimum saturation possible is $s = 0.50$. Since the porosity is $\varepsilon = 0.43$, the minimum humidity of the final cake is:

$$h = 100 \times \frac{\rho_f \varepsilon s}{\rho_f \varepsilon s + \rho_s (1 - \varepsilon)}$$

$$= 100 \times \frac{1{,}000 \times 0.43 \times 0.50}{1{,}000 \times 0.43 \times 0.50 + 3{,}870 \times (1 - 0.43)} = 8.9\ \%$$

Correlation for the residual saturation

According to Wakeman, see Eqs. (6.74) and (6.75) of this book, the residual saturation can be calculated in terms of the capillary number with the following equations:

$$s_\infty = 0.155\left(1 + 0.031\ N_{cap}^{-0.49}\right), \quad N_{cap} = \frac{\varepsilon_{av}^3 \bar{x}_{12}^2 \Delta p}{(1 - \varepsilon_{av})^2 \ell \gamma}$$

Table 9.9 Filtrate production in time during the blowing stage

Δp	Filtrate volume V_f	Liquid in the cake V	Porosity ε	Saturation s
0.00	16.47	26.3	0.43	1.00
0.20	17.07	25.7	0.43	0.98
0.40	22.84	19.9	0.43	0.76
0.45	24.32	18.5	0.43	0.70
0.47	26.17	16.6	0.43	0.63
0.50	27.09	15.7	0.43	0.60
0.55	27.79	15.0	0.43	0.57
0.60	28.39	14.4	0.43	0.55
0.65	28.56	14.2	0.43	0.54
0.70	28.57	14.2	0.43	0.54
0.80	28.57	14.2	0.43	0.53
0.90	28.86	13.9	0.43	0.53
1.00	28.89	13.9	0.43	0.51
1.50	29.32	13.5	0.43	0.50
2.00	29.58	13.2	0.43	0.50
2.50	28.58	13.2	0.43	0.50
3.00	29.58	13.2	0.43	0.50

where N_{cap} is the capillary number, ε_{av} and ℓ are the average cake porosity and the thickness, $\bar{x}_{12}$ is the surface-volume average particle size, Δp is the pressure drop across the cake and γ is the liquid surface tension.

Problem 9.10 Determine the residual saturation for the filter cake of the previous problem. The experimental information is:

$$\varepsilon_{av} = 0.43, \quad \bar{x}_{12} = 54.18\ \mu\text{m}, \quad \Delta p = 2.5\ \text{bars}, \quad \ell = 3.5\ \text{cm}, \quad \gamma = 72\ \text{dyns/cm}$$

$$N_{cap} = \frac{\varepsilon_{av}^3 \bar{x}_{12}^2 \Delta p}{(1-\varepsilon_{av})^2 \ell \gamma} = \frac{(0.43)^3 \times (54.18 \times 10^{-4})^2 \times (2.5 \times 10^6)}{(1-0.43)^2 \times 3.5 \times 72} = 0.0713$$

$$\begin{aligned} s_\infty &= 0.155 \times \left(1 + 0.031 \times N_{cap}^{-0.49}\right) = 0.155 \times \left(1 + 0.031 \times 0.0713^{-0.49}\right) \\ &= 0.173 \end{aligned}$$

The value obtained with Wakeman's correlation is about 35 % of the experimental value of $s_\infty = 0.50$.

(d) Relative permeability

We defined the *relative permeability* of the water and that of the air in a porous medium at time *t*, as the ratio of the respective water and air permeability at time *t* and the permeability of the saturated cake. As we know, this last parameter is a property solely of the porous medium.

To calculate the permeabilities of the water and the air during the blowing filtration stage, the flow of water and air as a function of time must be known (Table 9.10).

Problem 9.11 Determine the relative permeabilities of water and air for a copper concentrate filtration process. Experiments were made with the FILTRATEST on a copper concentrate pulp with 73 % solid by weight. Experimental conditions were: $\Delta p = 6$ bars for the cake formation stage, filtration area $S = 6.55\ \text{cm}^2$, temperature $T = 20\ °\text{C}$, filtrate viscosity $\mu = 1$ mPa-s, air viscosity 0.0187 mPa-s, concentrate density $\rho_s = 4.50\ \text{g/cm}^3$. At the end of the filtration cycle, the humid cake weighed 46.57 g and after drying, the weight was 43.27 g. The cake formation stage took 37 s and liberated 5.79 g of filtrate. The expression stage eliminated 4.56 g of filtrate during 38 s at $\Delta p = 7.5$ bars. Finally, during 121 s, air blown at $\Delta p = 6$ bars liberated 1.99 g of additional filtrate. Table 9.11 shows data for the several filtration stages. The rate of filtrate production is given in the first two columns of Table 9.12, for the cake formation stage, and the first three columns of Table 9.13 for the blowing stage.

Table 9.10 Summary of the mass balance results

Data	Symbol	Unit cgs
Solid density	ρs	4.5
Temperature	T(°C)	20
Filter area	S	6.5
Pressure drop in cake formation	ΔP	6.00E+06
Pressure drop in expresion	ΔP	8.50E+06
Pressure drop in blowing	ΔP	6.00E+06
Filtrate viscosity	μ_f	0.01
Air viscosity	μa	0.000187
Weight of humid cake	Wh	4.66E+01
Weight of dry cake	Wd	4.33E+01
Cake formation time	t_1	3.70E+01
Cake expresion time	t_2	3.80E+01
Cake blowing time	t_3	1.21E+02
Filtrate from formation	V_1	5.79E+00
Filtrate from expresion	V_2	4.56E+00
Filtrate from blowing	V_3	1.99E+00
Parameter t/Vf versus Vf	a	4.73E+00
Parameter t/Vf versus Vf	b	2.85E−01
Results		
Weight of humid cake	Wh	4.66E+01
Weight of dry cake	Wd	4.33E+01
Filtate in saturate cake	V_{fsc}	9.85E+00
Filtrate after expresion	V_{faexp}	5.29E+00
Filtrate in final cake	V_{ffc}	3.30E+00
Total filtrate produced	V_f	1.23E+01
Water in suspension	V_{wat0}	1.56E+01
Humidity of saturated cake	h_{sc}	18.54
Humidity of final cake	^{h}fc	7.09
% solid in suspension	w_0	73.5
Volume fraction in suspension	$\varphi 0$	0.381
Cake volume of saturated cake	$^{V}csat$	19.47
Cake volume after expression	$^{V}caexp$	14.91
Cake porosity in cake formation	ε_0	0.506
Cake thickenes in formation	^{L}exp	2.97
Cake thickeness after blowing	L_b	2.28
Cake porosity after blowing in blowing	ε_{ab}	0.355
Cake final saturation	S	0.624
Cake permeability	$k(\varepsilon)$	2.29E−10
Filtermedia resistance	R_m	1.86E+10

Cake permeability

To calculate the cake permeability, we model the filtrate volume flow from Table 9.12.

Figure 9.29 shows a plot of the filtrate volume versus time for the cake formation stage.

Table 9.11 Data for different filtration cycles

Stage	Time s	Δp bars	Filtrate production at stage, g
Formation	36.69	6.0	5.79
Expression	38.31	7.5	4.56
Blowing	120.57	6.0	1.99
Total	195.57		12.34

Table 9.12 Filtrate production in time during the cake formation stage

Cake formation stage			
time t, s	V_f cm^3	V_f (sim) cm^3	t/V_f (sim) s/cm^3
2.14	0.01	0.0491	43.550
4.18	0.25	0.2932	14.255
6.21	0.68	0.6570	9.452
8.24	1.14	1.0682	7.714
10.27	1.56	1.4913	6.887
12.31	1.93	1.9115	6.440
14.34	2.34	2.3170	6.189
16.37	2.65	2.7064	6.049
18.4	3	3.0788	5.976
20.43	3.35	3.4339	5.950
22.47	3.75	3.7740	5.954
24.5	4.06	4.0966	5.981
26.53	4.48	4.4042	6.024
28.56	4.7	4.6979	6.079
30.6	4.99	4.9798	6.145
32.63	5.27	5.2481	6.217
34.66	5.48	5.5051	6.296
36.69	5.79	5.7514	6.379

From the figure, the following correlation are obtained:

$$V_f = -0.0012\, t^2 + 0.2189\, t - 0.5716$$

Using this correlation, a graph $t/V_f \text{versus} V_f$ yields Fig. 9.30.

From the correlation $t/V_f = 4.7322 + 0.2848 V_f$, the following parameters a and b are:

$$a = 4.7322 \quad \text{and} \quad b = 0.2848$$

With these values, the permeability and the filter medium resistance are:

Table 9.13 Permeability data and results

Blowing stage

t_2 s	Q_{air} l/h	$V_{f\ (tot)}$ cm^3	$V_{f\ (blowing)}$ cm^3	$Q_{air\ (correl)}$ l/h	$V_{f\ (correl)}$ cm^3	$Q_{f\ (correl)}$ cm^3/s	$Q_{a\ (correl)}$ cm^3/s	k_{ef} cm^2	k_{ea} cm^2	k_{rf}	k_{ra}	Satur., s	Rel. sat., s_r $(s-s_\infty)/(1-s_\infty)$
0.00	0	10.35	0.00	0.0000	0.0000	0.0000	0.0000					1.000	1.000
2.04	67.2	10.83	0.48	40.2219	0.6150	0.0904	11.1728	5.23255E−11	1.20981E−10	0.228	0.528	0.909	0.859
4.07	56.49	11.11	0.76	45.4920	0.7554	0.0549	12.6367	3.17789E−11	1.36832E−10	0.139	0.597	0.856	0.777
6.10	50.64	11.18	0.83	46.2319	0.8511	0.0410	12.8422	2.37278E−11	1.39058E−10	0.103	0.606	0.843	0.757
8.13	48.31	11.24	0.89	46.3770	0.9259	0.0333	12.8825	1.92833E−11	1.39494E−10	0.084	0.608	0.832	0.739
10.17	46.54	11.34	0.99	46.4441	0.9885	0.0283	12.9012	1.64052E−11	1.39696E−10	0.072	0.609	0.813	0.710
12.20	46.54	11.39	1.04	46.5008	1.0423	0.0248	12.9169	1.43851E−11	1.39866E−10	0.063	0.610	0.803	0.695
14.23	46.92	11.44	1.09	46.5560	1.0899	0.0222	12.9322	1.28722E−11	1.40033E−10	0.056	0.611	0.794	0.681
16.26	47.04	11.44	1.09	46.6111	1.1329	0.0202	12.9475	1.16906E−11	1.40198E−10	0.051	0.611	0.794	0.681
18.29	47.04	11.49	1.14	46.6662	1.1722	0.0185	12.9628	1.07386E−11	1.40364E−10	0.047	0.612	0.784	0.666
20.33	47.17	11.54	1.19	46.7216	1.2086	0.0172	12.9782	9.94926E−12	1.40531E−10	0.043	0.613	0.775	0.651
22.36	46.41	11.54	1.19	46.7766	1.2423	0.0160	12.9935	9.28854E−12	1.40696E−10	0.041	0.614	0.775	0.651
24.39	46.92	11.61	1.26	46.8317	1.2738	0.0151	13.0088	8.72367E−12	1.40862E−10	0.038	0.614	0.762	0.631
26.42	46.79	11.68	1.33	46.8868	1.3035	0.0142	13.0241	8.23437E−12	1.41027E−10	0.036	0.615	0.749	0.610
28.46	46.41	11.68	1.33	46.9421	1.3318	0.0135	13.0395	7.80384E−12	1.41194E−10	0.034	0.616	0.749	0.610
30.49	47.04	11.73	1.38	46.9972	1.3585	0.0128	13.0548	7.42514E−12	1.41360E−10	0.032	0.617	0.739	0.596
32.52	47.04	11.73	1.38	47.0523	1.3839	0.0122	13.0701	7.08751E−12	1.41525E−10	0.031	0.617	0.739	0.596
34.55	46.41	11.78	1.43	47.1073	1.4082	0.0117	13.0854	6.78432E−12	1.41691E−10	0.030	0.618	0.730	0.581
36.58	46.67	11.78	1.43	47.1624	1.4315	0.0112	13.1007	6.51035E−12	1.41857E−10	0.028	0.619	0.730	0.581
38.62	47.17	11.78	1.43	47.2178	1.4540	0.0108	13.1160	6.26019E−12	1.42023E−10	0.027	0.619	0.730	0.581
40.65	47.04	11.83	1.48	47.2728	1.4756	0.0104	13.1313	6.03288E−12	1.42189E−10	0.026	0.620	0.720	0.566
42.68	46.92	11.83	1.48	47.3279	1.4963	0.0101	13.1466	5.82430E−12	1.42354E−10	0.025	0.621	0.720	0.566
44.71	46.41	11.88	1.53	47.3830	1.5164	0.0097	13.1619	5.63215E−12	1.42520E−10	0.025	0.622	0.711	0.552
46.75	47.93	11.93	1.58	47.4383	1.5360	0.0094	13.1773	5.45361E−12	1.42686E−10	0.024	0.622	0.701	0.537

(continued)

Table 9.13 (continued)

Blowing stage													
t_2 s	Q_{air} l/h	$V_{f\ (tot)}$ cm^3	$V_{f\ (blowing)}$ cm^3	$Q_{air\ (correl)}$ l/h	$V_{f\ (correl)}$ cm^3	$Q_{f\ (correl)}$ cm^3/s	$Q_{a\ (correl)}$ cm^3/s	k_{ef} cm^2	k_{ea} cm^2	k_{rf}	k_{ra}	Satur., s	Rel. sat., s_r $(s-s_\infty)/(1-s_\infty)$
48.78	47.67	12.02	1.67	47.4934	1.5548	0.0091	13.1926	5.28878E−12	1.42852E−10	0.023	0.623	0.684	0.511
50.81	47.93	12.02	1.67	47.5485	1.5731	0.0089	13.2079	5.13536E−12	1.43018E−10	0.022	0.624	0.684	0.511
52.84	48.05	12.02	1.67	47.6035	1.5908	0.0086	13.2232	4.99214E−12	1.43183E−10	0.022	0.624	0.684	0.511
54.87	47.8	12.01	1.66	47.6586	1.6081	0.0084	13.2385	4.85810E−12	1.43349E−10	0.021	0.625	0.686	0.514
56.91	48.05	12.02	1.67	47.7140	1.6250	0.0082	13.2539	4.73173E−12	1.43516E−10	0.021	0.626	0.684	0.511
58.94	48.31	12.06	1.71	47.7690	1.6414	0.0080	13.2692	4.61349E−12	1.43681E−10	0.020	0.627	0.677	0.499
60.97	48.31	12.06	1.71	47.8241	1.6573	0.0078	13.2845	4.50207E−12	1.43847E−10	0.020	0.627	0.677	0.499
63.00	48.05	12.06	1.71	47.8792	1.6729	0.0076	13.2998	4.39686E−12	1.44013E−10	0.019	0.628	0.677	0.499
65.04	47.93	12.06	1.71	47.9345	1.6883	0.0074	13.3151	4.29685E−12	1.44179E−10	0.019	0.629	0.677	0.499
67.07	47.93	12.11	1.76	47.9896	1.7032	0.0073	13.3304	4.20255E−12	1.44345E−10	0.018	0.630	0.667	0.484
69.10	48.18	12.11	1.76	48.0447	1.7177	0.0071	13.3457	4.11304E−12	1.44510E−10	0.018	0.630	0.667	0.484
71.13	48.05	12.11	1.76	48.0998	1.7320	0.0070	13.3610	4.02795E−12	1.44676E−10	0.018	0.631	0.667	0.484
73.16	47.93	12.11	1.76	48.1548	1.7460	0.0068	13.3763	3.94694E−12	1.44842E−10	0.017	0.632	0.667	0.484
75.20	48.05	12.11	1.76	48.2102	1.7598	0.0067	13.3917	3.86934E−12	1.45008E−10	0.017	0.632	0.667	0.484
77.23	47.93	12.11	1.76	48.2652	1.7732	0.0066	13.4070	3.79564E−12	1.45174E−10	0.017	0.633	0.667	0.484
79.26	48.44	12.11	1.76	48.3203	1.7864	0.0064	13.4223	3.72520E−12	1.45339E−10	0.016	0.634	0.667	0.484
81.29	48.05	12.11	1.76	48.3754	1.7993	0.0063	13.4376	3.65779E−12	1.45505E−10	0.016	0.635	0.667	0.484
83.33	47.55	12.11	1.76	48.4307	1.8121	0.0062	13.4530	3.59292E−12	1.45671E−10	0.016	0.635	0.667	0.484
85.36	47.67	12.15	1.80	48.4858	1.8246	0.0061	13.4683	3.53102E−12	1.45837E−10	0.015	0.636	0.660	0.473
87.39	48.05	12.15	1.80	48.5409	1.8369	0.0060	13.4836	3.47161E−12	1.46003E−10	0.015	0.637	0.660	0.473
89.42	47.67	12.15	1.80	48.5960	1.8489	0.0059	13.4989	3.41452E−12	1.46168E−10	0.015	0.637	0.660	0.473
91.46	48.05	12.2	1.85	48.6513	1.8609	0.0058	13.5143	3.35936E−12	1.46335E−10	0.015	0.638	0.650	0.458
93.49	49.03	12.2	1.85	48.7064	1.8726	0.0057	13.5296	3.30654E−12	1.46501E−10	0.014	0.639	0.650	0.458
95.52	49.47	12.2	1.85	48.7615	1.8841	0.0056	13.5448	3.25565E−12	1.46666E−10	0.014	0.640	0.650	0.458
97.55	49.17	12.2	1.85	48.8165	1.8954	0.0055	13.5601	3.20659E−12	1.46832E−10	0.014	0.640	0.650	0.458

(continued)

Table 9.13 (continued)

Blowing stage													
t_2 s	Q_{air} l/h	$V_{f\ (tot)}$ cm^3	$V_{f\ (blowing)}$ cm^3	$Q_{air\ (correl)}$ l/h	$V_{f\ (correl)}$ cm^3	$Q_{f\ (correl)}$ cm^3/s	$Q_{a\ (correl)}$ cm^3/s	k_{ef} cm^2	k_{ea} cm^2	k_{rf}	k_{ra}	Satur., s	Rel. sat., s_r $(s-s_\infty)/(1-s_\infty)$
99.58	49.17	12.24	1.89	48.8716	1.9065	0.0055	13.5754	3.15926E−12	1.46998E−10	0.014	0.641	0.643	0.446
101.62	49.47	12.24	1.89	48.9270	1.9176	0.0054	13.5908	3.11334E−12	1.47164E−10	0.014	0.642	0.643	0.446
103.65	49.17	12.25	1.90	48.9820	1.9284	0.0053	13.6061	3.06920E−12	1.47330E−10	0.013	0.643	0.641	0.443
105.68	49.61	12.24	1.89	49.0371	1.9391	0.0052	13.6214	3.02652E−12	1.47495E−10	0.013	0.643	0.643	0.446
107.71	49.61	12.28	1.93	49.0922	1.9497	0.0052	13.6367	2.98523E−12	1.47661E−10	0.013	0.644	0.635	0.434
109.75	49.32	12.29	1.94	49.1475	1.9601	0.0051	13.6521	2.94506E−12	1.47827E−10	0.013	0.645	0.633	0.432
111.78	49.03	12.29	1.94	49.2026	1.9704	0.0050	13.6674	2.90635E−12	1.47993E−10	0.013	0.645	0.633	0.432
113.81	49.32	12.34	1.99	49.2577	1.9805	0.0050	13.6827	2.86882E−12	1.48159E−10	0.013	0.646	0.624	0.417
115.84	49.32	12.34	1.99	49.3127	1.9905	0.0049	13.6980	2.83244E−12	1.48324E−10	0.012	0.647	0.624	0.417
117.87	49.76	12.34	1.99	49.3678	2.0004	0.0048	13.7133	2.79713E−12	1.48490E−10	0.012	0.648	0.624	0.417
119.91	49.03	12.34	1.99	49.4232	2.0102	0.0048	13.7287	2.76269E−12	1.48657E−10	0.012	0.648	0.624	0.417
120.57	49.03	12.34	1.99	49.4411	2.0133	0.0048	13.7336	2.75177E−12	1.48710E−10	0.012	0.649	0.624	0.417

$$k(\varepsilon) = \frac{1}{b} \times \frac{\mu}{2S^2\Delta p} \times \frac{\varphi_0}{1-\varphi_0-\varepsilon_0}$$
$$= \frac{1}{0.2848} \times \frac{0.01}{2 \times 6.55^2 \times 6 \times 10^6} \frac{0.381}{1-0.381-0.506}$$
$$= 2.293 \times 10^{-10}\,\text{cm}^2$$

$$R_\text{m} = a \times \frac{\Delta pS}{\mu} = 4.7322 \frac{6 \times 10^6 \times 6.55}{0.01} = 1.860 \times 10^{+10}\,\text{cm}^{-1}$$

Relative permeability

The first four columns of Table 9.13 show data for the blowing stage. The plot of the filtrate volume versus time for the blowing stage gives: (see Figs. 9.31, 9.32, 9.33, 9.34):

$$V_f = -0.0482 + 0.544\, t^{0.278} \quad Q_f(t) = 0.1512\, t^{-0.722}$$
$$Q_a(t) = 46.17(1 - 1/\exp(t)) + 0.02713 \times t$$

The effective and relative permeabilities of the filtrate and of the air are given by:

$$k_{ea}(t) = \left(\frac{\ell}{\Delta p}\right) \times \left(\frac{\mu_f Q_f(t)}{S}\right) \quad \text{and} \quad k_{ra} = \frac{k_{ea}}{k(\varepsilon)}$$
$$k_a(t) = \left(\frac{\ell}{\Delta p}\right) \times \left(\frac{\mu_a Q_a(t)}{S}\right) \quad \text{and} \quad k_{ra}(t) = \frac{k_{ra}}{k(\varepsilon)}$$

The saturation is given by:

$$s = \frac{V_{\text{f-suspension}} - V_{\text{f-total}}(t)}{V_{\text{f-suspension}} - \left(V_{\text{f-cakeformation}} + V_{\text{f-expression}}\right)} = \frac{15.64 - V_{\text{f-total}}(t)}{15.64 - (5.79 + 4.56)}$$
$$= \frac{15.64 - V_{\text{f-total}}(t)}{5.29},$$

The residual saturation is $s_\infty = 0.535$, so that the reduced saturation becomes:

$$s_r = \frac{s - s_\infty}{1 - s_\infty} = \frac{s - 0.535}{1 - 0.535} = \frac{s - 0.535}{0.465}$$

The relative permeabilities can now be correlated with the reduced saturation in the form (see Fig. 9.35):

$$k_{rf}(s_r) = \exp\left(-5.50\left(1 - s_r^2\right)\right) \quad k_{ra}(s_r) = \frac{0.67(1 - s_r)}{1 - 0.92 \times s_r - 0.045 \times s_r^2}.$$

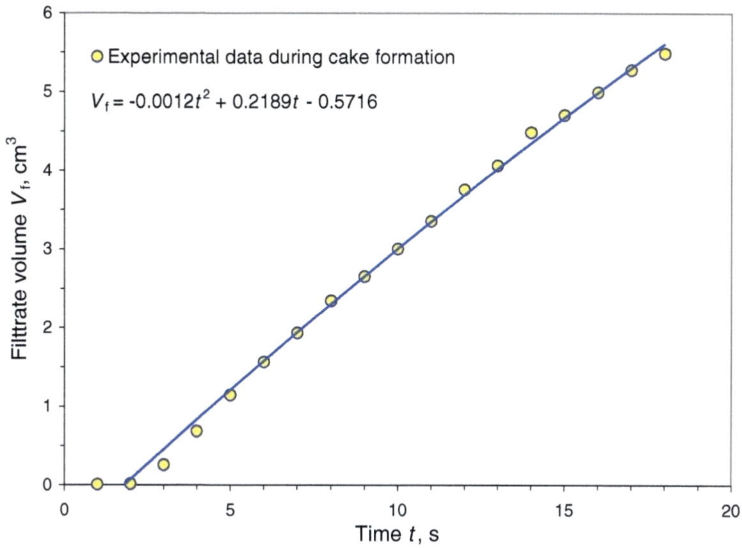

Fig. 9.29 Filtrate volume versus time for the cake formation stage

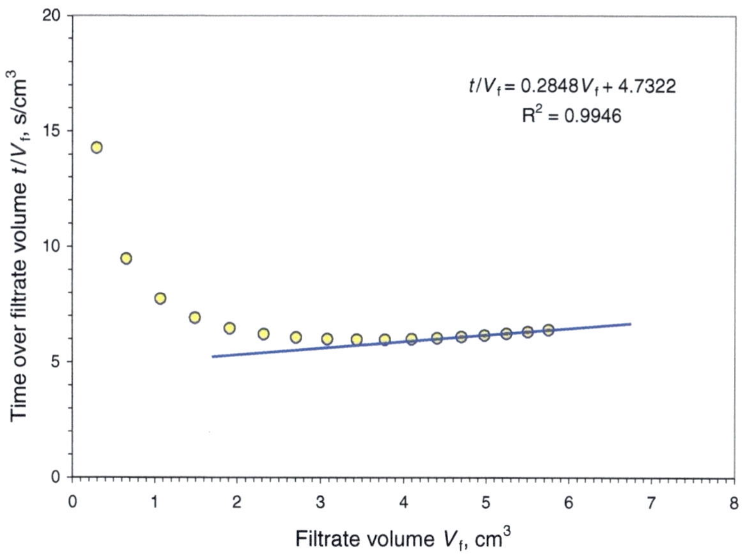

Fig. 9.30 Plot of t/V_f versus V_f for a copper tailings

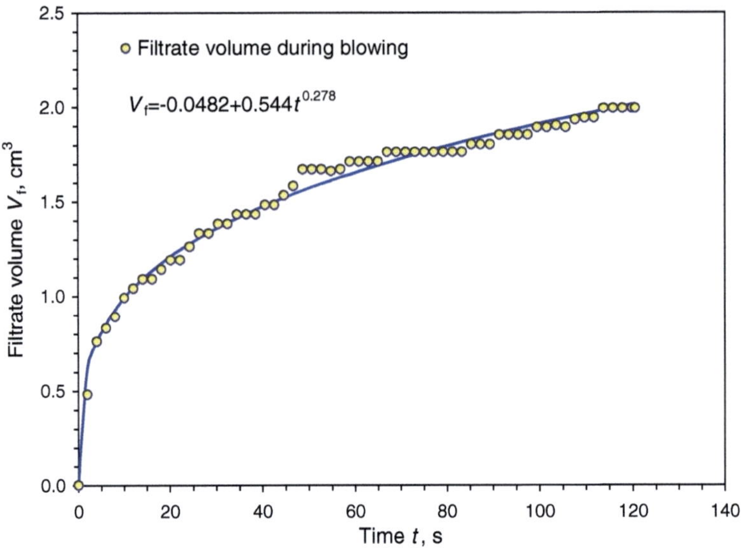

Fig. 9.31 Plot of V_f versus time for the blowing cycle

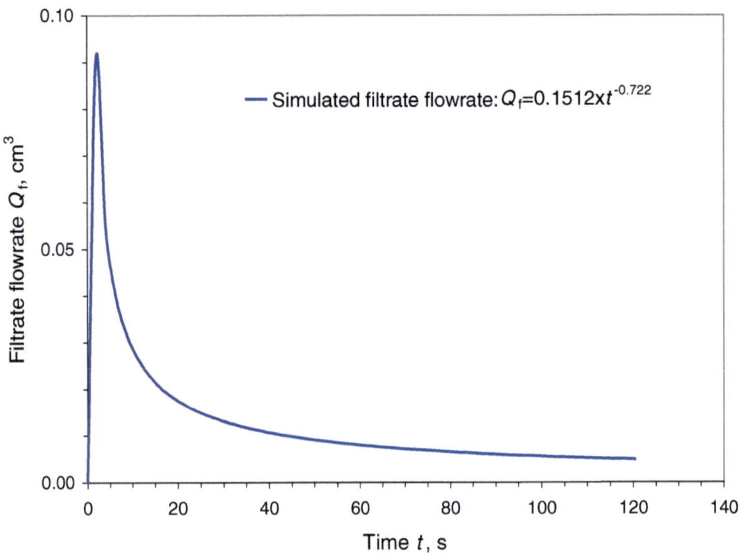

Fig. 9.32 Filtrate flowrate during the blowing stage

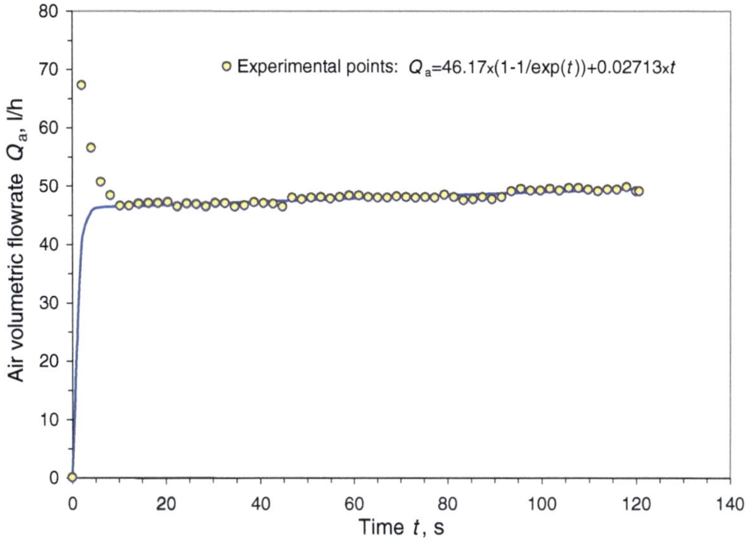

Fig. 9.33 Air flowrate in the blowing stage

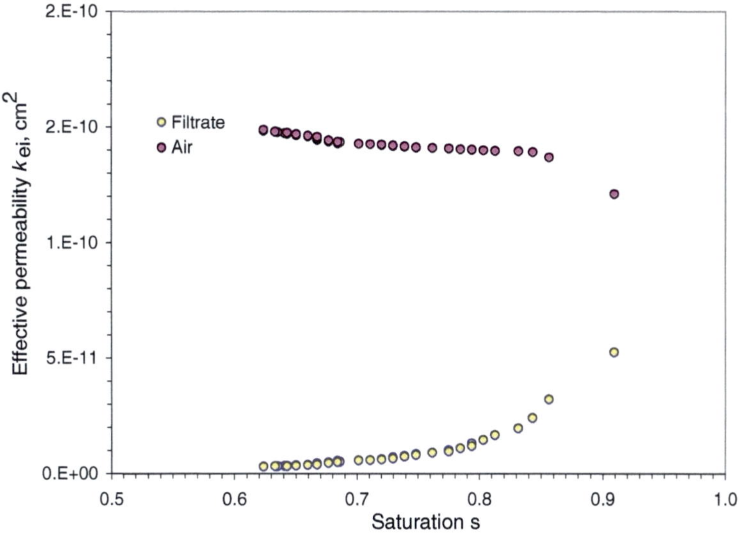

Fig. 9.34 Effective filtrate and air permeabilities

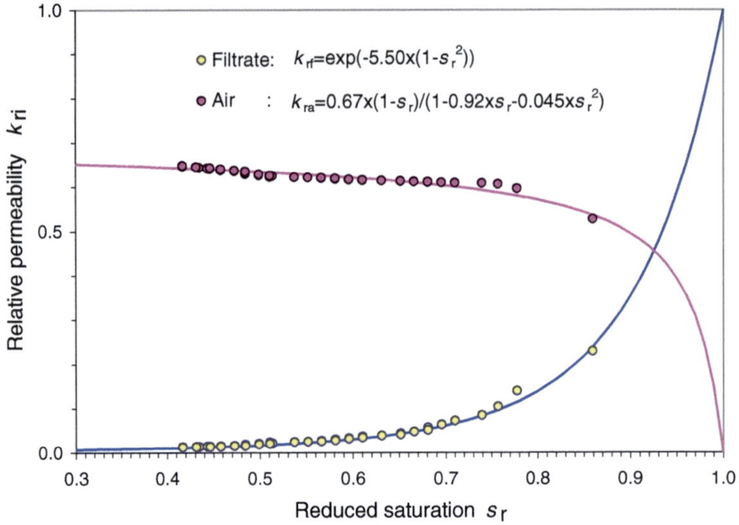

Fig. 9.35 Relative permeability versus reduced saturation

9.5 Continuous Modeling

9.5.1 Vacuum Filters

Vacuum filtration is an important process in the mining industry and for the first half of the 20th century it was the only filtration process in practice.

In this type of filtration, pulp feeds a tank in which a filter is submerged. The side of the filter in contact with the feed is at atmospheric pressure while the other side connects to a vacuum pump that provides a pressure drop across the filter. The difference in pressure drives the filtrate across the filter medium leaving the solid adhering in the form of a cake. The maximum theoretical pressure available for this type of filter is one atmosphere, but the pressure drop in the piping system and, especially, the height of the location with respect to sea level drastically diminishes this value, often reaching 0.6–0.8 bars. Thus vacuum filtration operates at low and constant pressure drop and therefore the filter cake produced is incompressible.

Rotary drum, disc, band and pan filters operate under vacuum. The selection of the appropriate filter for a given duty depends principally on the material to be filtered. Pulps with large solid particles, such as potassium chloride or potassium sulphide crystals, are difficult to maintain in suspension and therefore band filters are best, while disc or drum filters are appropriate for pulps of fine material.

One of the main costs of using filter cloth is the air consumption with power demands on the order of 2–15 kW/m^2 of filter area (Henriksson 2000). An alternative to filter cloth is ceramic plates made of alumina. These filters are homogeneous and have small pores on the order of 2 mμ, which produce capillary suction according to the Young–Laplace equation (see Sect. 6.5.1 in this book).

This action permits fluid flow through the filter medium with minimum help from a vacuum, requiring only about 0.05 kW/m^2 of filter area (Henriksson 2000). The principal problem with ceramic filters is blockage of the pores with small solid particles that must be removed with acid leaching.

Rotary Filter Model

Consider a rotary vacuum filter, for example a disc filter, rotating at N rpm with cake formation and dehumidification stages. The time t_R, in minutes, for each revolution is $t_R = 1/N$. If the fraction of this time used during the cake formation is I, then, the time for the cake formation stage t_0 and the time for dehumidification t_3 are:

cake formation stage:

$$t_0 = I \times t_R = \frac{I}{N}, \quad \min/\text{rev} \tag{9.47}$$

dehumidification stage

$$t_3 = (1 - I) \times t_R = \frac{1 - I}{N}, \quad \min/\text{rev} \tag{9.48}$$

Assume that the filter is submerged in the suspension forming an arc with an angle of θ, as is shown in Fig. 9.36. The fraction of time I in cake formation must be equal to the fraction of filter area submerged in the suspension, that is $I = \theta/360$, then, Eqs. (9.47) and (9.48) become:

$$t_0 = \frac{\theta}{360} \times \frac{1}{N} \quad \text{and} \quad t_3 = \left(1 - \frac{\theta}{360}\right) \times \frac{1}{N}, \quad \min/\text{rev} \tag{9.49}$$

(a) **Cake formation**

The relationship between the production of filtrate and the cake formation time is given by Eq. (9.16), therefore:

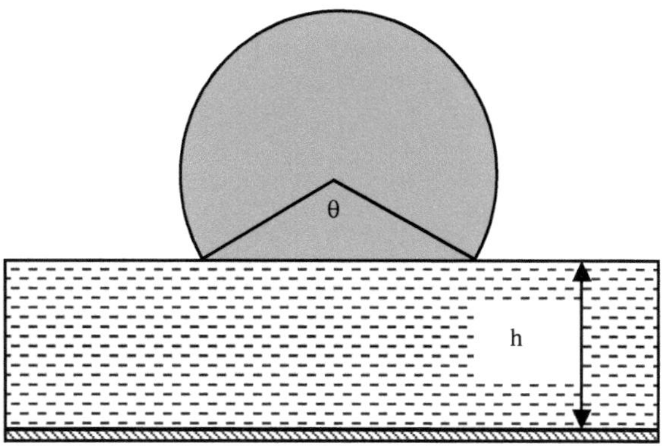

Fig. 9.36 Rotary vacuum filter

$$V_f^2(t_0) + 2Sk(\varepsilon_0)R_m \frac{1-\varphi_0-\varepsilon_0}{\varphi_0} V_f(t_0) - 2S^2k(\varepsilon_0)\frac{\Delta p}{\mu}\frac{1-\varphi_0-\varepsilon_0}{\varphi_0} t_0 = 0$$

Substituting (9.49) in this equation yields the volume of filtrate per revolution:

$$V_f^2(N) + 2Sk(\varepsilon_0)R_m \frac{1-\varphi_0-\varepsilon_0}{\varphi_0} V_f(N) - 2S^2k(\varepsilon_0)\frac{\Delta p}{\mu}\frac{1-\varphi_0-\varepsilon_0}{\varphi_0}\frac{\theta}{360\ N}$$

Rearranging:

$$\left(\frac{S}{N}\right)^2\left(\frac{V_f(N)N}{S}\right)^2 + 2Sk(\varepsilon_0)R_m\frac{1-\varphi_0-\varepsilon_0}{\varphi_0}\left(\frac{S}{N}\right)\left(\frac{V_f(N)N}{S}\right) - 2S^2k(\varepsilon_0)\frac{\Delta p}{\mu}\frac{1-\varphi_0-\varepsilon_0}{\varphi_0}\frac{\theta}{360\ N} = 0$$

Dividing by S^2/N^2 gives:

$$\left(\frac{V_f(N)N}{S}\right)^2 + 2Nk(\varepsilon_0)R_m\frac{1-\varphi_0-\varepsilon_0}{\varphi_0}\left(\frac{V_f(N)N}{S}\right) - \frac{2\ k(\varepsilon_0)}{\mu}\frac{1-\varphi_0-\varepsilon_0}{\varphi_0}\frac{\theta\ N}{360} = 0$$

Filtrate flow rate

The term $V_f(N)N/S = q_0$ is the filtrate flow rate produced per unit area during N revolutions. Substituting the previous equation yields an equation for the production of filtrate during the cake formation time

$$q_0^2(t) + 2Nk(\varepsilon_0)R_m\frac{1-\varphi_0-\varepsilon_0}{\varphi_0}q_0(t) - \frac{k(\varepsilon_0)N\Delta p}{\mu}\frac{1-\varphi_0-\varepsilon_0}{\varphi_0}\frac{\theta}{180} = 0 \quad (9.50)$$

The solution to this equation is:

$$q_0(t) = Nk(\varepsilon_0)\frac{1-\varphi_0-\varepsilon_0}{\varphi_0}\left\{\left(R_m^2 + \frac{\Delta p}{\mu k(\varepsilon_0)}\frac{\varphi_0}{1-\varphi_0-\varepsilon_0}\frac{\theta}{180\ N}\right)^{1/2} - R_m\right\} \quad (9.51)$$

Mass flow rate

The mass flow rate relates to the filtrate flow rate by Eq. (9.14), that is

$$m_s(t) = \frac{\rho_s\varphi_0(1-\varepsilon_0)}{(1-\varphi_0-\varepsilon_0)}Sq_0(t)$$

Substituting Eq. (9.51) yields:

$$m_s(t) = \rho_s(1-\varepsilon_0)k(\varepsilon_0)SN\left\{\left(R_m^2 + \frac{\Delta p}{\mu k(\varepsilon_0)}\frac{\varphi_0}{1-\varphi_0-\varepsilon_0}\frac{\theta}{180\,N}\right)^{1/2} - R_m\right\} \quad (9.52)$$

Neglecting the filter media resistance, Eqs. (9.54) and (9.52) become:

$$q_0(t) = \left(\frac{k(\varepsilon_0)\Delta p}{\mu}\frac{1-\varphi_0-\varepsilon_0}{\varphi_0}\frac{\theta\,N}{180}\right)^{1/2} \quad (9.53)$$

$$m_s(t) = \rho_s(1-\varepsilon_0)S\left(\frac{k(\varepsilon_0)\Delta p}{\mu}\frac{1-\varphi_0-\varepsilon_0}{\varphi_0}\frac{\theta\,N}{180}\right)^{1/2} \quad (9.54)$$

Equations (9.51)–(9.54) permit the calculation of a rotary vacuum filter capacity. In these equations φ_0 is the suspension volume fraction, ε_0 and $k(\varepsilon_0)$ are the filter cake porosity and permeability during the cake formation stage, ρ_s is the solid particle density, μ is the filtrate viscosity, Δp is the total pressure drop across the filter, N is the filter velocity in rpm and θ is the angle subtended by the part of the filter submerge in the suspension.

It is useful to separate the variables in Eqs. (9.53) and (9.54) that are properties of the suspension, the cake and the filtrate from operating variables:

$$q_0(t) = \underbrace{\left(\frac{1-\varphi_0-\varepsilon_0}{\varphi_0}\right)^{1/2} \times (k(\varepsilon_0))^{1/2}}_{\text{Suspension and cake properties}} \times \underbrace{\left(\frac{1}{\mu}\right)^{1/2}}_{\text{Filtrate property}} \times \underbrace{\left(\Delta p\frac{\theta}{180}N\right)^{1/2}}_{\text{Operational variables}} \quad (9.55)$$

$$m_s(\varepsilon_0, t) = S\underbrace{\left(\frac{1-\varphi_0-\varepsilon_0}{\varphi_0}\right)^{1/2}(k(\varepsilon_0))^{1/2}}_{\text{Suspension and cake properties}} \times \underbrace{\left(\frac{1}{\mu}\right)^{1/2}}_{\text{Filtrate property}} \times \underbrace{\rho_s}_{\text{Solidproperty}} \times \underbrace{\left(\Delta p\frac{\theta}{180}N\right)^{1/2}}_{\text{Operational variables}} \quad (9.56)$$

Cake thickness

The cake thickness is related to its mass by Eq. (9.13):

$$\ell = \frac{m_s(\varepsilon_0, t)}{\rho_s S(1-\varepsilon_0)} \quad (9.57)$$

Replacing Eq. (9.57) with Eqs. (9.56) or (9.52), depending whether or not the filter medium resistance is omitted, yields:

$$\ell(\varepsilon_0, t) = k(\varepsilon_0) \times \left\{\left(R_m^2 + \frac{\Delta p}{\mu k(\varepsilon_0)}\frac{\varphi_0}{1-\varphi_0-\varepsilon_0}\frac{\theta}{180\,N}\right)^{1/2} - R_m\right\} \quad (9.58)$$

$$\ell(\varepsilon_0, t) = \left(\frac{\varphi_0}{1 - \varphi_0 - \varepsilon_0}\right)^{1/2} \times (k(\varepsilon_0))^{1/2} \times \left(\frac{\Delta p \theta}{180\,N}\right)^{1/2} \tag{9.59}$$

(b) **Cake dehumidification**

The cake dehumidification takes place by vacuum suction of the filtrate in the interior of the equipment from the atmosphere at the surface of the filter cake. The suction time t_3 is given by Eqs. (9.36) and (9.49), then:

$$\left(1 - \frac{\theta}{360}\right)\frac{1}{N} = \frac{\mu_\ell \varepsilon}{\Delta p}\left(R_m(1 - s) + \frac{\ell}{k(\varepsilon_0)}\int_s^1 \frac{d\eta}{k_{r\ell}(\eta, s_\infty)}\right) \tag{9.60}$$

This equation relates the rotation speed to the final saturation. The air flow during the dehumidification stage is given by Eq. (9.37):

$$Q_a(t) = \left(\frac{S\Delta p}{\mu_a}\right)\left(R_m + \frac{\ell}{k(\varepsilon_0)k_{ra}(s, s_\infty)}\right)^{-1} \tag{9.61}$$

In these equations $k_{r\ell}(s)$ and $k_{ra}(s)$ are the relative permeabilities of the liquid and air respectively.

Problem 9.12 A rotary drum filter with 3 m^2 of area operates at a speed of 0.5 rpm, an internal pressure of 30 kN/m^2 and 30 % of its surface area submerged. The atmospheric pressure is 1.013×10^5 N/m^2. Calculate the filter capacity and the final humidity of the cake. The cake is incompressible and the filter cloth resistance is equivalent to the resistance of a 1 mm filter cake. The following data are known:

Submerged angle	$\theta = 120°$
Surface area	$S = 3\text{m}^2$
Rotational speed	$N = 0.5$ rpm
Atmospheric pressure	1.013×10^5 N/m^2
Internal pressure	30 kN/m^2
Suspension concentration	$N = 0.5$ rpm
Cake porosity	$\varepsilon_0 = 0.40$
Cake permeability	$k(\varepsilon_0) = 5 \times 10^{-13}\text{m}^2$
Cake thickness	$\ell = 2.3$ cm
Solid density	$\rho_s = 2{,}000\ \text{kg/m}^3$
Filtrate density	$\rho_\ell = 1{,}000\ \text{kg/m}^3$
Filtrate viscosity	$\mu_a = 1 \times 10^{-3}$ Pa - s
Air viscosity	$\mu_a = 1.85 \times 10^{-5}$ Pa - s
Liquid relative permeability;	$k_{r\ell} = \exp(4.466(s_r^2 - 1))$
Air relative permeability	$k_{ra} = \frac{0.5521 \times (1 - s_r)}{1 - 0.9155 \times s_r - 0.07643 \times s_r^2}$

Results:

Pressure drop	$\Delta p = 1.013 \times 10^5 - 0.3 \times 10^5 = 0.713 \times 10^5\ \mathrm{N/m^2}$
Suspension concentration	$\varphi_0 = 20 \times 1{,}000/(2{,}000 \times (100-20)+20) = 0.111$
Filter medium resistance	$R_m = \frac{1\times10^{-3}}{k(\varepsilon_0)} = \frac{1\times10^{-3}}{5\times10^{-13}} = 2 \times 10^9\ \mathrm{m^{-1}}$
Rotation time	$t_R = \frac{1}{N} = \frac{60}{0.5} = 120\ \mathrm{s}$

Filtration time:

Cake formation time	$t_1 = \frac{\theta}{360} t_R = \frac{120}{360} \times 120 = 40\ \mathrm{s}$
Dehumidification time	$t_3 = \left(1 - \frac{\theta}{360}\right) \times t_R = \left(1 - \frac{120}{360}\right) \times 120 = 80\ \mathrm{s}$

Capacity:

Filter capacity given by (9.52)

$$m_s = \rho_s \times (1-\varepsilon_0) \times k(\varepsilon_0) \times S \times N \left\{ \left(R_m^2 + \frac{\Delta p}{\mu_\ell \times k(\varepsilon_0)} \times \frac{\varphi_0}{1-\varphi_0-\varepsilon_0} \times \frac{\theta}{180 \times N} \right)^{1/2} - R_m \right\}$$

$$= 2{,}000 \times (1-0.4) \times 5 \times 10^{-13} \times 3 \times (0.5/60)$$

$$\times \left\{ \left((2\times10^9)^2 + \frac{0.713 \times 10^5}{1\times10^{-3} \times 5 \times 10^{-13}} \frac{0.111}{1-0.111-0.4} \times \frac{120}{180 \times (0.5/60)} \right)^{1/2} - 2\times10^9 \right\}$$

$$= 0.731\ \mathrm{kg/s}$$

$$= 0.731 \times 3{,}600/1{,}000 = 2.64\ \mathrm{tph}$$

$$= 2.64 \times 24 = 63.4\ \mathrm{tpd}$$

Cake thickness:

From Eq. (9.57): $\ell = \frac{m_s}{\rho_s(1-\varepsilon_0)SN} = \frac{0.792}{2{,}000 \times (1-0.4) \times 3 \times 0.5/60 \times 100} = 2.64\ \mathrm{cm}$

Filtrate flow: $Q_f = \frac{m_s}{\rho_s} \frac{1-\varphi_0-\varepsilon_0}{\varphi_0} \frac{1}{1-\varepsilon_0} = 0.0025\ \mathrm{m^3/s}$

Cake humidity:

Equation (9.34) relates the filtrate flow to the relative permeability:

$$\frac{\Delta p}{\mu_\ell} = \left(R_m + \frac{\ell}{k(\varepsilon_0) \times k \times k_{r\ell}(s)} \right) q_\ell$$

The relative permeability is then:

$$k_{r\ell}(s) = \frac{\ell}{(\Delta p/\mu_\ell Q_\ell - R_m) \times k(\varepsilon)} = \frac{2.64}{\left(\frac{7.13 \times 10^5 \times 3}{10^{-3} \times 2.5 \times 10^{-3}} - 2 \times 10^9\right) \times 5 \times 10^{-13}}$$

$$= 0.635$$

If the constitutive equation of the relative permeability of the filtrate is given by:

$$k_{r\ell}(s_r) = \exp\{4.4662 \times (s_r^2 - 1)\}$$

Then, the reduced saturation is:

$$s_r = \left(\frac{\ln k_{r\ell}(s_r)}{4.4662} + 1\right)^{1/2} = \left(\frac{\ln 0.635}{4.4662} + 1\right)^{1/2} = \left(\frac{-0.4547}{4.4662} + 1\right)^{1/2} = 0.948$$

The final saturation is:

$$s = s_\infty + s_r \times (1 - s_\infty) = 0.535 + 0.948 \times (1 - 0.535) = 0.976,$$

and the final humidity is:

$$\begin{aligned} h &= 100 \times \frac{\rho_f \varepsilon s}{\rho_f \varepsilon s + \rho_s(1-\varepsilon)} \\ &= 100 \times \frac{1{,}000 \times 0.4 \times 0.976}{1{,}000 \times 0.4 \times 0.976 + 2{,}000 \times (1 - 0.4)} = 24.5\ \% \end{aligned}$$

Air flow

The relative air permeability is given by:

$$\begin{aligned} k_{ra} &= \frac{0.5521 \times (1 - s_r)}{1 - 0.91554 \times s_r - 0.076429 \times s_r^2} \\ &= \frac{0.5521 \times (1 - 0.948)}{1 - 0.91554 \times 0.948 - 0.076429 \times (0.948)^2} = 0.453 \end{aligned}$$

The airflow through the filter cake is given by:

$$\begin{aligned} Q_a &= \left(\frac{S\Delta pN}{\mu_a}\right) \times \left(R_m + \frac{\ell}{k(\varepsilon)k_{ra}(s)}\right)^{-1} \\ &= \left(\frac{3 \times 7.1 \times 10^5 \times 0.5/60}{1.85 \times 10^{-4}}\right) \times \left(2 \times 10^9 + \frac{2.30}{5 \times 10^{-13} \times 0.453}\right)^{-1} \\ &= 9.31 \times 10^{-4} \mathrm{m}^3/\mathrm{s} \\ &= 3.35\ \mathrm{m}^3/\mathrm{h} \\ &= 80.4\ \mathrm{m}^3/\mathrm{d} \end{aligned}$$

9.5.2 Pressure Filters

As we said in Sect. 9.2.2, pressure filters work in a semi-continuous manner, that is, with filter cycles.

Pulp is fed to the filtration chambers from a manifold with as many rubber hoses as there are filtration chambers. The chambers are closed spaces between the filtration cloth and the rubber diaphragm. As soon as all the chambers are closed, the pulp is pumped and distributed evenly across the horizontal surface of the filter cloth in all the chambers. When the chambers are full, air is pumped to the other side of the diaphragms. Although filtration starts when the pulp begins to enter the chambers, the cake formation continues until no further filtrates comes out of the cake. If the material is compressible, the diaphragm pressure reduces the cake porosity in an expression stage. Once cake formation and expression are finished, air is blown through the cake to displace the water retained in the pores. Finally, automatic mechanisms open the set of filtration plates to discharge the cake. The table below shows a typical filtration cycle for a Larox pressure filter (Droguett 2000).

Stage	Function	Time (s)
1	Closing the plates	50
2	Filling the filtration chambers	85
3	Washing the feeding tube	13
4	Washing the feeding hoses	50
5	Pressing and expression	80
6	Compressed air release	5
7	Opening and closing the outlet tubes	30
8	Cake blowing	100
9	Pressure drainage	10
10	Opening the feeding valve	1
11	Cake discharge	18
12	Filter cloth washing	72
14	Total cycle	514

The table shows that time is divided into *effective filtration time,* cake formation t_1, cake pressing t_2 and cake blowing t_3 and time that is not used for filtration, which is called *dead time* t_4. The filtration time is 265 s and the dead time is 249 s. Optimizing a filtration system should shorten filtration time and especially dead time.

Problem 9.13 With the data from problem 9.11, calculate the capacity, in tpd, of a horizontal pressure filter with 144 m^2 in filter area, to produce a copper concentrate cake, $\rho_s = 4{,}500\ kg/m^3$, with 8.5 % humidity. The feeding and cake formation takes $t_1 = 85$ s at $\Delta p = 6$ bars producing a cake of 3.0 cm. An expression stage at 7.5 bars takes $t^2 = 75$ s and is followed by a blowing stage at 6 bars.

From problem 9.11 we have the following data and parameters:

Cake formation stage:

$$\Delta p = 6 \times 10^6 \text{ N/m}^2$$
$$\ell = 3.0 \text{ cm}$$
$$\varepsilon_0 = 0.520$$
$$R_m = 9.80 \times 10^9 \text{ cm}^{-1}$$
$$k(\varepsilon_0) = 1.55 \times 10^{-10} \text{ cm}^2$$
$$S = 144\text{m}^2$$

Expression stage:

$$\Delta p = 74.5 \times 10^6 \text{N/m}^2$$

Reduction of porosity from $\varepsilon_0 = 0.520$ to $\varepsilon = 0.362$

Blowing stage:

$$\Delta p = 6 \times 10^6 \text{ N/m}^2$$
$$\varepsilon_1 = 0.362$$

Reduced permeability correlations:

$$k_\ell = \exp\{4.4662(s_r^2 - 1)\}$$
$$k_a = \frac{0.5521(1 - s_r)}{1 - 0.91554001 \times s_r.0.07642878 \times s_r^2}$$
$$s_\infty = 0.535$$

Since the solid volume is the same before and after pressing, we can write:

$$V_s = V_0(1 - \varepsilon_0) = V_2(1 - \varepsilon_1)$$
$$V = \ell \times S$$
$$\ell_1 = \ell \times \frac{1 - \varepsilon_0}{1 - \varepsilon_1} = 3 \times \frac{1 - 0.400}{1 - 0.362} = 2.82$$

Blowing time

To obtain the desired humidity of 8.5 %, the saturation must be:

$$\text{s} = \frac{\rho_\text{s}}{\rho_\text{f}} \frac{1 - \varepsilon_0}{\varepsilon_0} \frac{\text{h}}{100 - \text{h}} = \frac{4.5}{1} \frac{1 - 0.362}{0.362} \frac{8.5}{100 - 8.5} = 0.737$$
$$\text{s}_\text{r} = \frac{\text{s} - \text{s}_\infty}{1 - \text{s}_\infty} = \frac{0.737 - 0.535}{1 - 0.535} = 0.4344$$

From expression (9.35) and from the relative permeability of the liquid, with parameters $a = -4.466256$ y $b = 4.4661532$, we obtain the blowing time to get a saturation of 0.4344:

$$t_3 = \frac{\mu \ell \varepsilon_0}{\Delta p_e}\left\{R_m(1-s) + \frac{\ell(1-s_\infty)}{k(\varepsilon_0)}\right\}\int_{s_r}^{1}\frac{d\eta}{\exp(a+b\eta^2)}$$

$$= \frac{\mu \ell \varepsilon_0}{\Delta p_e}\left\{R_m(1-s) + \frac{\ell(1-s_\infty)}{k(\varepsilon_0)}\right\}\left\{\frac{1}{2b \times \exp(a+b\eta^2)}\right\}$$

$$t_3 = \frac{0.01 \times 2.26 \times 0.362}{6 \times 10^6} \times \left\{9.80 \times 10^9 \times (1-0.737) + \frac{2.26 \times (1-0.535)}{10.55 \times 10^{-10}}\right\}$$

$$\times \left\{\frac{\exp(4.466256 - 4.4661532 \times \eta^2)}{-2 \times 4.4661532 \times \eta}\right\}_{0.43}^{1} = 91.7 \text{ s}$$

Airflow:

The airflow is calculated from (9.36):

$$Q_a = \left(\frac{S\Delta p_e}{\mu_a}\right) \times \left(\frac{\ell}{k(\varepsilon_0)f_a(\varepsilon_0, s)}\right)^{-1}$$

$$Q_a = \left(\frac{S\Delta p_e}{\mu_a}\right) \times \left(\frac{\ell}{k(\varepsilon_0) \times \left(\frac{a+cs_r}{1+bs_r+ds_r^2}\right)}\right)^{-1} = 1.24 \text{ m}^3/\text{s}$$

Air consumption:

$$V_a = Q_a \times t_3 = 1.24 \times 91.7 = 113.8 \text{ m}^3$$

Filter capacity:

The time for one cycle is:

$$t = t_1 + t_2 + t_2 + t_4$$
$$= 85 + 80 + 92 + 249 = 506 \text{ s}$$

Therefore, for each cycle of 506 s, a cake is formed in 85 s.

Mass per cycle:

$$m(\varepsilon_0, t) = \rho_s(1-\varepsilon_0)S\left(\frac{2}{\mu}\right)^{0.5} \times \left(\frac{\varphi_0}{1-\varphi_0-\varepsilon_0}\right)^{0.5} \times (k(\varepsilon_0))^{0.5} \times \Delta p_e^{0.5} \times t^{0.5}$$

$$= 2.356 \times 10^7 \text{ g/cycle}$$
$$= 2.356 \times 10^7 \times 3600/506 \text{ ton/cycle}$$
$$= 167.6 \text{ tph.}$$

References

Concha, F. (1990). *Suspension Rheology*. Universidad de Concepción, (in Spanish).

Droguett, M. H. (2000). Optimization of the filtration system of the Coloso Plant, Minera Escondida. *Engineering Thesis*. University of Concepción, (in Spanish).

Henriksson, B. (2000). Focus on separation in the mining industry. *Filtration + Separation, 37*(7), 26–29.

Holdich, R. (1996). Simulation of compressible cake filtration. *Filtration + Separation, 31*, 825–829.

Massarani, G. (1978). *Problems in Particulates Systems.* COPPE/UFRJ, 21. (in Portuguese).

Massarani, G. (1997). *Fluodynamics of Particulate Systems.* UFRJ (in Portuguese).

Tiller, F. M. (1953). The role of porosity in filtration. Numerical method for constant rate and constant pressure filtration based on Kozeny's law. *Chemical Engineering Progress, 49*(9), 467–479.

Tiller, F. M. (1958). The role of porosity in filtration, part III, variable-pressure-variable rate filtration. *AIChE Journal, 6*(4), 170–174.

Tiller, F. M., & Cooper, H. R. (1958). The role of porosity in filtration. Part IV. Constant pressure filtration. *AIChE Journal, 6*(4), 595–601.

Tiller, F. M., & Cooper, H. R. (1962). The role of porosity in filtration. Part V. Porosity variations in filter cakes. *AIChE Journal, 8*(4), 445–449.

Tiller, F. M., & Shirato, M. (1964). The role of porosity in filtration. Part VI. New definition of filter resistance. *AIChE Journal, 10*(1), 61–67.

Tiller, F. M., & Lu, W. (1972). The role of porosity in filtration. Part VIII, cake non-uniformity in compression-permeability cells. *AIChE Journal, 18*(3), 569–572.

Tiller, F. M., & Yeh, C. S. (1987). The role of porosity in filtration. Part XI, filtration followed by expression. *AIChE Journal, 33*(8), 1241–1256.

Tiller, F. M., Hsyung, N. B., & Cong, D. Z. (1985). The role of porosity in filtration. Part XII, filtration with sedimentation. *AIChE Journal, 41*(5), 1153–1164.

Wakeman, R. J., & Tarleton, E. S. (1999a). *Filtration: Equipment selection, modeling and process simulation* (pp. 81–82). Oxford: Elsevier Sci.

Wakeman, R. J., & Tarleton, E. S. (1999b). *Filtration: Equipment selection, modeling and process simulation*. Oxford: Elsevier Sci. 23.

Chapter 10
Suspension Rheology

Abstract The importance of rheology in the mining industry derives from the fact that all materials being processed are suspensions, that is, mixtures of solid particles and fluids, usually in water. In mineral processing plants, water is mixed with ground ore to form a pulp that constitutes the mill feed. The mill overflow is mixed again with water to adjust the solid content for classification in hydrocyclones. Pulp characteristics are essential in the transport of products to their final destination. A suspension, like all types of materials, must obey the laws of mechanics under the application of forces. The flow patterns of suspensions in tubes depend on their concentrations and transport velocities. In diluted suspensions at low velocities particles will settle. The suspension is termed a settling suspension and the flow regime is considered heterogeneous. At a velocity beyond a value at which all particles are suspended gives a non-settling suspension and the flow regime is homogeneous with Newtonian behavior. Concentrated suspensions are usually homogenous with non-Newtonian behavior. The variables and field equations for all types of fluids are presented and constitutive equations differentiate between Newtonian and non-Newtonian behavior. Empirical models of non-Newtonian behavior are presented, including pseudo-plastic and dilatant behavior with Cross and Carreau and Power-law models, and yield-stress models with Bingham and Hershel-Bulkley models. The study of the operational effect on viscosity includes variable such as solid particle size and concentration, temperature, pressure, time and pH. Rheometry provides experimental methods to determine rheological parameters such as viscosity and yield stress.

E. C. Bingham introduced the word *rheology* in 1929 to describe the study of deformation and flow of all types of materials. The axioms of mechanics and the mass and momentum balances are valid for all macroscopic bodies and the distinction among different materials is established by constitutive equations, that is, the response of materials to applied stresses. Strictly speaking, rheology covers the mechanical study of all matter considered as continua, but it is usually reserved for those with non-linear constitutive equations, therefore leaving out Hooken solids and Newtonian fluids. Rheology can be considered a description, with constitutive equations also called rheological equations of state, of material *behavior* and not of

F. Concha A, *Solid–Liquid Separation in the Mining Industry*,
Fluid Mechanics and Its Applications 105, DOI: 10.1007/978-3-319-02484-4_10,

materials. A rheological study includes the formulation of constitutive equations and the experimental methods to determine corresponding parameters, which is called rheometry.

The importance of rheology in the mining industry derives from the fact that all materials being processed are *suspensions*, that is, mixtures of solid particles and fluids, usually water. In mineral processing plants these suspensions are termed *pulps.* In a grinding plant, water is mixed with ground ore to form a pulp that constitutes the mill's feed. The mill overflow is mixed again with water to adjust the solid content required to be classified in hydrocyclones. Pulp characteristics are essential in the transport of products to their final destination in the flotation plant.

A suspension, like all types of materials, must obey the laws of mechanics under the application of forces. The flow patterns of suspensions in tubes depend on their concentrations and transport velocities. In diluted suspensions at low velocities particles will settle. The suspension is termed a *settling* suspension and the flow regime is considered *heterogeneous.* At a velocity beyond a value at which all particles are suspended gives us a *non-settling* suspension and the flow regime is *homogeneous* with Newtonian behavior. Concentrated suspensions are usually homogenous but with non-Newtonian behavior. Generally, mineral pulps have non-Newtonian behavior, therefore their rheological characteristics are essential in the different unit operations in a mineral processing plant.

10.1 Introduction to Rheology

The incompressible stationary shear flow of a fluid can be described with the following variables, (1) material density $\rho(\boldsymbol{r},t)$, (2) velocity $\boldsymbol{v}(\boldsymbol{r},t)$ and (3) the stress tensor $\boldsymbol{T}(\boldsymbol{r},t)$, where $\boldsymbol{r}$ and t are the position vector and time respectively. These three field variables must obey the mass and linear momentum field equations:

$$\nabla\cdot\boldsymbol{v}=0 \tag{10.1}$$

$$\rho\nabla\boldsymbol{v}\cdot\boldsymbol{v}=\nabla\cdot\boldsymbol{T}+\rho\boldsymbol{g} \tag{10.2}$$

where $\boldsymbol{g}$ is the gravitational constant vector.

Since there are three field variables and only two field equations, a constitutive equation must be postulated for the stress tensor:

$$\boldsymbol{T}=-p\boldsymbol{I}+\boldsymbol{T}^E(\boldsymbol{r}) \tag{10.3}$$

where p is the pressure and $\boldsymbol{T}^E$ is the shear stress tensor or extra stress tensor.

The extra stress tensor defines the type of fluid, for example, a Newtonian fluid is $\boldsymbol{T}^E$ given by:

$$\boldsymbol{T}^E=\mu\left(\nabla\boldsymbol{v}+\nabla\boldsymbol{v}^T\right) \tag{10.4}$$

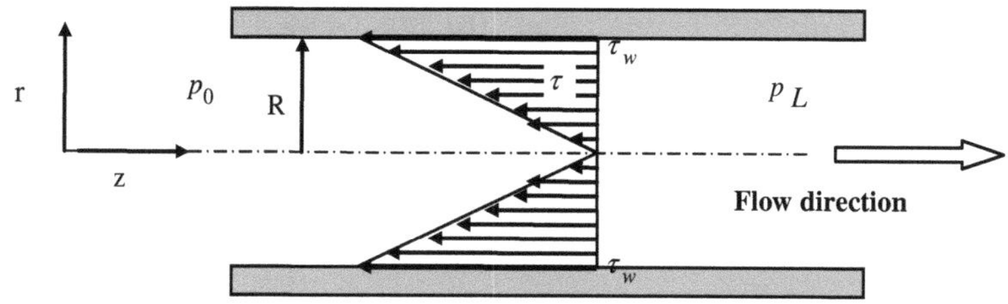

Fig. 10.1 Shear stress distribution for the flow in a cylindrical tube for $p_0 > p_L$

where μ is a constant called shear viscosity and $\nabla \boldsymbol{v}$ is the shear rate tensor.

For a two-dimensional axi-symmetrical flow of a Newtonian fluid in the x_2 or z direction in case of a cylindrical tube, the shear stresses $T_{12}(x_1, x_2) < 0$ or $T_{r,z}(r, z) < 0$ reduces to:

$$T_{12}^E = \mu \frac{\partial v_2}{\partial x_1} \quad T_{rz}^E = \mu \frac{\partial v_z}{\partial r} \tag{10.5}$$

The stresses are usually written in the form $T_{12}^E \equiv \tau$ or $T_{rz}^E \equiv \tau$ and the velocity gradient as $\partial v_2/\partial x_1 = \dot{\gamma}$ or $\partial v_z/\partial r = \dot{\gamma}$, then Eq. (10.5) is used in the form:

$$\tau = \mu \dot{\gamma} \tag{10.6}$$

where the shear stress τ is measured in Pascal (Newton per meter) (Pa = N/m^2), the shear rate $\dot{\gamma}$ in (s^{-1}) and the viscosity in (Pa s).

Figure 10.1 represents the shear stress for the flow in a cylindrical tube in the direction z, where τ_w is the shear rate at the wall of the tube (see Chap. 11).

10.2 Constitutive Equations

Materials with a constant viscosity behave as Newtonian fluids. Common fluids like water and air have Newtonian behavior. For these types of fluids, the shear stress is a liner function of the shear rate.

10.2.1 Suspensions with Newtonian Behavior

Diluted non-settling suspensions have *Newtonian* behavior, that is, the viscosity is constant and the relationship between shear stress and the shear rate is represented by a straight line called a *rheogram*, see Fig. 10.2 for a suspension with 0.01 volume fraction of solids. Einstein's constitutive equation applies; $\eta = \eta_s \times (1 + 2.5\varphi)$, where η_s is the viscosity of the continuous phase.

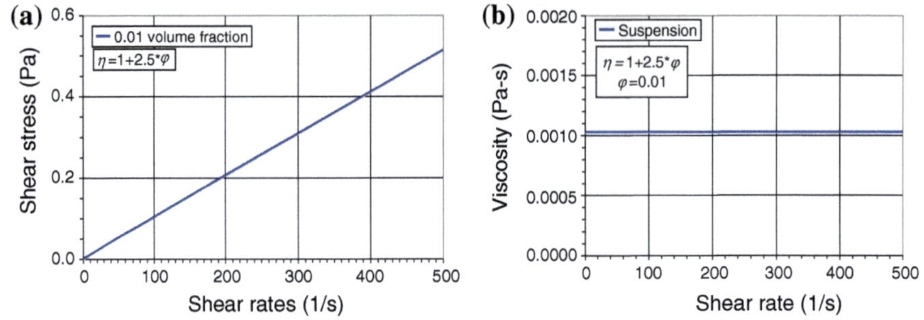

Fig. 10.2 Rheogram for a diluted non-settling suspension with Newtonian behavior. **a** Flow curve. **b** Viscosity curve

10.2.2 Non-Newtonian Behavior

Figure 10.3 shows a general viscosity curve for suspensions with non-Newtonian behavior. At a low shear rate, the shear stress shows a constant viscosity region followed by a drastic fall, then a new constant viscosity region and finally, in some cases, an increase in viscosity at very high shear rates.

If we consider Newtonian behavior as a reference, see the red line in Fig. 10.4, non-Newtonian behavior present two additional rheograms: *pseudo-plastic*, *also known as shear thinning* behavior, typical of mineral suspensions and polymer solutions (see the blue line in Fig. 10.4), and *dilatant,* also known *as shear thickening* behavior, where viscosity increases with shear rate, see the magenta line.

A copper flotation tailing has *non-Newtonian* behavior, that is, the constitutive equation of the stress is a non-linear function of the shear rate. These types of constitutive equations are written the same as Newtonian equations. However, in

Fig. 10.3 General representation of the *flow curve* of non-Newtonian suspension

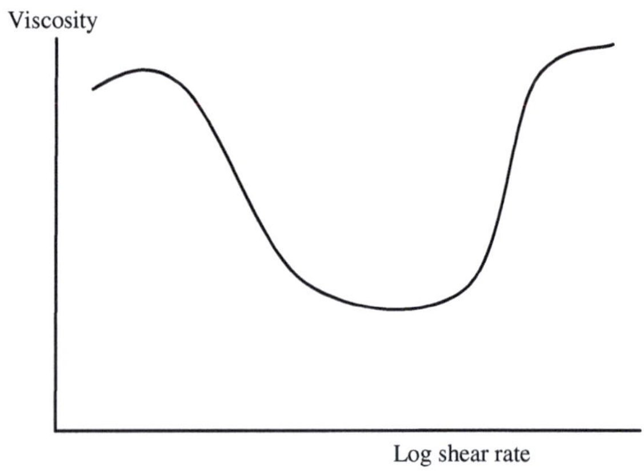

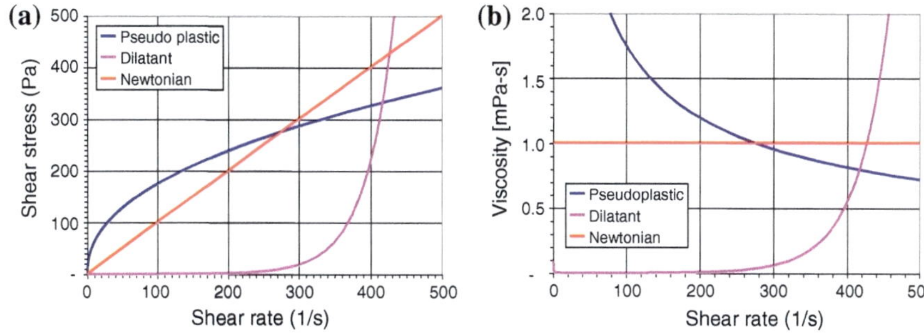

Fig. 10.4 Rheogram of a fluid with a typical non-Newtonian behavior (Schramm 2000). **a** Flow curve. **b** Viscosity curve

this case, viscosity is not constant but rather is a function of the shear rate. Figure 10.5 shows an example.

$$\tau = \eta(\dot{\gamma})\dot{\gamma} \tag{10.7}$$

where η is the shear viscosity.

(a) **Pseudo Plastic and Dilatant Behavior**

In general, materials with pseudo-plastic behavior present two Newtonian plateaus (constant viscosities; see Fig. 10.7), a *first Newtonian plateau*, with constant viscosity η_0 at low shear rates and a *second Newtonian plateau*, with viscosity η_∞ at high shear rates. Sometimes the first Newtonian plateau is so high that it cannot be measured, in which case, the low shear rate behavior is described as an apparent *yield stress* τ_y. Sometimes, the second Newtonian plateau is short and viscosity increases as the shear rate increases, which is termed dilatant behavior.

In mineral processing, we find discrete or agglomerate particle suspensions with different concentrations. At low concentrations, discrete particle suspensions have Newtonian behavior, as shown in Fig. 10.2, but with higher concentrations their

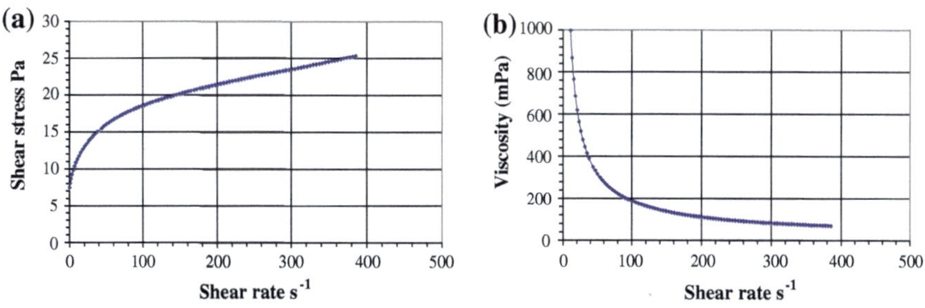

Fig. 10.5 Rheogram for flotation tailings. **a** Flow curve. **b** Viscosity curve

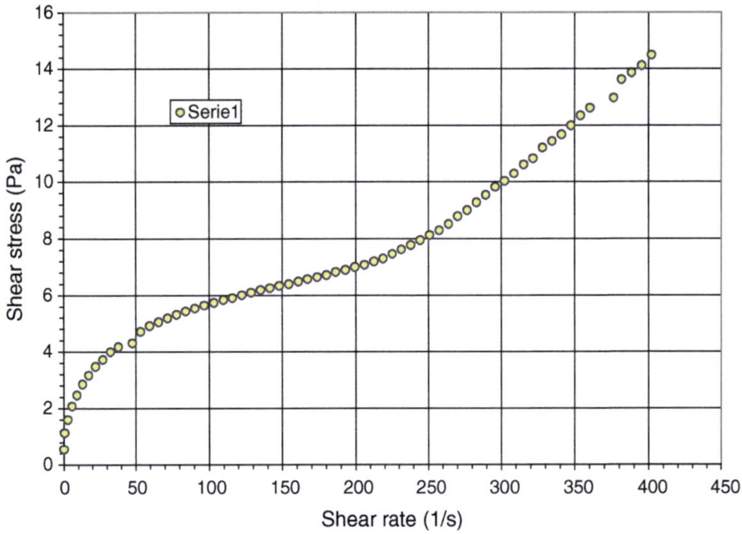

Fig. 10.6 Rheological behavior of a rougher flotation tailing with 56 % solids and pH 9.2

behavior is viscoplastic, see Fig. 10.5. In general, mineral particles at rest present a negative surface charge when suspended in water (see Chap. 7) and consequently become hydrated. In slow motion particles stay hydrated and present a certain resistance to flow but with an increase in shear rate, the hydration layer is striped away and particles become oriented in the direction of the flow, causing decreases in flow resistance and viscosity, approaching an optimum constant orientation.

Dilatant flow behavior is found in highly concentrated suspensions and depends on the solid concentration, the particle size distribution and the continuous phase viscosity. The region of shear thickening generally follows that of shear thinning.

Densely packed particles have enough fluid inside to fill the void between particles. At rest or at low shear rates, water lubricates particle surfaces, allowing an easy positional change of particles when forces are applied and the suspension behaves as a shear thinning liquid. At critical shear rates, packed particles lose water, which causes an increase in interior concentration. Particle–particle interaction increases drag, causing dilatant behavior as shown in Fig. 10.6.

10.2.3 Empirical Rheological Models

Empirical constitutive equations are quantified with different mathematical models. We will describe Cross and Carreau models; Ostwal-de Waele, commonly known as the power law model, the Herschel-Bulkley model and Bingham model.

(a) **Cross and Carreau Models**

Cross and Carreau models are represented by Eqs. (10.8) and (10.9), respectively. Given that $\dot{\gamma} = \tau/\eta$, the relationship between viscosity and shear rate is:

$$\frac{\eta - \eta_\infty}{\eta_0 - \eta_\infty} = \frac{1}{1 + (\lambda\dot{\gamma})^m} \qquad \tau = \frac{\eta}{\lambda} \times \left(\frac{\eta_0 - \eta_\infty}{\eta - \eta_\infty} - 1\right)^{1/m} \tag{10.8}$$

$$\text{Carreau} \quad \frac{\eta - \eta_\infty}{\eta_0 - \eta_\infty} = \frac{1}{\left(1 + (\beta\dot{\gamma})^2\right)^{(1-n)/2}} \qquad \tau = \frac{\eta}{\beta} \times \left[\left(\frac{\eta_0 - \eta_\infty}{\eta - \eta_\infty}\right)^{\frac{2}{1-n}} - 1\right]^{0.5} \tag{10.9}$$

where η_0 and η_∞ are the viscosities at low and high shear rates plateaus, and λ, β, m and n are experimental constants. Figure 10.7 represents the two models in terms of $\eta = f(\dot{\gamma})$, where λ and β are curve fitting parameters with the dimension of time and n as a constant.

(b) **Power Law Model (Ostwal-de Waele)**

Power law models represent pseudo-plastic and dilatant behavior with great accuracy. Equation (10.10) represents the viscosity and shear stress for material obeying the power law model:

$$\eta = m\dot{\gamma}^{n-1} \quad \text{and} \quad \tau = m\dot{\gamma}^n \tag{10.10}$$

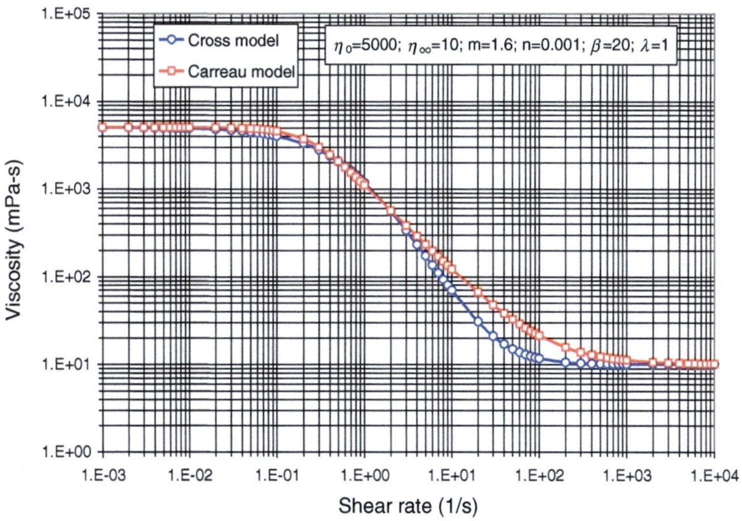

Fig. 10.7 Viscosity versus shear rate for a given material

where m is the *consistency index*, with units in Pa × s^2 and n is the *power index*. Values of the power index $n < 1$ represent pseudo-plastic behavior and values of $n > 1$ dilatant behavior.

Problem 10.1 Determine the rheogram of the thickener underflow of a copper flotation tailing from the experimental data in Table 10.1, obtained with a rotational viscometer. Plot the rheogram and determine the parameters of the power law model.

Results are shown in Eq. (10.11) and Figs. 10.8, 10.9, 10.10, where the points are the experimental values and the lines represent the simulation with the power law model.

$$\eta = 1.723\dot{\gamma}^{0.727}\ [\text{Pas}] \quad \tau = 1.723\dot{\gamma}^{0.2468}\ [\text{Pa}] \tag{10.11}$$

(c) Models with Yield Stress

In concentrated flocculated suspensions, particles aggregate as flocs, which interact with each other forming a network maintained by surface interaction forces extending throughout the entire volume of the suspension. The application of stresses to this structure deforms it elastically until the structure breaks down. This break down is related to the *yield stress* of the material and can be considered as the minimum shear stress at which the solid structure becomes liquid. Knowledge about yield stress is essential in transporting suspensions, especially in resuspending particles when they have settled in a pipeline or channel. See Chap. 11 for details.

There are two methods to measure the yield stress: (1) extrapolating the flow curve to a zero shear rate and (2) directly measuring shear stress when the flow begins. The first method depends on the rheological model in use, for example *Bingham or Hershel-Bulkley models,* which provide different values that are in both cases different from the yield stress determined by measuring with the vane method. We conclude that shear rate should be determined by the method that gives the best value for the application required.

1. **Extrapolation from flow curves**

Bingham Model

mineral pulps in tubes and channels the range of shear rates is usually high, on the order of hundreds of seconds to minus one. At these ranges, viscosity is constant and equal to the slope of the line of the shear values. In this case, the extrapolation of this line to a zero shear rate gives an appropriate yield stress that, together with the constant viscosity, provides the required rheological parameters. Bingham proposed this method in 1922 with the constitutive equation:

$$\tau = \tau_y + K\dot{\gamma} \tag{10.12}$$

Table 10.1 Rheological experiment

γ	τ	η
0.01	0.490	49021
0.02	0.592	29616
0.03	0.662	22055
0.04	0.716	17893
0.05	0.761	15213
0.06	0.799	13325
0.07	0.834	11912
0.08	0.865	10810
0.09	0.893	9923
0.1	0.919	9191
0.2	1.111	5553
0.3	1.241	4135
0.4	1.342	3355
0.5	1.426	2853
0.6	1.499	2498
0.7	1.563	2234
0.8	1.622	2027
0.9	1.675	1861
1	1.723	1723
2	2.082	1041
3	2.326	775
4	2.516	629
5	2.674	535
6	2.811	468
7	2.932	419
8	3.040	380
9	3.140	349
10	3.231	323
20	3.904	195
30	4.361	145
40	4.718	118
50	5.014	100
60	5.270	88
70	5.497	79
80	5.701	71
90	5.887	65
100	6.059	61
200	7.321	37
300	8.178	27
400	8.846	22
500	9.402	19

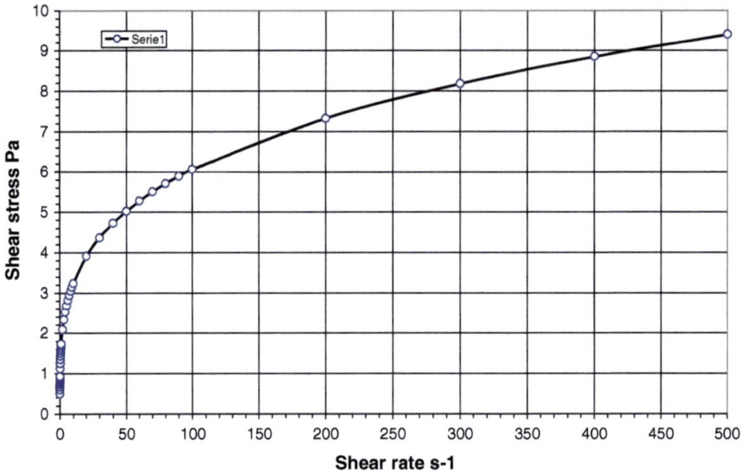

Fig. 10.8 Shear stress versus shear rate of a thickener underflow of copper flotation tailings modeled by the power law

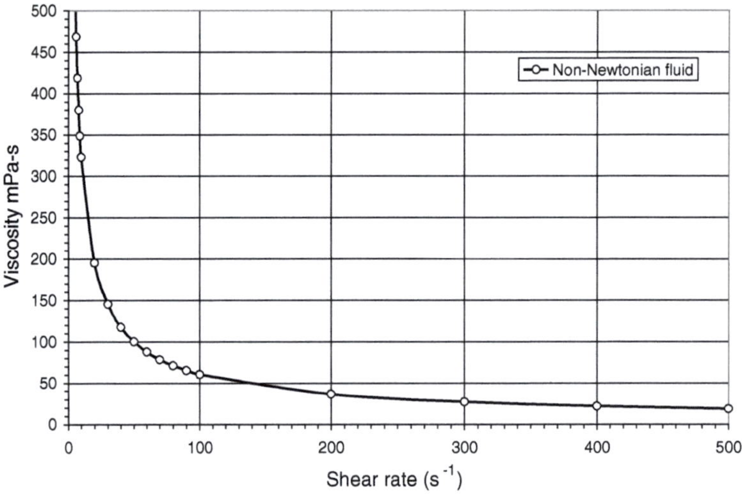

Fig. 10.9 Viscosity versus shear rate of thickener underflow of a copper flotation tailings modeled by the power law

where τ_y is the yield stress and K is the constant plastic viscosity. Equation (10.12) shows that Bingham's model is the combination of a yield stress τ_y with a Newtonian viscosity K. This model has the advantage of giving the result of the modeling in one plot.

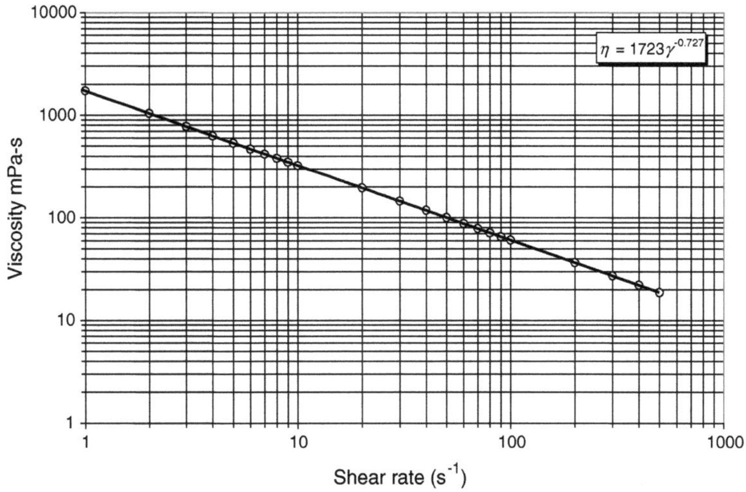

Fig. 10.10 Rheogram of a fluid with non-Newtonian behavior. *Circles* are experimental data and the line is the potential model

Problem 10.2 Determine the parameters of Bingham's model for the experimental data in Table 10.2. For high shear stress values, a tangent drawn to the shear stress-shear rate curve gives τ as the intercept of the tangent with the vertical axis and the viscosity as its slope. See Fig. 10.10.

$$\tau_y = 12.5 \text{ Pa} \text{ and } K = 0.0196 \text{ Pa s} \tag{10.13}$$

Herschel-Bulkley Model

For processes at low shear rates, the stress-shear function is curved and Bingham's model is inadequate since viscosity is not constant. In this case, the Herschel-Bulkley model can be used with the constitutive equation:

$$\tau = \tau_y + k\dot{\gamma}^z \tag{10.14}$$

where τ_y is the yield stress, k is the consistency index, similar to the power law model, and n is the power index.

Problem 10.3 For the experimental data in Table 10.2, determine the rheological parameters of the Herschel-Bulkley model.

The following values were obtained by non-linear curve fitting for the Herschel-Bulkley rheological parameters: $\tau_y = 1.25$, $k = 2.37$, $z = 0.343$, see Figs. 10.11 and 10.12.

Table 10.2 Data of a rheogram for a copper flotation tailing

γ (1/s)	τ (Pa)	η (m Pa s)	γ (1/s)	τ (Pa)	η (m Pa s)
0.0000	3.3	0.0000	186.5000	8.7	46.5300
0.4660	3.8	8109.0000	193.0000	8.7	45.0100
2.6380	4.5	1690.0000	199.4000	8.7	43.7300
5.5050	5.1	927.8000	205.8000	8.7	42.4100
8.9820	5.7	632.1000	212.3000	8.7	41.1800
12.9600	6.2	478.7000	218.7000	8.8	40.0700
17.3500	6.6	380.9000	225.2000	8.8	39.0100
22.1100	6.9	313.6000	231.5000	8.8	38.0400
27.1500	7.2	266.6000	238.0000	8.8	37.0300
32.4400	7.5	230.1000	244.4000	8.9	36.2200
37.8900	7.7	202.1000	250.8000	8.9	35.4400
47.6700	7.7	162.0000	257.3000	8.9	34.7500
53.4800	8.1	150.8000	263.7000	8.9	33.9300
59.5000	8.2	137.0000	270.2000	9.0	33.3700
65.5900	8.2	125.4000	276.6000	9.1	32.9500
71.8300	8.3	115.5000	283.1000	9.2	32.4500
77.9900	8.3	106.8000	289.5000	9.3	32.2100
84.2300	8.4	99.2600	295.9000	9.5	32.2300
90.5400	8.4	93.1500	302.4000	9.8	32.5300
96.8500	8.5	87.5300	308.8000	10.2	32.8800
103.2000	8.5	82.5300	315.3000	10.4	33.1200
109.5000	8.5	77.8900	321.7000	10.8	33.6200
115.9000	8.5	73.7100	328.2000	11.2	34.1300
122.4000	8.6	69.9300	334.6000	11.6	34.7200
128.7000	8.6	66.6400	341.1000	11.9	34.8800
135.1000	8.6	63.6700	357.0000	12.1	33.9600
141.5000	8.6	60.8100	362.2000	12.6	34.7700
148.0000	8.6	58.1600	369.1000	12.8	34.7800
154.4000	8.6	55.9700	375.8000	13.0	34.6500
160.9000	8.7	53.8000	382.4000	13.3	34.8500
167.3000	8.7	51.8400	389.1000	13.5	34.7300
173.6000	8.7	49.9900	395.7000	13.9	35.0800
180.1000	8.7	48.3700	402.3000	14.1	35.1100

For the same experimental data, the power-law model better describes the shear stress in the whole range of shear rates. The drawback is that this model requires knowledge of the variable viscosity obtained from the viscosity plot versus shear rate, while Bingham's model requires only the shear stress plot. We conclude that the power-law model is better for processes requiring low shear rate (lower than 150 s^{-1} in Fig. 10.11). Engineers designing and operating pipelines in the mining industry prefer Bingham's model because it gives them a constant viscosity and a yield stress value, which are important in transporting mineral pulps.

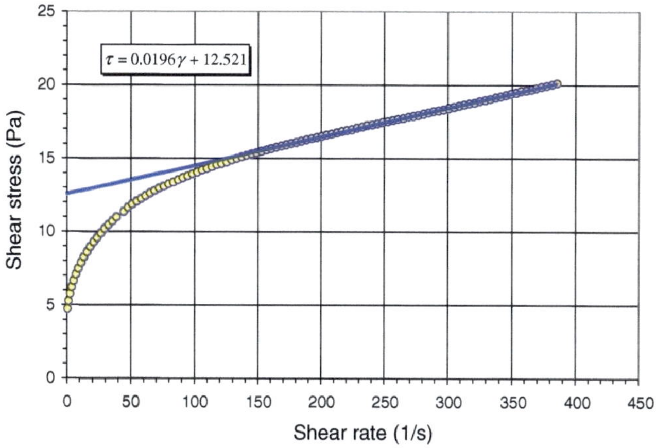

Fig. 10.11 Bingham model for data of Table 10.2

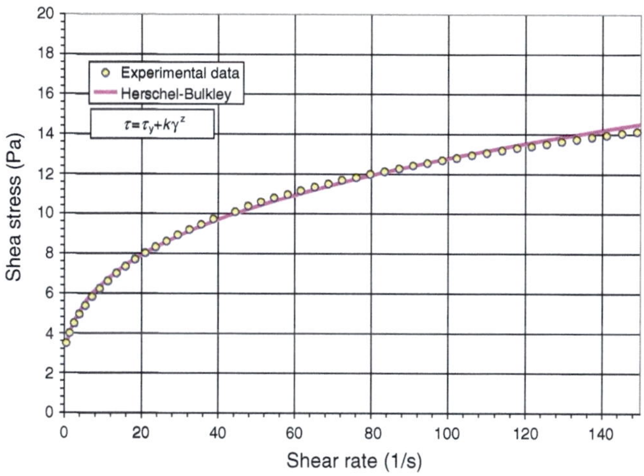

Fig. 10.12 Herschel-Bulkley model for data from Table 10.2

(d) **Pseudo-Plastic-Dilatant behavior**

Under some conditions, copper flotation tailings present pseudo-plastic-dilatant behavior similar to that of Fig. 10.6. There are no models for this type of behavior, but a polynomial of degree three or four can describe the entire rheogram in the full range of shear rates.

Problem 10.4 Determine the parameters for a rougher flotation tailing with data given in Table 10.3 and represented in Fig. 10.13. The result with a four-power polynomial is:

$$\tau = -2.000 \times 10^{-9}\dot{\gamma}^4 + 2.000 \times 10^{-6}\dot{\gamma}^3 - 6.000 \times 10^{-4}\dot{\gamma}^2 + 7.940 \times 10^{-2}\dot{\gamma} + 5.212 \tag{10.15}$$

10.2.4 Operational Effects on Viscosity

(a) **The effect of concentration**

Solid concentration has the most important effect on suspensions. In general, properties such as yield stress and viscosity increase with solid concentration. The viscosity of suspensions at low concentration can be modeled by a polynomial extension of Einstein's equation.

$$\eta = \eta_0\left(1 + k_1\varphi + k_2\varphi^2 + k_3\varphi^3 + \cdots\right) \tag{10.16}$$

where η_0 is viscosity at zero concentration, φ is the volume fraction of solids, $k_1 = 2.5$ is Einstein's parameter and $k_2, k_3, \ldots k_n$ are fitting parameters. For concentrations of less than 0.01 the suspension behaves Newtonian. See Fig. 10.14.

Krieger-Daugherty

At higher concentrations, suspensions have non-Newtonian behavior. Several equations describe this behavior; one of the most commonly used is the Krieger-Daugherty equation (Krieger 1972), in which viscosity depends on maximum particle packing φ_{max}. See Eq. (10.17) and Fig. 10.14.

$$\eta = \eta_0\left(1 + \frac{\varphi}{\varphi_m}\right)^{a\times\varphi_m} \tag{10.17}$$

Table 10.3 Dimension of the sensors

Sensor system	MV
Internal cylinder (Rotor)	18.4
Length L_1 (mm)	60.0
Radius R_1 (mm)	
External cylinder	21.0
Length L_C mm)	85.0
Radius R_2 (mm)	
Gap (mm)	2.60
Temperature range °C	−30 °C a 100
Sample volume cm^3	46
Viscosity range mPa s	20 a 4 × 10^5

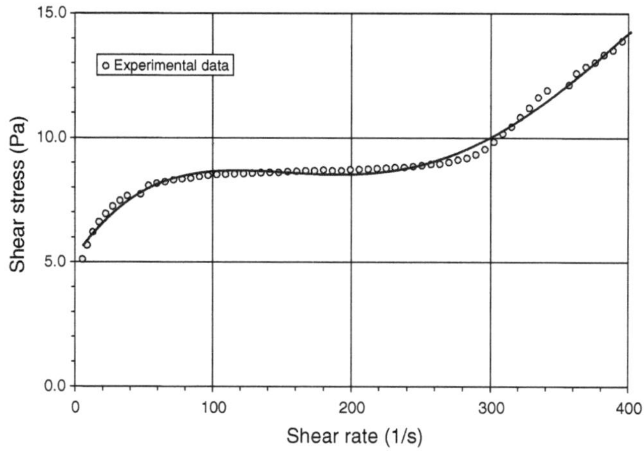

Fig. 10.13 Rheogram of a rougher flotation tailings of a copper ore at pH = 9.2 and 4 % solids modeled by a four power polynomial

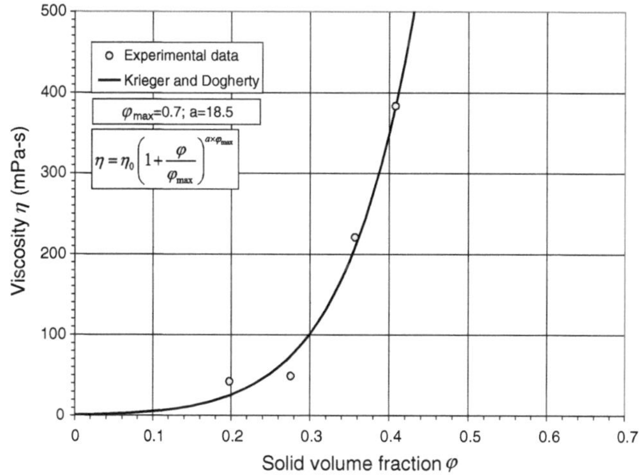

Fig. 10.14 Shear stress versus shear rate plot for a copper tailing at several solid concentrations

Exponential function

When using the power-law model for the rheogram, an exponential function sometimes gives a good fit for the effect of solid concentrations. Figures 10.15 and 10.16 give an example of a copper flotation tailing at a shear rate of 200 s^{-1}.

$$\eta = 0.838 \times \exp(0.142\ \%)\dot{\gamma}^{-0.670} \tag{10.18}$$

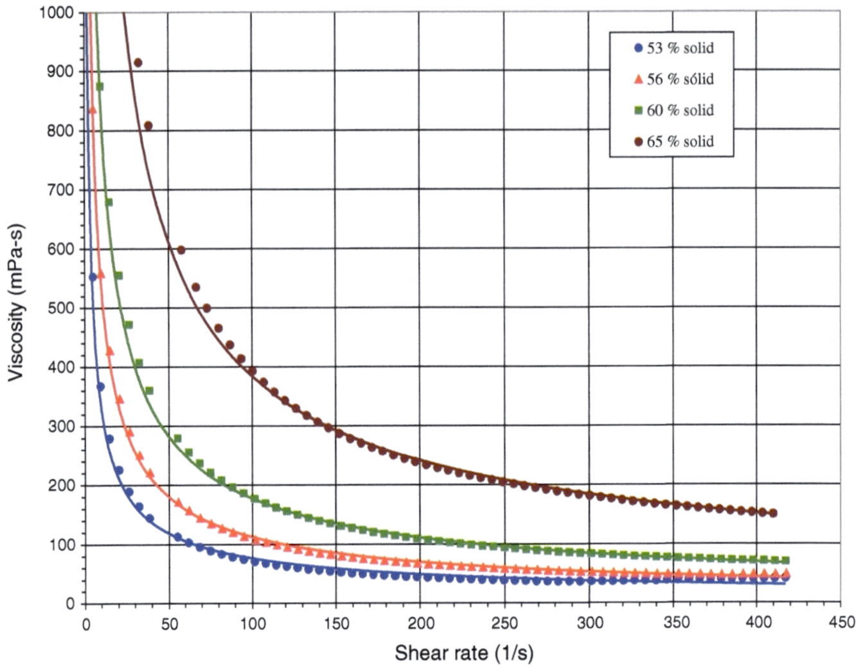

Fig. 10.15 Shear viscosity versus shear rate for a copper flotation tailing with % solid by weight as a parameter. *Symbols* are experimental values and the *lines* are simulations with Eq. (10.18)

Shaheen (1972) presented an alternative model to describe the concentration effect:

$$\eta = \eta_0 a \times \exp\left(\frac{b\varphi}{1 - (\varphi/\varphi_m)}\right) \tag{10.19}$$

where a, b sand φ_m are constant.

(b) **Effect of particle size**

Particle size distribution affects viscosity in three ways: (1) through the maximum particle packing φ_m; (2) the presence of very small particles; and (3) particle size distribution. Fine particles can fit in a packed bed between larger particles, increasing the density and affecting the relative concentration φ/φ_m. Another effect of small particles is to transform the continuous phase, usually water, into a viscous suspension that directly affects overall viscosity. Finally, particle size distribution contributes to shear thickening of mineral pulps, as shown in Figs. 10.6 and 10.13.

Unfortunately, there is no theoretical information on how these variables influence suspension viscosity. Consequently, the maximum packing density φ_m is usually obtained by curve fitting.

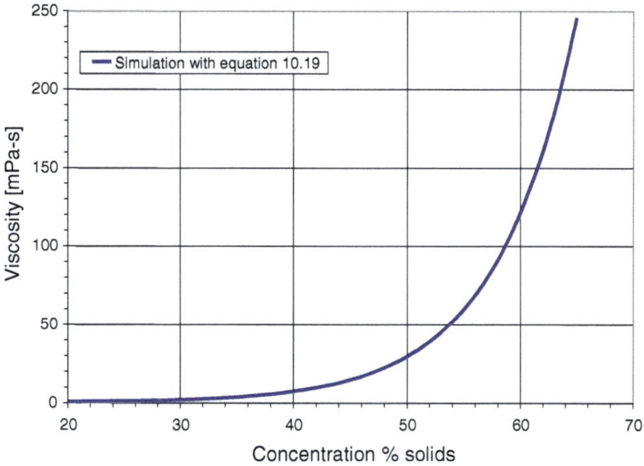

Fig. 10.16 Plot of viscosity versus particle concentration for data from Fig. 10.13

(c) The effect of temperature

An exponential function (Tanner 1988) fits the effect of temperature on Newtonian liquids.

$$\eta = A\,\exp(B/(T - T_0)) \tag{10.20}$$

where T is the absolute temperature and A, B and T_0 are characteristic constants. In the absence of information on the effects of temperature on suspensions, the equation for Newtonian fluids is used.

In Barrientos et al. (1994) performed numerous rheological experiments with quartz samples of different sizes and concentrations. They proposed a general equation based on the Shaheen model (1972) for the suspension viscosity in terms of four dimensionless variables x/x_0, φ, Re and °C, where x and x_0 are the average and a reference particle size, φ is the solid volume fraction, $\mathrm{Re} = \rho_f \dot{\gamma} x^2/\eta$ is the flow Reynolds number, ρ_0 and η_0 are the fluid density and viscosity, respectively, $\dot{\gamma}$ is the shear rate and °C is the temperature in Celsius. They separated the functional form of this equation into three terms, one for the effect of temperature, a second for the effect of concentration and a third for the interaction between these variables.

$$\begin{aligned} \eta/\eta_0 &= \mathrm{p}_1 \exp\left(\frac{3.462 \times 10^3}{T} +\right) \times \exp\left(\frac{p_2 \varphi}{1 - (\varphi/p_3)^{p_4}}\right) \\ &\quad \times \left(1 + \frac{p_5 \exp(-p_6 \varphi^{p_7}) \times (x/x_0)^{p_8}}{\mathrm{Re}^{1/p_9}}\right)^{p_9} \end{aligned} \tag{10.21}$$

Concha et al. (1999) performed 70 experiments with underflow material from the feed, overflow and underflow of a copper ore grinding-classification circuit and with ground underflow for several time ranges, temperatures from 5 to 25 °C and concentrations from 15 to 40 % solid by weight. After obtaining nine parameters of the four dimensional groups by non-linear curve fitting, they concluded by simulation that particle size has a significant effect on rheograms solely for shear rate values below $\dot{\gamma} \approx 200\ \mathrm{s}^{-1}$.

(d) **Effect of pressure**

In general, the effect of pressure on viscosity is small, except for materials such as oil subjected to very high pressures, where an exponential equation can be used (Tanner 1988).

$$\eta = \eta(0)\exp(p/\beta)$$

where p is the applied pressure, $\eta(0)$ is the viscosity at zero pressure and β is a constant.

(e) **pH Effect**

It is well know that plant operators add lime to thickener underflow of copper flotation tailings when it is too viscous for hydraulic transport. This does not always solve the problem because pH affects the slurry in a complicated way. Figure 10.17 show the viscosity of a copper flotation tailing for several particle concentrations at a shear rate of 200 s^{-1}.

Two minimum viscosities were present for this material at all concentrations, one at pH 7.5 and the second at pH 9.0, with more pronounced values at high concentrations. Several copper tailings present this behavior. To establish if this behavior is due to the silica content of the tailings, experiments were made with silica in distilled water. Figures 10.17 and 10.18 give the results. The two minimums are also shown but with somewhat higher pH values.

Fig. 10.17 Effect of pH on the viscosity of a copper flotation tailing at a shear rate of 200 (s^{-1})

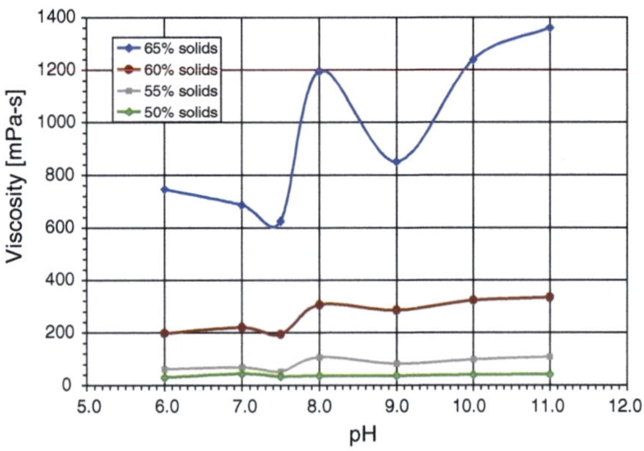

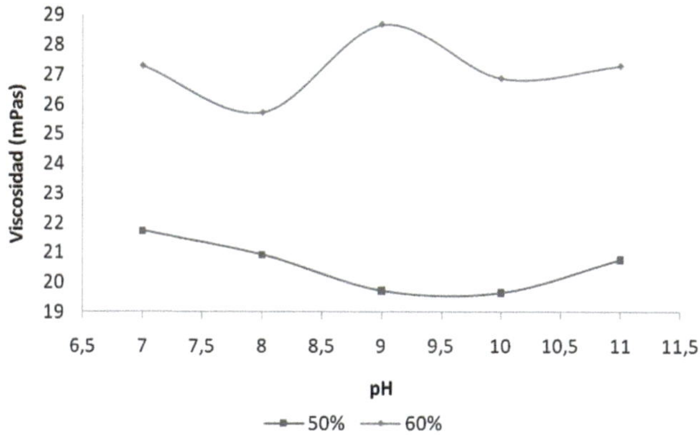

Fig. 10.18 Effect of pH on the viscosity of a suspension of silica of 2 % solids in distilled water at a shear rate of 200 (s^{-1})

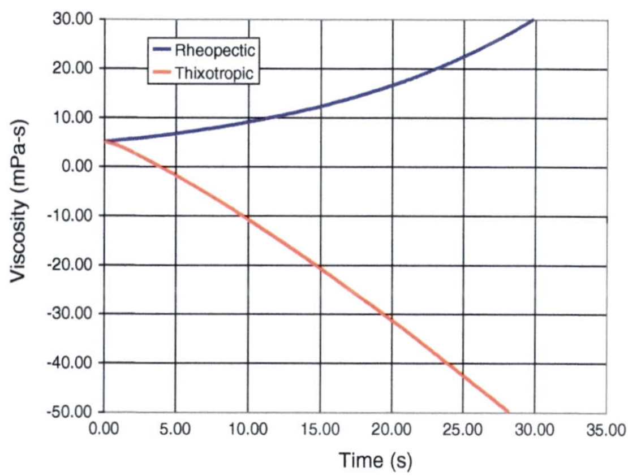

Fig. 10.19 Material with thixotropic or rheopectic behaviors

(f) **The effect of time**

Time is important in materials that suffer structural changes during measurement. For example, some flocculated suspensions change structure while sheared, which produces a change in viscosity. This behavior may be thixotropic or rheopectic depending if the viscosity diminishes or increases with time. A schematic drawing of these behaviors is shown in Fig. 10.19.

10.3 Rheometry

Rheometry is that part of Rheology which provides experimental methods to determine rheological parameters such as viscosity and yield stress, that is, it establishes the methods to determine the constitutive equation of a fluid material. Simple shear flows permit obtaining exact solutions of the Navier–Stokes equations, which in turn provides the rheological parameters.

10.3.1 Simple Shear Stationary Flows

Simple shear are flows produced by a unidirectional shear rate, an example of which are the flows in a circular tube, rotational flows in the annular gap of concentric cylinders, torsion flow between two flat plates and flow between a cone and a plate, among other. These flows are of interest to mineral processing because they provide the tools to calculate pipes and pumps (see Chap. 11) and experimental methods to determine rheological properties; shear stress versus shear rate plots; yield stress and viscosity versus shear rate.

(a) **Flow in concentric cylinders**

Consider the stationary rotational flow of a suspension of non-settling particles in two concentric cylinders of radius R_1 and R_2 produced by the rotation of the cylinders with angular velocities Ω_1 and Ω_2 respectively. Assume the cylinders are open to the atmosphere at one end. Figure 10.20 shows the cylinder.

Considering a constant fluid density, variables for this problem are the viscosity of the suspension and the tangential velocity of the fluid. The field equation in cylindrical coordinates in laminar flow is:

$$\eta \frac{\partial}{\partial r}\left(\frac{1}{r}\frac{\partial}{\partial r}(r v_\theta)\right) = 0 \tag{10.22}$$

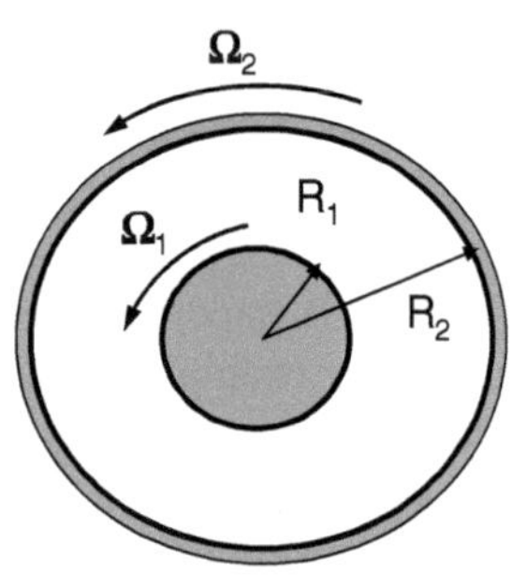

Fig. 10.20 Rotational flow between two concentric cylinders

with boundary conditions:

$$v_\theta(R_1) = \Omega_1 R_1\,,\ v_\theta(R_2) = \Omega_2 R_2\,,\ v_r = v_z = 0$$

where v_θ, v_r and v_z are the components of the velocity vector and Ω_1 and Ω_2 are the angular velocities of the cylinders with radius R_1 and R_2.

Tangential velocity

Integrating Eq. (10.22) twice results in:

$$v_\theta = \frac{C_1}{2} r + \frac{C_2}{r} \tag{10.23}$$

Applying boundary conditions, the constant C_1 and C_2 are:

$$C_1 = \frac{2\left(R_2^2\Omega_2 - R_1^2\Omega_1\right)}{\left(R_2^2 - R_1^2\right)} \quad C_2 = \frac{R_1 R_2^2(\Omega_1 - \Omega_2)}{\left(R_2^2 - R_1^2\right)} \tag{10.24}$$

and

$$v_\theta(r) = \frac{1}{\left(R_2^2 - R_1^2\right)}\left(\left(R_2^2\Omega_2 - R_1^2\Omega_1\right) r + \frac{R_1 R_2^2(\Omega_1 - \Omega_2)}{r}\right) \tag{10.25}$$

Shear Rate

Integrating Eq. (10.22) once yields:

$$\frac{\partial v_\theta}{\partial r} = C_1 - \frac{v_\theta}{r} \tag{10.26}$$

Substituting C_1 gives:

$$\frac{\partial v_\theta}{\partial r} = \frac{R_2^2\Omega_2 - R_1^2\Omega_1}{\left(R_2^2 - R_1^2\right)} - \frac{v_\theta}{r} \tag{10.27}$$

To obtain the average shear rate calculate the average of (10.27) for radius R_1 and R_2:

$$\bar{\dot{\gamma}} = \frac{1}{2}\left(\left.\frac{\partial v_\theta}{\partial r}\right|_{r=R_1} + \left.\frac{\partial v_\theta}{\partial r}\right|_{r=R_2}\right) \tag{10.28}$$

Applying boundary conditions:

$$\left.\frac{\partial v_\theta}{\partial r}\right|_{r=R_1} = \frac{R_2^2\Omega_2 - R_1^2\Omega_1}{\left(R_2^2 - R_1^2\right)} - \Omega_1$$
$$\left.\frac{\partial v_\theta}{\partial r}\right|_{r=R_2} = \frac{R_2^2\Omega_2 - R_1^2\Omega_1}{\left(R_2^2 - R_1^2\right)} - \Omega_2 \tag{10.29}$$
$$\bar{\dot{\gamma}} = \frac{R_2^2\Omega_2 - R_1^2\Omega_1}{\left(R_2^2 - R_1^2\right)} - (\Omega_2 + \Omega_1),\ \mathrm{s}^{-1}$$

(b) **Flow in a capillary**

Consider the stationary laminar axial flow of a fluid in a cylindrical tube, see Fig. 10.21. From Chap. 11 the flow rate and the shear stress at the wall of the tube are given by:

Since the shear rate is linear in r, the average value of $\bar{\dot{\gamma}}$ is given by:

$$\bar{\dot{\gamma}} = \frac{4 \times \bar{v}_z}{D}, \quad \bar{\dot{\gamma}} = \frac{5.1 \times Q}{D^3} \tag{10.30}$$

10.3.2 Types of Viscometers

There are two types of viscometers used in mineral processing, rotational and capillary. Searle-type rotational viscometers are used for mineral pulps while capillary viscometers are used for polymers.

(a) **Rotary viscometers**

The relative rotation of two concentric cylinders of a viscometer induces shear rate in the fluid. Usually one cylinder rotates while the other is fixed. In a Searle viscometer, the inner cylinder rotates while the outer cylinder is fixed. The system is called Couette. See Figs. 10.22 and 10.23.

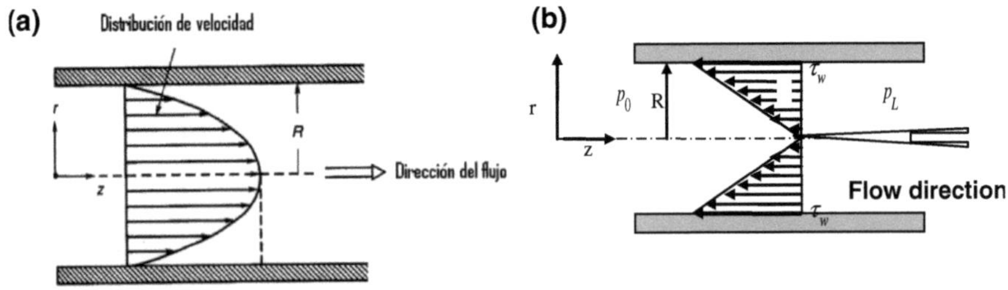

Fig. 10.21 Axial flow in cylindrical tube. **a** Velocity distribution. **b** Shear stress distribution

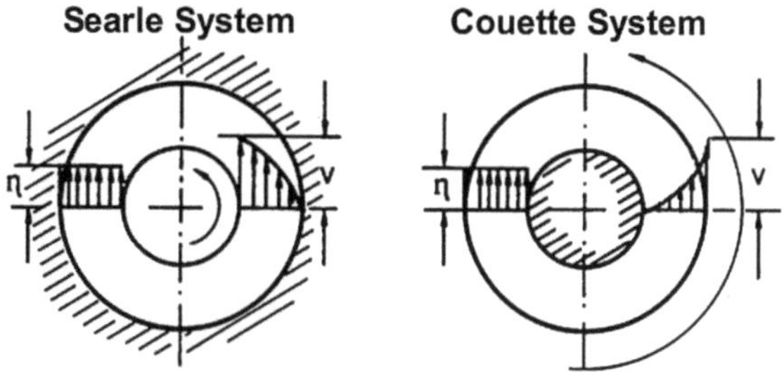

Fig. 10.22 Systems types used in rotary viscometers

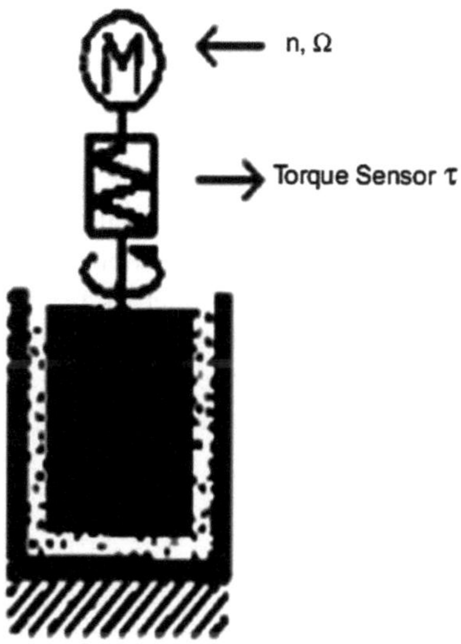

Fig. 10.23 Searle type of measuring system. Shear rate is measured on the rotor axis and the outer cylinder is fixed

During measurement, the fluid to be tested is allowed into the gap between the two cylinders. The relative motion of the cylinder induces a simple shear to the fluid, which produces a torque in the other cylinder that is measured by a suitable device. If the gap between the cylinders is small, the viscosity in the gap is constant as shown in Fig. 10.22.

The Searle system is the most commonly used for mineral pulps.

For a Searle system from Eq. (10.29), the shear rate is given by:

$$\dot{\gamma} = \frac{\Omega_1 R_2^2}{\left(R_2^2 - R_1^2\right)} \tag{10.31}$$

To determine a rheogram, a given shear rate $\dot{\gamma}$ is established in the equipment by imposing a rotational speed N_1 given in terms of $\dot{\gamma}$ by Eq. (10.32)

$$N_1 = \frac{60}{2\pi}\left(1 - \left(\frac{R_1}{R_2}\right)^2\right)\dot{\gamma} \ \text{[rpm]} \tag{10.32}$$

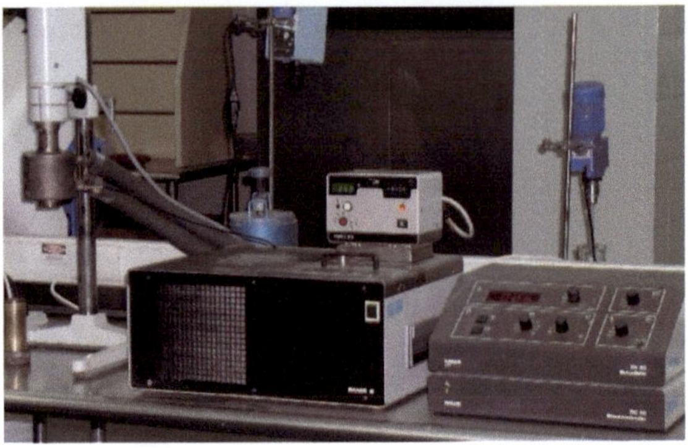

Fig. 10.24 Haake RV-20 rotational viscometer

where N_1 is the rotational speed in rpm of the inner cylinder with a radius of R_1 and R_2 is the radius outer cylinder.

A good example of a robust rotational viscometer for mineral pulps is the Haake RV-20 viscometer under ISO standard 3219. Figure 10.24 shows this instrument.

Figure 10.25 shows typical sensors for suspensions. The grooves on the outside of the inner cylinder and on the inside of the outer cylinder avoid slippage of particles along the walls.

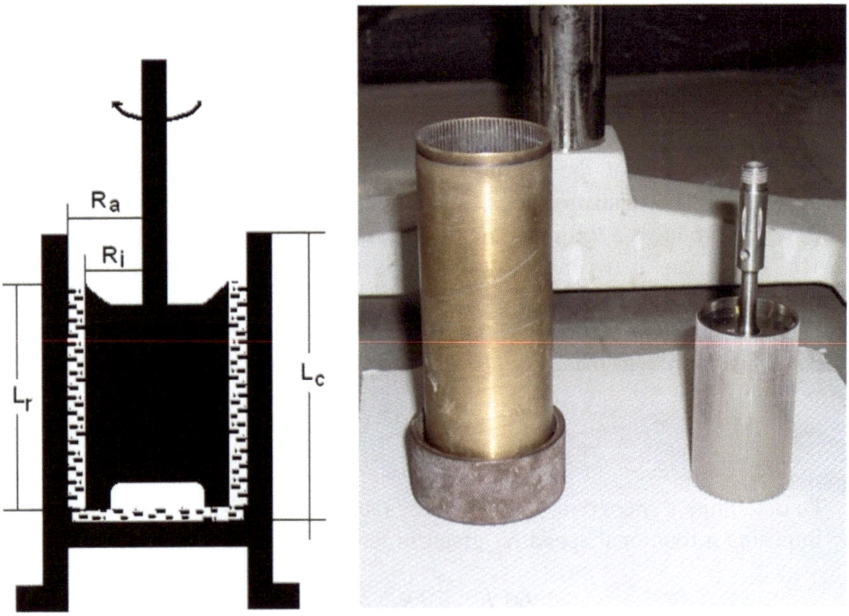

Fig. 10.25 Grooved sensors type MV

10.3.3 Standard Rheological Measurement (Rheogram ISO 3219)

1. Select the correct measuring system.
2. Fill the cup with a representative sample of slurry to the designated point.
3. Place the cup on the laboratory jack centered below the viscometer bob.
4. Slowly raise the jack so that the bob completely penetrates the slurry sample.
5. Fix the cup to the viscometer using the designated mounting screw.
6. Use the software from the rotational rheometer to produce a rheogram of the slurry sample.
7. Present a rheogram of shear stress versus shear rate between 0 and 450 s^{-1}.
8. Repeat for different concentrations.

Problem 10.5 The measurement of copper flotation tailings with a rotational viscometer at 65 % solids gave the data in Table 10.4. Obtain the complete rheogram and determine the rheological parameters with Bingham, Power-law and Herschel-Bulkley model.

Bingham Model

The Bingham model is characterized by a yield stress τ_y and a constant plastic viscosity K. Results are given in Fig. 10.26.

$$\tau_y = 40 \text{ Pa} \quad \text{and} \quad \text{K} = 0.197 \text{ Pa s}$$

Power law model

The power law model is described by the constitutive equation: $\tau = m\dot{\gamma}^n$

$$m = 6.958 \text{ Pa and } n = 0.457$$

The result is shown in Fig. 10.27.

Herschel-Bulkley Model

Herschel-Bulkley model combines a yield stress with a power-law model. Here Fig. 10.28:

$$\tau_y = 1.25 \text{ Pa}; \quad k = 2.37 \text{ Pa s}^z; \quad z = 0.343$$

Determination of the yield stress with vanes

Given the importance of yield stress in transporting mineral pulps, it needs to be determined accurately. The best way determine yield stress for values above 10 Pa is direct measurement at shear rate tending to zero. Measuring yield stress of mineral pulps at very low shear rates with rotary viscometers presents the problem of particle slip at the rotating cylinder. To avoid this problem, the *vane method* is used. This method consists of using a rotating vane, as shown in Fig. 10.29, to measure the yield stress under static conditions. The vane is submerged in the pulp,

Table 10.4 Experimental data of a copper mineral at 65 % solids

γ (1/s)	τ (Pa)	η (m Pa s)	γ (1/s)	τ (Pa)	η (m Pa s)	γ (1/s)	τ (Pa)	η (m Pa s)
0.00	7.45	0.00	90.54	54.72	604.40	226.10	83.63	369.90
0.09	7.56	81140.00	93.33	55.39	593.50	229.20	84.24	367.50
0.44	8.54	19200.00	96.27	56.31	584.90	232.40	84.79	364.90
0.90	9.65	10680.00	99.14	57.05	575.40	235.70	85.22	361.60
1.51	10.85	7174.00	102.20	57.78	565.60	238.90	85.77	359.10
2.26	12.05	5334.00	104.90	58.52	557.60	241.90	86.26	356.60
3.15	13.29	4223.00	108.00	59.31	549.40	245.20	86.76	353.80
4.07	14.53	3569.00	111.00	60.11	541.70	248.40	87.25	351.20
5.10	15.86	3111.00	113.80	60.90	535.00	251.50	87.80	349.00
6.24	17.09	2740.00	116.80	61.58	527.30	254.70	88.23	346.40
7.49	18.32	2446.00	119.80	62.31	520.20	257.90	88.72	344.00
8.82	19.53	2213.00	122.80	63.05	513.40	261.10	89.14	341.40
10.21	20.75	2033.00	125.80	63.66	506.00	264.30	89.70	339.40
11.72	21.96	1874.00	128.80	64.40	499.90	267.50	90.19	337.10
13.32	23.18	1741.00	131.80	65.13	494.10	270.80	90.55	334.40
14.96	24.41	1631.00	135.60	65.99	486.80	273.90	91.04	332.40
18.57	26.60	1433.00	138.60	66.66	481.10	277.10	91.53	330.30
20.61	28.01	1359.00	141.70	67.34	475.10	280.40	91.90	327.80
22.58	29.11	1289.00	144.70	68.01	469.90	283.50	92.33	325.70
24.54	30.15	1228.00	147.70	68.75	465.30	286.70	92.76	323.60
26.62	31.26	1174.00	150.80	69.36	460.10	290.00	93.19	321.40
28.75	32.36	1126.00	154.00	70.09	455.20	293.10	93.62	319.40
30.88	33.40	1082.00	157.00	70.64	450.00	296.30	94.04	317.30
33.09	34.50	1043.00	160.10	71.32	445.30	299.50	94.41	315.20
35.33	35.48	1004.00	163.20	71.99	441.10	302.80	94.78	313.00
37.61	36.58	972.90	166.50	72.60	436.20	306.00	95.21	311.10
39.92	37.57	941.00	169.50	73.22	432.10	309.10	95.58	309.20
44.73	39.46	882.20	172.60	73.83	427.70	312.40	95.94	307.10
47.38	40.69	858.70	175.70	74.50	424.00	315.60	96.37	305.40
49.96	41.61	832.80	178.90	75.12	419.80	318.70	96.74	303.50
52.47	42.59	811.60	181.90	75.73	416.20	321.90	97.11	301.60
54.98	43.57	792.40	185.20	76.34	412.30	325.20	97.54	300.00
57.56	44.55	773.90	188.20	76.95	408.80	328.40	97.78	297.80
60.22	45.34	753.00	191.40	77.51	404.90	331.50	98.27	296.40
62.87	46.26	735.90	194.50	78.12	401.70	334.70	98.52	294.30
65.59	47.12	718.40	197.60	78.67	398.10	338.00	98.88	292.60
68.24	48.04	704.00	200.90	79.28	394.70	341.20	99.25	290.90
70.97	48.96	689.90	204.00	79.83	391.30	344.40	99.62	289.30
73.69	49.88	676.80	207.20	80.45	388.30	347.60	99.80	287.10
76.34	50.61	663.00	210.30	81.00	385.20	350.80	100.30	285.90
79.14	51.53	651.10	213.50	81.49	381.70	354.00	100.70	284.40
82.01	52.27	637.30	216.60	82.10	379.10	357.20	100.90	282.50
84.87	53.12	625.90	219.80	82.59	375.80	360.40	101.30	281.00
87.67	53.92	615.00	222.90	83.14	372.90	363.70	101.60	279.30
						366.90	101.90	277.70

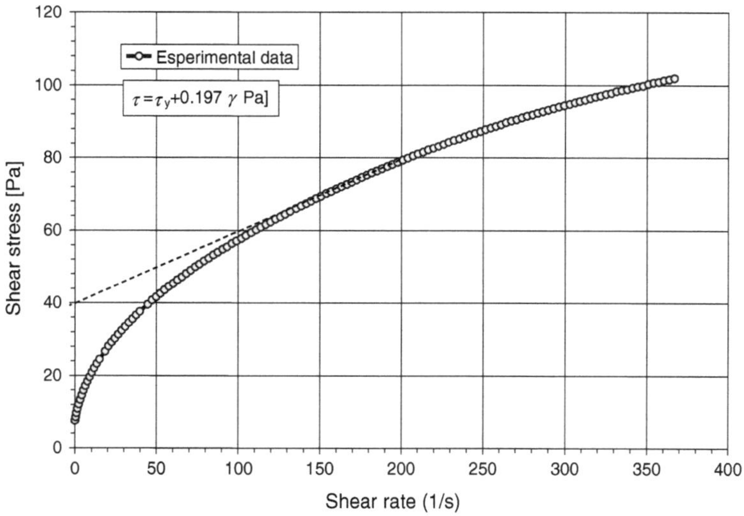

Fig. 10.26 Bingham parameters for data from Table 10.4 by extrapolation of the rheological curve

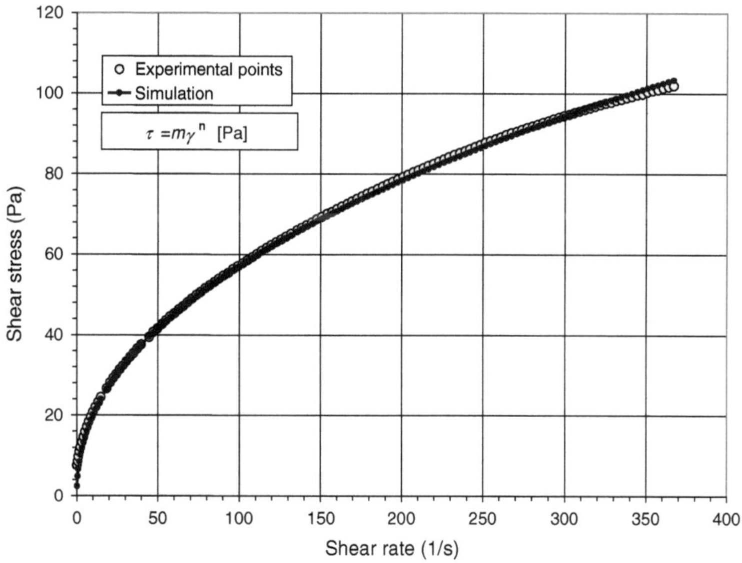

Fig. 10.27 Simulation with power law model of data of Table 10.4

rotated at a speed of less than 10 [rpm] and torque is slowly increased. After a linear elastic deformation of the shear surface formed, a maximum torque T_M is reached as shown in Fig. 10.30. Appropriate operating conditions are $D_T > 3d$ and $N < 10$ [rpm]. Three (kg) are needed for each test.

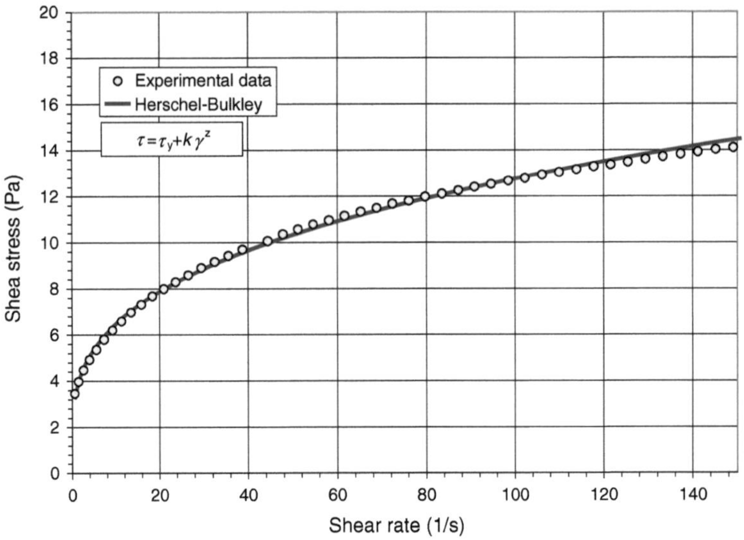

Fig. 10.28 Simulation with Herschley-Bulkley model for data from Table 10.4

Fig. 10.29 Vane rotor to determine yield stress

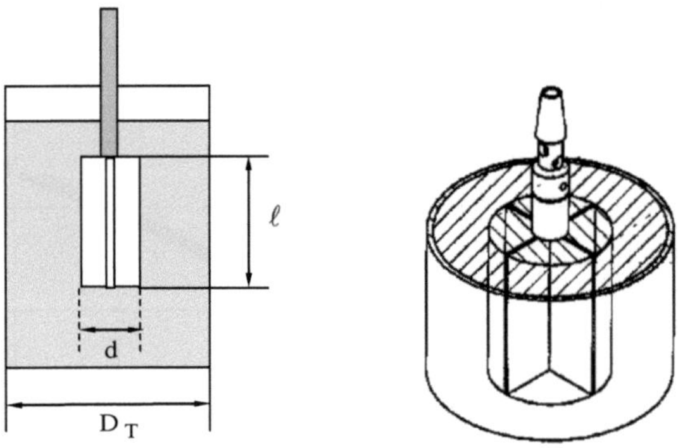

With vane measurements an approximate value of the yield stress is obtained from Eq. (10.33) with values 20–30 % lower than the real value. This is because the shear distribution is not uniform, the sides of the shear surface having different values from each other. In the absence of theoretical knowledge, Nguyen and Boger (1985) assumed a potential distribution with power *m*, and obtained the equation:

$$T_m = \frac{\pi d^3}{2}\left(\frac{\ell}{d} + \frac{1}{3}\right)\tau_y \tag{10.33}$$

Due to the presence of two unknowns, τ_y and m, in equation, it is necessary to perform more than one test, usually three, with vanes of different shapes ℓ/d to obtain the values of these unknowns simultaneously.

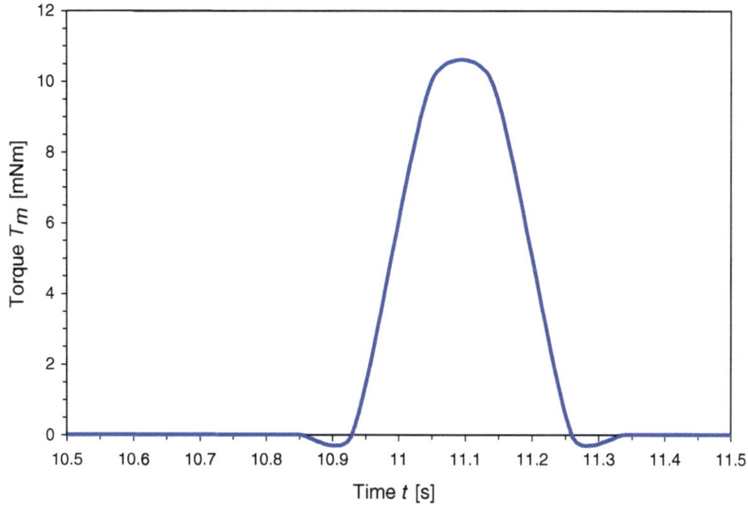

Fig. 10.30 *Torque curve* versus time showing the maximum torque reached

Writing Eq. (10.33) in the form:

$$\frac{2T_m}{\pi d^3} = \frac{\ell}{d}\tau_y + \frac{\tau_y}{m+3} \tag{10.34}$$

a plot of $2T_M/\pi d^3$ vs. ℓ/d gives a straight line. The slope of the line is τ_y and the intercept with the vertical axis is $\tau_y/(m+3)$, which gives the value of m. See Fig. 10.31. In the case of Fig. 10.31:

$$\frac{\tau_y}{m+3} = 2.8551 \;\rightarrow\; \tau_y = 32.444 \;\rightarrow\; m = 8.36 \tag{10.35}$$

(b) **Capillary viscometers**

A capillary viscometer is a straight cylindrical tube with diameter D and length L, through which the sample to be tested flows with constant velocity v. The time t for a given volume Q to flow between levels of the tube at a constant pressure gradient is measured. If the material has a Newtonian behavior, the Hagen-Poiseuille equation relates these variables. See Eq. (10.36).

$$Q = \frac{1}{8}\frac{\pi R^4}{\eta}\frac{\Delta p}{L}\;\left[\mathrm{m}^3/\mathrm{s}\right] \tag{10.36}$$

Since $Q = \bar{v}_z/t$, where $\bar{v}_z$ is the average velocity. The flow is gravity driven with $\Delta p/L = \rho g$, and the kinematical viscosity is $\nu = \eta/\rho$, we have:

$$\nu = \frac{\pi}{32}\frac{gD^2}{\bar{v}_z}t\;\left[\mathrm{cm}^2/\mathrm{s}\right] \tag{10.37}$$

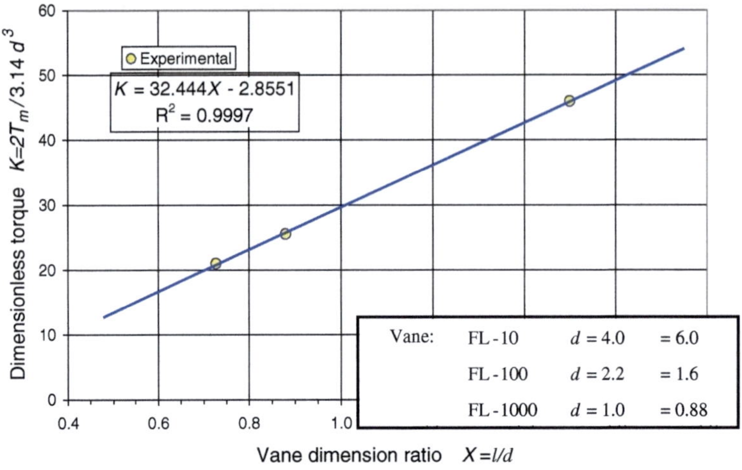

Fig. 10.31 Yield stress determination with the vane method

For a determined capillary and constant velocity, Eq. (10.37) is written in the form:

$$\nu = K \times t\ \mathrm{cm}^2/\mathrm{s}; \quad \text{where} \quad K = \frac{\pi}{32}\frac{gD^2}{\bar{v}_z} \tag{10.38}$$

Manufacturers have automated and standardized capillary viscometers and give the constant K for each capillary to facilitate its use. An example is the Cannon–Fenske capillary viscometer with Lauda control. See Fig. 10.32.

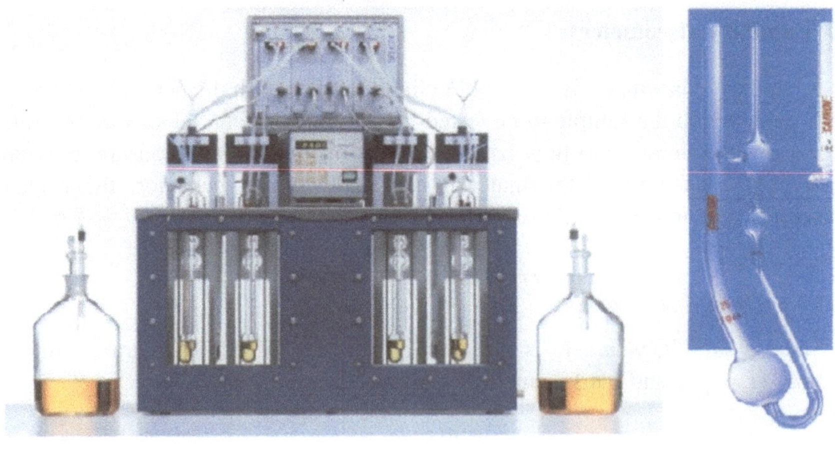

Fig. 10.32 Cannon-Fenske Capillary Viscometer with Lauda Control

Selection of capillary diameters

It is important that the material to be tested behave as a Newtonian fluid, because of which the above equation was developed. To ensure this requirement, the flow in the capillary must give a shear rate $\dot{\gamma}$ within the Newtonian range. In Chap. 11 we establish the following equation for the average shear rate in the flow in a tube:

$$\bar{\dot{\gamma}} = 6.8\frac{Q}{D^3} \tag{10.39}$$

For example, the flow of 15 mm/s in a capillary of 1.01 mm gives a shear rate of $\bar{\dot{\gamma}} = 99\,\text{s}^{-1}$, which is in the Newtonian range (see Chap. 11) and corresponds to a Cannon–Fenske capillary N° 200.

References

Barrientos, A, Concha, F., & León, J. C. (1994). A mathematical model of solid-liquid suspensions. IV Meeting of the Southern Hemisphere on mineral technology, vol. I. Mineral Processing and Environment (pp. 189–199). Concepción: University of Concepción.

Concha, F., Castro, O., & Muñoz, L. (1999). Shear viscosity of copper ore pulp in a grinding-classification circuit. *Mineral Processing and Extractive Metallurgy Review: An International Journal, 20*, 155–165.

Krieger, I. M. (1972). Colloid Interface Science. *Advances in Colloid and Interface Science, 3*, 111.

Nguyen, Q. D., & Bolger, D. V. (1985). Direct yield stress measurement with the vane method. *Journal of Rheology, 29*(3), 345–347.

Schramm, G. (2000). *A practical approach to rheology and rheometry* (p. 21). Gebrueder Haake, BmbH: Karlsruhe, Federal Republic of Germany.

Shaheen, I. (1972). Rheological study of viscosities and pipeline flow of concentrated slurries. *Powder Technology, 5*, 245–256.

Tanner, R. I. (1988). *Engineering rheology* (p. 348). Oxford: Oxford Science Publications, Clarendon Press.

Chapter 11
Transporting Concentrates and Tailings

Abstract Ore, water and mineral pulps are transported among the different operational units of a mineral processing plant. Water is pumped through pipelines to the grinding plant to be mixed with the ore to form the pulp that constitutes the mill feed. The mill overflow is again mixed with water to adjust the solid content and is sent through pipes to be classified in hydrocyclones. Cyclone underflow with coarse material is sent back to the mill and the overflow goes to the flotation plant. Transport in the flotation plant and between flotation sections and solid–liquid separation units is through pipelines, and finally flotation tailings are transported to tailing ponds through pipelines or channels. This chapter of the book is related to the transport of pulps in mineral processing plants. Starting from the continuity equation and the equation of motion for a continuous medium, the expression for the pressure drop during fluid flow in a tube is obtained. Newtonian fluid behavior is used to treat cases of laminar and turbulent flows. The concepts of friction factor and Reynolds number are introduced and the distribution of velocity, flow rate and pressure drop in a tube are obtained. The transport of suspensions in pipelines is then treated, defining the different regimes separated by the limiting deposit velocity. First, the flow of heterogeneous suspensions is introduced and the form to calculate head loss is presented. Next, homogeneous suspensions modeled by different rheological approaches are discussed. Finally equations for the transport of suspensions in open channel are dealt with.

Ore, water and mineral pulps must be transported among the different operational units of a mineral processing plant. In the crushing plant, where the ore is essentially dry, it is transported efficiently by conveyor belts. Water is pumped through pipelines to the grinding plant to be mixed with the ore to form the pulp that constitutes the mill feed. The mill overflow is again mixed with water to adjust the solid content and is sent through pipes to be classified in hydrocyclones. Cyclone underflow with coarse material is send back to the mill and the overflow goes to the flotation plant. Transport in the flotation plant and between flotation sections and solid-liquid separation is through pipelines, and finally flotation tailings are transported to tailing ponds through pipelines or Channels.

F. Concha A, *Solid–Liquid Separation in the Mining Industry*,
Fluid Mechanics and Its Applications 105, DOI: 10.1007/978-3-319-02484-4_11,

Pipelines in mineral processing plants enable transporting maximum loads with a minimum of space using conventional centrifugal pumps and pipes that in most cases do not exceed 24 inches in diameter. Pipelines are extremely flexible and can be used for short distance tailing disposal and long distance concentrate transportation, No matter how complex the topography; a pipeline can always be laid out.

Slurries can be classified as *homogenous* and *heterogeneous* suspensions. Homogenous suspensions behave like fluids with increased density and particular rheology, while in heterogeneous suspensions, also called mixed slurries; solid particles settle and form a solid vertical concentration profile and some bed formation while being transported.

A suspension at low solid concentration with particle sizes of less than 270 mesh (50 μm) behaves heterogeneously and requires high transport velocity to prevent particles from settling. The same suspension for high concentrations behaves homogeneously at any transport velocity. The latter suspension behaves as a mono phase fluid with particular rheological behavior. Knowledge about the rheological properties of dense slurries is fundamental to design pipeline systems. The power consumption to pump 100 (tph) of homogeneous slurry horizontally is between 0.1 and 0.2 (kW/ton-km) (Condolios and Chapus 1967). Gravity transport of homogeneous slurries is possible if there is a gradient of at least 1.5 (m) per 100 (m).

Slurries with particles larger than 270 mesh (50 μm) form heterogeneous or mixed slurries that produce vertical concentration profiles and bed formations while being transported. Particles in these suspensions are transported by saltation, by bed movement or with concentration gradients that depend on the size of the particles and the flow velocity. Higher velocities must be used to prevent settling. Pulp with particle sizes under 9 mesh (2 mm) and at least 20 % of material under 270 mesh can be transported by centrifugal pumps with a power consumption of about 3–4 kW/ton-km for a capacity of 100 tph (Condolios and Chapus 1967). Materials with sizes over 9 mesh (2 mm) require more power, in the range of 6–12 kW/ton-km (Condolios and Chapus 1967), and subject pipes to severe wear.

11.1 Transporting Fluids in Pipelines

Incompressible stationary flow in a horizontal circular tube can be described by the following variables, the fluid (1) density $\rho(\boldsymbol{r}, t)$, (2) velocity $\boldsymbol{v}(\boldsymbol{r}, t)$ and (3) stress tensor $\boldsymbol{T}(\boldsymbol{r}, t)$, where r and t are the position vector and time respectively. These three field variables must obey the mass and linear momentum field equations:

$$\nabla \cdot \boldsymbol{v} = 0 \tag{11.1}$$

$$\rho \nabla \boldsymbol{v} \cdot \boldsymbol{v} = \nabla \cdot \boldsymbol{T} + \rho \boldsymbol{g} \tag{11.2}$$

where $\boldsymbol{g}$ is the gravitational constant.

Since there are three field variables and only two field equations, a constitutive equation must be postulated for the stress tensor:

$$\boldsymbol{T} = -p\boldsymbol{I} + \boldsymbol{T}^E(\boldsymbol{r}) \tag{11.3}$$

where p is the pressure and $\boldsymbol{T}^E$ is the extra stress tensor.

Cylindrical tubes have axial-symmetry and cylindrical coordinates can be used. Thus for the horizontal tube shown in Fig. 11.1, the following equations are valid:

$$\text{Continuity} \quad \frac{\partial v_z(r,z)}{\partial z} = 0, \Rightarrow v_z = v_z(r) \tag{11.4}$$

Since the velocity varies in the r direction only, T_{rz}^E must be a function solely of r.

$$\text{Momentum component r:} \quad 0 = -\frac{\partial p}{\partial r} + \frac{\partial T_{rz}^E(r)}{\partial z} + \rho g_r \tag{11.5}$$

$$\text{Momentum component } \theta\text{:}\ 0 = -\frac{\partial p}{\partial \theta} + \rho g_\theta \tag{11.6}$$

$$\text{Momentum component z:}\ 0 = -\frac{\partial p}{\partial z} + \frac{1}{r}\frac{\partial}{\partial r}\left(rT_{rz}^E(r)\right) \tag{11.7}$$

Equation (11.7) can be written in the form:

$$\frac{\partial p}{\partial z} = \frac{1}{r}\frac{\partial}{\partial r}\left(rT_{rz}^E(r)\right) = K \tag{11.8}$$

Integrating by parts and writing the pressure drop $\Delta p = p_0 - p_L > 0$, the left side of (11.8) yields:

$$\begin{aligned} &\int_{p_0}^{p_L} dp = \int_0^L K dz, \quad \Rightarrow p_L - p_0 \equiv -\Delta p = KL \\ &K = -\frac{\Delta p}{L} \end{aligned} \tag{11.9}$$

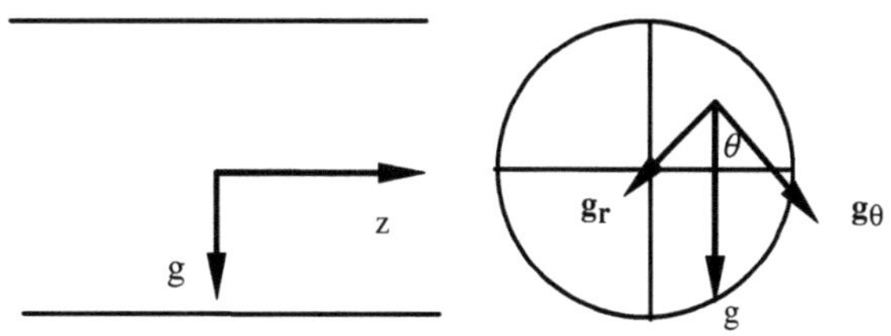

Fig. 11.1 Flow in a *horizontal* tube

Fig. 11.2 Shear stress distribution for the flow in a *cylindrical* tube

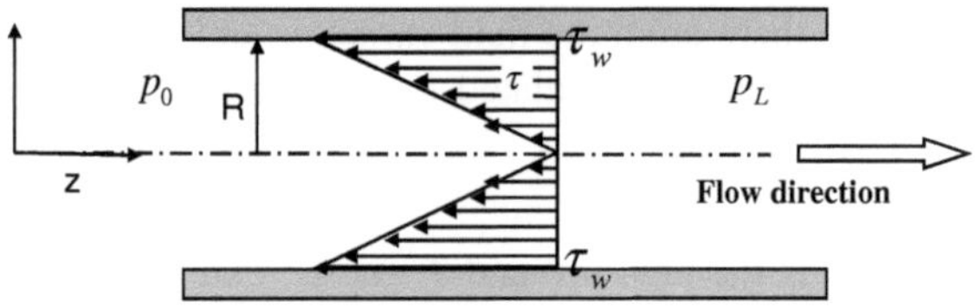

For the right side of (11.8), integrating by parts yields:

$$\int d\left(rT^E_{rz}(r)\right) = \int Krdr$$
$$T^E_{rz}(r) = \frac{1}{2}Kr + \frac{C}{r}$$

Since the stress is finite at the tube axis, $T_{rz}(0) \neq \infty$, for $r = 0\, C = 0$, then:

$$T^E_{rz}(r) = \frac{1}{2}Kr$$

Substituting K from (11.9) yields the distribution of shear stress in a cylindrical tube:

$$T^E_{rz}(r) = -\frac{1}{2}\frac{\Delta p}{L}r \tag{11.10}$$

Designating $T^E_{rz}(r) \equiv \tau(r)$, Eq. (11.10) is usually written in the form (See Fig. 11.2):

$$\tau(r) = -\frac{1}{2}\frac{\Delta p}{L}r \quad \text{with}\, \Delta p = p_0 - p_L > 0 \tag{11.11}$$

If we call τ_w the shear stress at the wall, from Eq. (11.10) we can write:

$$\tau_w = -\frac{1}{2}\frac{\Delta p}{L}R \tag{11.12}$$

The ratio of shear stress at r and at the wall is:

$$\frac{\tau(r)}{\tau_w} = \frac{r}{R} \tag{11.13}$$

It is important to realize that Eqs. (11.10–11.13) are valid for all types of fluids, since we have not invoked any type of constitutive equation for $T^E_{rz}(r)$.

11.2 Newtonian Fluids

11.2.1 Laminar Flows

For a Newtonian fluid, the constitutive equation for the extra stress $T_{rz}^E(r)$ is:

$$T_{rz}^E(r) = \mu\left(\frac{\partial v_z}{\partial r} + \frac{\partial v_r}{\partial z}\right) \quad (11.14)$$

Using the continuity Eq. (11.4) and substituting (11.10) gives:

$$-\frac{1}{2}\frac{\Delta p}{L}r = \mu\frac{\partial v_z}{\partial r} \quad (11.15)$$

Velocity distribution

Integrating (11.15) with boundary condition $v_z(R) = 0$ at the wall gives:

$$v_z(r) = \frac{1}{4}\frac{\Delta p R^2}{\mu L}\left(1 - \left(\frac{r}{R}\right)^2\right) \quad (11.16)$$

The velocity distribution is parabolic as shown in Fig. 11.3.

Volume flow rate

The volume flow rate is given by $Q = \int_0^R 2\pi v_z(r) r dr$, then:

$$\begin{aligned} Q &= \frac{1}{2}\frac{\pi \Delta p R^4}{\mu L}\int_0^1 \left(1 - \left(\frac{r}{R}\right)^2\right)\frac{r}{R}d\left(\frac{r}{R}\right) \\ Q &= \frac{1}{8}\frac{\Delta p \pi R^4}{\mu L} \end{aligned} \quad (11.17)$$

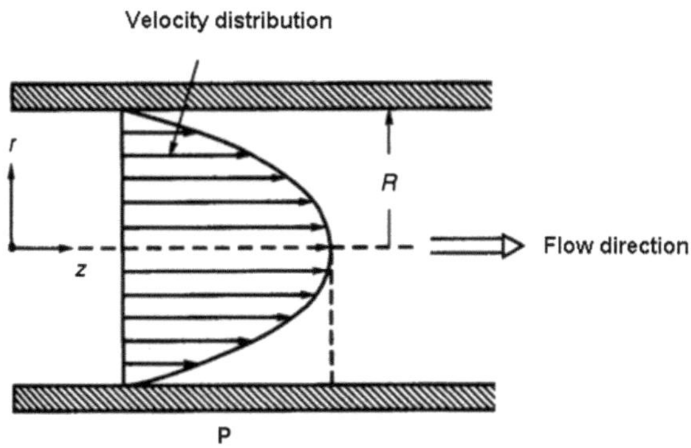

Fig. 11.3 Velocity distribution for the flow of a Newtonian fluid in a *circular* tube

Pressure gradient

$$\frac{\Delta p}{L} = \left(\frac{8\mu Q}{\pi R^4}\right) \tag{11.18}$$

Average velocity

The average velocity can be obtained from the volume flow rate by $\bar{v}_z = Q/A$, where $A = \pi R^2$ is the cross sectional area of the tube:

$$\bar{v}_z = \frac{1}{8}\frac{\Delta p R^2}{\mu L} \tag{11.19}$$

Shear rate at the wall

Defining the shear rate $\dot{\gamma}_w = \partial v_z/\partial r|_{r=R}$ at the wall as $\tau_w = \mu\dot{\gamma}_w$, from Eq. (11.12) for τ_w, we get:

$$\dot{\gamma}_w = \frac{1}{2}\frac{\Delta p R}{\mu L} \tag{11.20}$$

and using (11.19) we can write:

$$\dot{\gamma}_w = \frac{8\bar{v}_z}{D} \tag{11.21}$$

where $D = 2R$ is the tube diameter.

Maximum velocity

The maximum velocity is obtained from (11.16) for $r = 0$:

$$v_m = \frac{1}{4}\frac{\Delta p R^2}{\mu L} \tag{11.22}$$

Problem 11.1 Calculate the velocity distribution of three fluids with different viscosities, $\mu = 0.001,\ 0.002$ and 0.008 [Pa-s] in a tube 1 inch in diameter and 50 m in length, subjected to a pressure drop of 172 (Pa). As well, calculate the flow rate, average and maximum velocity, wall shear stress and shear rate and the Reynolds flow number.

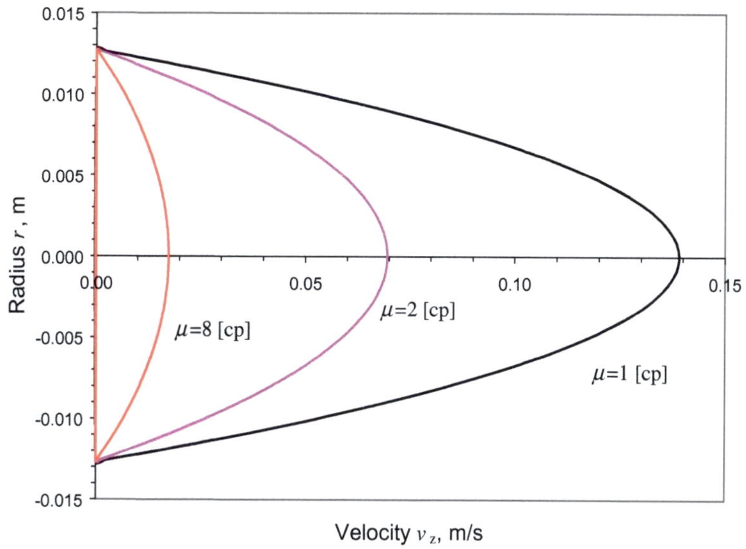

Fig. 11.4 Velocity distribution for the flow of three fluids; with viscosities 1, 2 and 8 cp, in a cylindrical tube 1 inch in diameter and 500 m in length

Data are: $R = 0.0127\,\text{m}$, $L = 50\,\text{m}$, $\Delta\text{p} = 172\,\text{Pa}$, $\mu = 0.001;\ 0.002,\ 0.008\,\text{Pa-s}$
As an example, calculate with $\mu = 0.001$ Pa-s (see Fig. 11.4).

$$v_z(r) = \frac{1}{4}\frac{\Delta p R^2}{\mu L}\left(1 - \left(\frac{r}{R}\right)^2\right) = \frac{172 \times 0.0127^2}{4 \times 0.001 \times 50}\left(1 - \left(\frac{r}{R}\right)^2\right)$$

$$v_z(r) = 0.14\left(1 - \left(\frac{r}{R}\right)^2\right)\ \text{m/s}$$

Maximum velocity:

$$v_m = 0.14\ \text{m/s}$$

Volume flow rate:

$$Q = \frac{1}{8}\frac{\Delta p \pi R^4}{\mu L} = \frac{172 \times \pi \times 0.127^4}{8 \times 0.001 \times 50} = 3.52 \times 10^{-5}\,\text{m}^3/\text{s}$$

Average velocity:

$$\bar{v}_z = \frac{Q}{\pi R^2} = \frac{3.52 \times 10^{-5}}{3.14 \times 0.0127^2} = 0.035\,\text{m/s}$$

Shear rate and shear stress at the wall

$$\dot{\gamma}_w = \frac{8\bar{v}_z}{D} = \frac{8 \times 0.035}{0.0127 \times 2} = 21.9\,\mathrm{s}^{-1}$$
$$\tau_w = \mu\dot{\gamma}_w = 0.001 \times 21.9 = 0.022\,\mathrm{Pa}$$

Reynolds number $\mathrm{Re} = \frac{\rho D\bar{v}}{\mu} = \frac{1000 \times 2 \times 0.0127 \times 0.035}{0.001} = 1.77 \times 10^3$

Summary:

Newtonian fluid			
μ (Pa-s)	0.001	0.002	0.008
L (m)	50	50	50
R (m)	0.0127	0.0127	0.0127
Δp (Pa)	172	172	172
Q (m^3/s)	3.52E−05	1.76E−05	4.40E−06
v_{av} (m)	0.070	0.035	0.009
v_m (m)	0.14	0.07	0.02
γ_w (s^{-1})	21.9	10.9	2.7
τ_w (Pa)	0.022	0.022	0.022
ρ (kg/m^3)	1.00E+03	1.00E+03	1.00E+03
Re	1.77E+03	4.41E+02	2.76E+01

Friction factor for Newtonian fluids

The dimensionless solid-fluid resistance coefficient, called the *Fanning friction factor,* is defined as the ratio of friction at the wall to the dynamic pressure:

$$f = \frac{-\tau_w}{1/2\rho\bar{v}_z^2} \tag{11.23}$$

From Eq. (11.12) $\tau_w = -\frac{1}{2}\frac{\Delta p}{L}R$, substituting (11.23) yields:

$$f = \frac{\Delta p}{L}\frac{D}{2\rho\bar{v}_z^2} \tag{11.24}$$

Equation (11.24) shows that the Fanning friction factor can also be interpreted as the ratio of the pressure gradient to halve the dynamic pressure. Substituting the value of $\Delta p/L$ from (11.19) with (11.24) results in:

$$f = \frac{16}{\rho D\bar{v}_z/\mu}$$

Using the definition of the Reynolds number $\mathrm{Re} = \rho D\bar{v}_z/\mu$, the Fanning friction factor for the laminar flow of a Newtonian fluid is:

$$f = \frac{16}{\mathrm{Re}} \tag{11.25}$$

Another definition of the friction factor is the ratio of head loss to velocity head: $\lambda = h_L / \left(\frac{\bar{v}_z^2}{2g}\right)\left(\frac{L}{D}\right)$ [m], and since $h_L = \frac{\Delta p}{\rho g}$, $\lambda = \frac{\Delta p}{L}\frac{D}{(1/2)\rho \bar{v}^2}$, then $\lambda = 4f$. This version is called the Darcy-Weisbach friction factor. In terms of λ, the friction factor for Newtonian fluids is:

$$\lambda = \frac{64}{\mathrm{Re}} \tag{11.26}$$

11.2.2 Turbulent Flows

The transport of suspensions occurs in laminar or turbulent regimes. The parameters defining the transition between laminar and turbulent flows are the *Fanning friction factor f* and the *Reynolds number* Re.

Due to the overriding effect of viscosity forces in the laminar flow of Newtonian fluids, even flows over asperous surfaces appear smooth. Therefore, the roughness of the walls, unless it is very significant, does not affect flow resistance. Under these flow conditions the friction coefficient is always a function of the Reynolds number alone.

As the Reynolds number increases, inertia forces, which are proportional to velocity squared, begin to dominate. The turbulent motion is characterized by the development of transverse component of the velocity, giving rise to agitation of the fluid throughout the stream and to momentum exchange between randomly moving masses of fluid. All this causes a significant increase in the resistance to the motion in turbulent flow compared to laminar flow.

When the surface of the wall is rough, separation occurs in the flow past the rough section and the resistance coefficient becomes a function of the Reynolds number and the relative roughness ε^*, defined as the ratio of the roughness height and the tube diameter:

$$\varepsilon^* = \frac{\varepsilon}{D} \tag{11.27}$$

where ε is the average height of the asperities and D is the tube diameter. While for low velocity flows in smooth tubes the friction factor decreases with higher Reynolds numbers, in rough tubes the friction factor increases with the Reynolds number and constant relative roughness. This is because at low flows the viscous sublayer δ is greater than the roughness protuberances $\delta > \varepsilon$ and the fluid moves smoothly past irregularities, while at higher velocities the sublayer becomes thinner than the roughness protuberances, $\delta < \varepsilon$, which enhances the formation of vortices and increases the friction factor and pressure drop. Tubes are considered

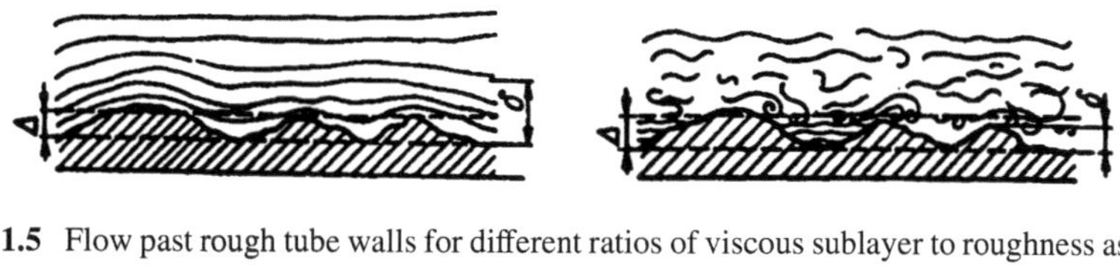

Fig. 11.5 Flow past rough tube walls for different ratios of viscous sublayer to roughness asperity

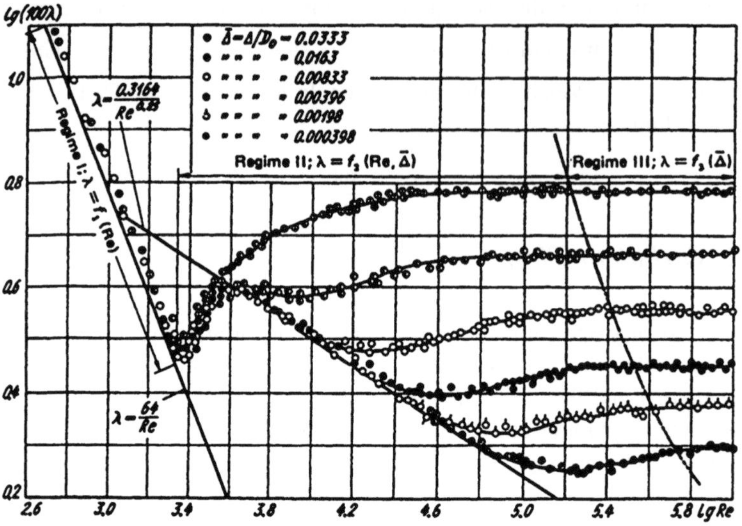

Fig. 11.6 Friction factor λ versus Reynolds number Re for tubes with uniform roughness, with $\bar{\Delta} = \varepsilon^*$, according to Nikuradse (1933), Idelchik et al. (1986)

smooth as long as the height of the asperity is less than the thickness of the laminar sublayer. See the following Fig. 11.5.

Nikuradse (1933) made flow experiments with tubes covered with different sizes of sand to simulate uniform roughness. His results are given in Fig. 11.6, which can be interpreted as consisting of three regimes (Tamburrino 2000): (1) laminar flow, (2) transition to turbulence and (3) rough walls regime.

First regime. In the first regime, with Reynolds numbers lower than 2,100, f is independent of the roughness of the tube and is given by:

$$\lambda = \frac{64}{\text{Re}} \tag{11.28}$$

Second regime. With $\text{Re} > 2,100$ and $\text{Re}_{\in} < 5$ the friction factor is given by the Blasius equation for all roughness (see Fig. 11.6):

$$\frac{1}{\sqrt{\lambda}} = -2\log\left(\frac{2.51}{\text{Re}\sqrt{\lambda}}\right) \tag{11.29}$$

For $\text{Re} > 2,100$ and $5 < \text{Re}_{\in} < 70$ the friction factor increases with the Reynolds number diverging to different lines for different degrees of constant relative roughness:

$$\frac{1}{\sqrt{\lambda}} = -2\log\left(\frac{2.5}{\text{Re}\sqrt{\lambda}} + \frac{\varepsilon^*}{3.7}\right) \quad \text{for} \quad \text{Re} > 2,100 \text{ and } 5 < \text{Re}_{\in} < 70 \tag{11.30}$$

where the Reynolds roughness number is $\text{Re}_{\in} = \varepsilon^* \sqrt{f/2}\text{Re}$.

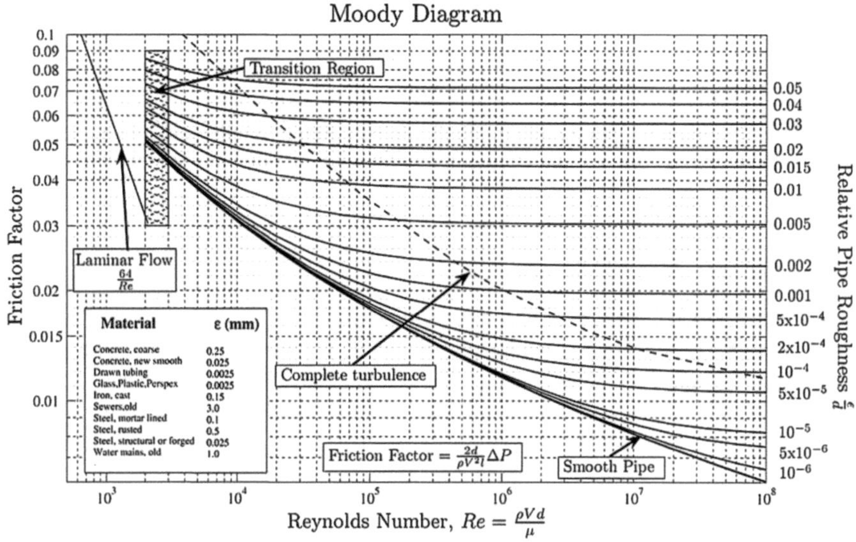

Fig. 11.7 Moody diagram

Third regime. In the third regime the friction factor becomes a different constant for each relative degree of roughness, independent of the Reynolds number:

$$\frac{1}{\sqrt{\lambda}} = -2\log\left(\frac{\varepsilon^*}{0.854}\right) \quad \text{for} \quad \text{Re} > 2{,}100 \text{ and } \quad \text{Re}_{\in} > 70 \tag{11.31}$$

Moody diagram for commercial pipes is a version of Eq. (11.30). See Fig. 11.7.

Problem 11.2 If D and $\bar{v}_z$ are known, calculate the pressure gradient due to friction when water flows through a 4-inch diameter pipe at 1.5 m/s for pipe roughness $\varepsilon = 0\,(\text{smooth}), 0.1, 0.5$ and 1 mm. Applying Eqs. (11.29–11.31) yields:

ε (mm)	0	0.1	0.5	1
D (inch)	4	4	4	4
ρ (kg/m^3)	1,000	1,000	1,000	1,000
μ (Pa-s)	0.001	0.001	0.001	0.001
v_{av} (m/s)	1.5	1.5	1.5	1.5
D (m)	0.1016	0.1016	0.1016	0.1016
E	0	0.00098425	0.00492126	0.00984252
Re	152,400	152,400	152,400	152,400
f	0.00413	0.00534	0.00771	0.51021
$1/f^{0.5} - 1/f^{0.5}$	(0.00)	(0.00)	(0.00)	(0.00)
Re_E	–	8	47	758
	183	237	342	22,598

Explicit equations for the friction factor

When the value of the average velocity or of the volume flow rate is not known, the Reynolds number and the friction factor cannot be calculated directly. To avoid using iterative calculations, Concha (2008) observed that $\mathrm{Re}\sqrt{f}$ is a dimensionless number independent of *average velocity*. From Eq. (11.24) $f = \left(D/2\rho\bar{v}_z^2\right) \times \Delta p/L$, so that

$$\mathrm{Re}^2 f = \left(\frac{\rho D\bar{v}_z}{\mu}\right)^2 \left(\frac{D}{2\rho\bar{v}_z^2}\right)\left(\frac{\Delta p}{L}\right) = \left(\frac{\rho}{2\mu^2}\frac{\Delta p}{L}\right) \times D^3 \tag{11.32}$$

Since the left-hand side of this equation is dimensionless, the right-hand side should also be dimensionless and a parameter Ξ, with dimensions of size, which can be defined as:

$$\Xi^3 = \left(\frac{2\mu^2 L}{\rho\Delta p}\right) \tag{11.33}$$

so that (11.32) can be written in the form:

$$\mathrm{Re}^2 f = \left(\frac{D}{\Xi}\right)^3 = D^{*3};\ \text{and}\ \mathrm{Re}\sqrt{f} = \left(\frac{D}{\Xi}\right)^{2/3} = D^{*2/3} \tag{11.34}$$

Similarly, Re/f is a dimensionless number independent of the *pipe diameter*. Consider the function:

$$\frac{\mathrm{Re}}{f} = \frac{\rho D\bar{v}_z}{\mu}\frac{2\rho\bar{v}_z^2}{D}\left(\frac{\Delta p}{L}\right)^{-1} = \frac{2\rho^2}{\mu}\left(\frac{\Delta p}{L}\right)^{-1}\bar{v}_z^3 \tag{11.35}$$

Defining the parameter Z, with the dimension of velocity by:

$$Z^3 = \left(\frac{\mu}{2\rho^2}\frac{\Delta p}{L}\right) \tag{11.36}$$

then Eq. (11.35) can be written in the following form:

$$\frac{\mathrm{Re}}{f} = \left(\frac{\bar{v}_z}{Z}\right)^3 = \bar{v}_z^{*3};\quad \text{and}\quad \frac{1}{\sqrt{f}} = \frac{\bar{v}_z^{*3/2}}{\mathrm{Re}^{1/2}} \tag{11.37}$$

Multiplying (11.32) and (11.37) yields:

$$\mathrm{Re} = D^*\bar{v}_z^* \tag{11.38}$$

From (11.34) $f = \dfrac{D^*}{\bar{v}_z^{*2}}$

and from (11.29) to (11.31) we get:

$$\frac{\bar{v}_z^*}{D^{*1/2}} = -4\log\left(\frac{1,26}{D^{*3/2}} + \frac{\in}{3.7}\right), \tag{11.39}$$

Since for Re < 2,100 $f = \frac{16}{\mathrm{Re}}$ and $\frac{\mathrm{Re}}{f} = \bar{v}_z^{*3}$, we finally obtain:and

$$\begin{aligned} \text{for Re2, 100} \quad \bar{v}_z^* &= \frac{1}{16} D^{*2} \\ \text{for Re2, 100} \quad \bar{v}_z^* &= -4 \log\left(\frac{A}{D^{*3/2}} + B\varepsilon^*\right) D^{*1/2} \\ \text{where} \quad \mathrm{Re}_\in 5 : A &= 1.25; \quad B = 0 \\ 5 \leq \mathrm{Re}_\in \leq 70 : A &= 1.25; \quad B = 0.270 \\ \mathrm{Re}_\in 70 : A &= 0; \quad B = 0.171 \end{aligned} \tag{11.40}$$

$\mathrm{Re}_\in = \varepsilon^* \sqrt{f/2}\mathrm{Re}.$

Problem 11.3 (If D and $\Delta p/L$ are known) Calculate the flow rate that will be achieved when water is forced through a pipe 8 inches in diameter under a pressure gradient of 200 Pa/m, if the pipe roughness is $\varepsilon = 0.25$ mm.

ε (mm)	0.25	0.25	0.25	0.25
D (inch)	8	8	8	8
ρ (kg/m^3)	1,000	1,000	1,000	1,000
μ (Pa-s)	0.001	0.001	0.001	0.001
	200	200	200	200
D (m)	0.2032	0.2032	0.2032	0.2032
$\varepsilon^* = Z$	0.0000508	0.00123031	0.00123031	0.001230315
Ξ	0.00021544	0.00021544	0.00021544	0.000215443
Z	0.00464159	0.00464159	0.00464159	0.00464159
D*	943.17	943.17	943.17	943.17
v^*_{av}	7.6778	11.4969	420.7164	349.0550
v_{av}	0.036	0.053	1.953	1.620
Re	7,241	10,844	396,807	329,219
f	16.00000	7.13560	0.00533	0.00774
$\mathrm{Re_E}$	1	25	25	25
Q (m^3/s)	0.00115568	0.00173055	0.06332765	0.052540943

The 4th column with $\mathrm{Re}_\in = 25$ gives the correct result.

Problem 11.4 (If D and $\Delta p/L$ are known) Calculate the flow rate that will be achieved when water is forced through a 4-inch diameter pipe under a pressure gradient of 180 Pa/m, if the pipe roughness is $\varepsilon = 0$ (smooth), $0.1, 0.5$ and 1 mm.

ε (mm)	0	0.1	0.5	1
D (inch)	4	4	4	4
ρ (kg/m^3)	1,000	1,000	1,000	1,000
μ(Pa − s)	0.001	0.001	0.001	0.001
$\Delta p/L$(Pa/m)	180	180	180	180
D (m)	0.1016	0.1016	0.1016	0.1016
$\varepsilon^* = Z$	0	0.00098425	0.00492126	0.00984252
Ξ	0.000223144	0.00022314	0.00022314	0.00022314
Z	0.00448140	0.00448140	0.00448140	0.00448140
D*	455.31	455.31	455.31	455.31
v^*_{av}	331.7712	290.4213	242.0344	218.0251
v_{av}	1.5	1.3	1.1	1.0
Re	151,059	132,232	110,201	99,269
f	0.00414	0.00540	0.00777	0.00958
Re_E	–	6.76166	33.80828	67.61657
Q (m^3/s)	0.012053971	0.01055164	0.00879364	0.00792133

11.3 Mechanical Energy Balance

In an open flow the mechanical energy balance reads:

$$p_1 + \frac{1}{2}\rho v_1^2 + \rho g z_1 = p_2 + \frac{1}{2}\rho v_2^2 + \rho g z_2$$

Table 11.1 Friction head losses

Fitting	X
45° elbow	0.3
90° elbow	0.7
90° square elbow	1.2
Exit from leg of T-piece	1.2
Entry into leg of T-piece	1.8
Unions and couplings	Small
Globe valve fully open	1.2–6.0
Gate valve fully open	0.15
Gate valve 3/4 open	1.0
Globe valve 1/2 open	4.0
Globe valve 1/4 open	16
Sudden expansion	$\left(1 - (D_1/D_2)^2\right)^2$
Discharge into a large tank	1
Sudden contraction	$X = 0.7867(D_2/D_1)^6 - 1.3322(D_2/D_1)^4 + 0.1816(D_2/D_1)^2 + 0.363$
Outlet of a large tank	0.5

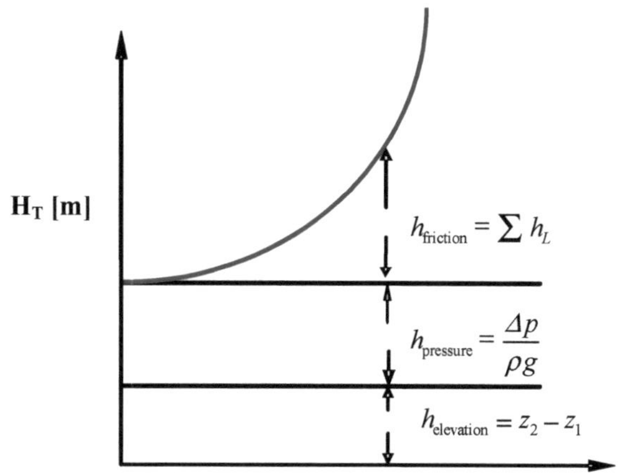

Fig. 11.8 Total head versus flow demand

that is:

$$\frac{\Delta p}{\rho g} = \frac{1}{2g}\left(\bar{v}_{z2}^2 - \bar{v}_{z1}^2\right) + (z_2 - z_1) \tag{11.41}$$

where the first term is the *pressure head* with $\Delta p = p_2 - p_1 > 0$, the second term is the *velocity head* and the third term is the *head.*

The basis to calculate flow in conduits is the mechanical energy balance in open flows to which two additional terms are added, one for the positive head H_T imposed by the pump and one for the loss $\sum h_L$ due to the friction within the fluid, on the pipe walls and on the fittings.

$$\begin{aligned} p_1 + \frac{1}{2}\rho v_1^2 + \rho g z_1 + H_T - \sum h_L &= p_2 + \frac{1}{2}\rho v_2^2 + \rho g z_2 \\ H_T &= \frac{\Delta p}{\rho g} + \frac{1}{2g}\left(\bar{v}_{z2}^2 - \bar{v}_{z1}^2\right) + (z_2 - z_1) + \sum h_L \end{aligned} \tag{11.42}$$

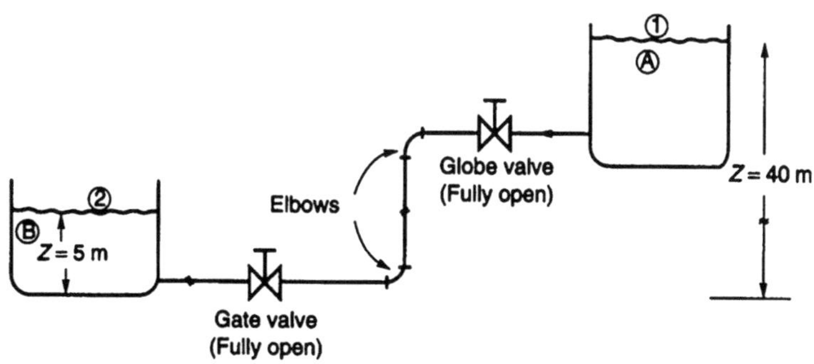

Fig. 11.9 Figure for problem 11.5

In Eq. (11.42) $H_T = \frac{P_o}{\rho g Q_f}$ and $\sum h_L = \frac{\dot{E}_v}{\rho g Q}$, where P_o and Q_f are the power and the flow rate delivered by the pump and $\dot{E}_v$ is the speed of energy dissipation by friction and $\sum h_L$ is the sum of the head loss in the pipe line $h_{pipe} = \left(L\bar{v}_z^2/gD\right) \times f$ and pipe line fittings given as the numbers X of velocity heads, $X \times \bar{v}_z^2/2\,g$. Table 11.1 gives the head loss for different fittings.

Figure 11.8 is a graphic description of the total head that a pump must deliver to a given flow rate of a Newtonian fluid.

Problem 11.5 Water flows under gravity from reservoir A to reservoir B, both of which are of large diameters. Estimate the flow rate through a 6-inch diameter pipe, with a roughness $\varepsilon = 0.4\,\text{mm}$, and 75 m length. See Fig. 11.9.

Apply Eq. (11.42):

$$0 = 0 + 0 + (z_2 - z_1) + \sum h_L$$
$$\sum h_L = z_1 - z_2 = 40 - 5 = 35\,[\text{m}]$$

Head loss:

$$\sum h_L = h_{friction}(\text{pipe}) + h_{fitting}(1\,\text{gate valve}) + h_{fitting}(1\,\text{globe valve}) + h_{fitting}(2\text{elbows}) + h_{entrance} + h_{\text{outlet}}$$

$$\frac{1}{\sqrt{\lambda}} = -2\log\left(\frac{\varepsilon^*}{0.458}\right)\text{forRe} > 2{,}100 \text{ and } \text{Re}_\in > 70$$

$$\lambda = \left(-2\log(\varepsilon^*/0.854)\right)^{-2}$$

$$h_{friction} = \lambda\left(\frac{\bar{v}_z^2}{2\,g}\right)\frac{\text{L}}{\text{D}}$$

$$\text{Re}_\in = \varepsilon^*\sqrt{\lambda/8}\text{Re}$$

Data		X_{outlet}	0.5
D (in]	6	X_{inlet}	1
L (m)	75	X_{elbow}	0.7
ε (mm)	0.4	$X_{\text{globe valve}}$	6
z_1 (m)	40	$X_{\text{gate valve}}$	0.15
Z_2 (m)	5	λ	0.039606674
P_f (kg/m^3)	1.000	assume v [m/s]	5.11
μ (Pa-s)	0.001	h_{friction}	25.94999706
Results		h_{fittings}	9.05
D (m]	0.1524	h_L [m]~	34.99999706
ε (m)	0.0004	h_L–h_L~0	2.94393E−06
ε* [-]	0.002624672	Re~	778,898
hL = $Z_1 - Z_2$	35	Re_ε>70	143.85

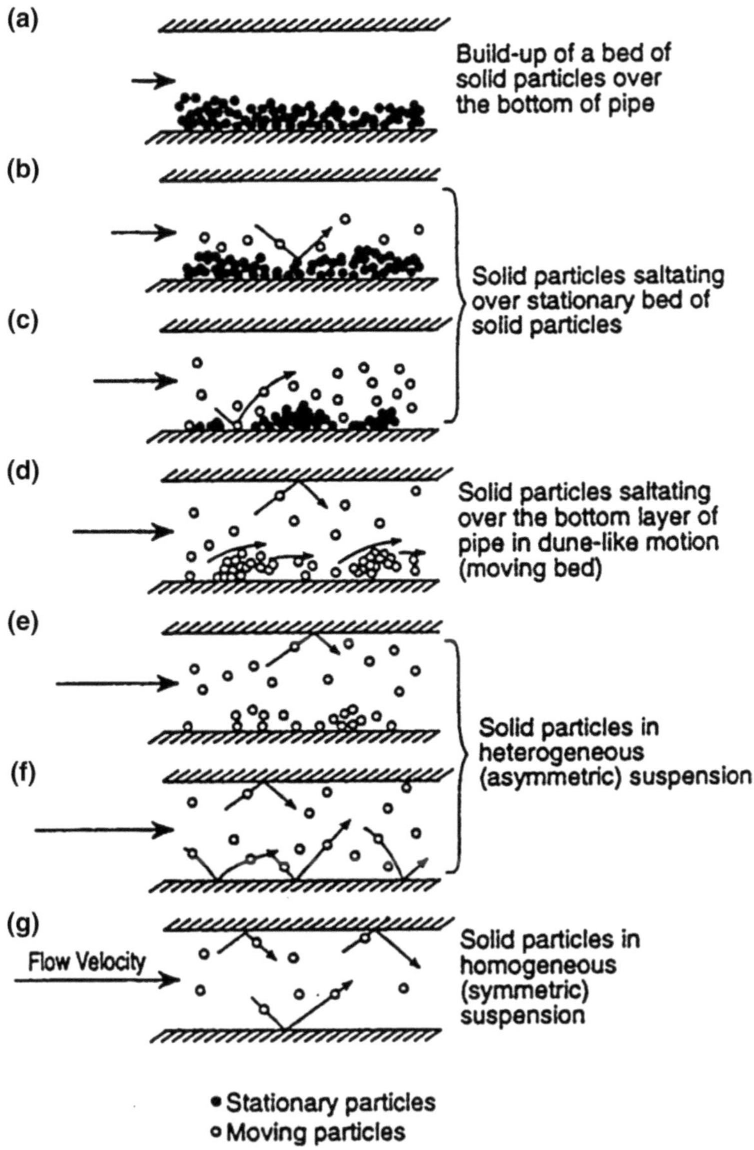

Fig. 11.10 Particle behavior for the flow of a suspension through a tube according to Chien (1994)

Table 11.2 Relationship between flow patterns and solid concentrations

Mixture velocity v_{Mi}	Flow pattern	Fraction of concentration φ/φ_F
v_{M1}	Homogeneous suspension	1.0
v_{M2}	Asymmetric suspension	0.7–1.0
v_{M3}	Living bed with asymmetric suspension	0.2–0.7
v_{M4}	Stationary bed with some particles in suspension	0–0.2

11.4 Transporting Suspensions in Pipelines

The flow patterns of suspensions in tubes depend on the transport velocity. See Fig. 11.10. At low velocities, the particles form a bed at the bottom of the tube and are not transported by the fluid. As the velocity increases, particles at the surface of the bed start moving. At higher velocities, the sediment moves as a cloud in saltatory motion, and some particles are suspended and carried away with the fluid. If the velocity increases, most particles are suspended but some settle. Under this condition the suspension is termed a *settling suspension* and the flow regime is *heterogeneous.* Increasing the velocity further, all particles are suspended and particles and fluid behave as a homogeneous mixture, the suspension is non-settling and the flow regime is *homogeneous*. Each of these behaviors corresponds to a pressure drop and the type of motion can be controlled by the pressure gradient.

The flow pattern for transporting suspensions in a tube is closely related to the suspension concentration. When particles begin to move above a stationary bed, the fractions of the feed concentration φ_F in suspension is in the fraction range of 0.7–12.0. Motions of the bed yield fractions of the feed concentration between $0.2<\varphi/\varphi_F<0.7$. Partial suspension gives fractions of feed concentration of $0.7<\varphi/\varphi_F<1$ and a complete suspension of particles gives $\varphi/\varphi_F=1$ (Table 11.2).

Settling velocity

Since particles will settle from a flowing suspension, it is important to be able to calculate the settling velocity of the particles at several concentrations. This can be obtained from laboratory experiments or by calculations from sedimentation models. A useful model was proposed by Concha and Almendra (1979a), which was discussed in Sect. 4.1.6.

For a suspension of spherical particles, Concha and Almendra (1979b) proposed using the same equation as for single particles, but with the P and Q parameters depending on the particle concentration. See Sect. 4.1.7.

11.4.1 Flow of Heterogeneous Suspensions

The flow of a suspension is heterogeneous if some particles segregate and settle. This happens when the average flow velocity is not fast enough to maintain the

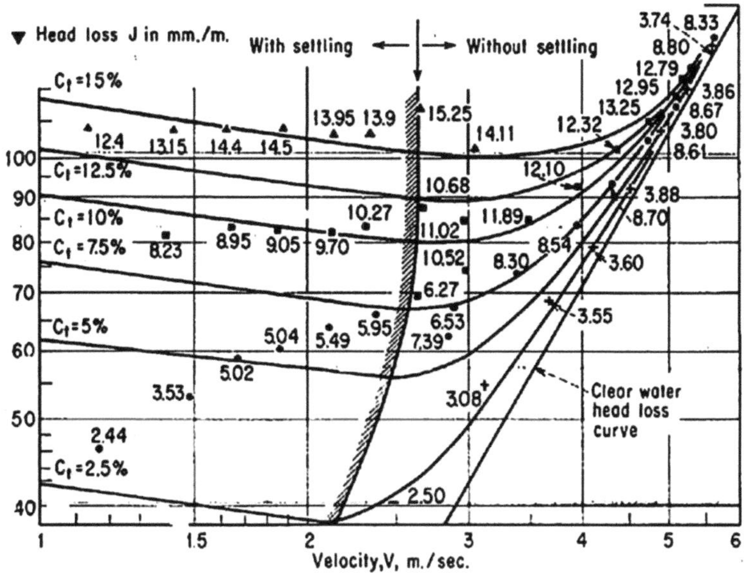

Fig. 11.11 Head loss versus average transport velocity for 0.44 mm fine sand at several concentrations (Condolios and Chapus 1963)

largest particles in suspension. In a heterogeneous regime, particles will form a concentration gradient without sediment at the bottom of the tube.

Figure 11.11 shows the pressure drop for the flow of a suspension as a function of the average flow velocity, with concentration as a parameter. For each suspension concentration there is a minimum flow velocity for which the pressure drop has a minimum. This is the lowest velocity that avoids sediment formation in the tube. Joining the minimum points for each concentration, a curve representing the *limiting settling velocity* is obtained.

Limiting deposit velocities

As discussed in the previous section, between the initiation of particle motion and the complete suspension at several concentrations, there is a small range of velocities at which the pressure drop is minimal. This range of velocities $v_L(\varphi)$ is termed *limiting deposit velocities*. The limiting deposit velocity is the lowest velocity at which no particles are deposited. The optimal transport velocity for a suspension produces the lowest pressures drop without depositing particles, that is, slightly to the right of the limiting deposit velocity in a pressure drop versus flow velocity graph.

In practice, the most common regime is with heterogeneous suspensions where the largest particles in the suspension settle, but this does not change the rheological characteristics of the pulp, although a concentration gradient is present in the pipe.

Correlations for the limiting deposit velocities

The simplest way that particles do not settle in a heterogeneous regime is to ensure that the regime is turbulent and that the Reynolds number for the largest particles is in Newton's regime (Faddick 1986):

$$\mathrm{Re} = \frac{D\bar{v}_z}{v} > 4,000 \text{ and } \mathrm{Re}_p = \frac{du}{v} > 1,000 \tag{11.43}$$

where D and $\bar{v}_z$ are the diameter of the pipe and the average velocity of the flow, d and u are the diameter of the largest particle in the suspension and its settling velocity and v is the kinematic viscosity of the fluid.

Problem 11.6 Design a pipe for the flow of 600 tph of magnetite mineral slurry that behaves as a Newtonian fluid with density 1,667 kg/m^3 and viscosity 5 mPa-s. The magnetite density is 5,000 kg/m^3 and its maximum particle size is 5 mm. Make sure that the flow regime is heterogeneous with an average velocity of 2.00 m/s.

Pulp volume flow is $Q = \frac{F}{\rho} = \frac{600}{1{,}667/1{,}000} = 0.100\ \mathrm{m}^3/\mathrm{s}$
Particle size: $d = 0.005\,\mathrm{m}$
Magnetite density

$$\rho_s = 5,000\,\mathrm{kg}\ /\mathrm{m}^3$$

Water density

$$\rho_f = 1{,}000\ \mathrm{kg/m}^3$$

% of solids:

$$\begin{aligned} w &= \frac{100 \times \rho_s \times (\rho - \rho_f)}{\rho \times (\rho_s - \rho_f)} = \frac{100 \times 5,000 \times (1,667 - 1{,}000)}{1{,}667 \times (5{,}000 - 1{,}000)} \\ &= 50\,\%\ \text{solid by weight} \end{aligned}$$

Volume fraction:

$$\varphi \quad = \frac{\rho_f \times w}{\rho_s \times (100 - w) + \rho_f w} = \frac{1,000 \times 50}{5,000 \times (100 - 50) + 1,000 \times 50} = 0.167$$

From Eq. (4.45):

$$P = \left(\frac{3}{4}\frac{\mu_f^2}{\Delta\rho \times \rho_f \times g}\right)^{1/3}\left(\frac{3}{4}\frac{0.001^2}{(5,000 - 1,000) \times 1,000 \times 9.81}\right)^{1/3} = 2.674 \times 10^{-5}\ \mathrm{m}^{1/3}$$

$$Q = \left(\frac{4}{3}\frac{\Delta\rho \times \mu_f \times g}{\rho_f^2}\right)^{1/3} = \left(\frac{4}{3}\frac{(5,000 - 1,000) \times 0.001 \times 9.81}{1,000^2}\right)^{(1/3)} = 0.0374\,(\mathrm{m/s})^{1/3}$$

$$d^* = \frac{d}{P} = \frac{0.005}{2.674 \times 10^{-5}} = 187.01$$

From Eq. (4.51):

$$u^* = \frac{20.52}{d^*}\left(\left(1+0.0921\times d^{*3/2}\right)^{1/2}-1\right)^2$$

$$=\frac{20.52}{187}\left(\left(1+0.0921\times 187^{*3/2}\right)^{1/2}-1\right)^2 = 22.698$$

$$u = u^* \times Q = 22.689 \times 0.03740 = 0.849 \text{ m/s.}$$

$$\text{Re}_p \frac{\rho_f u d}{\mu} = 4243.0 > 1,000$$

Select the average transport velocity: $\bar{v}_z = 2.0$ m/s

$$D = \left(\frac{4Q}{\pi \bar{v}_z}\right)^{1/2} = \left(\frac{4\times 0.100}{3.14\times 0.304}\right) = 0.2524 \text{ m} = 10.0 \text{ in}$$

$$\text{Re} = \frac{\rho_f \bar{v}_z D}{\mu} = \frac{1,667\times 2.00\times 0.2524}{0.005} = 1.6827\times 10^5 > 4,000$$

Reynolds number fulfills the conditions for a heterogeneous flow. Summary:

F (tph)	600	P	2.673688E−05
d (m)	0.005	Q	3.740152E−02
ρ_s (kg/m^3)	5,000	d*	1.870076E+02
ρ_f (kg/m^3)	1,000	u*	22.689
P (kg/m^3)	1,667	u (m/s)	0.849
η_f (Pa-s)	0.005	Rep	4243.0
μ_f (Pa-s)	0.001	v_z (m/s)	2.00
Q_F (m^3/s)	0.09998	D (m)	0.2524
w (% by weight)	50.0	D (inch)	9.94
ϕ	0.167	Re	1.6827E+05

It has not been possible to establish the limiting deposition velocity from fundamentals, but many correlations have been proposed in the range of particle sizes from 50 (mμ])to 5 (mm) and pipes from 50 mm (2 in) to 300 mm (12 in).

Chien (1994) reviewed the work of many researchers, among them Durand (1953), Durand (1953), Spells (1955), Newitt et al. (1955), Cairns et al. (1960), Govier and Aziz (1961), Schulz (1962), Sinclair (1962), Condolios and Chapus (1963), Yufin and Lopasin (1966), Zandi and Govatos (1967), Babcock (1968), Shook (1969), Bain and Bonnington (1970), Charles (1970), Wilson (1979), Thomas (1979), Oroskar and Turian (1980) and Gillies and Shook (1991).

Unfortunately these equations are valid for different particle sizes, densities and pipe diameters, and therefore give different values of limiting velocities that range from 0.5 to 7. Figure 11.12 shows the application of these equations to suspensions of particles 150 microns in size and 2,650 kg/m^3 in density, in an 8-inch pipe and volume fractions from 0 to 0.30.

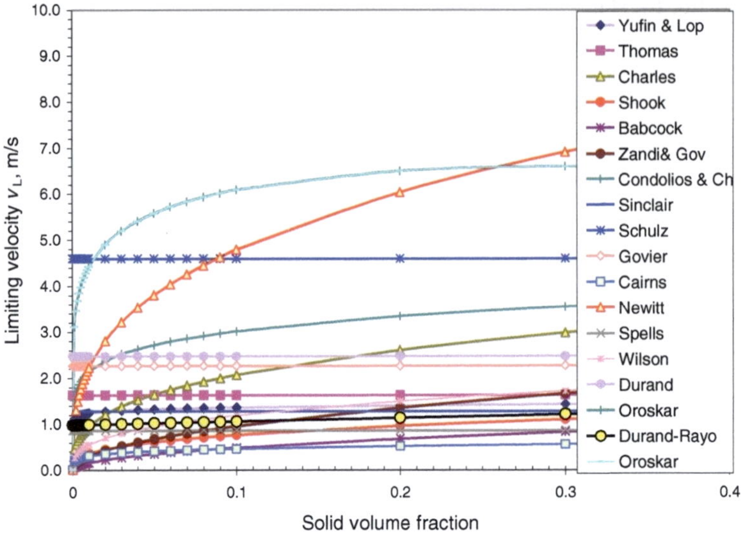

Fig. 11.12 Several correlations for the limiting velocities versus suspension concentration

In what follows, we use Durand's equation (1953) with parameters by McElvian and Rayo (1993):

$$v_L(\text{cm/s}) = F_L(\varphi)\sqrt{2\,gD\Delta\rho/\rho_f}\,; \quad \text{for } \varphi < 0.20 \tag{11.44}$$

$$F_L(\varphi) = \begin{cases} 1.1\left(\Delta\rho/\rho_f\right)^{1/5} F_{LM}(\varphi) & \text{for} \quad \text{small } d_{50} \text{ and small } D \\ 1.25\left(2\,gD\Delta\rho/\rho_f\right)^{-1/4} F_{LM}(\varphi) & \text{for} \quad \text{small } d_{50} \text{ and big } D \\ \left(d_{80}/d_{50}\right)^{1/10} F_{LM}(\varphi) & \text{for} \quad \text{big } d_{50} \text{ and extended distribution and small } D \end{cases} \tag{11.45}$$

$$\begin{aligned} F_{LM} = {} & (0.1248\varphi + 0.165)\ln(d_{50}) \\ & + (0.6458\varphi + 1.224) \text{ for } 0.005 < d_{50}(\text{mm}) < 0.5 \end{aligned} \tag{11.46}$$

In these equations φ is the volume fraction of solids in the suspension, F_{LM} is given by, with the particle diameter in mm and $F_L(\varphi)$ by Rayo (1993), where the units of the variables are v_L m/s, D m, d m; ρ kg/m^3 and $g = 9.81$ m/s^2. Rayo's equation is based on numerous years of experience designing pipelines for the copper mining plants in Chile.

Problem 11.7 Determine the limiting sedimentation velocity of quartz suspensions flowing in pipes 200 m long and 2 and 8 inches in diameter. The particle diameters are $d_{50} = 50\,\text{m}\mu$, $d_{80} = 374.5\,\text{m}\mu$ and $d_{50} = 1.5$ mm and concentration 20 % solid by weight. The solid density is $\rho_s = 2{,}650$ kg/m^3; water density $\rho_f = 1{,}000$ kg/m^{3}and suspension viscosity $\mu = 5$ cp. Use Durand's equation

(1953), with parameters by Rayo (1993). For the same data draw a figure of the limiting velocity versus concentration.

Utilizing Eqs. (11.44–11.46) and Fig. 11.13 yields:

Durand and Rayo	d_{small};D_{big}	d_{small}; D_{small}	d_{big}; D_{small}
ρ_s(kg/m^3)	2,650	2,650	2,650
ρ_f (kg/m$_3$)	1,000	1,000	1,000
L (m)	200	200	200
D (m)	0.2032	0.0508	0.0508
d_{50} (m)	1.500E−04	1.500E−04	1.500E−03
d_{50} (mm)	1.500E−01	1.500E−01	1.500E+00
d_{80} (m)	3.7450E−04	3.7450E−04	3.7450E−04
g (m/s^2)	9.81	9.81	9.81
X (%sol)	20	20	20
μ (Pa-s)	5.000E−03	5.000E−03	5.000E−03
φ	0.2284	0.2284	0.2284
P	1.050E−04	1.050E−04	1.050E−04
Q	4.761E−02	4.761E−02	4.761E−02
d*	1.428E+00	1.428E+00	1.428E+01
u*	0.082	0.082	2.994
u	3.923E−03	3.923E−03	1.426E−01
$\Delta\rho/\rho_f$	1.650	1.650	1.650
ρ (kg/m^3)	1377	1377	1377
v_L (m/s)	1.161	1.340	1.706
Re_p	1.177E−01	1.177E−01	4.277E+01
Re	4.717E+04	1.362E+04	1.733E+04
C_D	2.112E+02	2.112E+02	1.597E+00

Pressure drop in a heterogeneous regime

In a heterogeneous regime the head loss J_m has two contributions: J_L to maintain the turbulent fluid flow in a Newtonian fluid, and, J_S to maintain the particles in suspension in the fluid. Both values are measured in columns of water per meter of pipe length $(J = h/L = \Delta p/\rho g L)$, evaluated at the average mixture velocity:

$$J_m = J_L + J_S \tag{11.47}$$

There is no generally accepted equation for the head loss for the flow of suspensions. In a form similar to the limiting velocity, there are several empirical equations that give results with great scatter. We will use the Durand and Condolios equation (1953).

$$J_m = J_L\left(1 + 81\varphi A^{-3/2}\right); \quad \text{where } A = \frac{\bar{v}_z^2\sqrt{C_D}}{gD\,\Delta\rho/\rho_f} \tag{11.48}$$

If particle size is widely distributed, (Wasp et al. 1977) recommended calculating total head loss by weighing the individual head loss by its volume fraction;

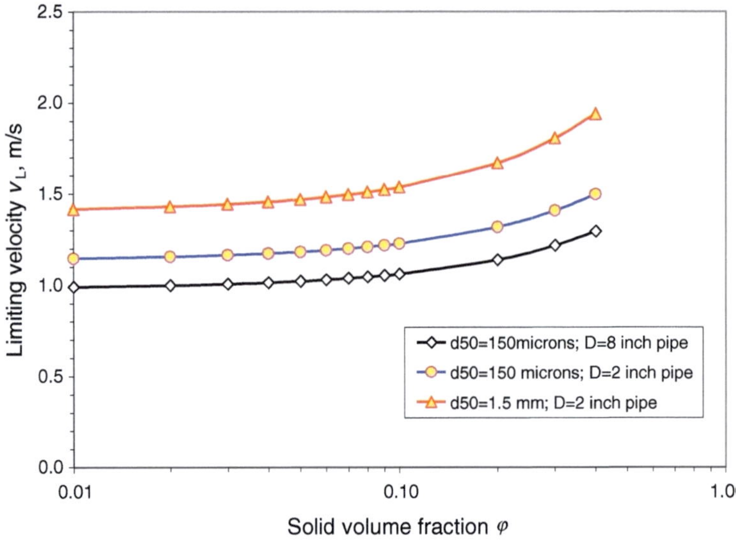

Fig. 11.13 Limiting sedimentation velocity for quartz particles of two different sizes and three different diameters, according to Durand's equation with Rayo's parameters

$$J_m = \sum_i J_i \varphi_i \tag{11.49}$$

where J_i and φ_i are the head loss associated with particle size x_i in a suspension with particle size distribution.

Problem 11.8 Calculate the pressure gradient due to friction when slurry, composed of 1 mm silica particles with a density of 2,700 kg/m^3, is pumped through a 5 cm diameter and 75 m pipeline at velocities of 3.5 m/s. The slurry contains 30 % silica by volume and the water has a density and viscosity of 1,000 kg/m^3 and 0.001 Pa-s.

$$\sum h_L = 23.96\left(\frac{\bar{v}_z^2}{2g}\right) \qquad J_L = 23.96/75\left(\frac{\bar{v}_z^2}{2g}\right) = 0.3195\left(\frac{\bar{v}_z^2}{2g}\right)$$

$$J_m = J_L\left(1 + 81\,\varphi A^{-3/2}\right); \quad \text{where } A = \frac{\bar{v}_z^2\sqrt{C_D}}{gD\Delta\rho/\rho_f}$$

$$\varphi = \frac{\rho_f w}{\rho_s(100 - w) + \rho_f w} = \frac{1{,}000 \times 30}{2{,}650(100 - 30) + 1{,}000 \times 30} = 0.139$$

$$P = \left(\frac{3}{4}\frac{\mu_f^2}{\Delta\rho\rho_f g}\right)^{1/3} = 3.59 \times 10^{-5} \text{ and } Q = \left(\frac{4}{3}\frac{\Delta\rho\mu_f g}{\rho_f^2}\right)^{1/3} = 2.7842 \times 10^{-2}$$

$$d_{50} = 1 \times 10^{-4}\ \mathrm{m} d^* = d_{50}/P = 2.78 f_p(\varphi) = (1-\varphi)^{-2.033} = 1.3563\,,\ f_q(\varphi)$$
$$= (1-\varphi)^{-0.167} = 1.0254 u^*$$
$$= \frac{20.52}{d^*} f_p(\varphi) f_q(\varphi)\left(\left(1 + 0.0921 f_p^{-3/2} d^{*3/2}\right)^{1/2} - 1\right)^2 = 0.7521 u = u^* \times Q$$
$$= 2.094 \times 10^{-2}\ (\mathrm{m/s})\mathrm{Re}_p = \frac{d_{50}\rho_f u}{\mu} = 2.094 C_D$$
$$= 0.28(1-\varphi)^{-2.01}\left(1 + \frac{9.08(1-\varphi)^{-1.83}}{\mathrm{Re}^{1/2}}\right)^2 = 32A = \frac{\bar{v}_z^2\sqrt{C_D}}{gD\,\Delta\rho/\rho_f} A/\left(\bar{v}_z^2/2g\right)$$
$$= \frac{2\sqrt{C_D}}{D\,\Delta\rho/\rho_f} = 2.31\,,\ A = 2.31\bar{v}_z^2 J_m = J_L\left(1 + 81\varphi A^{-3/2}\right)$$
$$= 0.3195\left(\frac{\bar{v}_z^2}{2g}\right)\left(1 + 81 \times 0.138 \times \left(10v_z^2\right)^{-3/2}\right)$$

Assuming that $v_z = 4.41\ \mathrm{m/s}$, with $J_m = 35/35 = 1$

$$1 - 0.9183\left(\frac{\bar{v}_z^2}{2g}\right)(1 + 81 \times 0.138 * (10v_z^2)^{-3/2}) = 0.0858$$

Using the solver results in:

$$1 - 0.9183\left(\frac{\bar{v}_z^2}{2g}\right)\left(1 + 81 \times 0.138 \times \left(10v_z^2\right)^{-3/2}\right) = 0.0858$$
$$v_z = 4.61\ \mathrm{m/s}\,,\ Q = \left(\pi D^2/4\right)v_z = 0.0842\ \mathrm{m^3/s}.$$

11.4.2 Flow of Homogeneous Suspensions

A homogeneous flow regime is such that suspensions are non-settling. According to Faddick (1985), if particles are small enough to be in a Stokes regime, their settling velocity will be low in relation to their transport velocity, and the suspension can be considered homogeneous.

Depending on the constitutive equation of the extra stress tensor, homogeneous suspensions can behave as Newtonian or no-Newtonian. If a suspension behaves Newtonian, the discussion and design criteria of Sect. 11.2 are valid. For non-Newtonian suspensions, we will consider flows of Bingham, power law and Herschel-Bulckley fluids in a tube.

(a) **Bingham Fluids**

Bingham fluids have the following constitutive equation for the shear stress in cylindrical coordinates:

$$T_{rz}^{E}(r) = \tau_y + K\frac{\partial v_z}{\partial r} \tag{11.50}$$

where K is a constant called *plastic viscosity*. From (11.10), $T_{rz}^{E}(r)$ is given by:

$$T_{rz}^{E}(r) = -\frac{1}{2}\frac{\Delta p}{L}r \tag{11.51}$$

Calling R_y the radius for which the stress is $T_{rz}^{E} = \tau_y$, we have:

$$T_{rz}^{E}(R_y) = \tau_y = -\frac{1}{2}\frac{\Delta p}{L}R_y, \tag{11.52}$$

Since the stress at the wall is given by (11.12), $\tau_w = -\frac{1}{2}\frac{\Delta p}{L}R$, the relationship between the yield stress τ_y and the wall shear stress τ_w is:

$$\frac{\tau_y}{\tau_w} = \frac{R_y}{R} \tag{11.53}$$

Velocity distribution

Substituting (11.50) with (11.51) yields:

$$\frac{\partial v_z}{\partial r} = -\left(\frac{1}{2}\frac{\Delta p}{KL}r + \frac{\tau_y}{K}\right)$$

Using (11.53) for τ_y results in: For

$$T_{rz}^{E}(r) > \tau_y \quad \frac{\partial v_z}{\partial r} = \frac{1}{2}\frac{\Delta p}{KL}(R_y - r) \tag{11.54}$$

Integrating this expression yields:

$$v_z = \frac{1}{2}\frac{\Delta p}{KL}\left(R_y r - \frac{1}{2}r^2\right) + C$$

For $r = R,\ v_z(R) = 0$, therefore:

$$C = -\frac{1}{2}\frac{\Delta p}{KL}\left(R_y R - \frac{1}{2}R^2\right)$$

and

$$v_z = \frac{1}{2}\frac{\Delta p}{KL}\left(R_y r - \frac{1}{2}r^2\right) - \frac{1}{2}\frac{\Delta p}{KL}\left(R_y R - \frac{1}{2}R^2\right)$$

For

$$T_{rz}^{E}(r) > \tau_y \qquad v_z(r) = -\frac{1}{2}\frac{\Delta p R^2}{KL}\left(\frac{R_y}{R}\left(1-\frac{r}{R}\right)-\frac{1}{2}\left(1-\left(\frac{r}{R}\right)^2\right)\right) \tag{11.55}$$

For

$$T_{rz}^{E}(r) \le \tau_y \qquad v_z(r) = -\frac{1}{2}\frac{\Delta p R^2}{KL}\left(\frac{R_y}{R}\left(1-\frac{R_y}{R}\right)-\frac{1}{2}\left(1-\left(\frac{R_y}{R}\right)^2\right)\right) \tag{11.56}$$

Using (11.53) we obtain the alternative expressions:

$$\begin{aligned} v_z(r) &= -\frac{1}{2}\frac{\Delta p R^2}{KL}\left(\frac{\tau_y}{\tau_w}\left(1-\frac{r}{R}\right)-\frac{1}{2}\left(1-\left(\frac{r}{R}\right)^2\right)\right); \quad \text{for} \quad \tau > \tau_y \\ v_z(r) &= -\frac{1}{2}\frac{\Delta p R^2}{KL}\left(\frac{\tau_y}{\tau_w}\left(1-\frac{\tau_y}{\tau_w}\right)-\frac{1}{2}\left(1-\left(\frac{\tau_y}{\tau_w}\right)^2\right)\right); \quad \text{for} \quad \tau \le \tau_y \end{aligned} \tag{11.57}$$

Volume flow rate

The volume flow rate is given by $Q = \int_0^R 2\pi v_z(r) r dr$, then:

$$Q = \int_{R_y}^{R} 2\pi v_z(r) r dr + \int_{0}^{R_y} 2\pi v_z(r) r dr$$

$$\begin{aligned} Q = &\int_{R_y}^{R} 2\pi\left(-\frac{1}{2}\frac{\Delta p R^2}{KL}\left(\frac{\tau_y}{\tau_w}\left(1-\frac{r}{R}\right)-\frac{1}{2}\left(1-\left(\frac{r}{R}\right)^2\right)\right)\right) r dr \\ &+ \int_{0}^{R_y} 2\pi\left(-\frac{1}{2}\frac{\Delta p R^2}{KL}\left(\frac{\tau_y}{\tau_w}\left(1-\frac{\tau_y}{\tau_w}\right)-\frac{1}{2}\left(1-\left(\frac{\tau_y}{\tau_w}\right)^2\right)\right)\right) r dr \end{aligned}$$

$$\begin{aligned} Q = -\frac{\pi \Delta p R^4}{KL}\Bigg(&\int_{R_y/R}^{1} \frac{\tau_y}{\tau_w}\left((1-\xi)-\frac{1}{2}\left(1-(\xi)^2\right)\right)\xi d\xi \\ &+ \int_{0}^{R_y/R} \left(\frac{\tau_y}{\tau_w}\left(1-\frac{\tau_y}{\tau_w}\right)-\frac{1}{2}\left(1-\left(\frac{\tau_y}{\tau_w}\right)^2\right)\right)\Bigg)\xi d\xi \end{aligned}$$

Integrating this expression we obtain:

$$Q = \frac{\pi \Delta p R^4}{8KL}\left(1-\frac{4}{3}\left(\frac{\tau_y}{\tau_w}\right)+\frac{1}{3}\left(\frac{\tau_y}{\tau_w}\right)^4\right) \tag{11.58}$$

Average velocity

The average velocity is given by $\bar{v}_z = Q/\pi R^2$, then:

$$\bar{v}_z = \frac{\Delta p R^2}{8KL}\left(1 - \frac{4}{3}\left(\frac{\tau_y}{\tau_w}\right) + \frac{1}{3}\left(\frac{\tau_y}{\tau_w}\right)^4\right) \tag{11.59}$$

Shear rate at the wall

Using a similar procedure as in the case of Newtonian fluids, we have:

$$\dot{\gamma}_w = \frac{8\bar{v}_z}{D} = \frac{1}{4}\frac{\Delta p D}{KL}\left(1 - \frac{4}{3}\left(\frac{\tau_y}{\tau_w}\right) + \frac{1}{3}\left(\frac{\tau_y}{\tau_w}\right)^4\right) \tag{11.60}$$

Maximum velocity

From (11.57) the maximum velocity is that for $0 \le r \le R_y \quad (\tau < \tau_y)$

$$v_m = \frac{1}{4}\frac{\Delta p R^2}{KL}\left(1 - \frac{\tau_y}{\tau_w}\right)^2 \tag{11.61}$$

Friction factor

Defining the Reynolds numbers $\mathrm{Re}_B = \rho\bar{v}_z D/K$ and $\mathrm{Re}_\in = \in/\sqrt{f}\,\mathrm{Re}_B$, the friction factor according to Eq. (11.24) can be written as $f = \Delta p D/2\rho L\bar{v}_z^2$:

$$f = \frac{16}{\mathrm{Re}_B}\left(1 - \frac{4}{3}\left(\frac{\tau_y}{\tau_w}\right) + \frac{1}{3}\left(\frac{\tau_y}{\tau_w}\right)^4\right) \quad \text{for} \quad \mathrm{Re}_B < 2,100 \tag{11.62}$$

$$f_{smooth} = \left(4.53\log\left(\mathrm{Re}_B\sqrt{f}\right) - 2.3 + 4.5\log\left(1 - \tau_y/\tau_w\right)\right)^{-2} \text{ for } \mathrm{Re}_B > 4,000;\ \mathrm{Re}_\in < 5 \tag{11.63}$$

$$f_{rough} = f_{mooth} \times \left(\frac{f_{water;\,rough}}{f_{water;\,smooth}}\right) \quad \text{for} \quad \mathrm{Re}_B > 4,000;\ 5 < \mathrm{Re}_\in < 70 \tag{11.64}$$

$$f = \left(4.07\log\left(\frac{D}{2\in}\right) + 3.36\right)^{-2} \quad \text{for } \mathrm{Re}_B > 4,000\ \mathrm{Re}_\in > 70 \tag{11.65}$$

Problem 11.9 For three suspensions of clay with density $\rho = 1,275\ \mathrm{kg/m^3}$ that can be represented by the Bingham model in the range $10 < \dot{\gamma} < 500\,[\mathrm{s^{-1}}]$ with

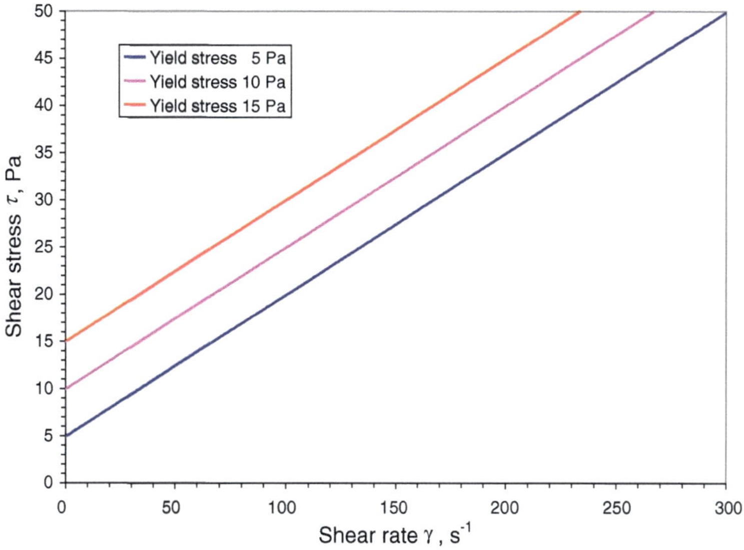

Fig. 11.14 Shear stress versus shear rate for a Bingham model of a material with plastic viscosity of 150 (mPa-s) and yield stresses of 5, 10 y 15 (Pa)

$\tau_y = 5,\ 10$ y 15Pa respectively and a plastic viscosity of $K = 150\,\text{mPa-s}$, flowing in a cylindrical tube 1 inch in diameter and 200 m in length. See Fig. 11.14. Calculate the pressure drop, the shear stress at the wall and the velocity distribution necessary to transport 100 l of suspension per minute.

Pressure drop:

$$Q_1 = 100\,\ell/\text{min} = 100/(60 * 1{,}000) = 1.6667 \times 10^{-3}\ \text{m}^3/\text{s}$$

$$Q_2 = \frac{\pi}{8}\frac{\Delta p R^4}{KL}\left(1 - \frac{4}{3}\left(\frac{\tau_y}{\tau_w}\right) + \frac{1}{3}\left(\frac{\tau_y}{\tau_w}\right)^4\right),\ \bar{v}_z = \frac{Q_f}{\pi R^2} = 0.923\,(\text{m/s});\ \tau_w = \frac{1}{2}\frac{\Delta p R}{L}$$

$$Q_2 = \frac{\pi}{8}\frac{\Delta p \times R^4}{K \times L}\left(1 - \frac{4}{3}\left(\frac{\tau_y}{0.5 \times \Delta p \times R/L}\right) + \frac{1}{3}\left(\frac{\tau_y}{0.5 \times \Delta p \times R/L}\right)^4\right)$$

Using solver to minimize the error $\Delta Q = Q_1 - Q_2$, by changing Δp leads to:

$$\Delta Q = 1.6667 \times 10^{-3} - \frac{\pi}{8}\frac{\Delta p \times R^4}{K \times L} \times \left(1 - \frac{4}{3}\left(\frac{5}{0.5 \times \Delta p \times R/L}\right) + \frac{1}{3}\left(\frac{5}{0.5 \times \Delta p \times R/L}\right)^4\right)$$
$$= 6.517 \times 10^{-9}$$

$$\Delta Q = 1.6667 \times 10^{-3} - \frac{\pi}{8}\frac{\Delta p \times R^4}{K \times L} \times \left(1 - \frac{4}{3}\left(\frac{5}{0.5 \times \Delta p/L}\right) + \frac{1}{3}\left(\frac{5}{0.5 \times \Delta p \times R/L}\right)^4\right)$$

$$= 6.517 \times 10^{-9}$$

$$\Delta p = 4.10 \times 10^5 \text{ Pa}$$

$$\Delta p = 4.10 \times 10^5 \times 1.45 \times 10^{-4} = 60 \text{ psi}$$

$$\tau_w = \frac{1}{2}\frac{\Delta p \times R}{L} = 26.09 \text{ Pa};\ R_y = R \times \frac{\tau_y}{\tau_w} = 0.00487 \text{ m};\ \text{Re} = \frac{\rho \times R \times \bar{v}_z}{\eta} = 1.40 \times 10^4$$

The velocity distribution:

$$v_z(r) = -\frac{1}{2}\frac{\Delta p R^2}{\eta L}\left(\frac{\tau_y}{\tau_w}\left(1 - \frac{r}{R}\right) - \frac{1}{2}\left(1 - \left(\frac{r}{R}\right)^2\right)\right)$$

For the three cases, calculations are in this excel sheet. See Fig. 11.15:

τ_y (Pa)	5	10	15
K (Pa-s)	0.150	0.150	0.150
Q (l/min)	100	100	100
R (inch)	1	1	1
L (m)	200	200	200
ρ (kg/m^3)	1,275	1,275	1,275
Q (m^3/s)	1.667E−03	1.667E−03	1.667E−03
R (m)	0.0254	0.0254	0.0254
τ_w (Pa)	26.09	32.76	39.42
τ_y/τ_w	0.19	0.31	0.38
γ_w (s^{-1})	173.94	218.39	262.83
V_{Zav} (m/S)	0.823	0.823	0.823
Δp (Pa)	4.109E+05	5.159E+05	6.209E+05
Δp (psi)	59.58	74.80	90.02
R_y (m)	0.00487	0.00775	0.00966
R_y (inch)	0.1916	0.3053	0.3805
Re	1.40E+04	1.40E+04	1.40E+04
ΔQ (m^3/s)	6.517E−09	9.775E−09	9.492E−09
v_m (m/s)	1.44	1.34	1.28

Problem 11.10 For three suspensions of materials with densities $\rho = 1,275\ \text{kg/m}^3$ that can be represented by the Bingham model in the range $10 < \dot{\gamma} < 500\ \text{s}^{-1}$ with $\tau_y = 15$ Pa respectively and a plastic viscosity of $K = 150$, 300 and 500 mPa-s, see Fig. 11.16, flowing in a cylindrical tube 1-inch in diameter and 200 m in length, Calculate the pressure drop and velocity distribution necessary to transport 100 l of the suspension per minute.

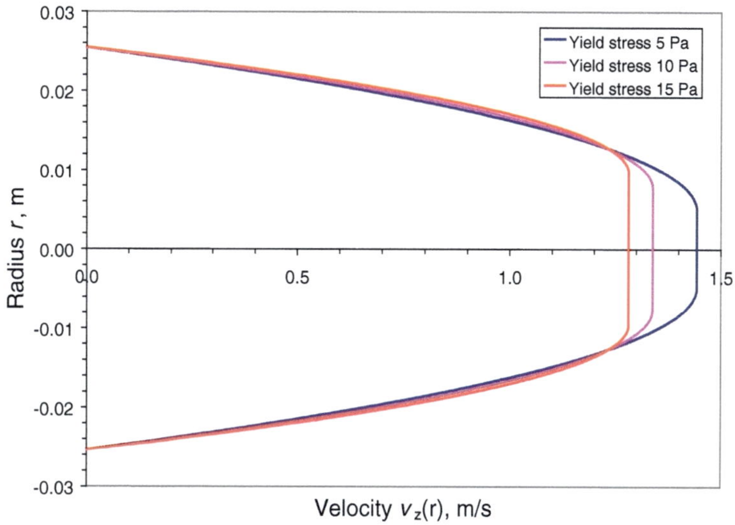

Fig. 11.15 Velocity distributions for a Bingham model of clays

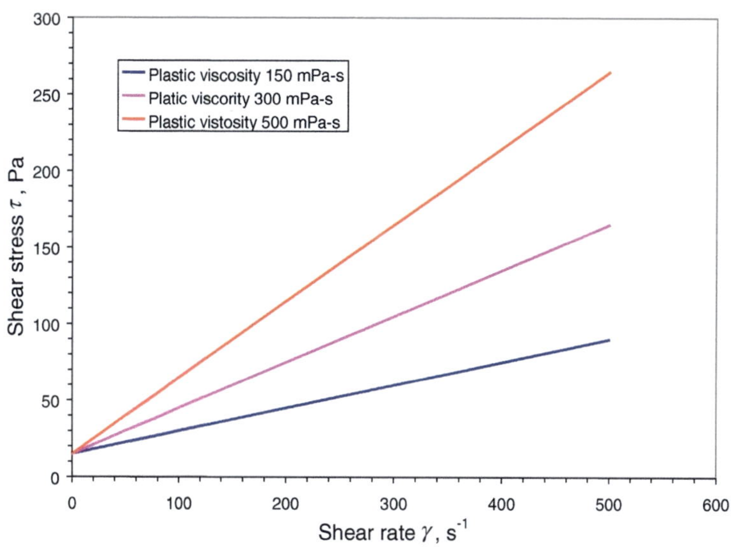

Fig. 11.16 Shear stress versus shear rate for a Bingham model of a material with plastic viscosity of 150, 300 and 500 (mPa-s) and yield stress of 15 (Pa)

$$Q = 100[\ell/\mathrm{m}] = 100/(60 \times 1{,}000) = 1.6667 \times 10^{-3}\,\mathrm{m}^3/\mathrm{s}$$

$$Q = \frac{\pi}{8}\frac{\Delta p R^4}{KL}\left(1 - \frac{4}{3}\left(\frac{\tau_y}{\tau_w}\right) + \frac{1}{3}\left(\frac{\tau_y}{\tau_w}\right)^4\right)$$

$$\tau_w = \frac{1}{2}\frac{\Delta pR}{L} = \frac{\Delta p \times 0.2}{2 \times 200} = 5 \times 10^{-4}\Delta p$$

$$Q = \frac{\pi}{8}\frac{\Delta p \times 0.2^4}{0.2 \times 200}\left(1 - \frac{4}{3}\left(\frac{15}{5 \times 10^{-4}\Delta p}\right) + \frac{1}{3}\left(\frac{15}{5 \times 10^{-4}\Delta p}\right)^4\right)$$

$$\text{Error} = 1.6667 \times 10^{-3} - \frac{\pi}{8}\frac{\Delta p \times 0.2^4}{0.2 \times 200}\left(1 - \frac{4}{3}\left(\frac{15}{5 \times 10^{-4}\Delta p}\right) + \frac{1}{3}\left(\frac{15}{5 \times 10^{-4}\Delta p}\right)^4\right)$$

$$= 1.000 \times 10^{-3}$$

Using solver minimizing the Error by changing Δp leads to:

$$\Delta p = 1.462 \times 10^6 \text{ Pa}$$

$$\tau_w = 5 \times 10^{-4}\Delta p = 5 \times 10^{-4} \times 1.462 \times 10^6$$

$$= 73.11 \text{ Pa}$$

The velocity distribution is given by:

$$v_z(r) = -\frac{1}{2}\frac{\Delta pR^2}{KL}\left(\frac{\tau_y}{\tau_w}\left(1 - \frac{r}{R}\right) - \frac{1}{2}\left(1 - \left(\frac{r}{R}\right)^2\right)\right)$$

$$= \frac{1}{2}\frac{0.2^2}{0.15 \times 200}\left(\frac{15}{73.11}\left(1 - \frac{r}{0.2}\right) - \frac{1}{2}\left(1 - \left(\frac{r}{0.2}\right)^2\right)\right)$$

For the three cases, see Fig. 11.17.

τ_y (Pa)	15	15	15
K (Pa-s)	0.150	0.300	0.500
Q (l/min)	100	100	100
R (inch)	1	1	1
L (m)	200	200	200
ρ (kg/m^3)	1,275	1,275	1,275
Q (m^3/s)	1.667E−03	1.667E−03	1.667E−03
R (m)	0.0254	0.0254	0.0254
τ_w (Pa)	39.14	58.87	84.79
τ_y/τ_w	0.38	0.25	0.18
γ_w (s^{-1})	260.95	196.24	169.57
v_{zav} (m/s)	0.823	0.823	0.823
Δp (Pa)	6.210E+05	9.271E+05	1.335E+06
Δp (psi)	90.05	134.43	193.61
R_y (m)	0.00973	0.00647	0.00449
R_y (inch)	0.3832	0.2548	0.1769
Re	1.40E+04	6.99E+03	4.20E+03
ΔQ (m^3/s)	9.995E−07	1.000E−06	1.000E−06
v_m (m/s)	1.27	1.38	1.46

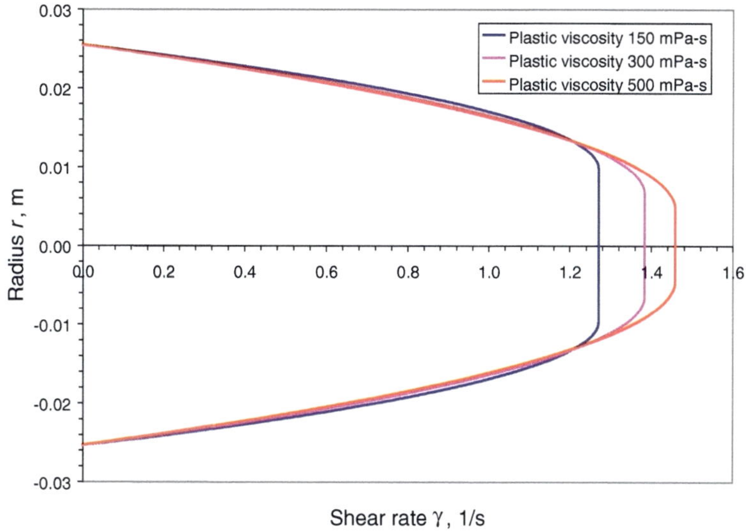

Fig. 11.17 Velocity distributions for a Bingham model of a material with plastic viscosity of 20, 50 and 100 mPa-s and yield stresses of 15 Pa

(b) **Power-Law Fluids**

The constitutive equation for the shear stress of power law fluids flowing in a circular tube is:

$$T_{rz}^{E}(r) = m\left(\frac{\partial v_z}{\partial r}\right)^n \tag{11.66}$$

where m is the consistency index and n is the power index.

Velocity distribution

Replacing (11.66) with (11.51) we have:

$$m\left(\frac{\partial v_z}{\partial r}\right)^n = -\frac{1}{2}\frac{\Delta p}{L}r$$

Integrating yields

$$\frac{\partial v_z}{\partial r} = \left(\frac{-\Delta p}{2\,mL}r\right)^{1/n} \tag{11.67}$$

$$v_z(r) = \left(\frac{\Delta p}{2\,mL}\right)^{1/n} \int (-r)^{1/n} dr + C$$
$$= \left(\frac{\Delta p}{2\,mL}\right)^{1/n} \frac{n}{n+1}\left(-r^{n+1/n}\right) + C$$

Using boundary condition $v_z(R) = 0$, results in:

$$v_z(r) = \frac{nR}{n+1}\left(\frac{\Delta pR}{2\,mL}\right)^{1/n}\left(1 - \left(\frac{r}{R}\right)^{(n+1)/n}\right) \tag{11.68}$$

Figure 11.18 of Problem 11.11 shows the velocity distribution for the flow of a power law fluid in a tube for different values of the power index.

Volume flow rate

The volume flow rate is given by $Q_f = \int_0^R 2\pi v_z r dr$, then substituting (11.68) and integrating:

$$Q_f = \int_0^R 2\pi \frac{nR}{n+1}\left(\frac{\Delta pR}{2\,mL}\right)^{1/n}\left(1 - \left(\frac{r}{R}\right)^{(n+1)/n}\right) r dr$$

$$Q = 2\pi \frac{nR^3}{n+1}\left(\frac{\Delta pR}{2\,mL}\right)^{1/n}\left(\int_0^1 \frac{r}{R} d\left(\frac{r}{R}\right) - \int_0^1 \left(\frac{r}{R}\right)^{(2n+1/n)} d\left(\frac{r}{R}\right)\right)$$

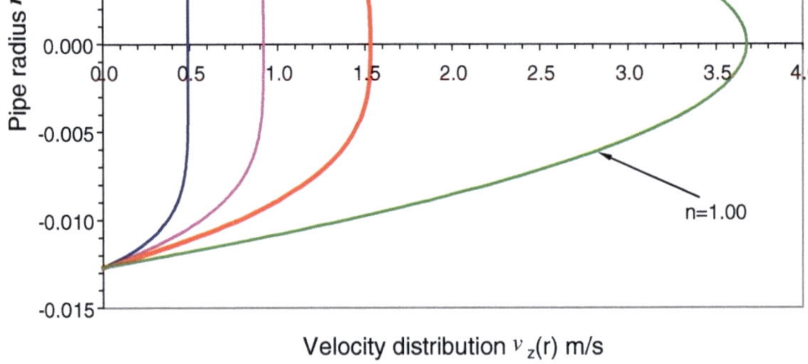

Fig. 11.18 Velocity distribution of power law fluids with consistency index $m = 3\text{Pa} \cdot \text{s}^n$ and power law indices 0.20; 0.33; 0.50; 1 and 3

$$Q_f = 2\pi R^3 \frac{n}{n+1}\left(\frac{\Delta pR}{2\,mL}\right)^{1/n}\left(\frac{1}{2}\left(\frac{r}{R}\right)^2\Big|_0^1 - \left(\frac{1}{((2n+1)/n)+1}\right)\left(\frac{r}{R}\right)^{((2n+1/n+1))}\right)$$

$$Q_f = \pi R^2 \frac{n}{(3n+1)}\left(\frac{\Delta pR^{n+1}}{2\,mL}\right)^{1/n}$$

Average velocity

The average velocity is given by $\bar{v}_z = Q_f/\pi R^2$, then:

$$\bar{v}_z = \frac{n}{(3n+1)}\left(\frac{\Delta pR^{n+1}}{2\,mL}\right)^{1/n} \tag{11.70}$$

$$\text{and} \qquad \frac{8\bar{v}_z}{D} = \frac{4n}{(3n+1)}\left(\frac{\Delta pR}{2\,mL}\right)^{1/n} \tag{11.71}$$

Shear stress and shear rate at the wall and maximum velocity

Since the shear rate and shear stress at the wall are given by:

$$\dot{\gamma}_w = \frac{\partial v_z}{\partial r}\Big|_{r=R} = \left(-\frac{\Delta pR}{2\,mL}\right)^{1/n} \qquad \tau_w = m\dot{\gamma}_w^n$$

from (11.71) we finally have:

$$\dot{\gamma}_w = \frac{(3n+1)}{4n}\frac{8\bar{v}_z}{D} \qquad \tau_w = m\left(\frac{(3n+1)}{4n}\frac{8\bar{v}_z}{D}\right)^n \tag{11.72}$$

The maximum velocity is obtained from (11.68) for $r = 0$, then:

$$v_m = \frac{nR}{n+1}\left(\frac{\Delta pR}{2\,mL}\right)^{1/n} \tag{11.73}$$

$$v_z(r) = v_m\left(1 - \left(\frac{r}{R}\right)^{(n+1)/n}\right) \tag{11.74}$$

Pressure drop

From (11.69)

$$\boxed{\Delta p = \frac{2\,mL}{R^{n+1}}\left(\frac{(3n+1)}{n}\frac{Q}{\pi R^2}\right)^n} \tag{11.75}$$

Problem 11.11 For a mass flow of $F = 1{,}000$ kg/h of a non-Newtonian fluid of the potential type with a density of $\rho = 1{,}074$ kg/m^3, consistency index of $m = 3$

and power law indices of $n = 1/5,\ 1/3,\ 1/2$ and $1,\ 3$, calculate the pressure drop Δp and draw a figure for the velocity distribution. See Fig. 11.18.

n	0.20	0.33333333	0.50	1.00
ρ (kg/m^3)	1,074	1,074	1,074	1,074
m (Pa-s^n)	3	3	3	3
d (inch)	1.00	1.00	1.00	1.00
L (m)	10	10	10	10
F (kg/h)	1,000	1000	1,000	1,000
Q (m^3/s)	9.311E−04	9.311E−04	9.311E−04	9.311E−04
R (m)	0.0127	0.0127	0.0127	0.0127
v_{zav} (m/s)	1.838	1.838	1.838	1.838
v_m (m/s)	0.49	0.92	1.53	3.68
Δp (Pa)	1.40E+04	3.12E+04	8.99E+04	2.73E+06

Wall shear stress and Reynolds number

Defining the friction coefficient in the same way as for Newtonian fluids in laminar flow, $f = 16/\mathrm{Re}$, we can define *a Reynolds number* Re_M for a power law fluid as the ratio of the wall shear stress τ_w to the dynamic pressure Metzner and Reed (1959). Then we have:

$$f = \frac{-\tau_w}{\frac{1}{2}\rho\bar{v}_z^2} = \frac{16}{\mathrm{Re}} \tag{11.76}$$

From (11.72) $\tau_w = m\left(\frac{(3n+1)}{4n}\frac{8\bar{v}_z}{D}\right)^n$

So that Re_M is:

$$\boxed{\mathrm{Re}_M = \frac{\rho\bar{v}_z^{2-n}D^n}{8^{n-1}m\left(\frac{3n+1}{4n}\right)^n}} \tag{11.77}$$

Transition to a turbulent regime

As in the case of fluids with Newtonian behavior, the friction factor gives the transition from a laminar to turbulent flow.

With the Reynolds number defined by (11.77), the roughness Reynolds number by $\mathrm{Re}_\in = \in\sqrt{1/f}\mathrm{Re}$ and the friction factor $f = -\Delta pD/2\rho L\bar{v}_z^2$, we have:

$$f = \frac{16}{\mathrm{Re}} \quad \text{for} \quad \mathrm{Re} < 2,100 \tag{11.78}$$

$$f_{rough} = \frac{f_{water-rough}}{f_{water-smooth}}\frac{16}{\mathrm{Re}} \quad \text{for} \quad \mathrm{Re} < 2,100,\ 5 < \mathrm{Re}_\in < 70 \tag{11.79}$$

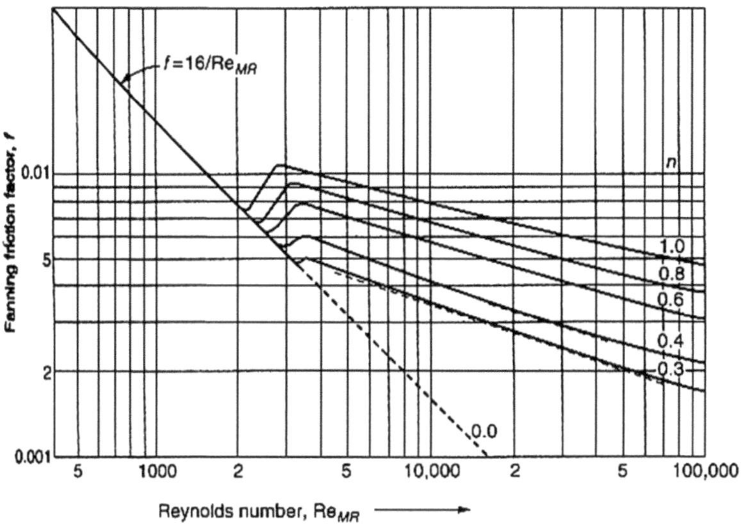

Fig. 11.19 Friction factor as a function of Metzner's Reynolds number for different values of the power function n (Chhabra and Richardson 1999)

$$\left(f = \frac{4.53}{n}\log\left(\mathrm{Re}\sqrt{f^{2-n}}\right) + \frac{2.69}{n} - 2.95 + 0.68\frac{5n-8}{n}\right)^{-2}, \quad \text{for} \quad \mathrm{Re} > 4,000\,,\ \mathrm{Re}_{\in} < 5 \tag{11.80}$$

$$f_{rough} = \left(4.07\log\left(\frac{1}{2\in}\right) + 6 - \frac{2.65}{n}\right)^{-2} \quad \text{for} \quad \mathrm{Re} > 4,000\,,\ \mathrm{Re}_{\in} > 70 \tag{11.81}$$

Figure 11.19 shows the friction factor as a function of Metzner's Reynolds number for different values of the power function n for smooth walls according to Chhabra and Richardson (1999).

Problem 11.12 A polyacrilamide solution of $\rho = 1,074\ \mathrm{kg/m^3}$ in density is to be pumped through a tube one inch in diameter and 10 m in length at a rate of 2.500 kg/h. Measurement in the laboratory showed that the fluid can be represented by the power law model with $m = 3\text{Pa-s}^n$ and $n = 0.5$. Calculate the necessary pressure to maintain the flow and calculate the velocity distribution, average and maximum velocity.

Volume flow

$$Q = \frac{F}{\rho \times 3,600} = 6.466 \times 10^{-4}\ \mathrm{m^3/s}$$

Pressure drop:

$$\Delta p = \frac{2\,mL}{R^{n+1}}\left(\frac{3n+1}{n}\frac{Q}{\pi R^2}\right)^n = 1.059 \times 10^5 \text{ Pa}$$

Average velocity

$$\bar{v}_z = \frac{Q}{\pi R^2} = 1.128\,\text{m/s}$$

Velocity distribution:

$$v_z(r) = \frac{nR}{n+1}\left(\frac{\Delta p R}{RmL}\right)^{1/n}\left(1 - \left(\frac{r}{R}\right)^{(n+1)/n}\right) = 2.128 \times \left(1 - \left(\frac{r}{R}\right)^{(n+1)/n}\right)$$

Maximum velocity:

$$v_z(r) = \frac{nR}{n+1}\left(\frac{\Delta p R}{RmL}\right)^{1/n} = 2.128$$

$$\text{Velocity distribution} \quad v_z = 2.128 \times \left(1 - \left(\frac{r}{R}\right)^{(n+1)/n}\right)$$

Problem 11.13 A non-Newtonian fluid with density equal to that of water flows in a tube 300 mm in diameter and 50 m long at a rate of 300 kg/s. Rheological measurements yield the following power law parameters: $m = 2.74\,\text{Pa-s}^{0.3}$ and $n = 0.30$. Determine the necessary power of a pump and the wall shear stress. See Fig. 11.20.

ρ (kg/m^3)	1,074
m (Pa-s^n)	3
n	0.5
D (inch)	1
L (m)	10
F (kg/h)	2,500
Q (m^3/s)	6.466E−04
R (m)	1.270E−02
v_{zav} (m/s)	1.277
Δ (Pa)	1.059E+05
v_m (m/s)	2.12787

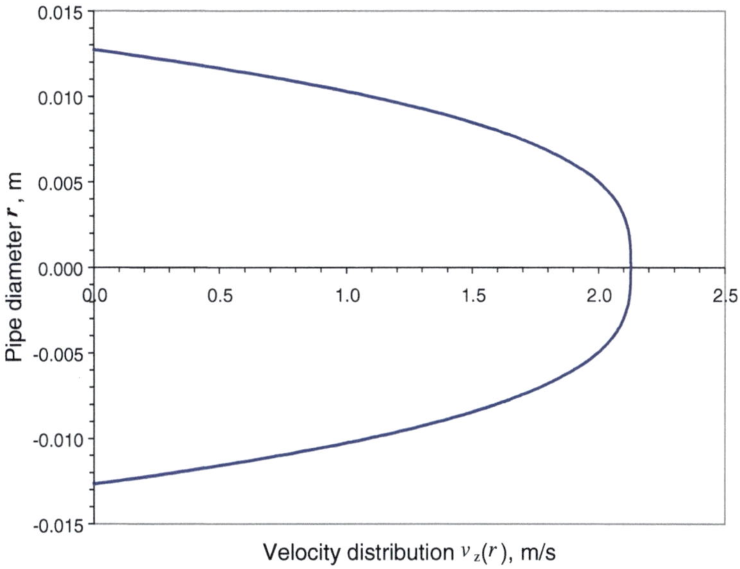

Fig. 11.20 Velocity distribution for a polyacrilamide solution with a power law model: $m = 3[\text{Pa-s}^{0.5}]$ y $n = 0.5$

$$\text{Average velocity}: \quad \bar{v}_z = \frac{Q_f}{\pi R^2} = \frac{300/1,000}{3.14 \times (0.15)^2} = 4.24\,(\text{m/s})$$

The transitional or critical Reynolds number is $\text{Re}_c \approx 2,100$.

$$\text{Re}_{MRc} = \frac{\rho v_c^{2-n} D^n}{8^{n-1} m \left(\frac{3n+1}{4n}\right)^n} = 2,100$$

then the critical velocity, that is, the velocity at which the flow changes from laminar to turbulent, is:

$$v_c = \left(\frac{8^{n-1} m \left(\frac{3n+1}{4n}\right)^n 2,100}{\rho D^n}\right)^{n-2} = 2.91\,(\text{m/s})$$

Since the average velocity 4.24 (m/s) is greater than the critical velocity 2.91 (m/s), the regime is turbulent. The actual Reynolds number is:

$$\text{Re}_{MR} = \frac{\rho \bar{v}_z^{2-n} D^n}{8^{n-1} m \left(\frac{3n+1}{4n}\right)^n} = 11090$$

With $\text{Re}_{\text{MR}} = 11,090$ and $n = 0.3$, from Fig. 11.16, we get a friction factor $f = 0.0033$.

The value of the pressure drop necessary to produce the flow is obtained from the friction factor definition: $f = -\tau_w/1/2\rho\bar{v}_z^2$; $\tau_w = -\frac{1}{2}\frac{\Delta p}{L}R \rightarrow f = \frac{\Delta p}{L}\frac{R}{\rho\bar{v}_z^2}$ therefore:

$$\Delta p = \frac{2\rho\bar{v}_z^2 L}{D} f = 1.9814 \times 10^4 \text{ (Pa)}$$

and, the pump power P*o* is:

$$Po = Q_f \times \Delta P = 6.0\,\text{kW}$$

The wall shear stress is

$$\tau_w = -\frac{1}{2}\frac{\Delta p}{L}R = 29.7\,\text{Pa}$$

D (m)	0.3
L (m)	50
ρ (kg/m^3)	1,000
F (kg/s)	300
m (Pa-s$^{0.3}$)	2.74
n	0.3
Re_{MRc}	2,100
R (m)	0.15
Q (m^3/s)	0.300
v_{zav} (m/s)	4.24
v_c (m/s)	2.921
Re_{MR}	11,090
f (11090; 0.30)	0.0033
Δp (Pa)	19,814
Po (W)	5,944
τ_w (Pa)	29.7

(c) **Herschel-Bulkley Fluid**

The constitutive equation for the stress tensor for Herschel-Bulkley fluids in a pipe has the form:

$$T_{rz}^E(r)\begin{cases} < \tau_y; & \frac{\partial v_z}{\partial r} = 0 \\ \geq \tau_y; & T_{rz}^E(r) = \tau_y + m\left(\frac{\partial v_z}{\partial r}\right)^n \end{cases} \tag{11.82}$$

Since for any fluid $T_{rz}^E(r) = -\frac{1}{2}\frac{\Delta p}{L}r$, for Herschel-Bulkley fluids we have:

$$T_{rz}^E(R_y) = \tau_y = -\frac{1}{2}\frac{\Delta p}{L}R_y$$

Then:

$$T_{rz}^{E}(r) - \tau_y = -\frac{1}{2}\frac{\Delta p}{L}\left(r - R_y\right)$$

Velocity distribution

For $T_{rz}^{E}(r) \geq \tau_y\,;\;\; R_y \leq r \leq R:$

$$\frac{\partial v_z}{\partial r} = \left(\frac{T_{rz}^{E}(r) - \tau_y}{m}\right)^{1/n} = \left(-\frac{\Delta p\left(r - R_y\right)}{2\,mL}\right)^{1/n} \tag{11.83}$$

Integrating with boundary condition $v_z(R) = 0$

$$v_z(r) = \left(-\frac{\Delta p}{2\,mL}\right)^{1/n} \int \left(r - R_y\right)^{1/n} dr$$

$$v_z(r) = \left(-\frac{\Delta p}{2\,mL}\right)^{1/n} \times \frac{n}{n+1}\left(r - R_y\right)^{(n+1)/n} + C_1$$

$$v_z(R) = \left(-\frac{\Delta p}{2\,mL}\right)^{1/n} \times \frac{n}{n+1}\left(R - R_y\right)^{(n+1)/n} + C_1 = 0 \tag{11.84}$$

$$\mathrm{C}_1 = -\left(-\frac{\Delta p}{2\,mL}\right)^{1/n} \times \frac{n}{n+1}\left(R - R_y\right)^{(n+1)/n}$$

$$v_z(r) = \left(-\frac{\Delta pR}{2\,mL}\right)^{1/n} \times \frac{nR}{n+1}\left(\left(\frac{r}{R} - \frac{R_y}{R}\right)^{(n+1)/n} - \left(1 - \frac{R_y}{R}\right)^{(n+1)/n}\right)$$

Using Eqs. (11.12) and (11.53) for the shear stress at the wall and the yield stress, Eq. (11.84) can be written in the form:

$$\boxed{v_z(r) = \left(\frac{\tau_w}{m}\right)^{1/n} \times \frac{nR}{n+1}\left(\left(\frac{r}{R} - \frac{\tau_y}{\tau_w}\right)^{(n+1)/n} - \left(1 - \frac{\tau_y}{\tau_w}\right)^{(n+1)/n}\right)}; \quad \text{for} \quad R_y \leq r \leq R \tag{11.85}$$

For $T_{rz}^{E}(r) < \tau_y\,;\quad 0 < r < R_y \quad$ where $\quad \tau_y = T_{rz}^{E}\left(R_y\right):$

$$\frac{\partial v_z}{\partial r} = 0 \qquad \rightarrow \qquad v_z(r) = v_z\left(R_y\right) \tag{11.86}$$

From (11.84),

$$v_z(r) = -\left(-\frac{\Delta pR}{2\,mL}\right)^{1/n} \times \frac{nR}{n+1}\left(1 - \frac{R_y}{R}\right)^{(n+1)/n} \quad \text{for} \quad 0 \leq r \leq R_y \tag{11.87}$$

$$\boxed{v_z(r) = -\left(\frac{\tau_w}{m}\right)^{1/n} \times \frac{nR}{n+1}\left(1 - \frac{\tau_y}{\tau_w}\right)^{(n+1)/n}} \quad \text{for} \quad 0 \leq r \leq R_y \tag{11.88}$$

Volume flow rate

The volume flow rate is given by $Q_f = \int_0^R 2\pi v_z r dr$, then substituting (11.84) and (11.87) into this equation and integrating yields:

$$Q_f = 2\pi\left(\frac{\tau_w}{m}\right)^{1/n} \frac{nR}{n+1}\left(\begin{array}{l} \displaystyle\int_0^{R_y} -\left(1 - \frac{\tau_y}{\tau_w}\right)^{(n+1)/n} rdr \\ \displaystyle + \int_{R_y}^{R}\left(\left(\frac{r}{R} - \frac{\tau_y}{\tau_w}\right)^{(n+1)/n} - \left(1 - \frac{\tau_y}{\tau_w}\right)^{(n+1)/n} rdr\right)\end{array}\right)$$

$$= 2\pi\left(\frac{\tau_w}{m}\right)^{1/n} \frac{nR^3}{n+1}\left(\begin{array}{l} \displaystyle\int_0^{R_y/R} -\left(1 - \frac{\tau_y}{\tau_w}\right)^{(n+1)/n} \xi d\xi \\ \displaystyle + \int_{R_y/R}^{1}\left(\xi\left(\xi - \frac{\tau_y}{\tau_w}\right)^{(n+1)/n} - \left(1 - \frac{\tau_y}{\tau_w}\right)^{(n+1)/n} \xi\right) d\xi\end{array}\right)$$

Integrating yields (Bird et al. 1987):

$$Q_f = \pi n R^3\left(\frac{\tau_w}{m}\right)^{1/n}\left(\frac{1}{3n+1}\left(1 - \frac{\tau_y}{\tau_w}\right)^2 + \frac{2}{2n+1}\frac{\tau_y}{\tau_w}\left(1 - \frac{\tau_y}{\tau_w}\right) + \frac{1}{n+1}\left(\frac{\tau_y}{\tau_w}\right)^2\right) \tag{11.89}$$

Average velocity

The average velocity is given by $\bar{v}_z = Q/\pi R^2$, then:

$$\bar{v}_z = nR\left(\frac{\tau_w}{m}\right)^{1/n}\left(\frac{1}{3n+1}\left(1 - \frac{\tau_y}{\tau_w}\right)^{(n+1)/n} + \frac{2}{2n+1}\frac{\tau_y}{\tau_w}\left(1 - \frac{\tau_y}{\tau_w}\right) + \frac{1}{n+1}\left(\frac{\tau_y}{\tau_w}\right)^2\right) \tag{11.90}$$

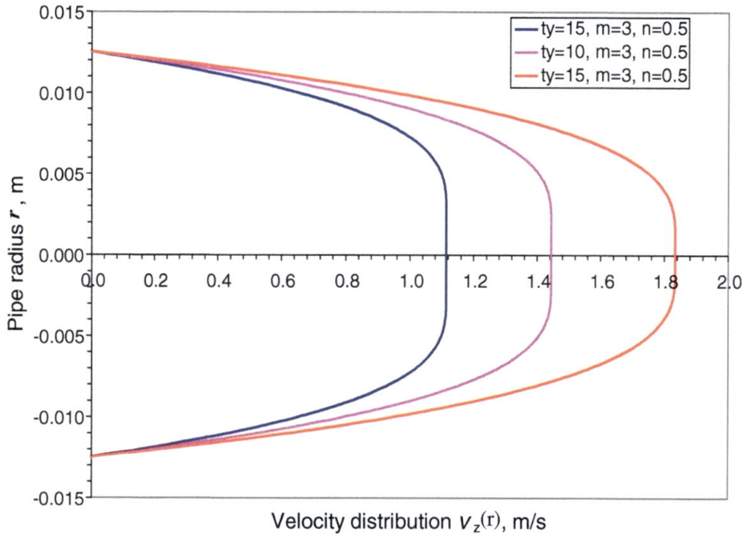

Fig. 11.21 Velocity distribution of Herschel-Bulkley fluids with yield stresses τ_y=5; 10 and 15; Pa, consistency index m = 3 Pa-s^n and power law index of 0.50

Maximum velocity

The maximum velocity is obtained from (11.88), then:

$$v_m = -\left(\frac{\tau_y}{\tau_w}\right)^{1/n} \times \frac{nR}{n+1}\left(1-\frac{\tau_y}{\tau_w}\right)^{(n+1)/n} \tag{11.91}$$

Problem 11.14 Figure 11.21 shows the velocity distribution for Herschel-Bulkley fluids with yield stress $\tau_y = 5,\ 10$ and15Paconsistency index m = 3 Pa-s^n and power law indices of 0.50. See Fig. 11.21.

τ_y (Pa)	15	10	5
n	0.5	0.5	0.5
m (Pa-s^n)	3	3	3
ρ (kg/m^3)	1,074	1,074	1,074
d (cm)	2.5	2.5	2.5
L (m)	10	10	10
F (kg/h)	2,500	2,500	2,500
Q (m^3/s)	6.466E−04	6.466E−04	6.466E−04
R (m)	1.250E−02	1.250E−02	1.250E−02
v_m (m/s)	1.318E+00	1.318E+00	1.318E+00
Δp (Pa)	1.125E+05	1.125E+05	1.125E+05
R_y	2.667E−03	1.778E−03	8.890E−04

Transition to turbulent regime

As in the case of Newtonian fluids, the friction factor gives the transition from laminar to turbulent flow. The Reynolds number is the same as that of pseudo plastic fluids. See Eq. (11.77):

$$\mathrm{Re}_{HB} = \frac{\rho \bar{v}_z^{2-n} D^n}{8^{n-1} m \left(\frac{3n+1}{4n}\right)^n} \; ; \; \mathrm{Re}_{\in} = \in \sqrt{1/f}\,\mathrm{Re}_{HB} \tag{11.92}$$

Now the friction factor $f = -\Delta p D / 2\rho L \bar{v}_z^2$ is given by:

$$f_{smooth} = \frac{1}{2^{2n-4}} \left(\frac{\pi R^3}{Q}\right) \left(\frac{m}{\tau_w}\right)^{1/n} \times \frac{1}{\mathrm{Re}_{HB}} ; \quad \text{for} \quad \mathrm{Re}_{\mathrm{HB}} < 2,100 \tag{11.93}$$

$$f_{rough} = \frac{f_{water-rough}}{f_{water-smooth}} \frac{1}{2^{2n-4}} \left(\frac{\pi R^3}{Q}\right) \left(\frac{m}{\tau_w}\right)^{1/n} \times \frac{1}{\mathrm{Re}_{HB}} ; \quad \text{for} \quad \mathrm{Re}_{PL} < 2,100\,,\; 5 < \mathrm{Re}_{\in} < 70 \tag{11.94}$$

$$\left(f_{smooth} = \frac{4.53}{n} \log\left(\mathrm{Re}_{PL} \sqrt{f^{2-n}}\right) + \frac{2.69}{n} - 2.95 + 0.68 \frac{5n-8}{n} \right)^{-2}, \quad \text{for} \quad \mathrm{Re}_{HB} > 4,000\,,\; \mathrm{Re}_{\in} < 5 \tag{11.95}$$

$$f_{rough} = \left(4.07 \log\left(\frac{1}{2\in}\right) + 6 - \frac{2.65}{n}\right)^{-2} \quad \text{for} \quad \mathrm{Re}_{PL} > 4,000,\ \mathrm{Re}_{\in} > 70 \tag{11.96}$$

Problem 11.15 The rheology of copper tailings is described by the values in the following table:

%	ty (Pa)	η (mPa-s)			
		10	100	150	200
55	0.678	166	32	25	21
60	1.035	230	44	34	29
65	1.579	318	61	46	39
70	2.409	441	83	63	53

Calculate the pressure drop necessary to transport 5,000 l per minute of a copper tailing with a density of 2,650 kg/m^3, at 55, 60, 65 and 70 % of solid by weight in a pipe 4 inches in diameter and 200 m long, if the rheological parameters of the pulp are those given in the table. Model the rheology of the tailing and draw the Rheological curves.

Pressure drop:

$$Q_f = 5,000\,\ell/\text{m} = 5,000/(60 \times 1,000) = 8.333 \times 10^{-2}\ \text{m}^3/\text{s}$$

$$Q_f = \frac{\pi}{8}\frac{\Delta pR^4}{KL}\left(1 - \frac{4}{3}\left(\frac{\tau_y}{\tau_w}\right) + \frac{1}{3}\left(\frac{\tau_y}{\tau_w}\right)^4\right)$$

$$\dot{\gamma}_w = \frac{8\bar{v}_z}{D} = \frac{32Q}{\pi D^3} = \frac{32 \times 1.6667 \times 10^{-3}}{3.14 \times 0.02^3} = 26.5\ 1/s$$

$$\tau_w = \frac{1}{2}\frac{\Delta pR}{L} = \frac{\Delta p \times 0.1016}{2 \times 200} = 2.54 \times 10^{-4}\Delta p$$

$$\tau_y = 0.678\ \text{Pa}$$

$$Q_f = \frac{\pi}{8}\frac{\Delta p \times 0.1016^4}{0.818 \times 200}\left(1 - \frac{4}{3}\left(\frac{0.678}{2.54 \times 10^{-4}\Delta p}\right) + \frac{1}{3}\left(\frac{0.678}{2.54 \times 10^{-4}\Delta p}\right)^4\right)$$

$$\text{Error} = 8.333 \times 10^{-2} - \frac{\pi}{8}\frac{\Delta p \times 0.1016^4}{0.818 \times 200}\left(1 - \frac{4}{3}\left(\frac{0.678}{2.54 \times 10^{-4}\Delta p}\right) + \frac{1}{3}\left(\frac{0.678}{2.54 \times 10^{-4}\Delta p}\right)^4\right)$$

$$= 6.998 \times 10^{-7}\Delta p =$$

The following excel sheet permits the calculation of all solid percentages:

% solids	55	60	65	70
ρ_s (kg/m^3)	2,650	2,650	2,650	2,650
ρ_f (kg/m^3)	1,000	1,000	1,000	1,000
τ_y (Pa)	0.678	1.035	1.579	2.409
η (Pa-s)	0.818	1.142	1.612	2.266
Q (l/min)	5000.00	5000.00	5000.00	5000.00
R (inchs)	4.00	4.00	4.00	4.00
L (m)	200	200	200	200
ρ (kg/m^3)	1520.80	1596.39	1679.87	1772.58
Q (m^3/s)	8.333E−02	8.333E−02	8.333E−02	8.333E−02
R (m)	0.1016	0.1016	0.1016	0.1016
τ_w (Pa)	83.66	116.92	165.19	232.46
$\tau_y/\ \tau_w$	0.01	0.01	0.01	0.01
γ_w (s^{-1})	102.28	102.38	102.48	102.59
v_{zav} (m/s)	2.571	2.571	2.571	2.571
Δp (Pa)	3.294E+05	4.603E+05	6.504E+05	9.152E+05
Δp (psi)	47.76	66.74	94.30	132.71
R_y (m)	0.00082	0.00090	0.00097	0.00105
R_y (inch)	0.0324	0.0354	0.0382	0.0415
Re	3.82E+04	2.88E+04	2.14E+04	1.61E+04
ΔQ (m^3/s)	9.998E−07	9.999E−07	1.000E−06	1.000E−06
v_m (m/s)	5.11	5.11	5.11	5.10

% solid	τ_y Pa	m	n − 1	n
55	0.678	818.26	−0.6963	0.3037
60	1.035	1141.8	−0.6996	0.3004
65	1.579	1611.7	−0.7072	0.2928
70	2.409	2266.1	−0.7134	0.2866
Average			−0.70	0.30

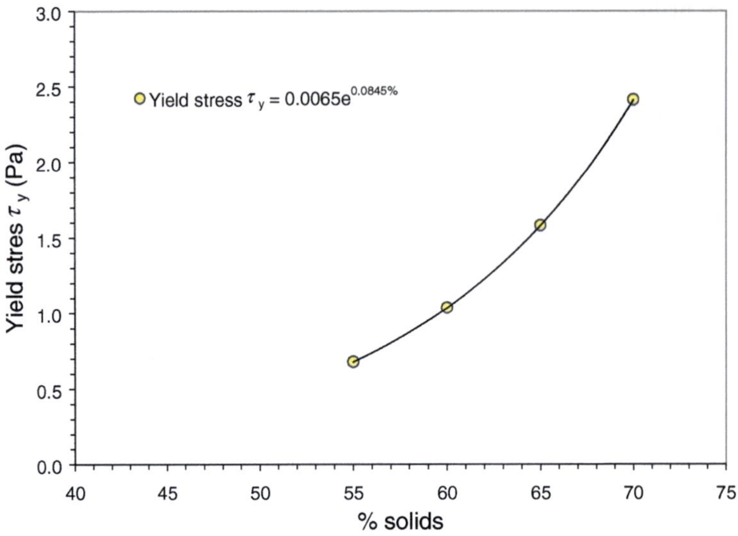

Fig. 11.22 Yield stress versus % solid by weight for a copper tailing

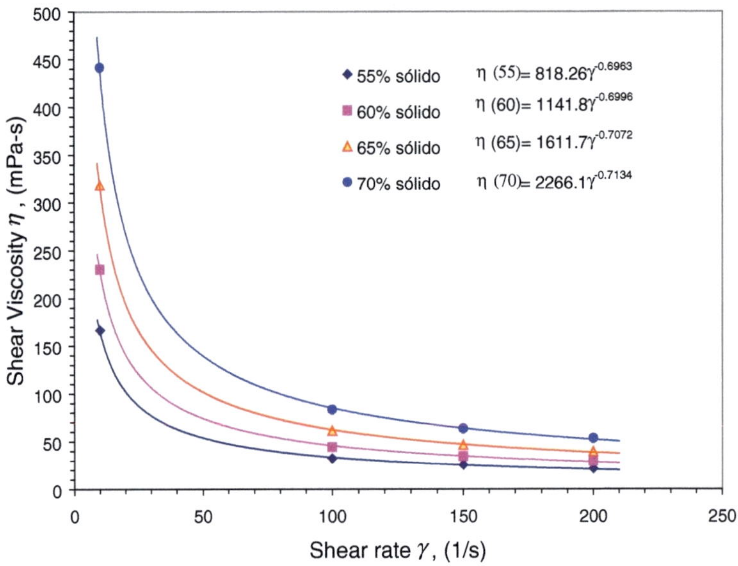

Fig. 11.23 Shear viscosity versus shear rate for a copper tailing with % solid by weight as parameter

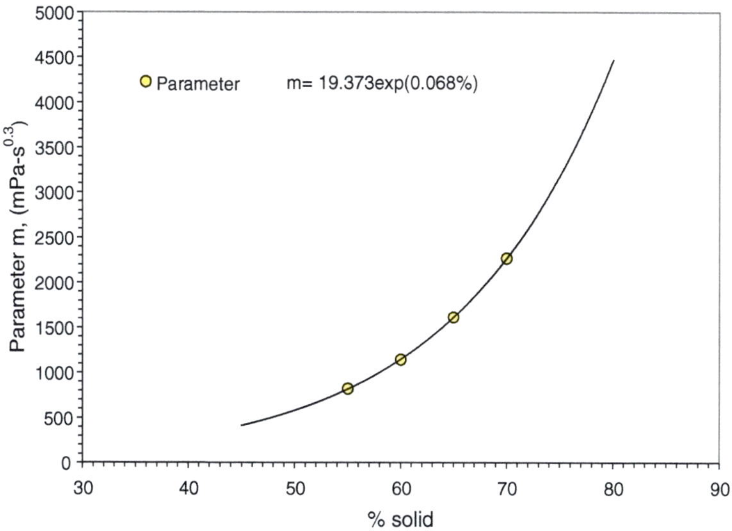

Fig. 11.24 Parameter m for a copper concentrate versus % solid

$$\eta(\text{mPa-s}) = 19.3673\ \exp(0.067\,\%)\dot{\gamma}^{-0.7}$$
$$\tau_y\,(\text{Pa}) = 0.0035\exp(0.0845\,\%)$$
$$\tau(Pa) = 0.0035\ \exp(0.0845\,\%) + 19.3673\ \exp(0.067\%)\dot{\gamma}^{0.3}$$

Figures 11.22, 11.23, 11.24, and 11.25 show the results graphically.

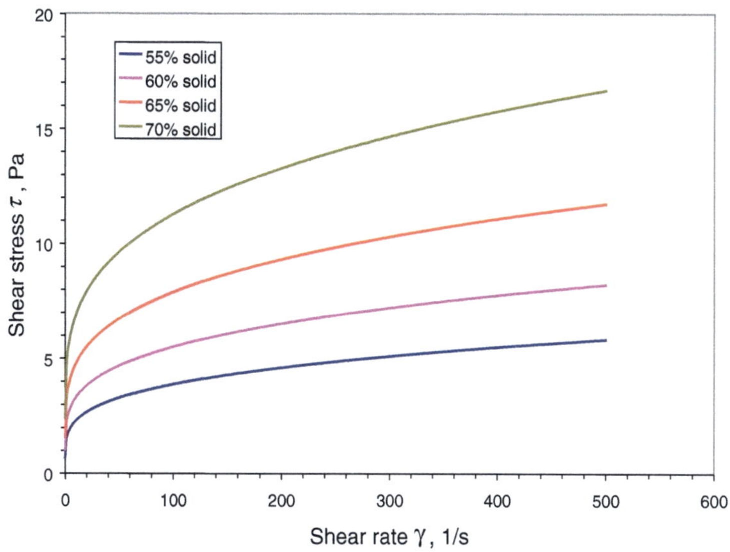

Fig. 11.25 Shear stress versus shear rate for a copper tailing

Table 11.3 Geometrical parameters of typical channels used for slurry conveyance

Geometry	Area	Wetted perimeter
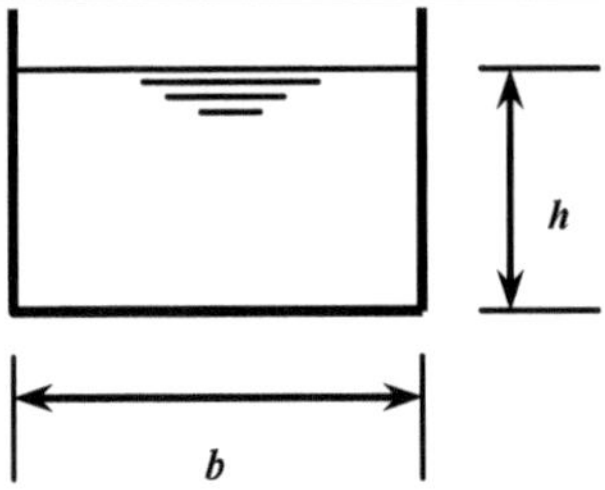	$A = bh$	$P = b + 2h$
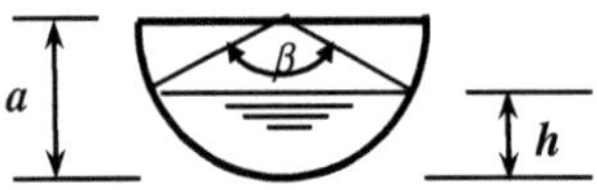 $h = a\left(1 - cos\left(\frac{\beta}{2}\right)\right)$	$A = \frac{a^2}{8}(\beta - sen(\beta))$	$P = \frac{a}{2}\beta$
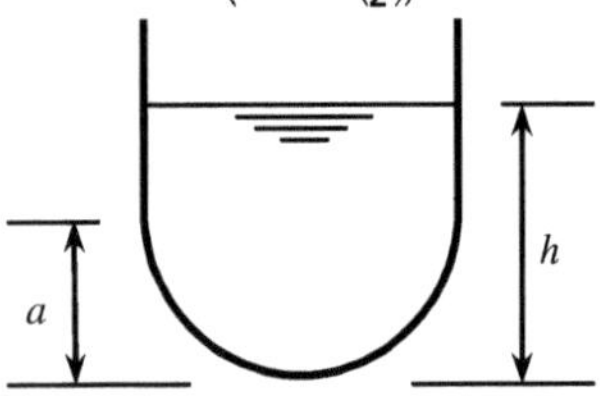	$A = 2a(h - a) + \frac{\pi}{2}a^2$	$P = \pi a + 2(h - a)$

11.5 Transporting Suspensions in Open Channels

Due to the natural slopes of the land around mines, it is often convenient to use Channels instead of pipelines to transport tailings. From a fundamental point of view, the problem of Channel flow is more complex than tube flow because the flow area is not known in advance and it can change while the flow is developing. The case is simpler if the flow is uniform.

Although several Channel geometries are used in open Channel flows, the most commonly used in slurry transport are the rectangular section, semicircular, and a composed semicircular-rectangular section. The relevant geometrical parameters are the cross-section flow area, A, and the wetted perimeter, P, as given in Table 11.3.

Steady open Channel flow is classified as uniform or non-uniform. A uniform flow is one for which the fluid depth h above the Channel bed is constant. Non-uniform flows are further classified into gradually varied flows, where the curvature of the free surface is small compared to the depth of the fluid, and rapidly varied flows, where the curvature is comparable to the fluid depth. Analysis of gradually varied flow is simpler because a hydrostatic pressure distribution can be assumed. Curvature in rapidly varied flows adds a radial acceleration to the fluid particles that must be added to the gravity effect to compute the pressure.

In studying steady state gradually varied flows in open Channels we want to determine the flow depth h as a function of the distance x for a given flow rate Q_f.

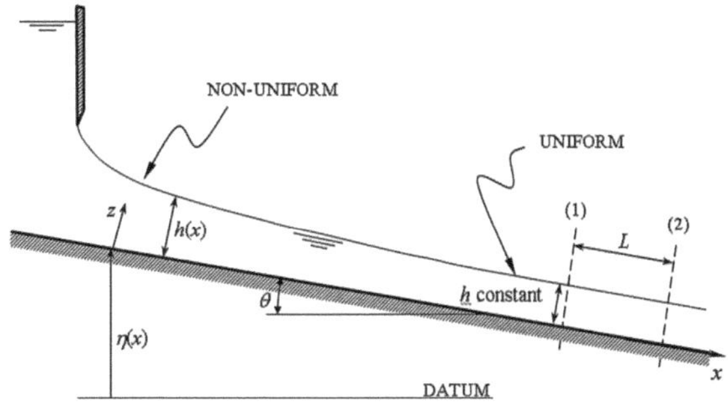

Fig. 11.26 Uniform and non-uniform flow in a channel

In this case the flow depth depends on the Channel characteristics (geometry, slope, wall roughness), and fluid properties (density and viscosity). See Fig. 11.26.

Flow of copper tailings in Channels has more favorable conditions than flow in pipelines. For example, the concentration of the pulp in Channels has no influence on the speed of the flow and whether or not the flow is turbulent, (velocities above 0.8 m/s will yield turbulent flow). The viscosity has no influence on the transport velocity but influences in limiting deposit velocity. On the other hand, the head loss can be calculated using the methods for water with similar wall conditions. These simplifications are not valid for pipe flow.

The slope of the Channel is important. As a rule of thumb, slight slopes, such as 0.3 %, need transport velocities greater than 1.2 m/s to avoid embankment. Velocities of 1.5 m/s are recommended for copper tailings (Kleiman 1960). If water is added to a developed flow with high solid content, such as 45 % by weight in a Channel with a slight slope and slow transport velocity, particles will settle. This is because the water dilutes the pulp and larger particles can segregate. Once a bed forms under this condition, it cannot be eliminated by washing with water. A flow with high concentration at velocities higher than 1.0 (m/s) will eventually removes the bed. Channels with slopes greater than 0.6 % and flows with high solid concentration will not segregate particles if water is added, and Channels with slopes greater than 0.9 % will not embank even with low flows.

11.5.1 Sub-Critical and Super-Critical Flow

Open Channel flows can be classified (Tamburrino 2000) in several ways, depending on the aspects we are interested in. We already distinguished between gradual and rapid flow, depending on the flow curvature. Other aspects we can consider are also found in pipe flows, such as the flow variation over time (steady or unsteady flow), and the importance of viscous effects with respect to inertia (viscous or turbulent regime). For non-homogeneous suspensions, a turbulent flow is required to avoid particle settling.

Another important classification arises from comparing mean flow velocity $\bar{v}_x$ to the speed of the small surface wave c. Assuming low wave amplitude and negligible surface tension effects, the speed of a surface wave is given from potential flow theory by:

$$\frac{c^2}{gh} = \frac{\lambda}{2\pi h}\operatorname{tgh}\left(\frac{2\pi h}{\lambda}\right) \tag{11.97}$$

where λ is the wave length and g is acceleration due to gravity. When $\lambda \gg h$, $\operatorname{tgh}(2\pi h/\lambda) \approx 2\pi h/\lambda$, and the speed of a small perturbation on the free surface of a flow having a finite depth is given by:

$$c = \sqrt{gh} \tag{11.98}$$

A further classification arises when $\bar{v}_x$ is compared to c, it being customary to work with the ratio $\bar{v}_x/c = \bar{v}_x/\sqrt{gh} \equiv Fr$, which is termed the Froude number. Thus, the following classification arises in open Channel flows:

$Fr < 1$, the flow is called sub-critical (or tranquil flow).
$Fr = 1$, the flow is called critical.
$Fr > 1$, the flow is called supercritical (or rapid flow).

When flow conditions are such that the Froude number moves in the range 0.8–1.2, the flow is called trans-critical. Design of open Channels usually avoids trans-critical flows due to the presence of water surface oscillations and flow depth variations. A supercritical regime is recommended for slurry transport in Channels.

The speed given in (0.91) is valid for two-dimensional flows. A general definition, valid for a Channel with any shape is as follows:

$$c = \sqrt{g\frac{A}{b_s}} \tag{11.99}$$

where A is the flow cross sectional area and b_s is the free surface width.

11.5.2 Steady Uniform Flow

In steady uniform flow there is equilibrium between the force generating the motion (gravity) and the resistance force opposing the flow. Theoretically, a gradually varied sub-critical flow will became uniform at an infinite distance upstream, and a super-critical flow reaches the condition of uniform flow at an infinite downstream. In practice however, as shown in Fig. 11.26, we can consider a finite distance for the uniform flow to develop.

In the figure, the sluice gate provides a control section that imposes a boundary condition for the flow downstream the gate. The flow that develops close to the gate is non-uniform, with h as a function of x. After a certain distance, variations of h with x are very small and of the order of the natural water surface fluctuations. Here we can consider that the flow has reached the uniform condition. The uniform flow is also called *normal flow* and its depth *normal depth*.

Mass balance

The macroscopic mass balance indicates that the volume flow rate Q is constant, then:

$$\begin{aligned} Q_2 &= Q_1 \\ A_2\bar{v}_2 &= A_1\bar{v}_1 \end{aligned} \tag{11.100}$$

where A is the wetted area and $\bar{v}_i$ is the average velocity. Since $A_2 = A_1$, we have:

$$\bar{v}_2 = \bar{v}_1 \tag{11.101}$$

Momentum balance

The macroscopic momentum balance at steady state applied to the control volume defined between sections (1) and (2) in Fig. 11.26 results in:

$$\oint_S \rho \boldsymbol{v}\boldsymbol{v} \cdot \boldsymbol{n} dA = \int_V \rho \boldsymbol{g} dV + \oint_S \boldsymbol{T} \cdot \boldsymbol{n} dA \tag{11.102}$$

The wall shear stress τ_w is defined by:

$$\tau_w = -\frac{1}{S}\int_S \left(\boldsymbol{T}^E \cdot \boldsymbol{n} dA\right) \cdot \boldsymbol{i}$$

where ρ is the pulp density and $\boldsymbol{i}$ is the unit vector in the direction of the flow. Since the velocity and the areas are constant, the first term of Eq. (11.102) vanishes and the other two terms become:

$$0 = \rho\,(\boldsymbol{g} \cdot \boldsymbol{i})\,AL - \tau_w S \tag{11.103}$$

$(\boldsymbol{g} \cdot \boldsymbol{i}) = g\sin\theta$ is the slope of the Channel, $S = LP$ is the wetted surface, that is, the Channel surface in which the shear stress is acting. Then,

$$\tau_w = \rho g \sin\theta \frac{A}{P} \tag{11.104}$$

Thus, Eq. (11.97) provides an expression for the average wall shear stress in terms of the Channel characteristics (A, P and θ), the density of the substance being conveyed and the acceleration of gravity. Note that the equilibrium does not discriminate between liquid and mixture, so that Eq. (11.104) is as valid for water as for slurries.

The ratio between the cross sectional flow area and the wetted perimeter is called *hydraulic radius* R_h, which is an important geometric parameter of the flow representing the ratio between the slurry (or liquid) volume, where gravity is acting, and the Channel surface where there is shear stress between the liquid (or slurry) and the wall.

$$R_h = \frac{A}{P} = \frac{bh}{P} \frac{\langle \text{cross sectional area} \rangle}{\langle \text{wetted perimeter} \rangle}$$

Note that Eq. (11.104) is valid for steady uniform flow in any geometry. The only restriction is that the Channel must be prismatic, that is, its shape must not change with distance in the flow direction:

$$\tau_w = \rho g \sin\theta R_h \tag{11.105}$$

For a rectangular Channel $A = bh$ and $P = b + 2h$

$$\tau_w = \rho g \sin\theta \frac{bh}{b + 2h} \tag{11.106}$$

Flow velocity

In terms of the dimensionless wall shear stress, known as Fannig friction factor, defined by $f = 4\tau_w/(1/2)\bar{v}_x^2$, we have:

$$f = 8\rho g \sin\theta \frac{bh}{b + 2h} \frac{1}{\bar{v}_x^2} \tag{11.107}$$

The most popular expression for the Fanning friction f factor is:

$$f = 116 \frac{\chi^2}{R_h^{1/3}} \tag{11.108}$$

so that the average flow velocity is:

$$\bar{v}_x = \sqrt{\frac{8\rho g \sin\theta}{116\chi^2} \left(\frac{bh}{b + 2h} \right)^{4/3}} \tag{11.109}$$

where χ is the roughness coefficient. Table 11.4 shows friction factors f for several Channels of uniform cross sections.

Volume flow rate

From Eq. (11.109), the flow rate for rectangular Channels is:

$$Q_f = bh \sqrt{\frac{8\rho g \sin\theta}{116\chi^2} \left(\frac{bh}{b + 2h} \right)^{4/3}} \tag{11.110}$$

Table 11.4 Friction factors for channels

Type of channel of uniform cross section	$\chi, \mathrm{ft}^{1/6}$	$\chi, \mathrm{m}^{1/6}$
Sides and bottom lined with wood	0.009	0.0074
Neat cement plaster; smoothest pipes	0.010	0.0082
Cement plaster; smooth iron pipes	0.011	0.0090
Unplanned timber evenly laid; ordinary iron pipes	0.012	0.098
Best brick work; well-laid sewer pipes	0.013	0.0170
Average brick work; foul iron pipes	0.015	0.0123
Good rubble masonry; concrete laid in rough form	0.017	0.0139

Fluid depth

Calculating h from (11.110) yields:

$$h = \frac{1}{b}\left(\frac{116(b+2\,h)^{4/3}\chi^2 Q_f^2}{8\,g\mathrm{sen}\theta}\right)^{3/10} \tag{11.111}$$

The height h is calculated from the implicit Eq. (11.111) by iteration. Another version of this equation is:

$$h = \frac{1}{b}\left((b+2\,h)\left(\frac{116\chi^2 Q^2}{8\,g\mathrm{sen}\theta}\right)^{\frac{3}{4}}\right)^{\frac{2}{5}} \tag{11.112}$$

Limiting velocity

Domínguez and Harambour (1989) proposed the following limiting deposit velocity to ensure that particles do not settle:

$$v_L = 0.6505\left(8g\left(\frac{\rho_s}{\rho}-1\right)d_{85}\right)^{0.5}\left(\frac{d_{85}}{4R_h}\right)^{0.342}\left(\frac{d_{99}}{d_{85}}\right)^{0.386} \tag{11.113}$$

where ρ_s and ρ are the solid particle and pulp densities, d_i are the sizes where i % of the material passes and R_h is the hydraulic radius.

Mechanical energy balance

The mechanical energy balance is:

$$\oint_S \left(1/2\rho\bar{v}_x^2 \boldsymbol{v}\cdot\boldsymbol{n}\right)dA = \oint_S \boldsymbol{v}\cdot\boldsymbol{T}\cdot\boldsymbol{n}dA - \oint_S \rho\phi\boldsymbol{v}\cdot\boldsymbol{n}dA - \dot{E}_v$$

$$\oint_S (1/2)\rho\bar{v}_x^2\boldsymbol{v}\cdot\boldsymbol{n}dA = -\oint_S p\boldsymbol{v}\cdot\boldsymbol{n}dA - \oint_S \rho\phi\boldsymbol{v}\cdot\boldsymbol{n}dA - \dot{E}_v$$

Since

$$\phi = g(\eta(x) + z\cos\theta)$$
$$p = \rho g[(h - z)\cos\theta]$$

$$(1/2)\rho\bar{v}_2^2 A_2 - (1/2)\rho\bar{v}_1^2 A_1 = -\rho g\big((\eta_2 + h\cos\theta)\bar{v}_2^2 A_2 - (\eta_1 + h\cos\theta)\bar{v}_1^2 A_1\big) - \dot{E}_v$$

but $\bar{v}_2 A_2 = \bar{v}_1 A_1, \bar{v}_2^2 A_2 = \bar{v}_1^2 A_1, p_2\bar{v}_2 A_2 = p_1\bar{v}_1 A_1, z_2\bar{v}_2 A_2 = z_1\bar{v}_1 A_1,$ so that this equation reduces to:

$$\eta_1 - \eta_2 = -\big(\dot{E}_v/\rho g Q\big)$$

Since:

$$\eta_1 - \eta_2 = L\mathrm{sen}\,\theta \text{ and } h_f = \big(\dot{E}_v/\rho g Q\big)$$

$$h_f = L\mathrm{sen}\,\theta \tag{11.114}$$

Thus, the *viscous dissipation*, or *head loss*, is just equal to the decrease in potential energy for uniform flow.

Problem 11.16 A uniform flow of copper flotation tailings takes place in a rectangular Channel constructed of concrete. If the angle between the Channel and the horizontal is 1.0° and the Channel is 0.9 m wide and water is 0.50 m deep, calculate the velocity and the volume flow rate.

$$\chi = 0.0139 \text{ m}^{1/6}$$

$$Q_f = bh\sqrt{\frac{8\,g\mathrm{sen}\,\theta(bh)^{4/3}}{116\chi^2(b+2h)^{4/3}}} = 0.9\times 0.5$$

$$\times\sqrt{\frac{8\times 9.8\times \mathrm{sen}(1\times\pi/180)\times(0.9\times 0.5)^{4/3}}{116\times 0.0139^2(0.9+2\times 0.5)^{4/3}}} = 1.347 \text{ m}^3/\text{s}$$

$$\bar{v} = \frac{Q}{b\times h} = \frac{1.346}{0.9\times 0.5} = 2.99 \text{ m/s}$$

b (m)	1.00
θ (°)	1
Q (m^3/s)	1.346
assume h* (m)	0.45
h*-h = 0	9.18E−06
g (m/s^2)	9.81
χ ($m^{1/6}$)	0.0139
v ((m/s)	146683.42

Problem 11.17 A Channel 0.9 m wide and 1 m high with a slope of 1.0° carries 1.347 m^3/s of copper flotation tailings. Calculate the height of the water in the Channel.

Using solver from Excel by assuming $h = 1$ m in Eq. (11.112) results in:

Problem 11.18 Design a Channel of rectangular cross section to transport a volume flow rate of 0.3 (m^3/s) of copper tailings. The Channel should have a slope of $\tan\theta = 0.0157$ and a height to width ratio of $h/d = 0.5$.

Assume $h = 1$ (m) in Eq. (11.112) and search for objective, with the result

$\tan\theta$ (°)	0.0157
Q (m^3/s)	0.300
h/b	0.5
θ (°)	0.01569871
assume h* (m)	0.27
b (m)	0.54
h*-h = 0	2.83E−04
g (m/s^2)	9.81
χ ($m^{1/6}$)	0.0139
v (m/s)	2.05

Problem 11.19 For a smooth concrete Channel 2 m wide with a slope of 0.001 has a volume flow rate of 1.0 m^3/s, determine the wall shear stress per unit length.

Using the solver of Excel and assuming $h^* = 1$ in Eq. (11.112) and calculating τ_w from (11.106) gives:

ρ (kg/m3)	1,000
$\tan\theta$ (°)	0.001
Q (m^3/s)	1.000
b (m)	2.00
g (m/s^2)	9.81
θ (°)	0.001
senθ	0.001
assume h* (m)	0.21
h*-h = 0	2.62E−06
χ ($m^{1/6}$)	0.0139
v (m/s)	2.41
τ_w (Pa/m)	1.68

References

Babcock, H.A. (1968). Heterogeneous flow of heterogeneous solids. *International Symposium on solid-liquid flow in pipes*, University of Pennsylvania, Philadelphia.

Bain, A. G., & Bonnington, S. T. (1970). *The hydraulic transport of solids by pipelines* (1st ed., pp. 125–148). Oxford: Pergamon Press.

Bird, R. B., Armstrong, R. C., & Hassager, O. (1987). *Dynamics of polymeric liquids: Vol I, fluid dynamics* (2nd ed.). New York: Wiley.

Cairns, R. C., Lawther, K. R., & Turner, K. S. (1960). Flow characteristics of dilute small particles suspensions. *British Chemical Engineering, 5*, 849–856.

Chhabra, R. P., & Richardson, J. F. (1999). *Non-Newtonian flow in the process industries.* Oxford: Butterworth-Heinemann.

Charles, M.E. (1970). Transport of solids by pipelines. In: *Proceedings of Hydrotransport-1*, (pp. 25–26). Warwick: British Hydromechanics Research Association, Paper A3.

Charles, M. E. (1979). Transport of solids by pipelines, *Proceedings of Hyrotransport-1* BHRA, paper A3, 25–26.

Chien, A. F. (1994). Critical velocity of sand-fluid mixtures in horizontal pipe flow. *ASME FED, 189*, 231–246.

Concha, F. (2008). Settling velocities of particulate systems 15: Velocities in turbulent Newtonian flows. *International Journal of Mineral Processing, 88*, 89–93.

Concha, F. & Almendra, E. R. (1979a). Seettling Velocities of Particulate Systems, 1. Settling Velocity of individual spherical. *International Journal of Mineral Processing*, *5*, 349–367.

Concha, F. & Almendra, E. R. (1979b). Seettling Velocities of Particulate Systems 2: Settling of Velocities of Suspensions of Spherical Particles. *International Journal of Mineral Processing, 6*, 31–41.

Condolios, E., & Chapus, E. E. (1963). Designing solids handling pipelines. *Chemical Engineering, 70*, 131–138.

Condolios, E., & Chapus, E. E. (1967). New trends in solid pipelines. *Chemical Engineering, 74*, 131–138.

Domínguez, B., SouyrisR., & Harambour, F. (1989). Caracterización global del fenómeno de la depositación de los sólidos en el flujo sólido líquido en Channeles. In *ninth Congreso Nacional de Hidráulica* (pp. 19–30). Santiago.

Domínguez, E. (1986). Análisis de los parámetros característicos de flujo sólido-líquido en Channeles. *Civil Engineer Thesis*, Pontificia Universida católica de Chile.

Durand, R. (1953). Basic relationships of the transportation of solids in pipes; Experimental research. In: *Proceedings of Minnesota International Hydraulics Convention*, pp. 89–103.

Faddick, R. R. (1985). *Slurry transport course.* Colorado School of Mines: Department of Civil Engineering.

Faddick, R. R. (1986). Slurry Flume design, Hydrotransport 10, BHRA Fluid Engineering, (pp. 143–147).

Gillies, R. G., & Shook, C. A. (1991). A deposition velocity correlation for water slurries. *Canadian Journal of Chemical Engineering, 69*, 1225–1227.

Govier, G. W. (1961). The flow of complex solid-liquid mixtures. *Journal of Engineering*, *44*, 50–57.

Govier, G. W., & Aziz, K. (1961). The flow of complex solid-liquid mixtures. *Engineering Journal (Canada), 44*, 50–57.

Idelchick, I. E., Malyavskaya, G. R. O. G. & Fried, E. (1986). *Handbook of Hydraulic Resistance*, (2nd ed.). Hemisphere Pub. Co., New York.

Metzner, A. B., & Reed, J. C. (1959). *AIChE Journal, 1*, 434.

Newitt, D. M., Richardson, J. F., & Gliddon, B. J. (1955). Hydraulic conveying of solids in horizontal pipes. *Transactions IChE London, 33*, 93–110.

Nikuradse, J. (1933). Srömungsgesetze in rauhen Röhren.

Oroskar, A. R., & Turian, R. M. (1980). The critical velocity in pipeline flow of slurries. *AIChE Journal, 26*(4), 550–558.
Schulz, L. (1962). Experiences of the Soviet Union in hydromechanization. *Bergbautechnik, 12*, 353–361.
Shook, C. A. (1969). Pipelining solids: The design of short-distance pipelines, Symposium on pipeline transport of solids, Canadian Society for Chemical Engineering.
Sinclair, C. G. (1962). The limit deposit velocity of heterogeneous suspensions, Proceedings of Symposium on interaction between fluid and particles, (pp. A68–A76). London: IChE.
Spells, K. E. (1955). Correlations for use in transport of aqueous suspensions of fine solids through pipes. *Transactions IChE London, 33*, 79–82.
Tamburrino, A. (2000). *Class Notes on Hydraulic Solid Transport*. Department of Civil Engineering, University of Chile (in Spanish).
Thomas, A. D. (1979). Predicting the deposit velocity for horizontal turbulent pipe flow of slurries. *International Journal Multiphase flow, 5*, 113–129.
Wasp, E. J., Kenney, J. P., & Gandhi, R. L. (1977). Solid-liquid flow slurry pipeline transportation, Trans. Tech. Pub., (1st ed.), Clausthal.
Wilson, K. C. (1979). Deposition limit nomograms for particles of various densities in pipeline flow. In *Proceedings of Hydrotransport-6* (pp. 1–12). Warwick: British Hydromechanics Research Association.
Yufin, A. P., & Lopasin, N. A. (1966). A summary and comparison of known correlations of critical velocity of solid-water mixtures ans some aspects of the optimization of pipelines. In *Proceedings Hydrotransport-2*, Warwick: British Hydromechanics Research Association.
Zandi, I., & Govatos, G. (1967). Heterogeneous flow of solids in pipelines. *Journal of Hyd. Division, ASCE, 93*(3), 145–159.

MIX
Papier aus verantwortungsvollen Quellen
Paper from responsible sources
FSC® C105338

If you have any concerns about our products,
you can contact us on
ProductSafety@springernature.com

In case Publisher is established outside the EU,
the EU authorized representative is:
Springer Nature Customer Service Center GmbH
Europaplatz 3, 69115 Heidelberg, Germany

Printed by Libri Plureos GmbH
in Hamburg, Germany